NONLINEAR EXCITATIONS
IN BIOMOLECULES

Les Houches School, May 30 to June 4, 1994

Editor

M. Peyrard

Springer-Verlag Berlin Heidelberg GmbH

ISBN 978-3-540-59250-1   ISBN 978-3-662-08994-1 (eBook)
DOI 10.1007/978-3-662-08994-1

© Springer-Verlag Berlin Heidelberg
and
© Les Editions de Physique Les Ulis 1995

Originally published by Springer-Verlag Berlin Heidelberg New York in 1995.

Preface

In his famous book "What is life?", addressing the question "how can we explain the basic phenomena of life with physics and chemistry", E. Schrödinger points out that one essential character of life is its ability to show *cooperative* behavior. Instead of the incoherent fluctuations of individual atoms or small molecules, living cells show coherent global dynamics. In the last few years, hopes have emerged that simple concepts could perhaps explain the extremely complicated biomolecular processes which are known to a greater and greater accuracy thanks to the extraordinary progress of biology. These expectations have been motivated by the progress of nonlinear sciences in the last few decades which have shown that the combination of *nonlinearity*, which characterizes most biological phenomena, and cooperative effects in a system having a large number of degrees of freedom, can give rise to coherent excitations with remarkable properties. New concepts such as solitons and nonlinear energy localisation have become familiar to physicists and applied mathematicians. Nonlinear excitations are extremely stable, immune to perturbations, and, more importantly, able to emerge spontaneously from thermal fluctuations or almost arbitrary excitations of a system. It is tempting to make an analogy between these coherent excitations and the exceptional stability of some biological processes, such as, for instance, DNA transcription, which require the coordination of many events in the ever changing environment of a cell. Physicists have made this bold move, sometimes without caution, and are now invoking nonlinear excitations to describe and explain many biomolecular processes. But biologists, who are in contact everyday with the extreme complexity of nature, often doubt that the seemingly infinite variety of phenomena that they are attempting to classify can be reduced to such simple concepts as coherent nonlinear excitations. While they are certainly right to point out the weaknesses of some models, they may also deny themselves the use of powerful tools simply because they do not look familiar. If one wants to go beyond classification of biomolecular processes, one needs to take advantage of all the available tools. This requires a close cooperation between physicists and biologists, which is often hindered by language and cultural barriers.

It was the aim of the workshop titled "Nonlinear Excitations in Biomolecules" to attempt to bridge the gap between these two communities. This is why, rather than simply reporting the latest research results, a large part of the meeting was devoted to tutorial lectures. The lecturers have agreed to write their contributions to the present volume in the same spirit, so that the book can provide a pedagogical introduction to the two topics forming the backbone of the meeting:

(i) the theory of nonlinear excitations and solitons, and their applications in biology,
(ii) the structure and function of biomolecules, as well as energy and charge transport in biophysics.

In order to emphasize the link between physics and biology, the volume is not divided along these two topics but according to biological subjects. Each chapter starts with a short introduction attempting to help the reader to find his or her way among the contributions and point out the connection between them.

The scientific committee of the meeting included M. Peyrard (ENS-Lyon), M. Barthès (University of Montpellier), G. Careri (University of Rome), P. Christiansen (Technical University of Denmark, Lyngby), J.A. Krumhansl (Cornell and University of Massachusetts), C. Reiss (Institut Jacques Monod, Paris), A.C. Scott (University of Tucson and Technical University of Denmark) and G. Tsironis (Research Center of Crete). I want to express a particular gratitude to G. Careri and J.A. Krumhansl who played a very important role in the preparation of the meeting. Discussions with H. Frauenfelder and J.A. Krumhansl, as the workshop was going on, also contributed significantly to shape the meeting.

The workshop would not have been possible without support from the European Union Euroconference Programme and from the French Centre National de la Recherche Scientifique, which are gratefully acknowledged. The International Science Foundation has also provided funds for the invitation of two scientists.

Finally I want to thank the Centre de Physique des Houches and its director Michèle Leduc for the opportunity of holding the meeting in a beautiful center, very favourable to contacts between participants. The efficiency and kindness of the administrative staff of the school have been very much appreciated by the participants, and even more by the director of the workshop who, thanks to them, could concentrate on the scientific aspects of the meeting.

Michel Peyrard

CONTENTS

LECTURE 4

Modelling the DNA double helix: techniques and results
by R. Lavery

LECTURE 5

Potential-of-mean-force description of ionic interactions and structural hydration in biomolecular systems
by G. Hummer, D.M. Soumpasis and A.E. García

LECTURE 6

Inelastic neutron scattering studies of oriented DNA
by H. Grimm and A. Rupprecht

LECTURE 7

Model simulations of base pair motion in B-DNA
by M.A. Collins and F. Zhang

LECTURE 8

A nonlinear model for DNA melting
by T. Dauxois and M. Peyrard

LECTURE 9

Dynamics of conformational excitations in the DNA macromolecule
by A.M. Kosevich and S.N. Volkov

VIII

CHAPTER II
Proteins, conformation and dynamics..... 175

LECTURE 13

Proteins and the physics of complexity
by H. Frauenfelder

LECTURE 14

Multi-basin dynamics of a protein in aqueous solution
by A.E. García

LECTURE 15

Nonlinear excitations in molecular crystals with chains of peptide bonds
by M. Barthes

LECTURE 16

Low temperature Raman spectra of acetanilide and its deuterated derivatives: comparison with normal mode analysis
by G. De Nunzio

LECTURE 17

Conformational dynamics of proteins: beyond the nanosecond time scale
by H. Grubmüller, N. Ehrenhofer and P. Tavan

LECTURE 18

Motions and correlations of the transmembrane domain of a protein receptor studied by molecular dynamics simulation
by N. Garnier, D. Genest and M. Genest

CHAPTER III

Energy and charge transport..... 247

LECTURE 19

Solitary waves in biology
by A.C. Scott

LECTURE 20

Exact two-quantum states of the semiclassical Davydov model and their thermal stability
by L. Cruzeiro-Hansson and V.M. Kenkre

LECTURE 21

Post-soliton quantum mechanics
by D.W. Brown

LECTURE 22

Dynamic form factor for the Yomosa model for the energy transport in proteins
by A. Neuper and F.G. Mertens

LECTURE 23

Energy and charge transfer in photosynthesis
by W. Mäntele

LECTURE 24

The role of nonlinearity in modelling energy transfer in Scheibe aggregates

by O. Bang, P.L. Christiansen, K.Ø. Rasmussen and Y.B. Gaididei

LECTURE 25

Protons in hydrated protein powders

by G. Careri, F. Bruni and G. Consolini

LECTURE 26

Nonlinear models of collective proton transport in hydrogen-bonded systems

by M. Peyrard

LECTURE 27

Proton-solitons bridge physics with biology
by G.P. Tsironis

LECTURE 28

Neutron scattering studies of biopolymer-water systems: solvent mobility and collective excitations
by H.D. Middendorf

CHAPTER IV

Beyond biological molecules..... 385

LECTURE 29

The cell's microtubules: self-organization and information processing properties
by J.A. Tuszynski, B. Trpisová, D. Sept, M.V. Sataric and S. Hameroff

LECTURE 30

Translation optimization in bacteria: statistical models
by F. Bagnoli, G. Guasti and P. Liò

LECTURE 31

Dynamics of vibrational dissociation of a pseudo-cluster
by D. Hennig and H. Gabriel

LECTURE 32

The step-potential model for π-electrons in hydrocarbon-systems
by C. Kuhn

The intersection of nonlinear science, molecular biology, and condensed matter physics. Viewpoints

J.A. Krumhansl

Department of Physics and Astronomy,
University of Massachusetts,
Amherst MA 01002, U.S.A.

1. INTRODUCTION

This conference, Nonlinear Excitations in Biomolecules (Les Houches, May 30 – June 4, 1994) has had a quite different purpose from the conventional scientific meeting; rather than focusing on a specific well developed topic or problems, the purpose here was to bring together scientists in three cultures – Molecular Biology, Condensed Matter Physics, and Nonlinear Science – as a first step toward addressing the question whether there are problems at the biomolecular level where new progress can be made through an interdisciplinary approach, particularly in theoretical and modeling methods.

A first prerequisite to progress was to address the language barrier between fields, so a key part of the program was a set of review lectures to convey a sense of the central problems and concepts in a way as to be generally understandable to each other. Thus, these talks were not research reports, rather an introductory "short course" in the current elements, language, and key methods in the three fields, concentrating on understandability and perspective. One very important issue has been the concern on the part of critical scientists that in many models and carryover of ideas between fields there has been considerable lack of realism; these reviews were very useful in remedying this situation for the participants, and hopefully this conference proceedings will carry on that mission. Indeed, in an examination of the literature (e.g. in Nature) one finds negligible referencing between fields, even when the subject of the reports purports to be interdisciplinary.

A second question quickly comes to mind: is there evidence that there are cross-disciplinary opportunities in this topic? It seems clear from the discussions that there is a good chance that the answer to that question is in the affirmative, but there is now much to be done to establish substantive foundations rather than circumstantial speculations. The latter do have their use, and there are many examples one could choose. For instance, most biological molecules are polymeric, polymer chemistry is obviously involved, but many aspects of configuration at the secondary level are in the domain of polymer physics (i.e. condensed matter physics), and the mesoscopic conformations and dynamics embrace nonlinear science. It seems clear that the overall biomolecular behavior generally involves concepts from all these fields. Further aspects of this matter will be discussed below.

There is a third and more subtle aspect, however, of this cross–disciplinary adventure which appears to me to be worth appreciating by anyone who takes up the challenges which this topic presents. This observation is based on my own experience, of many years in both condensed matter physics and nonlinear science (applied mathematics), and at least an earnest effort over the past decade to become acquainted with molecular biology. The point is that there are essential "cultural" differences in conceptual and experimental styles and methods between these fields, which fall under the general heading of *Viewpoints*. In my opinion it is important to realize this and to have some understanding of what those differences are.

Thus, and in the spirit of the meeting, I chose to make comments on viewpoints my major topic and I will do so here, rather than presenting a research report, hoping that this will be useful to those engaged in this interesting venture. Nor is it intended to be a review article. For these reasons I have not attempted to provide a bibliography; to do so and cover the fields would be a voluminous undertaking; only a few very specific references are given.

2. VIEWPOINTS

The aims and perspectives in nonlinear science, molecular biology, and condensed matter physics, respectively, differ in many respects.

The central concepts in **Nonlinear Science** are largely those of applied mathematics, now supplemented by computer simulations and numerical analysis. The terminology "nonlinear" in most general terms simply means that the effects of combined phenomena are not additive, rather are functionally interdependent in a more general way. Within this characterization almost any natural phenomena perturbed far from equilibrium will be described; excitations mix intrinsically. Many of the results are physically counterintuitive, and unexpected until recently. However the development of new mathematical techniques leading to a quantification of solitary and chaotic excitations have led to many new applications in physical and engineering systems. Their robustness suggest that they might be functionally relevant in conformational and dynamical questions in biomolecules.

The central scientific issues in **Molecular Biology** appear to be molecular constitution and structure (both primary and secondary), macromolecular properties, and interaction with other macromolecules (e.g. DNA transcription). In recent years computer simulations built on several decades of "first principles" development of interatomic molecular potentials have now become a standard tool of the trade; to an extent this has pushed aside phenomenological modeling, on which I will comment below.

Condensed Matter Physics encompasses an extremely wide variety of phenomena in a correspondingly wide variety of forms of matter – crystals, liquids, ceramics, polymers, glasses and special materials such as liquid crystals. The common feature is that unlike atoms or small molecules they have vast numbers of atomic units (i.e. 10^{23}) structurally coordinated and interacting with each other. Thus qualitatively distinct phenomena occur, such as extended ordering e.g. ferromagnetism, collective electronic excitations described by energy bands rather than discrete levels. These are of course subject to the laws of quantum mechanics, as are atoms and molecules but the point is that because of the extended connectivity of many atoms there are additional phenomena not found in atomic and chemical physics. Inasmuch as biomolecules also contain large numbers of atomic units and are connected in an extended coordinated manner they also qualify as "condensed matter", but at this point there has been minimal cross-disciplinary interaction. One reason for this might be the quite distinct nature of the questions being asked, as will be suggested below.

Given these briefs of the fields I offer several comparative observations about styles, motivations, and methods.

2.1 Scientific objectives differ intrinsically; thus do also the choices of problems and methods.

In biology, each biological genus or system must be "discovered" by experimental observation. This is due, at the very least to the nearly infinite diversity of biological systems; but it seems, more deeply, due to the absence of predicting principles such as theorems in mathematics or the quantum laws in physics. The only general principle that seems to encompass all biology is that of Evolution, which provides powerful logical connections, but says little predictively, say in the sense of determinism of Newton's law. Thus, *in most respects biology is an experimental science,* with theoretical principles (mostly from chemistry) applied piecewise. "Theoretical Biology" is not a well defined sub-discipline.

By contrast, in physics and nonlinear science a few general principles or key theorems and methods can lead to the understanding and prediction of a tremendously diverse number of experimentally observed phenomena. Theoretical physics or chemistry are well recognized as distinct activities. Reductionism is widely useful. So already at this very basic conceptual level the different nature of their problems is reflected in the way the practitioners of the respective fields look at their science.

2.2 The biologist is interested in **Differentiation;** physicists and those in nonlinear science search for **Generalization.**

Differentiation is the essence of biology; *Species* is a key word in all of biology. At the heart of molecular biology is, of course, the amazing concept that specificity in the (base pair) molecular sequence can translate into specific macroscopic species differentiation. Thus the targets in molecular biology are the sequencing of DNA/RNA specific conformation and shape as essential (differentiating) factors in both DNA/RNA and proteins, enzymes, etc. in the transcription process. *Details* are the essence in biology.

The attention of physicists, on the other hand, has largely been directed at finding general laws. One classical example in condensed matter is the Debye rule for specific heats; at sufficiently low temperatures for the three dimensional matter all heat capacities go like T^3, crystal, glasses, amorphous materials, etc. The *microscopic details are lost,* so to speak! No matter how elegant and important this law has been in physics it provides no information of the kind that molecular biologists need. What is needed is an understanding of phenomena which can be localized, either as probes or in biological processes themselves. An example of the former is the application of electron and nuclear magnetic resonance methods. And, for the latter the main point of this meeting is whether nonlinearly localized excitations, particularly if stabilized by locally specific structure, are important in molecular processing.

As in many other instances of the synergism between physics and mathematics, the development of nonlinear science has its examples. The earliest recognition of wave solitons by Scott-Russell in 1834 in large amplitude canal waves is a canonical example. The other general type of soliton, which may be static, not wavelike, appeared in the 1950's in a model by Skyrme for a nonlinear Schrödinger equation of the so called sine-Gordon form. And, of course the famous early computer simulation by Fermi, Pasta and Ulam in a nonlinear lattice dynamics model which produced persistent, localized excitations, led to the naming of the "soliton", and the "inverse scattering" generalization of linear transform methods. The resurrection of Poincare's mapping concepts, new demonstrations of integrability in some nonlinear classes, and the whole subject of chaos and fractal dimensionality have followed.

Now ask the question whether there is such a synergism with biomolecular science. Given the interests of biologists in particularization and the physicists/mathematicians in generalization it is easy to see that at present there is an inter cultural problem. I believe that there can and will be useful transfers of understanding. However, I strongly believe that at present there are both misunderstandings and seriously uncritical, unrealistic, models purported to demonstrate important biological behavior. A prime example is the study of one-dimensional ball-spring models, with nonlinear potentials, advertised to describe DNA; however, almost universally one will find no reference, either specific or general, to any biological literature in their reports. The criticism is that these simply do not contain factual rather than fictional variables, parameters, and potentials that represent real biomolecules.

At a far more sophisticated level over the past decade or so a remarkable development in statistical mechanics, the recognition of *universality* in phase transitions, not only of ordered but also disordered (under the rubric of "spin glass") extended systems, rationalized by the elegant Nobel winning application of the renormalization group theory, has been one of the great events in the 20th century physics. However, both theoretically and experimentally the focus is on *generalization*. Sorry, no matter how beautiful this theory may be, universality is of little interest to the molecular biologist! At the same time most biologists seem to feel that modern physics or nonlinear mathematical science have nothing to offer to real biology; that attitude is wrong as well. This cultural misunderstanding places a special obligation on whoever works in this interdisciplinary area to maintain a substantive dialog with the other cultures.

2.3 Biomolecules as Polymers are treated realistically these days.

Motion and Shape (i.e. **Conformation**) of biomolecules are central factors in the description or interpretation of processes such as transcription, protein folding, and enzyme activity. About 90% of biological matter is polymeric: polypeptides (proteins), polynucleotides (DNA/RNA), and polysaccharides (starch, glucose). This segment of bioscience is now highly developed and spans polymer chemistry, polymer physics, statistical mechanics, and applied mathematics. In particular realistic coordinates and potentials for treating conformational behavior have been extensively studied, especially in forms useful for computer simulations, based on experience gained in the evolution of first principles quantum chemical computations. The review of such computer simulations by Lavery at this meeting shows clearly how sophisticated the simulation of average molecular configurations has come. The motions, i.e. conformational changes, due to dynamic fluctuations or external perturbations, e.g. enzyme attachment, is a much more complex topic which I will return to further on.

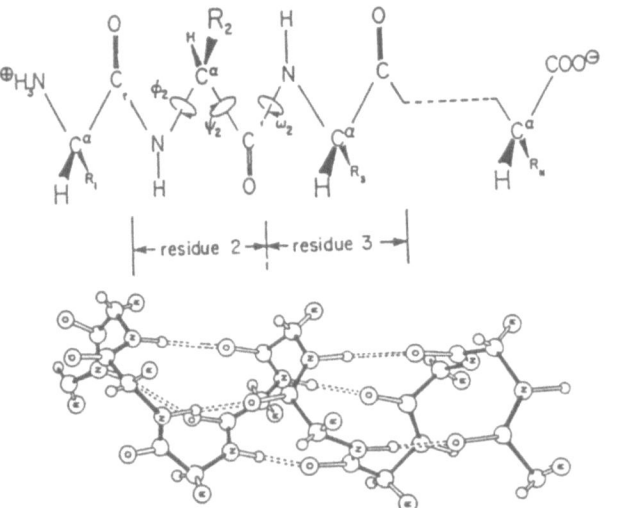

Figure 1. Schematics of the Construction of a Polypeptide Protein

For the present, however, the current treatments of biopolymers provide an opportunity to discuss the nature of the relevant and realistic coordinates for molecular dynamics or Monte Carlo studies. The book by McCammon and Harvey [1] is an excellent overview of this topic. It will become abundantly (painfully?) clear how greatly the biomolecular motions, and the governing potentials, differ from those which have evolved for discussing the lattice dynamics of crystal or glasses. In Fig. 1 we display the primary (molecular) and secondary (α-helical) structure of a polypeptide polymeric protein.

A tremendous amount of research on determining the structure of both proteins and DNA/RNA from model potentials has been done. Both the static, minimum energy structures and molecular dynamics have been and continue to be simulated to ever increasing detail, as several of the papers at this meeting demonstrate. In the present discussion I want to compare the features of the potentials which must be used for biopolymers with those usually used (and more or less adequate for pedagogical purposes) in condensed matter physics and nonlinear dynamics. The fundamental difference is that biopolymers are covalently bonded, and because of their open structures able to easily make large motions if potentials allow.

Thus, note that in the peptide residue at the molecular scale in Fig. 1 "dihedral" motions about the $N - C$, $C - C$, and $C - N$ bonds, denoted by ϕ, ψ and ω respectively are singled out for emphasis. The physical reason is that in this polymer those motions are "rotationally permissive"; put more quantitatively, the energy required for such distortions is much lower than for other distortions. How is dihedral rotation related to the cartesian displacements of the atoms making up the polymer? In Fig. 2 we show the atoms involved in the torsional motion around the $N - C$ bond. It is apparent that the positions of *no less than 4 atoms* must be specified simultaneously to determine the angle ω.

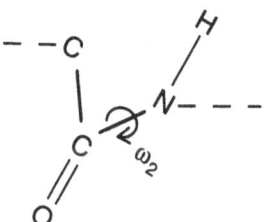

Figure 2

Using position vectors in obvious notation the direction cosine of that angle can be obtained from $[\{(\mathbf{R}_C - \mathbf{R}_O) \times (\mathbf{R}_N - \mathbf{R}_C)\} \cdot (\mathbf{R}_H - \mathbf{R}_N)]$ /(normalization). The potential for dihedral rotation is then a function of this argument, which explicitly involves 4–"centers". Analogously "bond bending" energies involve 3–centers, bond stretching 2–, and core and coulomb (screened) interactions also are 2–center. However the major motions of both protein and nucleic acid polymers are of rotational character; **the potential energy for these motions simply cannot be expressed as a sum of pairwise interactions.** For this reason many would-be simulations of biomolecules which have been carried out in nonlinear studies are not of much relevance.

Truly first principle quantum calculations for such extensive molecules as these are out of the question, but phenomenological potentials can be very quantitative, based on experimental data and the results therefrom. A typical representative model potential of the "molecular mechanics" type is strongly nonlinear in the dihedral rotations:

$$V = \frac{1}{2} \sum_{\text{bonds}} K_b (b - b_0)^2 + \frac{1}{2} \sum_{\text{bond angles}} K_\theta (\theta - \theta_0)^2$$

$$+ \frac{1}{2} \sum_{\text{dihedral angles}} K_\phi [1 + \cos(n\phi - \delta)] + \sum_{\text{nonbonded pairs}} \left[\frac{A}{r^{12}} - \frac{C}{r^6} + \frac{q_1 q_2}{Dr} \right]$$

Energetically, the bond terms may be ≈ 1 eV, the bond angles ≈ 0.1–0.25 eV, the dihedral terms ≈ 0.05 eV, and the nonbonded pairs also $\approx 10^{-2}$ eV. Since biological processes take place at energies near or slightly above $kT = 0.025$ eV it is clear that bond stretching or breaking is not significant in most conformational processes. The analogous development for representing DNA/RNA has many more degrees of freedom.

2.4 Models Play an Important Role in Science. Biology is no Exception.

Modeling is ubiquitous in science. Models are in a sense caricatures of the real situation and designed to extract those features which are the essence of the phenomena under investigation, and little more. The economy of description in good models is invaluable to both theoretical and experimental studies. It allows carrying out formal (i.e. theoretical) analysis in systems with otherwise intractable complexity, and it provides experimentalists with a method not only to interpret but also conceive new experiments. We have just discussed such a model in the form of the molecular mechanics potential for describing protein conformations.

... In thinking about modeling, comparing over fields, I suggested that they could be grouped into two classes: **Phenomenological** and **Deductive**. To illustrate, consider an oil drop falling through air in a gravitational field. At the phenomenological level the motion is simply described by one variable, the center of mass position of the drop, and it's law of motion is the familiar one, with the viscosity parameter introduced. In a more detailed, but still phenomenological description a Langevin random force is added introducing Brownian motion in addition to the steady dropping. This is a good model because its variables and parameters are quantitatively measurable, and consistency checks can be made. But this model description can also be obtained deductively. Consider together not only the oil drop but also the kinetic motions of the air molecules around it; in the final analysis, for the purpose at hand we do not care about their detailed behavior, but we recognize that they collide with the drop and transfer energy and momentum. Separating the variables into "relevant" (the drop) and "irrelevant" (molecular), then by suitable averaging over the latter motions, one can obtain not only the Langevin equation for the motion of the drop and the viscosity parameter quantitatively, but also obtain an important additional result, the fluctuation–dissipation relationship between the phenomenological viscosity coefficient and the Langevin fluctuation force. In this teaching example it was possible to carry out both methods of modeling; this, unfortunately is not often so; not enough is known about the underlying microscopic details to connect them directly to a macroscopic observable.

... In spite of this common state of affairs, it is well worth while to try to find a few representative systems in the class under consideration which can be modeled deductively, since even when this cannot be carried out explicitly for another member of the class understanding how the model relates to the underlying detailed variables, one can imagine how that goes for the system under consideration, even estimate parameters, or know what auxiliary experiments might be useful.

In spite of the difficulty, it turns out that in condensed matter, a number of important phenomena can be modeled deductively, in principle, and parameters deduced from microscopic data. the methodology for doing that was pioneered by the Russian physicist Landau [2] and bears his name, e.g. Landau Free Energy, and pervades the literature in this field.

... The procedure is straightforward in principle. Assuming that we wish to determine the free energy when a system has been macroscopically distorted from a reference state, one calculates the usual partition function, but with a *constraint* – the expectation over the system shall yield the prescribed value for the macroscopic distortion. As in the oil drop–Langevin example above fluctuations are integrated out. Why is this method for modeling so successful? The answer is simple: the resulting free energy is expressed directly in terms of laboratory measurable quantities. It has been applied widely – to phase transitions, to ferromagnetic and ferroelectric domain patterns, superconductivity, to a deductive approach to nonlinear elasticity, and many others. It is also general enough to provide nonlinear free energy functionals as a basis for nonlinear theory development. There is one exceptional situation, very special, under which it clearly is not a faithful representation; that is when fluctuations become large compared with the macroscopic average which occurs in the case of second order phase transitions near the transition. The difficulty is well understood and can be dealt with.

... What about molecular biology? I have not been able to find comparable developments to date. The comparable situation does occur in practice, however. Describing the macromolecular configurations of DNA, e.g. supercoiling, the familiar approach is to model the bare DNA as an elastic, twistable, bendable, biaxial cylinder, adding the effect of the solvent and ionic interactions [3]. This direct phenomenological model is clearly a good one, particularly because experimental measurements of the parameters as well as direct observations of the conformations can be made to check consistency. The Landau approach, based for instance on the detailed results of molecular dynamics simulations, would be difficult to carry out. On the other hands if localized kinks, bends, etc. are seen in the simulations a purely elastic macroscopic model is not sufficient. In that case nonlinear localized excitations may be relevant. This looks like an interesting direction for future study.

Summing up: **Phenomenological Models** are of great use in the many complex situations encountered in nature; their validity rests on consistent and quantitative relationship to experiment. **Deductive Models** are in a sense complementary to phenomenological models and are not often easy to come by; however, when possible, they can offer deeper insight into the foundation of the phenomenology. Both must be mutually consistent, and must be experimentally testable.

As a postscript here I also mentioned at the meeting, only partly in jest, what might be called "hot-air" models. Their basis is usually just circumstantial or semantic, and they would not survive experimental test. As an example I took the question: "What makes the wind blow?" Answer: Note that whenever the wind blows the leaves on the tree moves, complete correlation! The leaves derive energy from the sun by photosynthesis; they can then oscillate and push the wind. Conclusion: photosynthesis makes the wind blow. Over and above the fact that cause and effect can never be obtained from correlation only. From the modeling point of view any critical examination of this circumstantial logical chain would disclose multiple violations of the physics in the process proposed. Silly, yes. Nonetheless some purely circumstantial models, in physics and mathematics, of biomolecular behavior come close in their complete detachment from experimental reality.

2.5 Toward More and Finer Measurements.

There is another significant manner in which molecular biology differs from physics and mathematics, and it is not obvious on first examination. The amount of new knowledge and understanding which molecular biology has achieved for biology over the past several decades is nothing less than staggering, in both its content and often elegance. Yet when viewed by a physicist like myself there is a feeling that something is missing. The central point is that molecular biology has revolutionized the understanding of how biological process work but there does not seem to be the same level of understanding why the steps taken are what they are. Put another way I often wonder whether if examined in greater detail and more of the methods of sophisticated modern physics or mathematical analysis were applied one could see why a step was taken. Without in the least meaning to diminish their accomplishments this question is worth very serious examination. More or less at the same time as I was preparing for the meeting a lucid discussion of exactly the same issues appeared in the journal Nature, in a commentary by John Maddox entitled "Toward more measurement in biology" [4]. Speaking to an elegant explanation of a genetic switch he nonetheless raises several points, which I simply quote selectively:

"Yet molecular biology remains a largely descriptive science."

"Even the best known systems in biology may not be as well understood as is generally believed, which means that understanding is incomplete and may even be misplaced."

"Indeed the chances are that this delicate genetic switch is not a phenomenon of equilibrium thermodynamics at all, but a kinetic one. And if this is possible in this rudimentary case, how much more likely is it that, in the functioning of the factors that transcribe genes, kinetic factors will be dominant?" ... To me this suggests that the important gaps in the understanding of kinetics might be filled by applying concepts from nonlinear dynamics; a tough challenge but one that is open.

... And summing up: "The complaint that molecular biology is insufficiently quantitative at present is a modest and subtle complaint: molecular biology would be in even better condition were it otherwise."

I cannot add much to these points, but I hope that his and my messages are clear. There is more to do, and condensed matter physics and nonlinear science may provide some of the answers. But the first step is to become realistically aware of what is known already about any given biomolecular process. Opportunity, and tough challenges lie ahead.

2.6 Some Paradigms in Current Nonlinear Science.

This paper is concerned with viewpoints, and I continue in that vein with some general thoughts about the possible relationships between nonlinear science and molecular biology. Detailed reviews and reports of developments in nonlinear science are amply provided by many of the other speakers, as well as particular models for nonlinear phenomena in biomolecules.

Rather I would like to suggest an overview perspective of what paradigms of potential utility for the purposes at hand have been developed. There are three in my opinion: **Deterministic Chaos, Solitons and Solitary Excitations, and Intermittency.** In all three cases the phenomena are distinct, counterintuitive, and not contained in any form in harmonic concepts. They are quite beyond the reach of normal mode, even quasi–harmonic, analysis of experiments. As such they have hardly been considered by molecular biologists; but in the spirit of the remarks in **2.5** it seems clear that they should be.

Deterministic Chaos does not seem to relate to molecular level biology, indeed if present would be undesirable for transcription, etc. On the other hand there are phenomena at the higher biological level, for example the physiological, where chaotic behavior must be present. Cardiac fibrillation and epilepsy may be examples.

By contrast, Solitary phenomena at the molecular level might appear in many forms. There have been extended studies of dynamic, wave like, solitons as energy carriers; they could package much larger amounts of energy than simple harmonic wave packets. That application is still under active examination and is discussed by several speakers; quantitatively, their presence seems to be an open question at this point. But though historically first, wave solitons are not the only class; they are distinguished by the equilibrium state from which they are formed being unique. Any excitation always returns to the same (ground) state. Perforce, solitons in this system cannot be static. This kind of soliton is termed non–topological (i.e. does not produce a change of state with its passage). There is however, a quite different, large class of solitons found in systems with multiple metastable states. Transition from one state to another in these does not occur gradually in a distributed fashion but in many cases occurs in a localized, nonlinear manner in space–time. Moreover, because the states are topologically distinct the transition solitons can be perfectly static, and the wave way of thinking of them is inappropriate; they are referred to as topological solitons. One of the oldest example is the domain wall in ferromagnetism, but over the past decade many more examples have been recognized. One direct application to biomolecules, at least in principle, has been to possible localized transformations in a DNA molecule from the B to A form or vice versa. These are distinguished by the inversion of the ribose pucker between two forms; the DNA molecule has multiple metastable states so qualifies for the presence of topological, static solitons. These also produce localized kinks and bending, of considerable interest in conformational studies. However, given the specific variations in the stacking of base pairs, localized kinks may not be easily treated as solitons. As in many other cases in this discussion, and as noted in the previous section, more quantitative studies and careful analysis is needed. It could be very exciting and useful.

Finally there is Intermittency. It is intuitively quite straightforward in concept. An intermittent system has two classes of modes: nontopological and topological. The latter flip-flop back and forth between states, the former appear superimposed as fluctuations. In this case there are many established examples of intermittency in molecular biology. Hinge bending models, multiple fine structure in nuclear resonance measurements, multimodal dynamics in molecular dynamics simulations of both proteins [5] and DNA are only a few examples. Again what appears to be open for more specific application of the nonlinear science of intermittency is the determination of what controls the period of intermittency, how reproductible, how changed by interactions with other molecules, how important to attachment processes, and so forth.

One may sum up these thoughts, certainly somewhat speculative at this point, by suggesting that, as Maddox would put it, the aim of bringing in nonlinear science is not to diminish the accomplishments of molecular biology, but to further add to quantitative understanding. We will not know until we try.

3. CONCLUDING REMARKS

The main purpose of this meeting was to examine the interface between molecular biology and nonlinear science, with the many recent applications of the latter in condensed matter physics also providing perspective. All three fields have been remarkably fertile over the past half century. Rather than a particular technical contribution it seemed very important for me to provide some perspective on how the fields interrelated "culturally" because there clearly is a greater gap between biologists and physical scientists or mathematicians than between the latter two. Some of the more apparent characteristics I see have been set out under the rubric of Viewpoints, with the belief that an understanding of the differences will be useful to those of us interested in pursuing this interdisciplinary field.

Many of the Viewpoints are cautionary, so it may seem that I am pessimistic about a fruitful, especially productive interaction. That is definitely not the case; I strongly believe that there are exciting and important cross–disciplinary opportunities to be explored. However, as in all of meaningful science, plausibility alone is not sufficient; quantification and experimental test are ultimately necessary. the stage is set for that.

I would like to thank particularly the "real" molecular biologists here who have provided the rest of us with such lucid and useful overviews of current issues and accomplishments in their field. Thanks are also due to the the sponsors UE Euroconference and CNRS for their interest in this very exploratory meeting.

References

[1] McCammon, J. A. and Harvey, S. C., "Dynamics of Proteins and Nucleic Acids" (University Press, Cambridge, 1987)

[2] Landau, L. D. and Lifshitz, E. M. "Statistical Physics", (Addison–Wesley, Reading, MA, 1958) Chapter 14.

[3] Marko, J. F. and Siggia, E. D., *Macromolecules*, **27** (1994) 981–988.

[4] Maddox, J., *Nature*, **368** (1994) 98.

[5] Garcia, A. E., *Phys. Rev. Lett.* **68** (1992) 2696–2699.

Introduction to solitons and their applications in physics and biology

M. Peyrard

*Laboratoire de Physique, Ecole Normale Supérieure
de Lyon, URA 1325 du CNRS, 46 allée d'Italie,
69007 Lyon, France*

1. THE SPECIFICITY OF NONLINEAR SYSTEMS.

The response of *most* of the physical systems to combined excitations is *not a simple superposition* of their response to individual stimuli. Mathematically, this property means that almost all physical and biological systems are *nonlinear*. Although this has been recognized for a long time, most of the theoretical models are still relying on a *linear description*, corrected as much as possible for nonlinearities which are treated as small perturbations. It is now well known that such an approach can be *qualitatively wrong*. Although linear response theory can give useful approximate results to analyze some experimental data, the linear approach can sometimes miss completely some essential behaviors of the system. This is particularly true for biological systems in which the nonlinear effects are often the *dominant ones*. For instance many biological reactions would not occur without large conformational changes which cannot be described, even approximately, as a superposition of the normal modes of the linear theory.

In the last two decades, the *intrinsic* treatment of nonlinearities in mathematical models, and later in physical systems, has renewed completely our view of some phenomena. In particular, two concepts have emerged: chaos and solitons. Chaos, which can appear in a nonlinear system with a small number of degrees of freedom, has been extensively discussed, even in non scientific journals, because it challenges our belief that a system described by a small number of deterministic equations should be easy to understand and predictable. But the concept of *soliton*, relevant for systems with many degrees of freedom which can cooperate to form a coherent nonlinear excitation, is also very important, particularly in the context of biology. It is tempting to say that these self-organized dynamical structures, which can emerge in a complex system from a large variety of excitations, provide a paradigm for similar behavior in biological systems. Although this analogy must certainly be taken with caution, the soliton provides nevertheless a new tool which should not be ignored in the studies of biomolecules because it has no equivalent in the world of linear excitations. The aim of this tutorial is to present the main ideas that underline the soliton concept and to discuss some applications.

2. THE CONCEPT OF SOLITON.

The first observation of a soliton was made by J. Scott-Russel in 1834. He was observing a boat moving on a shallow canal and noticed that, when the boat suddenly stopped, the wave that it was pushing at its prow took its own life: "it accumulated round the prow of the vessel in a state of violent agitation, then suddenly leaving it behind, rolled forward with great velocity, assuming the form of a large solitary elevation, a rounded, smooth and well defined heap of water which continued its course along the channel apparently without change of form or diminution of speed". J. Scott-Russel followed the wave along the canal for several kms. He was so impressed by this "great solitary wave", as he called it, that he spent 10 years of his life to study experimentally the properties of this "singular and beautiful phenomenon". Whatever the actual experiment which is made to show a soliton, it always impresses the observer because the properties of the soliton are against our intuition built on the observation of small amplitude linear waves that spread over as they propagate. On the contrary the soliton is extremely robust. The analytical explanation of Scott-Russell's observations was only provided much later with the works of Boussinesq in 1972 and Korteweg-de Vries in 1895. The phenomenon was then forgotten until 1965 when N. Zabusky and M.D. Kruskal coined the word "soliton" to describe this solitary wave which has particle-like properties. They also provided a fundamental understanding of the origin of the soliton and initiated the development of mathematical methods which started a "soliton industry" in applied mathematics.

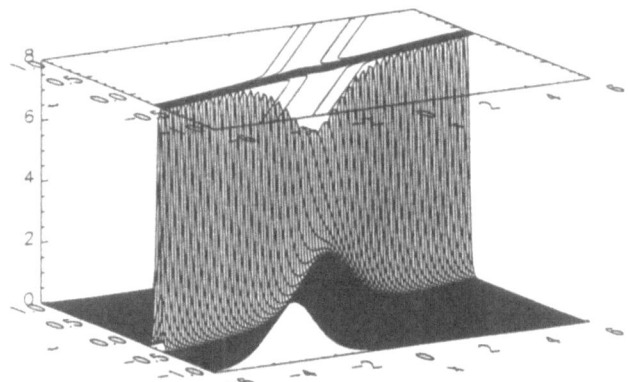

FIG. 1. Collision of two solitons, solutions of the Korteweg–de Vries equation describing shallow water waves.

Solitons are *solitary waves*, i.e. waves localized in space, which have very special properties:

- they propagate at constant speed without changing their shape,

- they are extremely stable to perturbations, and in particular to collisions with small amplitude linear waves,

- they are even stable with respect to collisions with other solitons as illustrated in fig. 1. In such a collision they pass through each other and recover their speed and shape after the interaction. This is surprising because one might think that the strong nonlinearity would break up the pulses during the interaction which is indeed not a simple superposition of the two waves. This can be observed directly on fig. 1 because, at the collision point, the amplitude of the wave is *not* the sum of the amplitudes of the two incoming waves. Moreover, after the collision the trajectories of the two excitations have been shifted with respect to trajectories without the collision. It is however the "soliton miracle" that finally, the outcome of the collision of two solitons is a simple phase shift of each excitation.

Solitons are not restricted to hydrodynamics and can appear in a large variety of systems. Another interesting example is provided by the chain of pendula shown in fig 2. If the first pendulum is displaced by a small angle θ_1, this disturbance propagates as a small amplitude linear wave from one pendulum to the next through the torsional coupling. As it moves along the chain, the small amplitude localized perturbation spreads over a larger and larger domain due to dispersive effects. The soliton is much more spectacular to observe. It is generated by moving the first pendulum by a full turn. This 2π rotation propagates along the whole pendulum chain and even reflects at a free end to come back unchanged. With such a simple mechanical device, it is easy to check the exceptional properties of the solitons. Launching a soliton and keeping on agitating the first pendulum we can test the ability of the soliton to propagate over a sea of linear waves. The particle-like properties appear clearly if one static soliton is created in the middle of the chain and then a second one is sent. The collision looks to the observer exactly similar to a shock between elastic marbles.

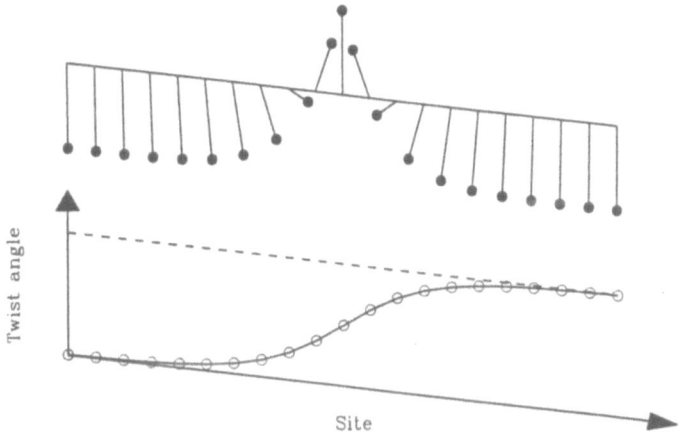

FIG. 2. Soliton in a chain of pendula coupled by a torsional spring. The soliton is a $2\,\pi$ rotation. The upper part shows the aspect of the pendulum chain and the lower figure shows the rotation angle of the pendula as a function of their position along the chain.

The pendulum chain involves only simple mechanics and it is easy to write its equations of motion. Its energy is the sum of the rotational kinetic energy, the elastic energy of the torsional spring connecting two pendula, and the gravitational potential energy. Denoting by θ_n the angle of deviation of pendulum n from its vertical equilibrium position, by m its mass and l the distance between the rotation axis and its center of mass, the expression of the energy is then

$$H = \sum_n \frac{1}{2}I\left(\frac{d\theta_n}{dt}\right)^2 + \frac{1}{2}C(\theta_n - \theta_{n-1})^2 + mgl(1 - \cos\theta_n)\,, \qquad (1)$$

where I is the moment of inertia of a pendulum around the axis, and C is the torsional coupling constant of the springs. The equations of motion which can be deduced from this hamiltonian are

$$\frac{d^2\theta_n}{dt^2} - c_0^2(\theta_{n+1} + \theta_{n-1} - 2\theta_n) + \omega_0^2 \sin\theta_n = 0 , \qquad (2)$$

with $c_0^2 = C/I$ and $\omega_0^2 = mgl/I$.

This simple looking set of differential equations has *no known analytical solution*. We encounter here a frequent situation in nonlinear science because nonlinear differential equations (or partial differential equations) are generally very difficult to solve. The common attitude in this case is to *linearize* the equation, i.e. to replace $\sin\theta_n$ by its small amplitude expansion $\sin\theta_n \approx \theta_n$. Then the system of equations is easy to solve *but essential physics has been lost*. The linearized equations have no localized solutions and they have no chance to describe the soliton even approximately because $\theta_n = 2\pi$ is not a small angle!

There is however another possibility to solve approximately the set of equations (2), while preserving their full nonlinearity, if the coupling between the pendula is strong enough. In this case, adjacent pendula have similar motions and the discrete set of variables $\theta_n(t)$ can be replaced by a single function of two variables, $\theta(x,t)$ such that $\theta_n(t) = \theta(x = n,t)$. A Taylor expansion of $\theta(n+1,t)$ and $\theta(n-1,t)$ around $\theta(n,t)$ turns the discrete set of equations (2) into the partial differential equation

$$\frac{\partial^2\theta(x,t)}{\partial t^2} - c_0^2 \frac{\partial^2\theta(x,t)}{\partial x^2} + \omega_0^2 \sin\theta = 0 . \qquad (3)$$

This equation is called the "sine-Gordon" equation and it has been extensively studied in soliton theory because it has exceptional mathematical properties. In particular, it has a soliton solution

$$\theta(x,t) = 4\tan^{-1}\exp\left[\pm\frac{\omega_0}{c_0}\frac{x - vt}{\sqrt{1 - v^2/c_0^2}}\right] , \qquad (4)$$

which is plotted in fig. 2. This solution provides a very accurate description of the 2π-torsion propagating as a soliton along the pendulum chain.

The example of the pendulum chain shows also clearly that the soliton is a localized packet of energy. Away from its center, the pendula are in their rest position which can be chosen as the reference energy level. But, inside the moving soliton, the three types on energy terms appearing in the hamiltonian (1) contribute to raise the energy: the rotating pendula have kinetic energy, the torsion is associated to an elastic energy in the springs and the pendula involved in the soliton have extra gravitational energy. The soliton solution introduced into the hamiltonian shows that the energy fall off exponentially away from the soliton center.

The pendulum chain has a double interest. First it provides an experimental device which convincingly demonstrates the properties of the soliton. But it also illustrates a new approach to a theoretical description of a physical problem which can be very fruitful. When one is confronted to a set of complicated equations describing a physical or biological phenomenon, *it is important to remember that linearizing the equations to get an approximate solution is not always a good answer*. The continuum limit approximation that we have discussed above is one alternative. Another one which is very fruitful is the multiple scale expansion [1,2] which can for instance be used to provide a description of the hydrodynamic soliton observed by Scott-Russel.

3. CONDITIONS TO HAVE SOLITONS.

The equation established by Korteweg and de Vries (now known as KdV equation) to describe the soliton observed by Scott-Russel illustrates clearly the conditions that a system must fulfill in order to sustain solitons. Calling $u(x,t)$ the height of the wave above the free surface at equilibrium, the KdV equation is

$$\frac{\partial u}{\partial t} + \frac{3}{2h} u \frac{\partial u}{\partial x} + \frac{h^2}{6} \frac{\partial^3 u}{\partial x^3} = 0 , \tag{5}$$

where h is the depth of the water in the canal. This equation is written in a frame moving at the speed of long-wave linear disturbances of the surface.

Let us suppose for a moment that the second term which contains the product $u(\partial u/\partial x)$ can be ignored. This term is the nonlinear term of the KdV equation, and, without it, the linearized KdV equation,

$$\frac{\partial u}{\partial t} + \frac{h^2}{6} \frac{\partial^3 u}{\partial x^3} = 0 , \tag{6}$$

has plane wave solutions of the form

$$u = u_0 \exp[i(qx - \omega t)] . \tag{7}$$

Putting such a solution into equation (6), one gets the dispersion relation of the wave $\omega = -h^2 q^3 / 6$. Any initial disturbance of the surface can be decomposed into its Fourier components (7) and the velocity of each component in the moving frame is is $v(q) = \omega/q = -h^2 q^2/6$. Threfore each component has its own velocity, or, in other words, the medium is *dispersive*. Consequently, without the nonlinear term, a localized perturbation of the surface would tend to spread over as it propagates as shown schematically on fig 3 and the KdV equation would not support solitons.

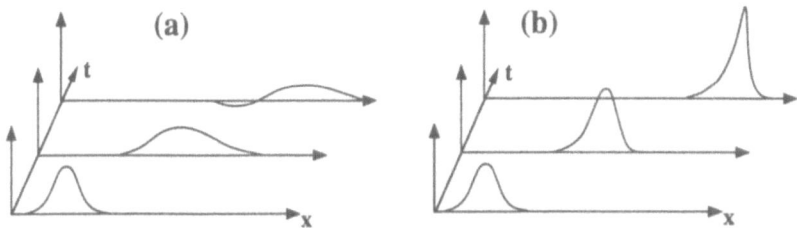

FIG. 3. Schematic picture of the time evolution of a localized pulse in a system showing dispersive effects (a) or nonlinear effects (b).

But the effect of the nonlinear term is exactly opposite. If we now forget temporarily the dispersive term, but keep the nonlinear term, the KdV equation reduces to

$$\frac{\partial u}{\partial t} + \frac{3}{2h} u \frac{\partial u}{\partial x} = 0 . \tag{8}$$

To understand the behavior of this equation, it is convenient to consider an even simpler equation

$$\frac{\partial u}{\partial t} + v_0 \frac{\partial u}{\partial x} = 0 \tag{9}$$

which has traveling wave solutions $u = f(x - v_0 t)$ propagating at speed v_0. This suggests that the solutions of the truncated KdV equation (8) can be viewed as waves for which the speed $v_0 = 3u/2h$ is proportional to the amplitude u. As shown in fig. 3b, the front of such a wave tends to steepen because the crest of the wave moves faster than the bottom. If nonlinearity were acting alone, this steepening would lead to the formation of a shock wave and then to a breakup of the wave.

But the KdV equation contains *both dispersion and nonlinearity. Their balance is responsible for the existence of the solitons.* While nonlinearity tends to localize energy, dispersion tends to spread it over and the "miracle of the soliton" is that this balance is stable: if a wave is too broad, its Fourier spectrum is very narrow and dispersion plays only a small role while nonlinearity tends to win and make the wave steeper until the dispersion, which grows as the wave gets more localized, balances the nonlinearity. Similarly, for a wave which is initially too narrow, the huge dispersion causes it to become wider until the dispersion is low enough to be balanced by the nonlinear effects. Therefore, what may appear as a first glance as a fragile equilibrium is in fact a mechanism which guarantees the robustness of the soliton. For surface waves, the KdV equation shows that the dispersion and nonlinearity are governed by the depth h of the water. The nonlinear term, proportional to $1/h$ increases for shallow water, while the dispersion, proportional to h^2 decreases. While small fluctuations of h do not perturb the soliton, the equilibrium between the two can however be achieved only is h is roughly constant. This is not true for waves approaching a beach because h decreases continuously in the frame of the wave and nonlinearity finally wins: the wave breaks and rolls over, a phenomenon well known of people practicing surf riding.

This discussion on stability shows that the balance between dispersion and nonlinearity can not only explain the *existence* of solitons but also their *formation* from a wide range of localized initial conditions. This is a very general property which can also be found in the pendulum chain for instance. Here the situation is however slightly more subtle because the same term $\omega_0^2 \sin\theta$ of the sine-Gordon equation contains dispersion and nonlinearity. Its nonlinear character comes from the sinusoidal function. Its role in the dispersion can by observed by linearizing the sine-Gordon equation which becomes

$$\frac{\partial^2 \theta(x,t)}{\partial t^2} - c_0^2 \frac{\partial^2 \theta(x,t)}{\partial x^2} + \omega_0^2 \theta = 0 \ . \tag{10}$$

A plane wave $\theta = \theta_0 \exp[i(qx - \omega t)]$ has the dispersion relation $\omega^2 = \omega_0^2 + c_0^2 q^2$, so that, for $\omega_0^2 \neq 0$, ω/q is not a constant and the medium is dispersive for the wave.

Besides the balance between dispersion and nonlinearity, equations having exact soliton solutions have many nice mathematical properties which have delighted many mathematicians [1,3]. They can have multi-soliton solutions and there exists a nonlinear analogous of the Fourier transform for linear equations which allows the decomposition of a given signal into its soliton content. This "inverse scattering transform" has been used for instance to analyze experimental observations of solitary waves created by the tide in the Andaman sea near the coast of Thailand [4]. These quasi-solitons can be observed from satellites and they propagate over hundreds of kilometers in a shallow sea. This example provides an impressive illustration of the soliton concept, and it also shows its power. Although the bottom of the Andaman sea is not flat, although the coast is far from being as straight as the side of the Scott-Russel's canal, the KdV equation provides a very good framework to analyze quantitatively the behavior of these large, extremely long lived, waves that any linear theory simply cannot explain.

4. THE DIFFERENT CLASSES OF SOLITONS.

Solitons can be divided into two main classes, topological and non-topological solitons.

The 2π-rotations of the pendulum chain provide an example of *topological* solitons. In order to understand this terminology, it is convenient to plot the gravity potential acting on the pendula, versus θ and the position x of a pendulum (fig. 4).

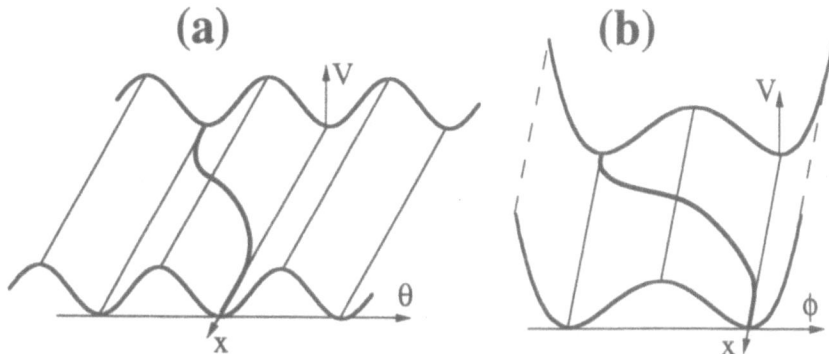

FIG. 4. Representation of topological solitons showing the shape of the on-site potential as a two-dimensional surface depending on the soliton variable and on position. The heavy line shows the trajectory of the soliton variable on this potential energy surface for a soliton solution. (a) The sinusoidal potential of the pendulum chain. (b) The double-well potential of the ϕ^4 model.

Since all the angles $\theta = 0, 2\pi, 4\pi, \ldots$ correspond to the minimal energy of the pendula, the physical system has *degenerate energy minima*. The topological soliton is an excitation which interpolates between these minima. It can exist at rest and is extremely stable because, in an infinite system, it can only be destroyed by moving a semi-infinite segment of the system above a potential maximum. This would require an infinite energy. The system can have solitons (going from one minimum θ_0 to another minimum θ'_0, but also anti-solitons going from θ'_0 to θ_0. The topological soliton could only be destroyed by a collision between a soliton and an antisoliton. In an integrable system having exact soliton solutions, solitons and anti-solitons simply pass through each other with a phase shift, as solitons do, but in a real system like the pendulum chain which has some dissipation of energy, the soliton–antisoliton equation may destroy the nonlinear excitations.

The example of the pendulum chain shows that the condition for the existence of the topological soliton is simply the existence of degenerate minima of the gravitational potential $V(\theta) = mgl(1 - \cos\theta)$. Two minima are sufficient for the existence of the topological soliton, as illustrated in fig 4b for another typical model, the "ϕ^4" model where the potential $V(\phi) = V_0(1 - \phi^2)^2$ replaces the gravitational energy (in this case, it is customary to call ϕ the soliton variable, instead of θ). Another remakable feature of the topological solitons is that they are Lorentz invariant with respect to the maximum group velocity c_0 of the linear waves in the system. This is reflected in the factor $\gamma = 1/\sqrt{1 - v^2/c_0^2}$ which appears in the solution of the sine-Gordon equation (4). Consequently the soliton cannot propagate faster than c_0 and fast solitons show a "relativistic" contraction which is easily visible in the experiments performed with the pendulum chain.

Contrary to the topological solitons of the pendulum chain, for the KdV solitons of water waves, the state of the system is the same on both sides of the soliton. Such *non topological* solitons are dynamical entities. They cannot exist at rest. The amplitude of the KdV solitons is related to their velocity v and to the maximum speed c_0 of linear waves by $u_{max} = \sqrt{v^2 - c_0^2}$ which shows that these solitons are necessarily supersonic. Moreover their equation of motion is galilean invariant instead of Lorentz invariant for the topological solitons.

The two examples that we have discussed up to now are *permanent profile solitons*. This is not always so and some solitons have an internal dynamics. An important example is provided by the solitons of the "Nonlinear Schrödinger" (NLS) equation

$$i\frac{\partial A}{\partial t} + P\frac{\partial^2 A}{\partial x^2} + Q|A|^2 A = 0 \, , \tag{11}$$

where P and Q are constant coefficients. This equation describes the time evolution of the complex amplitude A of a weakly nonlinear wave. If $PQ > 0$, it has soliton solution which are localized wavepackets as shown in fig. 5a.

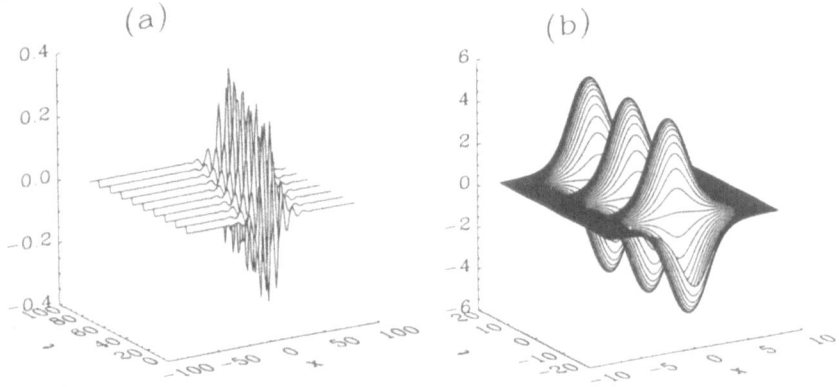

FIG. 5. (a) Propagation of a soliton solution of the NLS equation. (b) Time evolution of a breather, solution of the sine-Gordon equation.

The relative motion of the envelope and carrier wave is responsible for the internal dynamics of the NLS soliton. Another example of a soliton with internal dynamics is provided by the "breather" of the sine-Gordon equation. It is a large amplitude oscillation in the bottom of one of the potential valleys of fig. 4a. Its time evolution is shown in fig. 5b.

These examples show that the world of solitons includes various types of localized nonlinear excitations. Although they may look very different, they have in common their exceptional stability and particle-like properties which distinguish them from the linear waves that we are used to consider. In fact, real systems do not carry exact soliton solutions in the strict mathematical sense (which implies an infinite life-time and an infinity of conservation laws) but "quasi-solitons" which have most of the features of true solitons. In particular, although they do not have an infinite lifetime, quasi-solitons are generally so long-lived that their effect on the properties of the system are almost the same as those of true solitons. This is why physicists often use the word soliton in a loosly way which does not agree with mathematical rigor. Following this habit, we use henceforth the denomination soliton for quasi-solitons in real systems.

5. SOLITONS ARE EVERYWHERE!

This title has been used once in a scientific journal to show how the concept of soliton is ubiquitous and powerful. Examples of solitons can be found at all scales (from the hydrodynamic solitons of the Andaman sea to domain walls which are only a few crystal cell wide in ferroelectric materials) and in various domains of physics and chemistry. Let us consider a few typical examples where the soliton paradigm has been involved in the explanation and prediction of physical phenomena.

The sine-Gordon equation (3) does not only describe torsions in a pendulum chain. It provides also a very accurate description of "fluxons", i.e. quanta of magnetic flux, in Josephson transmission lines. A Josephson junction is a sandwich made of a thin dielectric film between two supraconductors (fig. 6a).

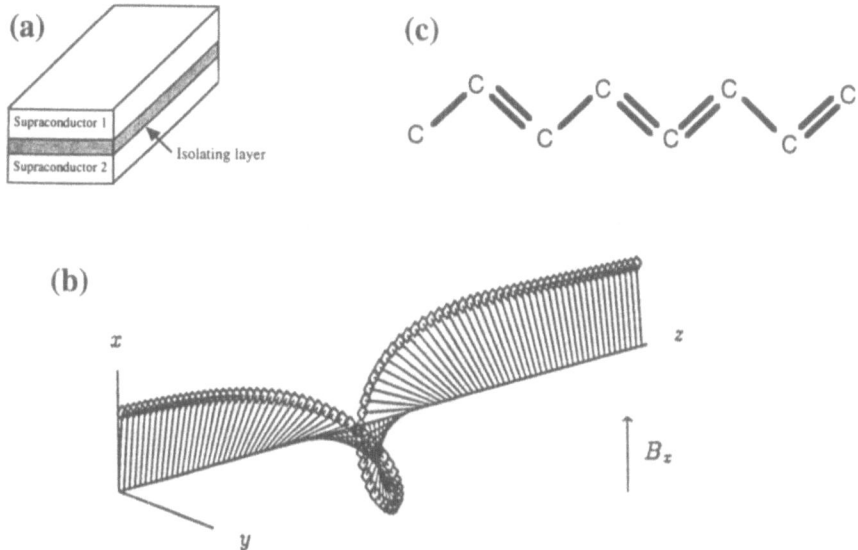

FIG. 6. (a) A long Josephson junction. (b) Magnetic soliton in a ferromagnetic spin chain. (c) Formula of polyacetylene showing the possibility to have a topological defect.

The potential difference V across the dielectric is related to the phase difference θ between quantum states of the Cooper pairs in the two supraconductors by

$$\frac{d\theta}{dt} = \frac{2e}{\hbar} V .$$

(12)

Moreover the tunneling of the Cooper pairs across the dielectric layer generates a current density j of the form $j = j_0 \sin \theta$, where j_0 is a current characteristic of a junction. Taking into account the inductance and the capacitance of the junction, one finally gets a sine-Gordon equation for $\theta(x,t)$, forced by a right hand side term which is associated to the current imposed through the junction by an electrical generator (bias current). The sine-Gordon solitons describe quanta of magnetic flux, expelled from the supraconductors, that travel back and forth along the junction. Their presence, and the validity of the soliton description, can be easily checked by the special shape of the current voltage characteristics of the junction as well as by the microwave emission which is associated to their reflections at the two ends of the junction.

While the solitons of the Josephson junctions are just beginning to be used in practical devices, optical solitons are moving fast to a multimillion dollar industry. They can be generated in optical fibers in which the optical index n depends on the amplitude of the electric field through high order terms of the dielectric tensor

$$n(E) = n_0 + \chi_3 |E|^2 \ . \tag{13}$$

Introducing this expression into the Maxwell equations describing the electric field in the fiber, one gets a NLS equation for the amplitude A of the electric field of a plane wave of frequency ω_0 and wavevector k_0.

$$i\frac{\partial A}{\partial x} - \frac{1}{2}\left(\frac{\partial}{\partial \omega}\frac{1}{v_g}\right)\frac{\partial^2 A}{\partial t^2} + \frac{\chi_3\omega_0^2}{2k_0 c^2}|A|^2 A = 0 \tag{14}$$

The coefficient of the second term is the dispersion of the group velocity v_g in the fiber while the nonlinear contribution is in the third term. In some frequency range, this equation has soliton solutions which can be used to carry information on extremely long distances at a very high rate. Moreover the exceptional robustness of the soliton to external perturbations can be used very efficiently to increase drastically the signal to noise ratio by the introduction of filters at regular intervals in the optical path so that the wavelength which they select changes slightly from one filter to the next. A small amplitude signal which is able to pass through one filter does not match the wavelength of the next one and is therefore stopped. On the contrary, the soliton is able to adjust itself to the perturbation and to propagate through the filters. It is remarkable that optical solitons, which are expected to be so widely used in applications in the near future, have been suggested from theoretical considerations. Nonlinear theory is still at the forefront of the industry for the development of new devices.

As mentioned above, solitons are not restricted to the macroscopic world. They show-up clearly in the properties of one dimensional magnetic materials. In compounds such as TMMC ($(CH_3)_4NMnCl_3$), the crystal structure is such that spins interact strongly along one axis of the crystal and very weakly along the other axes. These spin chains are qualitatively similar to the pendulum chain described above, and a torsion of the spin lattice can propagate as a soliton in the crystal. Figure 6b shows an example of such a magnetic soliton. These solitons are created thermally and their dynamics can be studied very accurately using magnetic resonance or neutron diffraction. In order to analyze the experimental results of such experiments, it is necessary to study the statistical mechanics of the magnetic chain. The soliton concept is again precious because the solitons can be treated as quasi-particles and the results can be obtained by investigating a "soliton gas".

Another example which has attracted a lot of attention is the case of conducting polymers. The simplest example is polyacetylene and, as shown in fig. 6c, its structure can exhibit solitons. The polymer chain has two degenerate states in which the positions of simple bonds and double bonds are exchanged. A defect interpolating between these two states can be treated as a topological soliton. It carries an electric charge and, due to the high mobility of the soliton, it can be responsible for a high electrical polymer conductivity. Following the original analysis [6], the model has been refined and it has been shown that the most probable defects are not the topological solitons but breathers (or polarons) which are simply another class of nonlinear excitations. The breathers generate electronic states in the gap between the valence and conduction energy bands of the electrons. These states have been observed in infra-red experiments, showing the validity of the nonlinear treatment of conducting polymers.

There are many other examples where the soliton concept has been used to analyze experimental results in macroscopic physics (for instance nonlinear electrical lines) or solid state physics (dislocations [7], domain walls in ferroelectrics [8], charge transport in hydrogen bonded chains [9], etc). Some examples in biology are discussed in the present volume [10] but, in this domain, a lot of work has still to be done to derive models which are sufficiently accurate while staying tractable, and which can be confronted with experiments.

6. THE FORMATION OF SOLITONS: NONLINEAR ENERGY LOCALIZATION.

Solitons would be merely mathematical curiosities if their creation in a system would require launching a wave with exactly the right profile. As shown in fig. 7a, it is very easy to generate solitons in a wavetank by dropping a large stone at one end. The initial disturbance evolves into one or several solitons due to the stable balance between dispersion and nonlinearity as discussed above. This experiment is a reduced model of the creation of a "tsunami" by an earthquake in the sea! The soliton can be considered as an attractor for localized perturbations in the system. Sometimes the situation is more complicated because there are several basins of attractions corresponding to different types of solitons that can coexist in the system.

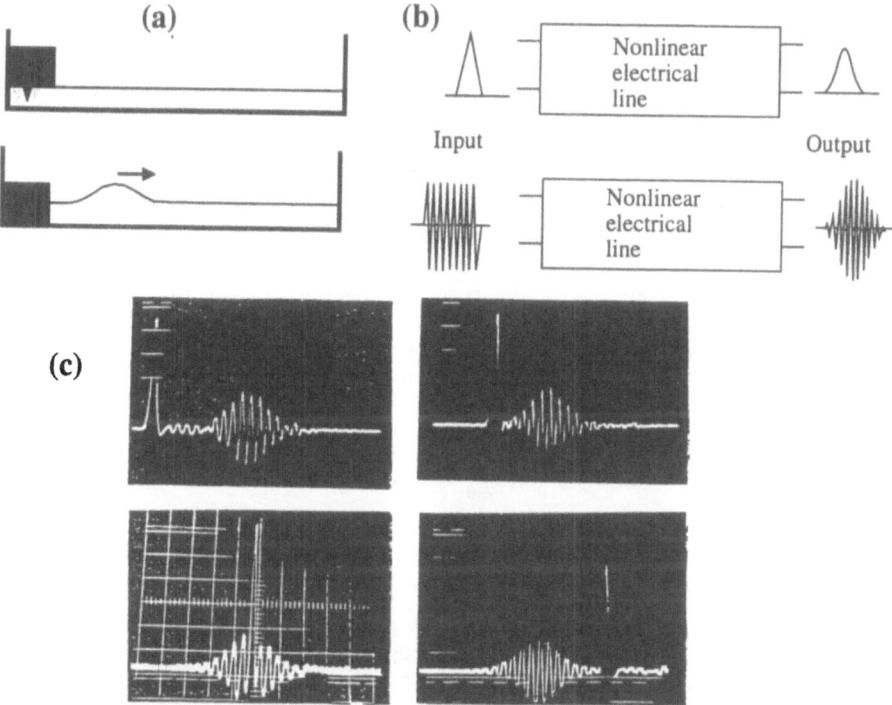

FIG. 7. a) Generation of solitons in a wavetank from an arbitrary initial condition. (b) Two classes of initial conditions generating two different solitons in an electrical line: a triangular signal generates a pulse soliton while a short wave-train generates an envelope soliton. (c) Experimental collision of the two types of solitons in the electrical line.

This is for instance the case for nonlinear electrical lines built as a sequence of filters including an inductance and a nonlinear capacitance made of a varicap diode [11]. As shown in fig. 7b, in such a device a triangular or square voltage pulse evolves into a pulse soliton well described by the KdV equation while a short wavetrain evolves into an envelope soliton well described by the Nonlinear Schrödinger equation. The oscillogrames shown in fig 7c show that the two types of solitons coexist in the system and can even pass through each other without being destroyed. This case illustrates an important property of nonlinear systems. The description of such a system by a nonlinear equation may not be unique and can depend on the particular type of excitation of interest. Even when only one equation can provide a good description of the system, the solutions of this equation can belong to different classes, showing the richness of nonlinear equations. Both the KdV equation and the NLS equation can be derived from the full equations of the electrical circuits, using different approximation schemes. And both provide a view of the system which is correct as long as the approximations that have been made to derive them stay valid.

In the cases that we have described up to now, solitons emerge from localized initial conditions. In these cases, the formation of the soliton is essentially a reshaping of the initial condition. But nonlinear phenomena can have a more important effect in physical and biological systems because they induce the *localization of energy* in the system. Nonlinear solitonlike excitations can also emerge from non-localized initial conditions or from thermal excitations. This property appeared from the first time in a numerical simulation performed in 1955 by Fermi, Pasta and Ulam (FPU) [12] but it was not recognized as such until 1965 because scientists we used to think in terms of linear excitations. In their work, FPU decided to investigate the behavior of a one-dimensional chain of 64 particles of mass m, interacting through forces that contain nonlinear terms, for times long compared to the characteristic periods of the corresponding linear problem. Their aim was to study "experimentally" the rate of approach to the equipartition of energy among the various degrees of freedom of the system. The hamiltonian of the chain is

$$H = \sum_{i=0}^{N-1} \frac{p_i^2}{2m} + \sum_{i=0}^{N-1} \frac{K}{2}(u_{i+1} - u_i)^2 + \frac{K\alpha}{3} \sum_{i=0}^{N-1}(u_{i+1} - u_i)^3 , \qquad (15)$$

where u_i and $p_i = m\dot{u}_i$ are the coordinate and momentum of the $i - th$ particle. They chose fixed boundary conditions $u_0 = u_N = 0$. The harmonic coupling constant is K, and α is a small parameter measuring the magnitude of the nonlinear term in the interaction potential. This hamiltonian can be expressed in terms of the normal coordinates A_k of the linearized hamiltonian

$$A_k = \sqrt{\frac{2}{N}} \sum_{i=0}^{N-1} u_i \sin\left(\frac{ik\pi}{N}\right) \qquad (16)$$

as

$$H = \frac{1}{2}\left(\sum \dot{A}_k^2 + m\omega_k^2 A_k^2\right) + \alpha \sum C_{klp} A_k A_l A_p , \qquad (17)$$

where $\omega_k = 2\sqrt{K/m}\sin(k\pi/2N)$ is the frequency of the $k - th$ normal mode and C_{klp} are constants. In the harmonic case ($\alpha = 0$), the energy stored initially in a given mode stays in that mode and the system does not approach thermal equilibrium. Fermi, Pasta, Ulam started their simulation by exciting the lowest mode ($k = 1$), i.e. by chosing a non localized initial condition having the shape of a plane wave with a wavelength equal to the size of the system. They thought that, for $\alpha \neq 0$, the mode coupling term in the expression (17) of the hamiltonian would cause energy to flow to the other modes, leading, in the long term to an equipartition of energy among the modes. At the beginning of the simulation, this is indeed what they observed: modes 2,3,4 became gradually excited. But, at their surprise, after about 157 periods of the fundamental mode, almost all the energy was back to the lowest mode. After this recurrence time, the initial state was almost restored. Much longer calculations performed later showed that this periodic recurrence of the initial state could be repeated many times and a "super-recurrence" which restores the initial state almost perfectly, was even found. This remarkable result, known as the FPU paradox, shows that introducing nonlinearity in a system does not guarantee an equipartition of energy. In order to understand the properties of the FPU system (and of most of the nonlinear systems), *it is essential to abandon the expansion on the linear modes* and to consider the full nonlinear excitations of the system. This was recognized by Zabusky and Kruskal in 1965 [13] who gave, ten years after the FPU discovery, an explanation of the FPU paradox in terms of solitons, solutions of the KdV equation. It is important to notice that Zabusky and Kruskal solved the FPU paradox because they plotted the displacements in *real space* instead of looking at the Fourier modes. As shown in fig. 8, the formation of the solitons is associated to a *localization in space* of the energy of the initial signal which evolves into rather narrow solitons. The FPU recurrence is simply a manifestation of the stability of the solitons. The initial condition can be expanded in terms of its soliton content. The solitons that were not apparent in the initial profile show up clearly at a later time (fig. 8). Because they survive in the system, they can later come back almost to the same positions that they occupied initially, to form again a quasi-sinusoidal disturbance of the system. This is the recurrence of FPU.

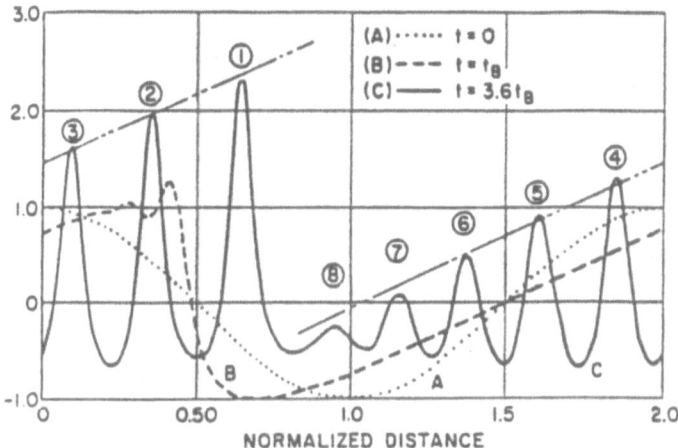

FIG. 8. Temporal evolution of the initial cosine condition introduced by Zabusky and Kruskal in the KdV equation (from ref [13]). (A) Cosine initial condition, (B) displacements after 1.0 period of the lowest mode, (C) displacements after 3.6 periods of the lowest mode.

The FPU simulation illustrates the localization of energy into pulse solitons. Another important phenomenon is the *modulational instability* of a plane wave in a nonlinear system. It is illustrated in fig. 9 which shows the time evolution of a plane wave in a model electrical line. Due to the nonlinear terms in the equation of propagation, a plane wave is unstable with respect to a modulation. A small amplitude modulation grows deeper and deeper and the wave spontaneously break up into wavepackets. The energy initially evenly distributed in the system concentrates itself into the wavepackets where the energy density can become significantly higher than in the initial wave. In the case presented in fig. 9, the perturbation is simply caused by the wavefront and the modulation starts from the front before spreading along the wave.

For application to biology, it is important to notice that nonlinear energy localization can also *occur from the thermal fluctuations* of the system. This is illustrated in the work presented by T. Dauxois in this workshop [14] on a simple DNA model. A thermalized nonlinear lattice can form spontaneously large amplitude localized oscillatory modes (i.e. breathers) which are extremely long lived. They can survive for thousands periods of the lattice oscillations. On timescales of millions of periods, there is an equipartition of energy in the lattice because breathers in contact with a thermal bath die out in one place and others form elsewhere, but during time intervals which are already long in comparison to typical periods of the dynamics of the lattice, the energy can self-localize in specific regions of the system. This could allow the large amplitude conformational changes which are necessary for some biological reactions without requiring a high temperature, because the thermal energy can be used very efficiently if it is essentially localized in the relevant place in the system. For instance the breathing of DNA, which can have important consequences when it traps small molecules that perturb the genetic code, is a an excitation which breaks temporarily several hydrogen bonds. This requires an energy which is well above the typical thermal energy $k_B T$. Nonlinear energy localization can perhaps provide a mechanism to explain this breathing [14].

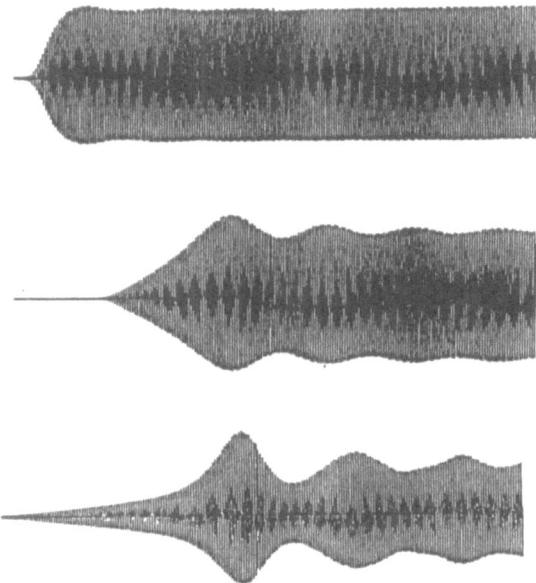

FIG. 9. Evolution of a sinusoidal signal sent at the input of a nonlinear electrical line. The upper picture shows the signal at the beginning of the line, and the lower picture shows the signal after propagation in the nonlinear medium. The perturbation due to the front seeds up a self modulation of the signal.

7. CONCLUSION.

It is interesting to notice that the original papers on solitons written by Scott Russel or Zabusky and Kruskal contained rather pictorial descriptions of the formation and dynamics of solitons. This is perhaps due to the unusual properties of the solitons for people who had been trained to think in terms of linear dispersive waves, so that they felt the need to explain in detail what they had seen to the sceptical reader. However, contrary to the title of our section 4, it is certainly not correct to claim that "solitons are everywhere". In their enthusiasm for these excitations with exceptional properties, physicists have sometimes overestimated the role of solitons. It is nevertheless important to realize that nonlinear excitations provide *new tools* that it would be a pity to ignore in physics and biology. Using simple examples, we have shown how linear expansions can miss completely essential physical properties of a system. This is particularly characteristic for the pendulum chain. Soliton theory offers alternative methods. Multiple scale approximations, or expansion on a soliton basis, can be very useful to provide a description of some physical phenomena. Nonlinear energy localization is also a very important concept valid for a large variety of systems.

These concepts are probably even more relevant for biological molecules than for solid state physics because these molecules are very deformable objects where large amplitude motions or conformational changes are crucial for function. These motions are fundamentally nonlinear. To what extend the soliton concept is relevant in this context is not yet established, but it would certainly be a pity to leave it out of our toolbox.

References

[1] R.K. Dodd, J.C. Eilbeck, J.D. Gibbon and H.C. Morris, *Solitons and nonlinear wave equations,* Academic Press, London 1982.
[2] A.C. Newell, *Solitons in mathematics and physics,* SIAM, Philadelphia (1985)
[3] P. Drazin and R.S. Johnson, *Solitons: an introduction,* Cambridge University Press, N.Y. (1988)
[4] A.R. Osborne and T.L. Burch, Science **208** (1980) 451-460
[5] J.P. Boucher, F. Mezzei, L.P. Regnault and J.P. Renard, Phys. Rev. Lett. **55** (1985) 1778-1781, and J.P. Boucher, Hyperfine Interactions **49** (1989) 423-438
[6] W.P. Su, J.R. Schrieffer and A.J. Heeger, Phys. Rev. B **22** (1980) 2099-2111
[7] J. Frenkel and T. Kontorova, J. Phys. Moscow **1** (1939) 137 and A. Seeger and P. Schiller, *Kink in dislocation lines and their effect on the internal friction in crystals,* in "Physical Acoustics" vol III A, ed. W.P. Mason, 361-512, New-York (1966)
[8] J.A. Krumhansl and J.R. Schrieffer, Phys. Rev. B **12** (1975) 3535-3545
[9] V. Ya. Antonchenko, A.S. Davydov and A.V. Zolotariuk, *phys. stat. sol. (b)* **115** (1983) 631-640.
[10] see for instance the papers by O. Bang et al, T. Dauxois et al., M. Peyrard, M. Salerno, A.C. Scott, G. Tsironis, S. Volkov et al. in the present volume.
[11] M. Remoissenet, *Waves called solitons: concepts and experiments,* Springer Verlag (1994)
[12] The Fermi Pasta Ulam work was never published as a paper because Fermi died before the paper was written. It appeared as a Los Alamos report which was later included in the collected works of Fermi. A reprinted version can be found in the book *The Many Body Problem,* by D.C. Mattis, World Scientific, Singapore 1993.
[13] N.J. Zabusky and M.D. Kruskal, Phys. Rev. Lett. **15** (1965) 240-243
[14] T. Dauxois and M. Peyrard, in the present volume.

Chapter I
DNA, structure and function.

DNA is a very important biological molecule since it carries the genetic code. The famous double helix structure has emphasised the relationship between structure and function. But static DNA would be useless in a cell. Large amplitude motions are essential for the exploitation of the code. This question has attracted the attention of many physicists who have developed nonlinear models for various biological phenomena involving DNA, such as base pair opening or conformational changes. But before proposing a model, it is essential to get acquainted with the molecule and its mode of operation. This is why this chapter starts by several lectures providing detailed pictures of DNA, both from the experimental and computer modelling point of views. **C. Reiss** introduces us to the fascinating aspects of the structure of DNA and its biological function. The lecture of **R. Lavery** shows how the magnifying glass of computer modelling can be used to obtain extremely detailed data on DNA fragments. It discusses a few examples such as base pair openings or backbone transitions which complete the general view given by C. Reiss and also presents the problems of modelling. One of the serious difficulties of molecular modelling is the role played by water in the force field. A biological molecule cannot be studied independently of the solvent in which it is embedded. **G. Hummer, D. Soumpasis and A. Garcia** discuss a formalism which describes quantitatively the solvent effect on the basis of an approximate statistical-mechanical representation and allows the study of various hydration effects.

Although computer modelling is essential to address questions which cannot be directly studied by experiments, it must be completed by observations. High resolution neutron scattering studies of DNA can teach us a lot, as shown by **H. Grimm** in his lecture.

However, if one attempts to go beyond description and classification obtained either directly form experiments or through detailed modelling, it is important to select the relevant dregrees of freedom and to build simple models that allow a deeper analysis. A first step in this direction has been made by **M. Collins and Fei Zhang** who describe a molecular mechanics model of DNA. The second half of the chapter considers even simpler models and illustrates the possibilities offered by nonlinear excitations to describe various aspects of DNA structure and dynamics. **T. Dauxois and M. Peyrard** show how nonlinear energy localisation could play a role in DNA thermal denaturation and perhaps transcription, and discuss the statistical mechanics of a nonlinear DNA model. Collective excitations and solitons are introduced by **A. Kosevich and S. Volkov** to describe conformational changes, and **M. Salerno** investigates the possibility to use a soliton model to explain the role of active promoter regions to initiate the motion of the transcription bubble. While most of the simple models ignore the helicoidal character of DNA, a first attempt to include helicoidal interaction is presented by **G. Gaeta**. Finally, **S. Flach and C. Willis** analyse from a fundamental point of view the role of nonlinearity to localise energy in a lattice. Their work shows the possible existence and stability of local modes, which could correspond to the "breathing" of DNA observed experimentally, and which can form efficient barriers for the transmission of phonons, or heat flux, along a molecule, opening a possibility for a dynamical control of thermal conductivity or thermal energy storage in a molecule.

Although the "soliton" and "local mode" models cannot yet claim to provide accurate descriptions of biological processes, the papers presented in this chapter illustrate new possibilities offered by nonlinear systems that could not be foreseen from a simple linear analysis, and which can perhaps provide new inspirations for biologists involved in detailed studies of DNA function.

Selected topics in molecular biology, in need of "hard" science

C. Reiss

Group "Structure and Dynamics of the Genome",
Institut Jacques Monod, Université Paris VII and CNRS,
2 place Jussieu, 75251 Paris, France

The almost exclusive repository of the information needed by all beings we know of, for birth, development, living, reproduction, and (most probably) setting their life-span, is DNA (deoxyribonucleic acid). Before reviewing how this information is organized, derived and expressed, we will first sketch the basic features of DNA: its chemical composition, its structures and structural flexibility, and the cooperativities which exist inside this macromolecule. In the second part, the salient basic features of the first two steps in gene expression, the transcription of the genetic message and the translation of the transcription product, will be summarized. These two parts are intended as an introduction to the two last parts of this contribution: control and regulation of gene expression, and protein folding, the third major step in gene expression. These two problems remain to date the major challenges in biology in general, and are badly in need of "hard" science.

Only basic, essential facts will be presented. They will be complemented by references to comprehensive reviews, comprehensible for interested, non-specialized readers. A rather mechanistic approach was choosen throughout this contribution, as it may be the easiest for scientists not familiar with molecular biology.

I. DNA structure.

Dramatic pictures of DNA were obtained by electron microscopy studies of phages (fig. 1), or bacteria (fig. 2), bursted by means of an osmotic shock. Most of the DNA is then displayed as a long, uninterrupted and more or less coiled thread about 20Å in diameter, the genome. Phages are virus-like particles infecting bacteria, primitive unicellular beings. The length of a phage or virus genome is in the µm range, that of bacteria measures from one to ten mm usually. Bacteria are prokaryotes, which distinguish from eukaryotes by the fact that their genome is loosely floating inside the cell, whereas eukaryotes have their genome tightly packed in a separate, intracellular body, the nucleus (eukaryotes are more evolved and appeared probably some $1.2 \ 10^9$ years ago; bacteria may have been on earth for over $4 \ 10^9$ years). Although the size of an eucaryotic cell compares to that of bacteria (µm range), genomes in simple eucaryotic cells, like yeast, are one to ten cm long. The size increases in more evolved species. Mammalians carry in each of their 10^{12} to 10^{14} cells, a genome of about one meter in length, but the genome of the salamander is over hundred times larger than ours... Actually, in vertebrates,

C. Reiss

less than 1% of the genome is "used", the remaining (often approximate copies of expressed parts) having unknown functions.

Figure 1: Redrawing from an electron micrograph by A.K. Kleinscmidt et al., Biochim. Biophys. Acta 61, p. 857, 1962. The thread is the duplex DNA chromosome released following osmotic shock from bacteriophage T2 (its envelop is the black item in the center of the picture). The two free ends are clearly seen.

Figure 2 : Redrawing from an electron micrograph by M. A. MacHattie et al., J. Mol. Biol. 11, p. 648, 1965. The duplex DNA chromosome of the bacterium Hemophilus influenza following osmotic shock (the bacterial wall is the structure at the center of the picture). Box: an enlarged view of the "huddle".

Let us zoom down at the Å range the picture of DNA in figs. 1 or 2. This can be done with the help of X-ray diffraction of fibers of purified DNA (which enabled Watson and Crick to discover the famous double helical structure of DNA), and more recently of oligomeric DNA in monocrystals, with resolution at the level of atoms.

Fiber diffraction studies yield a picture sketched in fig. 3. The DNA double helix is contained within a cylinder of 20Å in diameter. It is schematically made of two ribbons, winding along the cylinder and facing each other, supporting "stairs" with 3.4Å steps (the base-pairs), and 10 to 11 stairs per helix turn.

The ribbons are strands made of alternating phosphate (PO4-) and sugar (ribose) groups. The formula of the ribose is shown on fig. 3. Notice that two consecutive phosphate groups are linked by the C3'-C4'-C5' chain of the ribose. Comparing the orientations of the segment on the two strands, it appears that they run in opposite directions; the strands are antiparallel (1).

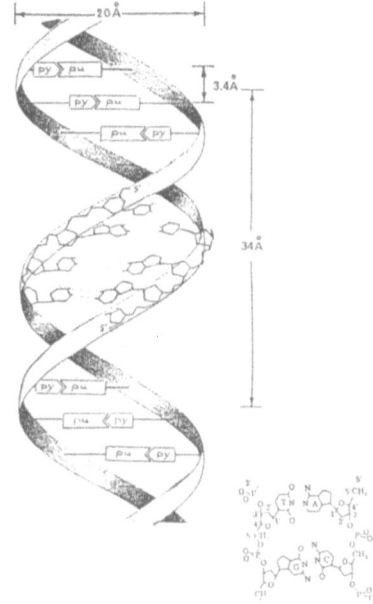

Figure 3 : Sketch of the double helix. In the central part, the sugar-phosphate backbone is drawn along the ribbons, with the bases attached. At the lower right corner, the chemical formula of these components.

The main internal degrees of freedom in the double helix.

The picture shown on fig. 3 is only an approximate sketch of the actual double helix structure. X-ray diffraction studies of oligomeric DNA monocrystals revealed that at atomic resolution, only the planar structures of the four bases are invariant. The conformation of the sugar-phosphate backbone, the shape of the base pairs, and the relative positions of the base pairs appear to change within limits which, however, maintain the overall helical appearance. The double helix is highly polymorphic. Its precise structure depends not only on the bases sequence, but also on external or environmental parameters, of physical or chemical nature. The polymorphism reflects the presence in the double helice of several, rather flexible degrees of freedom.

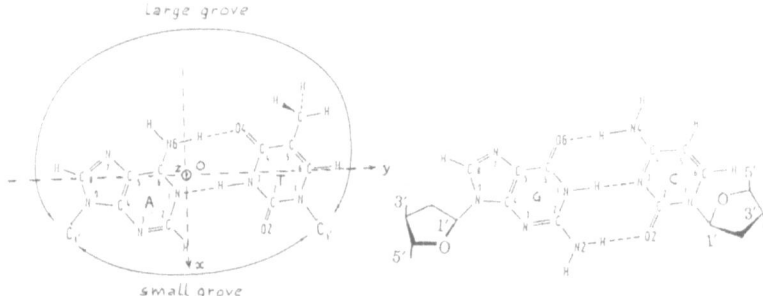

Figure 4: Detailed chemical formula of the two base pairs. Right: the Adenine-Thymin pair, with the Oxyz frame used to describe the degrees of freedom (see text), and the definition of the two groves. Left: the Guanin- Cytosine pair, with the attached sugar moieties (notice the opposed orientations of the 3' to 5' links, also shown in fig. 3). The base pairs expose in both groves potential H-bond donors(d) and acceptors (a): for GC, in the large grove C6 (d) of C, O6 (a) and N7 (a) of G; in the small grove: O2 (a) of C, C2 (d) and N3 (a) of G; for AT, in the large grove O4 (a) of T, C6 (d) and N7 (a) of A; in the small grove, O2 (a) of T, C2 (d) and N3 (a) of A.

First, the base- pair is seldom planar. Let the yy' axis of the pair run from C6 (pyrimidine) to C8 (purine). The zz' axis (fig. 4) is taken normal to the pair (assumed planar for the moment) and intercepts the yy' axis at O, the "center"of the pair (near the N1 of the purine). In monocrystals, it is observed that, within certain limits, the bases of a pair can rotate and translate with respect to each other. Intrapair rotations around zz', yy' and xx' are termed "opening", "propeller twist" and "buckle", respectively. Intrapair translations along the same axes are respectively called "stagger", "stretch" and "shear". Furthermore, it is observed that a given base pair can also rotate and translate as a whole with respect to its neighbour pair. "Twist", "roll" and "tilt" are interpair rotations around zz', yy' and xx', "rise", "slide" and "shift" are interbase translations along these axes, respectively. For some of these degrees of freedom, the observed variations can be rather ample: for interpair degrees, the twist can be found between 28 and 40° (average 36°), the roll between 20 and -10°, the tilt may be as large as 20°, the rise may deviate from its mean (3.4A) by 25%, the slide may be found from -1A to +2A. The propeller twist may be found in the +/- 20° range (2).

Close inspection of the crystal structure of the double helix reveals that the degrees of freedom just mentionned help to avoid, or overcome, potential steric clashes between consecutive base pairs. Steric clashes depend on the base sequence. In order for instance to accomodate in the double helix a sequence of two or more bulky purines present on the same stand, neighbouring bases or base pairs are forced to make use of their degrees of freedom, to escape the clash (). The positional freedom granted to bases or base pairs, serves also to keep at a minimum   the contacts between the rather hydrophobic bases and water surrounding the double helix.

The bases are attached to the sugar-phosphate backbone, hence their degrees of freedom must correspond to the flexibility of the backbone. Indeed, rotational freedom is present around all the bond chains linking the base to the phosphate in 3', i. e. N (base)-C1'-C2'-C3'-O3'-P, N (base)-C1'-O4'-C4'-C3'-O3'-P, or to the phosphate in 5', i.e. N (base)-C1'-C2'-C3'-C4'-C5'-O5'-P and N (base)-C1'-O4'-C4'-C5'-O5'-P (see fig. 3). Details of these rotational freedoms can be found in (1).

Of particular interest is the flexibility of the sugar moiety, which plays a major role in the polymorphism of the double helix. Indeed, the sugar is not planar, but puckered. In the double helix, the five-membered sugar ring is usually found in the twisted form, with two of its atoms, C2' and C3', located on both sides of the plane defined by the three others, C1', O4' and C4'. The energy barriers between the puckered states are rather shallow (fig. 5).

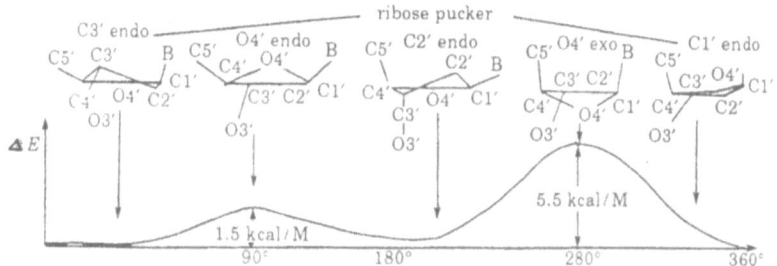

Figure 5: Nucleoside (i.e. the sugar-base ensemble) free energy as a function of the concise pucker descriptor P, the pseudo-rotational phase angle, defined by tgP= .32((v4+v1)- (v3+v0))/v2 (vi is the tortional angle of the bond linking Ci' to Ci'+1 for i=1 to 3, v4 and v0 correspond to bonds C4'-O4' and O4'-C1', respectively). The ribose pucker is characterized in short by the atom (usually C) of the sugar ring which is the most out of the plane, completed by "endo" or "exo" depending on the position of this atom on the side of, or opposite, C5'. Puckers corresponding to selected values of P are sketched on top of the curve. The C2' endo and C3' endo puckers, located in the minima, are those most frequently found in natural DNA.

Given the conformational flexibilities just described, and others of lesser importance (see ref. 1 and 2), the double helix may adopt quite different shapes, characterized by the handedness of the double helix and the depth of its grooves. Looking down the helix axis in either direction, in a right (or left)-handed helix each strand winds clockwise (or counter-clockwise) as it moves away from the observer. The helix has a small and large groove, found respectively on the side of the base-pair turned towards the small or the large angle made between the two C1'-N (base) bonds of the base-pair (see fig. 4).

The DNA helices have been classified into three main families: the B family, the most frequent one, with base-pairs almost perpendicular to the helix axis, a shallow wide groove and a deep small groove; the A family has a deep large groove, a shallow minor groove and its base-pairs are markedly non-perpendicular to the helix axis; both A and B helices are rigth- handed, in contrast to the members of the Z family, which are left-handed. A DNA of given sequence may be found in diverse conformational families, depending in particular on the presence and concentration of certain ions and the degree of hydration. A, B and Z conformations may be found coexisting on the same DNA segment (1,2).

Cooperativity inside the DNA molecule.

In vacuum, three major, non-covalent interactions take place in the double helix:

-the H-bonds exchanged between the bases in a base-pair;

-the stacking between consecutive base-pairs, due to dispersive interactions between the delocalized electrons associated with the heterocyclic base structures

-the electrostatic interactions beween the phosphate groups, intra- or inter-strand.

The interplay between these interactions gives rise to a host of interesting physical properties (3). The first, and to date the best studied, is cooperativity associated with nearest neighbour interactions. It can be modelled quite accurately by an adaptation of the Ising model proposed by Az'bel (4).

To give a general formulation (5,6), we assume a chain of two monomeric species (0 and 1), which may for instance stand for AT and GC base-pairs, or purines and pyrimidines, etc... Depending on an external parameter X, the chain can adopt two states (a and b), for instance the native state of DNA (double helix) and its denatured state (where over a sequence all H-bonds linking the base-pairs are disrupted), or A and B or Z conformations of DNA,... X stands for temperature, pressure, pH, type and concentration of ions or chemicals, etc...

The free energy of species (0) and (1), may be written as $\Delta F0=\Delta S(X-X0)$ and $\Delta F1=\Delta S(X-X1)$, assuming that ΔS is identical for both species. X0 and X1 are the values of X for which chains of pure (0) and (1) change their state. Introducing the dimensionless parameter $p=(X-X0)/\Delta X$, with $\Delta X=X1-X0$, the free energy of a chain built with n0 (0) and n1 (1) is $E=\Delta S\Delta X[n0p+n1(p-1)]= \Delta S\Delta XN(p-r)$, where $N=n0+n1$, $r=n1/N$. Each boundary between segments ("domain") in states (a) and (b) adds a boundary free energy V written as $V=W\Delta S\Delta X/2$. Therefore, the transition can be characterized by $N(p-r)=bW$, where the boundary coefficient b $=+1$, 0 or -1, depending on whether the transition appears within a domain in a given state, will extend a domain in a given state, or will bring to merging two domains in the same state. Given W, a particular base sequence is then divided into domains, each specified by its boundaries in the sequence and the value of p for which it changes its state.

W depends stringently on the DNA environment (ionic strength, pH, chemicals...). Therefore, the partition of a DNA of given sequence into cooperative domains depends on the environment also. A phase diagram can be traced (fig. 6), showing a hierarchical organization of the sequence partition as W changes (6).

This simple adaptation of the Ising model is surprisingly capable of accounting accurately for experimental observations. Thermal (X=T) denaturation curves (cooperative double-helix (a) <=> coil (b) transition) of DNA, considered a copolymer of AT (0) and GC (I) "monomers", measured in buffers of different ionic strengths, can be fitted to within experimental error from its known AT and GC sequence, with just one adjustable parameter (W) (7, 8). As the AT pair (two H-bonds) is less stable than the GC pair (three H-bonds), the cooperativity considered in this example involves the DNA "stability". The model predicts the partition of the sequence into domains and the value of the parameter p (temperature in the present example) at which a given domain changes its state from native double helix, to random single-strands.

The sequence of a given DNA segment can be analyzed for several types of cooperativities. As seen above, a row of two or more bulky purines (Pu) on one strand of the double helix introduces steric constraints, which can be eased by the inter- or intrapair degrees of freedom, but at the expense of constraints imposed to the sugar-phosphate backbones. As a result, the double helix structure may be affected (propensity to form A-type helices). Sequences of alternating purines and pyrimidines (Py) on one strand introduce no or less such constraints, and have a propensity to form B-type helices, and even Z-type helices if the purine G alternates with the pyrimidine C, under particular ionic conditions. The sequence of DNA, considered a copolymer of Pu and Py, can then be analyzed for cooperativity among alternating Pu and Py (B- or Z-type "conformability"), and non alternating Pu or Py (A-type conformability) (6). The transition from one type to the other can be induced by a change in ionic strength for instance, or the addition of selected chemicals (alcool, water,...). In addition, at the boundary between domains in A and B form for instance, it was shown that the helix axis experiences a more or less pronounced bend. An example of stability and conformability cooperativities of a given DNA sequence is shown in fig. 6.

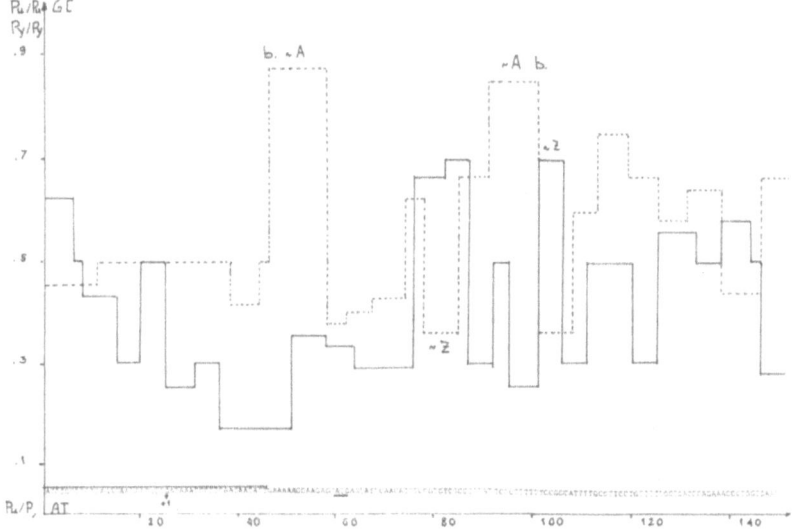

Figure 6. Two cooperativity maps drawn for the sequence at the bottom. . W=1.5. Abscissa: sequence in base pairs. Ordinate: p.
    Full line: DNA stability. p stands for temperature, i.e. a given domain changes it state at the temperature corresponding to its p value. The domains with the lowest p are the less stable, those with the highest p are the most stable. In the model, AT is (0), GC is (1), (a) is double-stranded, (b) is unwound.
    Dotted line: DNA conformability. p stands for ionic strength, i.e. a given domain changes its conformation at the ionic strength corresponding to its p value. The domains with the lower p values tend to adopt the B conformation, those with the higher p values have a propensity to adopt the A form. At the junction between domains of high and low p values, the helix axis mat be markedly bent. In the model, PuPy on the same strand is (0), PuPu or PyPy on the same strand is (1). (a) and (b) are B and A conformations, respectively.
    Segments cumulating p values corresponding to high stability and low conformability tend to be in the Z conformation. See (5-7).
    The predicted stability domains fit surprisingly well those determined by thermal denaturation experiments (5, 7, 8). The conformability domains are less well established, due to the lack of adapted experimental methods (work in progress in this lab.).

Bases in DNA cooperate also to provide hydration spines in the grooves of the double helix, or particular steric patterns of proton donors and acceptors exposed in both grooves, which may serve as docking facilities for specific ligands (see fig. 4 and (1)).

We will see later that, as a rule, vital genetic informations, or signals, are encoded in the DNA as sequences of variable size. The bases of a given sequence support together a dedicated signal, which is to instruct the enzymatic machinery in charge of gene expression. An ever growing body of evidences shows that a given signal identifies with a distinct property of the double helix (precise helix conformation, bend, stability...) sensed and processed by the machinery, provided the property meets the required characteristics (amplitude and spacial extension).

This strongly suggests that the property results from a selected cooperativity associated with the local base sequence. The amplitude of the property would be set by the strength of the local cooperativity, and the size of segment supporting the signal would be that of the local cooperativity domain. Amplitude and size would depend of course on the physiological DNA environment, including (during gene expression) the enzymatic machinery itself. Modifications of this environment could change the domain partition and cooperativity strength, with possibly important biological implications and consequences. Experimental results reported below support this view. We will also see that at certain places of the genome (promoters), several signals coexist. They could correspond to several types of cooperativities supported by the same base sequence.

II. Gene expression.

From the point of view of molecular biology (9), gene expression appears to be the primary and most fundamental event of life. Gene expression is the process whereby the information stored in DNA is transferred and materialized, most often by the production of proteins, to which we will restrict below. The basic information needed to produce a given protein is deposited in a particular base sequence of the of the genome, the gene (size from 150 to 6000, average about 1000 base pairs). Bacterial genomes carry a few thousand genes, eukaryots up to ten times more.

A gene can be identified and localized in the genome genetically or biochemically. For instance, modifying its sequence (mutation) often materializes in changing, or even abolishing its expressoin. Methods have been devised to sequence large segments of DNA. Data bases including several tens of thousand sequences are available. Most of the sequenced segments bear one or several genes, separated by sequences not coding for genes and of mostly unknown function.

The basic genetic information -the base sequence- is at present readily deciphered. However, the expression of a particular gene at the right time, at the appropriate place and in due amounts is of vital importance. Excesses or defaults in expression mean waste of energy, failure to coordinate the life processes at the cellular or higher levels, and may end with lethal penalty. Therefore, control and regulation of gene expression must obey to strict laws for smooth life of the cell and for performance of the dedicated cell duty in pluricellular organisms. The informations needed to achieve these tasks are encoded in the nucleic acid, mostly in the gene itself, but sometimes also at more or less remote places.

Loss of control and regulation may be inborn, or may result from environmental insults (radiations, chemicals including polluants, viruses), or from seemingly spontaneous acquired disorders (aging). These laws, and the underlaying mechanisms, are at present largely unknown.

Furthermore, even if control and regulation of gene expression are normal, the product of the gene, the protein, must be able to perform its role properly. The activity of a given protein is tightly linked to its structure. Understanding the former requires at least the knowledge of the latter. Here is a second important challenge in molecular biology, namely the determination of the steric structure of a protein, and specially its prediction, knowing the aminoacid sequence of the protein. This so-called protein folding problem has been frustrating so far and awaits a comprehensive solution.

Control and regulation of gene expression, and protein folding, are at present major challenges in molecular biology. The description of their main features will be presented in some detail in the remaining of this contribution.

Transcription.

The expression of a given gene occurs in two major steps: transcription and translation. Transcription is the process whereby the (linear) genetic information is copied into an auxiliary nucleic acid, the messenger ribonucleic acid (mRNA). Translation is the transfer of the information stored on mRNA into a linear sequence of aminoacids, the protein, using the genetic code. We will first focus on transcription.

RNA distinguishes from DNA by having an OH attached to the C2' of all ribose rings, instead of an H in DNA (hence the prefix "deoxy"). In contrast to DNA, which almost always comes in two complementary strands, only a single strand of RNA is usually produced.

mRNA is produced by a dedicated, well characterized machinery, RNA polymerase (RNAP), a key enzyme of all cells (fig. 8). It can be easely extracted from bacterial cells (each contains some 10^4 RNAP particles) and purified in large quantities. In the following, we will refer to the RNAP from the bacterium Escherichi coli (E.coli), which has 5 subunits (of the order of 100 kD each) and a "diameter" of about 150A. Purified RNAP is able to perform transcription in the test tube with efficiency and fidelity comparable to those it has in vivo.

If pure RNAP, and a segment of bacterial DNA bearing a gene, are mixed in a test tube containing an appropriate salt buffer (say .1M K+, .01M Mg++), RNAP immediatly binds to DNA. Binding results from electrostatic interactions mainly, as it cancels above .2M K+. The bound RNAP immediately starts sliding along the double helix at high speed, equivalent to a one- dimensional random walk at a rate of $10^6 bp^2/s$, in search of a specific sequence, usually located in front of the gene, the promoter. Promoter location proceeds at an effective rate of 1000bp/s.

The genome of E.coli contains about 4.7 10^6 base pairs. It bears of the order 4000 genes and a comparable number of promoters, many of them sequenced. Although almost all sequenced promoters are recognized by the same RNAP, not two of the sequenced promoter share the same base sequence. To the best, a few "prefered" base pairs are found at certain places in the promoters, but these are by no means mandatory (see fig. 7 for a few promoter sequences, chosen at random in the Genebank data base). To date, promoter identification cannot be made from the base sequence, and relays exclusively on biochemical or genetic informations.

In the test tube, once an RNAP has met the promoter sequence, it will remain there for hours, until the monomeric units necessary for the synthesis of mRNA (the triphosphates (TP) of the four bases, XTP below) are provided.

```
                        -35                   -10        ~~~→
gal-P2/mut-2 m  TAATTATTCCAT GTCACA CTTTTCCC   ATTTTGT TATGCT ATPCTTaTTTCATAC
glnL      b  CAATTCTCTGATGC TTGCCC CTTTTTATC   CGTAAAAGC TATAAT GCACThAATGCTCC
glnS      b  TAAAAACTAACAG TGTCA GCCTGTCC      CGCTTATAA GATCAT ACCCCgttaTAGCTT
gltA-P1   b  ATTCATTGGGACA CTTATT AGTGGTAG      ACAAGTTT AATAAT TCCCALTCCTAACTA
gltA-P2   b  AGTTGTTACAAACA TTAGCA GGAAAAGCA    TATAATGCG TAAAG TTAtGAAGTCGGT
glyA      b  TGCGTTGTCAAGAC CTGTTA TGGCACAA     TCATTCGGT TATACT GTTCgGCGTTGTCC
glyA/geneX b ACACCAAAGAAGCCA TTTACA TTGCACGG     CTATTTTTA TAACAT CCATTCGACATACAT
gnd       b  CCATGCATAAGCTA TTTATA CTTTAATA     AGTAGTTTG TATACT TATTTGCgAACATTCCA
groE      b  TTTTTCGGCCC TTGAAG GGGGCAAG        CCATGGGCA TTTCTC TCGTCaCCAGCCCCGAA
gyrB      b  CCGAACCAAAA TTCGAA GATGTTtAGCGTGAAAAGG TAAAAT AACGGATtAAGCCAAGTT
his       b  ATATAAAAGTTC TTCGTT TGTAACGTG       AAAGTGGTT TACGTT AAAAGACaTCAGTTGAA
hisA      b  GATCTACAAAGTAA TTAATA AATAGTTA     ATTAAGCGT CATCAT TGTACAATGCAaCTGTAC
hisBp     b  CGTCCAGTGCGGTG TTTAAA TCTTTGTG      GGATCAGCG CATTAT CTTacGTGATCAG
hisJ(St)  b  TAGAATGCTTTGCC TTGTCC GGCTGATT      AATGCGCAC GATAGT CGCATCGGATCTG
hisS      b  AAATAATAACGTCA TGGGAA GGGGCTCG     CTTCGGGTG TATCAT TGAACccgCATGCGTC
htpR-P1   b  ACATTAGCGCAGTT AGGCGT GAATAATA      AAAGGCTGT TATACT CTTTCCtGCAATCGTT
htpR-P2   b  TTCACAAGCTTGCA TTGCAG TTGTGGCGATA   AAAATCACGG TCTGAT AAAACAgTGAATG
htpR-P3   b  AGGCTTGCATTGCAAC TTGTCG ATAAAATC    ACGGTGTCA TAAAAC AGTGAATgATAAGCCTGCT
ilvGEDA   b  GGCAAAAAATATCT TGTAGT ATTTACAA     AACGTATCG TAACTG TTTACGCaTTGCTTGGA
ilvIH-P1  b  CTCTGGCTGGCAA TTGGTT AAGCAACA       TCGGACGGT TAATGT GTTttacacatttTTTC
ilvIH-P2  b  GAGGATTTTATGGT TTGTTT TCACCTTT      CCTCGTGTT TATTGT TATtACCCCGTGT
ilvIH-P3  b  ATTTTAGGATTAA TTAAAA AAATACAG       AAATTTGCTG TAAGTG GTGGGATTcAGCGGATT
ilvIH-P4  b  TGTAGAATTTTATT GTCAAT GTCTGGGC      TGTGTATTT TAGGAT TAATTAAAAAAATACAG
ISlins PL p  CCAGCGCCGGTGATG CTGCCA ACTTACTG     ATTTAGTG TATGCAT GGTGtttTTCAGGTCGT
ISlins PR p  ATATATAGCCTTA TGGTAA TGACTCCA        AGTTATTCA TAGTGT TTTATGTtCACATAAT
IS21-II   M  ATGTC TGGAAA TATAGCGG              CAAATCCAC TAGTAT TAACAGCtaTCACTTATT
lacI      b  CACACCATGCAATG GGGCAA AAGCTTTC     CGGGTATCG CATCAT AGGGGCGgCAAGAGACGT
lacP1     b  TAGGGCAGGGCAGCGG TTTACA GTTTATGCT  TCGGGGCTGG TATGTT GTGTGGaATTGTGAGCG
lacP115   M  TTTACAGTTTATG CTTGGG GCTGGTATG      TGTGTGG TATTGT GACCggataacaATTT
lacP2     b  AATGTGAGCTTAGCT CAGTCA TTAGGCAC     CCTAGGGTT TACACT TTATGGCtTCGGGCTCG
lep       b  TCGTCGGCTCAAATG TTGTAG TGTAGAAT     CGGGGGGTT TCTATT AATAcaCAGCTTAAT
leu       b  G TTGACA TGGGTTTT                  TGTATGCAG TAACTG TAAAAGCCATATGGCATT
leultRNA  b  TCGATAATTAACTA TTGCAG AAAAGCTG     AAAACCAC TAGAAT GGGGCTCCGgTGGTAGCA
lex       b  TGTGCAGTTTATGG TTCCAA ATAGGCCT     TTTGGTGTA TACACT CACACGCaTAACTGTAT
livJ      b  TGTCAAAATAGCTA TTCCAA TATCATAA      AAATCGGGCA TATGTT TTAGCaCAGTATGCT
lpd       b  TGTTG TTTAAA AATGTTA               ACAATTTG TAAAAT ACCGACGGtagAACGA
lpp       b  CCATCAAAAAATA TTCTCA ACATAAAAA      ACTTTGTG TAATAC TTGTAACgCTACATGGA
lpp/P1    m  ATCAAAAAAATA TTGTCA ACATAAAAA       ACTTTGT TATACT TGTAACgCTACATCGA
lpp/P2    m  ATCAAAAAAATA TTGTCA ACATAAAAA       ACTTTGTGT TATAAT TGTAACgCTACATCGA
lpp/R1    m  ATCAAAAAAATA TTCACA ACATAAAAA       ACTTTGTG TAATAC TGTAACgCTACATCGA
Kl.rna    b  ATCGGCAACGCCGG GTCACA AGGGCCGG      CAAACCCTG TATACT GGGGGGCgAAGCTCACC
mac11     M  CCCCGGCAGGGAT GAGCAA GGTGGTGCG       CGGGGCTCG TATGTT GTGTGGaATTGTCAACC
mac12     M  CCCCGGCAGGGAT GAGCAA GGTGGTGCG      ACGGGCTCG TATGTT GTGTGGaATTGTCAACC
mac21     M  CCCCGGCAGGGAT GAGCAA GGTGGCAGCCT    TCGGGGCTCG TATGTT GTGTGGaATTGTCAACC
mac3      M  CCCCGGCAGGGAT GAGCAA GGTGGTC         CACGGGCTCG TATGTT GTGTGGaATTGTCAACC
mac31     M  CCCCGGCAGGGAT GAGCAA GGTGCGTC       CACGGGCTCG TATATT GTGTGGaATTGtTCAACC
malEFG    b  AGGGCCAAGCAGCA TGGAAA GACGTTCC      CGTATAAA GAAACT AGAGTGCgTTTAGGTGT
malK      b  CAGGGGGTCGAGCGA TTTAAG CCATGTCC     TGATGACG CATGAT CACGCaTCATCGAATG
malPQ     b  ATCCCGGCAGGATG AGGAAG GTCAACAT        CCAGGCTGG CAAACT AGCGATaACCTTCGTGT
```

Figure 7. Sequences of E. coli promoters. Except for the propensity to find part of the hexameric sets (E.coli "promoter consensus sequences"), TATAAT and TTGACA, about 10 and 35 base pairs upstream the transcription start, there is no obvious sequence conservation among these promoters, although all of them are processed by the same enzyme. The wavy line on top indicates the transcription start area, the arrow the direction of transcription.

In the absence of XTP, the complexation of RNAP with the promoter sequence occurs in several steps. First, a "closed" complex forms, in which the promoter sequence remains in its double-stranded state (10). This complex can form at low temperature (5-10°C) and withstands the challenge of up to .7M K+. The electrostatic interactions, involved in the binding of RNAP to non-specific DNA sequences, are complemented on the promoter by specific and strong non covalent interactions, probably of hydrophobic nature. The formation of the closed complex results in the removal of 12-15 monovalent cations (K+), which shielded the phosphates of the promoter sequence, prior to its complexation (11). RNAP "covers" now a promoter sequence of some 70-80 base pairs, about 250A long. RNAP (150A uncomplexed) must have undergone a large conformational change upon complexation. Also, biochemical probing reveals that about one out of four bases covered by RNAP are in close, stable contact with the aminoacids of the enzyme subunits. The base contact patterns of the same RNAP species on different promoters are not conserved.

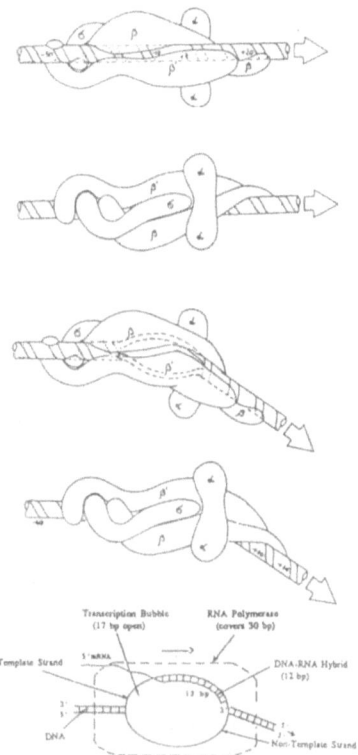

Figure 8: Sketch of the RNAP-DNA complex. From top to bottom: the two sides of the closed promoter complex, with the 5 subunits, the same ofter transconformation into the open promoter complex, and the elongation complex after leaving the promoter region, at non-pause sites. The position of the five subunits was determined by chemical contact experiments. Drawn after Chenchick et al, FEBS Lett. 128, p. 46-50, 1981.

E.Coli grows optimally at 37°C, but keeps growing at reduced rate below 30°C. In the test tube and for a selected promoter, if the closed complex is kept below 25°C, it will remain in that state for hours (12). In contrast, if the temperature exceeds 27°C, the closed complex soon transconforms into a new state, the "open" complex. A row of 10-20 bases on each strand of the promoter sequence, which were not accessible to chemical probes in the closed complex because they were engaged in base pairs, can now be probed. The corresponding base pairs have disrupted in the process, that is the promoter DNA has been unwound over one to two helix turns, as is indeed observed by direct experiments. The precise extention of unwinding depends on the particular promoter studied. In the open complex, RNAP keeps the position that it had in the closed complex. It has been shown in addition that on both sides of the unwound sequence, the helix axis is now bent by some 45°, whereas the uncomplexed promoter DNA is not bent (13).

The formation of the open complex (fig. 8) has some rather intriguing characteristics. In addition to the energy required for bending the helix axis, the disruption of 10-20 base pairs corresponds to that of several tens of H-bonds, amounting to invest some 50-100kcal/mole. Yet no chemical energy has been supplied to the system, which includes only purified RNAP and promoter DNA in a salt buffer. Hence the energy required for bending and opening must have been borrowed from the thermal reservoir and localized in the complex. This typical non linear effect has arroused the interest of several teams of physicists (see accompagnying contributions).

The open complex is quite stable. It can be kept at room temperature for hours and days, without loosing its ability to start transcription as soon as the XTP monomers are added. As the latter are supplied, RNAP begins the production of mRNA (10).

The genetic information is written on one strand only of the gene, the "coding" strand. The complementary strand is called the "template" strand, as the RNAP uses this strand as the template to produce the mRNA. The mRNA sequence is therefore complementary to that of the template strand, that is it is identical to that of the coding strand. The information on the coding strand always reads in the 5' to 3' direction, and transcription proceeds in the same direction. (Both strands of a genome carry coding sequences, reading of course in opposite directions)

As XTPs are supplied to the open complex, RNAP selects the template strand and begins synthesizing its complement. Transcription starts at a precise base located inside the unwound region of the promoter DNA. For a particular promoter studied, if +1 designates the first transcribed base, the promoter sequence covered by RNAP extends from about -55 to +23, the unwound region from -11 to +3 at least (- and + stand respectively for upstream, or non transcribed, and for downstream, or transcribed, bases)

In vitro, the enzyme makes several attempts to start transcription (the energetically rich triphosphates (XTP) are split to monophosphates (XMP) and pyrophosphate), but synthesis aborts as short runs of oligomers (<10 bases) are synthesized. The oligomer is released and the process resumes at the start site. The open complex does not change its structure during this abortive initiation step.

Following several attempts (their number depends on the promoter), abortion is overcome. The RNAP loses one of its subunit (sigma, stringently required for promoter recognition, which is bound to RNAP by fairly weak electrostatic forces), progressively covers shorter segments of the DNA (about 50, then 30 base-pairs, as 12, respectively 18mer mRNA are synthesized) and starts moving in the 5' to 3' direction of the coding sequence. The unwound DNA sequence ("bubble") found in the open promoter complex rewinds, probably as 15-20mers are synthesized. Meanwhile, the bubble has moved with the RNAP, i.e. in the sequence covered now by the enzyme, 10-20 base pairs became disrupted.

RNAP will continue "walking" on the gene and dissociates only as it meets the termination signal, up to several thousand base pairs downstream. It may then enter a new round of transcription, on the promoter of the same, or of another gene. At the places where RNAP actively elongates the nascent mRNA, DNA is found unwound (bubble of 10-20 bases), and the helix axis is bent by 45°, a few bases downstream of the base about to be transcribed (see fig. 8 for structural details of the elongation complex).

It can be observed in vitro that, at well-specified sites of the DNA template, the transcription complex pauses for characteristic time intervals, then resumes the synthesis of the nascent mRNA (the pause delays, but does not abort mRNA synthesis). At the pause sites, the transcription "bubble" (unwound DNA segment) appears to temporarely collapse, then reforms as transcription resumes (14).

Pause sites are on the average a few tens base pairs apart, but pausing on several consecutive base pairs has been observed, as well as gene segments of several hundred base pairs devoid of a pause site. The pausing signals sensed by RNAP on a given site of the DNA (or on the nascent mRNA?) are largely unknown. To date, the only identified and probed signal is a row of six adenines, known to have the potential to make a permanent bend in the DNA helix axis (15). It has been proposed that the overlap of this permanent bend with the bend associated with the elongation complex, could cancel each other. This would arrest the elongation complex, until thermal fluctuation opening (of the type experienced by the closed complex?) would restore the transcription bubble and the bend.

Kinetics of the various transcription steps have been studied in vitro. For the particular promoters studied, the rate limiting step in transcription initiation is the closure of the open promoter sequence (16). The rate of elongation from base to base between pause sites (of the order of a few milliseconds in vivo) is three to four orders of magnitude higher than that through a pause site, which compares to the rate of formation of the open complex. RNAP spends at pause sites some 80% of the total time taken to transcribe the gene (15). Transcription kinetics can be envisionned schematically as an alternance of slow steps (initiation and restart after pausing), which involve DNA opening, and productive steps (the addition of bases to the nascent mRNA between pause sites), at least thousandfold faster, which occur in already open transcription bubbles.

Translation of the messenger RNA into protein (9).

Transcription and translation are strickingly illustrated on the drawing (fig. 9), sketched from an electron micrograph of bacterial DNA in the process of gene expression. Several RNAP particles are attached to the central DNA strand, each in the process of synthesizing a mRNA. The picture looks much like a christmas tree, with DNA as the stem, and the branches (nascent mRNAs) covered with a series of balls, each with a pending coil with a length increasing , as the balls come closer to the stem.

An enlarged drawing of a branche is seen on fig. 10. The coils correspond to nascent proteins. Each ball is identified as a ribosome, the largest device found in the bacterial cell, which harbours a total of some 50.000 of them (about ten times more than RNAPs). A ribosome is made of about 50 different proteins, assembled in two independent subunits, a smaller one, which catches the mRNA (about 50 bases of the mRNA are covered by the ribosome), and a larger one, which assembles the protein, according to the information read on the mRNA.

The mRNA is functionnally divided in three parts. Usually, the 15-30 first bases (5' end) of the mRNA are not translated. They support the assembly of the two ribosome subuits, and set the assembly rate. The final (3' end) tens to hundreds bases are not translated either, and seem to be involved mainly in setting the life-time of the mRNA, usually of the order of 1-3 minutes. The middle part is the coding sequence, which bears the information needed to built the protein. This information is encoded as an ordered, linear array of the four bases. It is translated into a linear sequence of aminoacids via the genetic code. This code

Figure 9. Transcription- translation process. Box: an enlarged view. Redrawn from electron micrographs by O.L. Miller et al., Science 164, p.955, 1969, and 169, p. 392, 1977 (box)

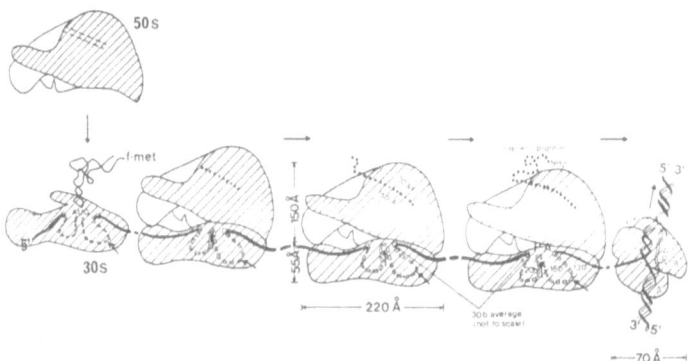

Figure 10. Close-up of transcription and translation. The double-stranded DNA is transcribed by RNA polymerase (far right). The mRNA is the thick black thread running from left to right. At its left end (the 5' end), the small subunit of the ribosome (30S, hatched) is attached, and the special tRNA (f-Met) binds its anticodon to the initiation codon AUG. The large Ribosomal subunit (50S) is about to complex to the small subunit. Notice the tunnel (dotted) in the 50S subunit. The ribosome which attached just befor has moved to the right of the ribosome about to assemble. It has already sythesized a short protein (dots in the large subunits, the tunnel is omitted in the drawing). The ribosomes to the right have initiated translation earlier, the one closest to the RNA polymerase was the first to translate the mRNA, and therefore has synthesized the largest protein portion.

associates each (except 3) of the $4^3=64$ possible arrangements of the four bases into triplets or codons, with one of a total of 20 different aminoacids. The general formula of aminoacids is NH2(amino)-CHR-COOH(acid), R standing for the 20 different "side chains" almost universally found in proteins. 8 aminoacids have polar side chains, 7 hydrophobic and 4 charged ones; the 20st (R=H) is ambivalent (see fig 11 for details). Because 20 aminoacids are specified by 61 codons, the genetic code is degenerate. Indeed, most aminoacids are specified by 3 or more (up to 6) codons, two only by one codon. The importance of the code degeneracy in the control and regulation of gene expression appeared quite recently and will be discussed later.

The coding sequence begins usually with an AUG "initiation" codon (U (uridine) replaces T in RNA; its formula is that of T, except for the methyl group). AUG specifies also an aminoacid, methionine, one of the two aminoacids encoded by one codon only. The first AUG on the 5' end of the mRNA is not necessarily the initiation codon, and in 10% of the cases, the initiation codon reads GTG, illustrating the fact that the initiation codon cannot be at present firmily determined from the sequence alone. Obviouly, additional, yet unidentified informations are stored on the local mRNA sequence, which qualify the initiation codon.

The initiation codon is crucial in setting the reading frame of the gene. Indeed, this codon sets the three base (codon) frame in the base sequence downstream (5' to 3'), which specifies the codon sequence of the gene. This "coding" sequence ends with a stop codon (one of the three codons which do not specify an aminoacid). Hence a coding sequence of 3n bases, followed by a stop codon, specifies a linear protein of n aminoacids: $NH_2\text{-CHR1-CO-(NH-CHRi-CO)}_{n-2}\text{-NH-CHRN-COOH}$.

To perform its task of producing the protein according to the coding sequence, the ribosome needs, at the site of the mRNA-born codon it is about to translate, the delivery of the corresponding aminoacid. Aminoacids are synthesized and stored at a remote places of the cell.

The transfer of the aminoacids from that pool to the ribosomes is performed by specialized devices, the "transfer" RNAs (tRNAs). E.coli for instance has 43 different tRNAs. tRNAs are short RNAs (about 80 bases long), sharing a common CCA sequence at their 3' end, to which the transported aminoacid is attached. tRNAs adopt in general a clover-leaf like structure. Close to its center, each tRNA exposes a specific 3 base sequence, the "anticodon". For a given tRNA, the anticodon sequence is complementary to at least two bases of the codon(s) specifying the aminoacid attached to its CCA end. Some aminoacids are transferred by two or three different tRNAs, which may each "recognize" the same, or different, codons,all specifying this aminoacid. For other aminoacids, one and the same tRNA recognizes all codons (as much as four) of its cognate aminoacid.

Once the small ribosomal subunit has attached to the untranslated 5' end of the mRN, it moves to the initiation codon, where a special tRNA enters the ribosome and binds to that codon by its anticodon. The large ribosomal subunit binds then to the small one, the ribosome is now ready for translation The special aminoacid (modified methionine, most often removed from the protein later) is discharged from the CCA end of the tRNA. The latter is then released and enters a new transfer cycle. Meanwhile, the ribosome has moved one codon downstream (5' to 3'), the second codon has been bound by its cognate tRNA, the attached aminoacid is removed and fused to the modified methionine, and so on until the stop codon is reached. There the ribosome dissociates into its two subunits, which are ready for another round, and the protein just synthesized is released. Its half-live is of the order of tens of minutes, and may last for several hours.

The kinetics of translation received little attention so far, but appear quite important in controlling and regulating gene expression. The tRNAs are in general in short supply, 3 to 4 per ten ribosomes for the most abundant species, as few as 1 per hundred ribosomes for the less abundant (17). Therefore, the rate-limiting step in translation elongation at a given codon is the availability of the cognate tRNA. From codon to codon, the rate of elongation was indeed observed to vary hundredfold, depending mainly on the cellular concentration of the cognate tRNAs, and to a lesser degree on codon-specific rate constants (18).

Assuming that the elongation rate is inversely proportional to the relative cellular concentrations of the tRNAs, which are known for E.coli, and with the help of a simple and plausible hypothesis on the codon-specific rate constants (see fig 11), the translation kinetics can be computed from the gene sequence (19). They compare quite well to those measured experimentally for a set of genes. In addition, parameters at present not accessible to experiment can be computed also, like for instance the translation rate for any codon, and the ribosome traffic charcteristics. Indeed, in the sequence of about 50 bases, or 17 codons, covered by the ribosome, the codon that it is about to translate is codon number 10 from the 5' end. This allows in particular to compute the largest number of ribosomes which can be present at a given instant on the mRNA in the absence of jam (traffic density), the distance and time interval between consecutive ribosomes at any place of the mRNA, the places where jam occurs, the number of queuing ribosomes,... It is observed for instance that for 40% of over 2000 sequenced genes of E.coli, the place of the slowest translation is right at the beginning of the coding sequence. This warrants automatically that the traffic will never experience jam, since upon translation initiation, the slowly moving ribosome has to clear the start place before the next can enter, and can never be caught up by the latter.

These computations gave insight into species-specific strategies, which exploit the genetic code degeneracy to control and regulate gene expression. As already mentioned, most aminoacids are specified by several (synonymous) codons. A given aminoacid may be transferred by up to four different tRNAs, each recognizing one or more synonymous codons. Large differences (up to 20-fold in E.coli) are observed in the relative cellular concentrations of tRNAs charging a given aminoacid. Thus a given aminoacid can be encoded either by a codon recognized by an abundant tRNAs, or by a synonymous codon corresponding to scarce tRNAs, and is accordingly translated fast or slowly.

It is observed that the most abundant codons are almost systematically used in coding sequences for strongly expressed proteins. In lesser expressed proteins, clusters of codons recognized by abundant and scarce tRNAs alternate. In one experiment, a few frequent codons were exchanged for synonymous but rare codons. The result in vivo was dramatic (20,21). The expression of the protein was reduced almost hundredfold, due to the fact that many proteins aborted during translation, mostly because of shifts in the reading frame. Transcription aborted frequently also and the life-time of the mRNA was reduced fourfold. Similar, though less dramatic effects, are observed upon exchanging unfrequent for frequent synonymous codons. These observations can be readily understood in terms of perturbations of the translation kinetics and ribosome traffic.

In several species, the relative intracellular concentrations of tRNA have been measured. It is observed that for a given codon, the relative concentrations of the cognate tRNA may differ significantly among species. It is further observed that, in a given species, a linear relationship between the frequency of codon usage, and the relative concentrations of the corresponding tRNAs, holds reasonably well.

Character	Name, abrev	R formula	codon	anticod	tRNA%	k .01/ms
Hydrophobic	alanine, Ala	CH3	GCT	GGC	4.20	6.12
			GCC	GGC	4.20	7.63
			GCA	UGG	6.06	7.28
			GCG	UGG	6.06	5.54
	valine, Val	CH-CH3 CH3	GTT	GAC	2.33	4.60
			GTC	GAC	2.33	6.76
			GTA	UAC	6.12	6.25
			GTG	UAC	6.12	5.99
	leucine, Leu	CH2-CH-CH3 CH3	CTT	GAG	1.75	2.61
			CTC	GAG	1.75	9.79
			CTA	?	.58	4.35
			CTG	CAG	5.83	4.44
			TTA	AAA	1.46	4.73
	isoleucine, Ile	CH-CH2-CH3 CH3	TTG	AAA	1.46	4.73
			ATT	GAU	5.83	7.15
			ATC	GAU	5.83	6.79
			ATA	NAU	.29	3.09
	phenylalanine, Phe	CH2-phenyl	TTT	GAA	2.04	6.09
			TTC	GAA	2.04	5.64
	methonine, Met	CH2-CH2-S-CH3	ATG	CAU	1.75	6.60
	proline, Pro	(N-CH-COO) CH2-CH2-CH2	CCT	?	.99	3.93
			CCC	?	.99	3.06
			CCA	?	3.32	6.31
			CCG	?	3.32	6.50
Polar	serine, Ser	CH2-OH	TCT	GGA	2.39	4.25
			TCC	GGA	2.39	4.03
			TCA	UGA	1.46	3.77
			TCG	UGA	1.46	4.17
			AGT	GCU	1.46	3.96
			AGC	GCU	1.46	5.31
	threonine, Thr	CH-OH CH3	ACT	GGU	4.67	4.35
			ACC	GGU	4.67	6.54
			ACA	UGU	1.40	3.83
			ACG	UGU	1.40	4.83
	tyrosine, Tyr	CH2-phenyl-OH	TAT	QCA	2.92	5.44
			TAC	QCA	2.92	4.93
	cystene, Cys	CH2-SH	TGT	GCA	1.17	2.96
			TGC	GCA	1.17	3.35
	asparagine, Asn	CH2-CO NH2	GAT	QUC	4.67	5.70
			GAC	QUC	4.67	6.47
	glutamine, Gln	CH2-CH2-CO NH2	CAA	UUG	1.75	6.76
			CAG	CUG	2.33	4.70
	histidine, His	CH2-C=CH-NH NH=CH	CAT	QUG	2.33	4.99
			CAC	QUG	2.33	7.41
	tryptophane, Trp	CH2-C=CH-NH phenyl	TTT	GAA	2.04	6.09
			TTC	GAA	2.04	5.64
Charged	aspartic acid, Asp	CH2-C-O- O	GAT	QUC	4.67	8.66
			GAC	QUC	4.67	6.02
	glutamic acid, Glu	CH2-CH2-C-O- O	GAA	UUC	5.25	4.12
			GAG	UUC	5.25	7.66
	lysine, Lys	CH2-CH2-CH2-CH2-NH3+	AAA	UUU	5.83	8.02
			AAG	UUU	5.83	4.83
	arginine, Arg	CH2-CH2-CH2-NH-C=NH2+ NH2	CGT	ICG	5.25	2.54
			CGC	ICG	5.25	3.22
			CGA	ICG	5.25	4.31
			CGG	?	.64	6.38
			AGA	?	.70	2.25
			AGG	?	.70	1.80
Ambivalent	glycine, Gly	H	GGT	GCC	6.53	3.86
			GGC	GCC	6.53	6.54
			GGA	UCC	.88	5.15
			GGG	CCC	.58	6.89

Figure11. List of the 20 aminoacids, classified according to their main physico-chemical characteristics, their chemical structure and the corresponding codons. The anticodons of the cognate tRNAs in E. coli are listed also, together with the relative abundance of the tRNAs in the E. coli cell and the codon- characteristic translation rate constant, k. The latter was computed from the frequency of usage of the codon in the 50 most expressed genes in E. coli, supposed to determine the optimal usage of the tRNAs

Obviously, a species has the option to exploit the degeneracy of the genetic code in its own way, by producing a specific distribution of the concentrations of its tRNAs, and by using the cognate codons accordingly. If it is assumed that the linear relationship between codon frequency and tRNA concentration applies for a species with unknown cellular tRNA distribution, translation kinetics and traffic on a coding sequence of this species can be computed from the frequencies of occurrence of the codons, which can be simply derived from the sequence database of that species.

III. Control and regulation of gene expression: physical aspects.

We just mentionned the effect of synonymous codon selection in translation kinetics and ribosome traffic, showing that the flexibility allowed by the genetic code degeneracy is actually used to actively modulate within large limits the production of proteins. Obviously, the modulation can occur only at places where the genetic code applies, i.e. inside coding sequences. However, important steps in gene expression, like transcription, or assembly of the ribosome at the 5' end of the mRNA, take place also outside coding sequences, or inside (RNAP pausing for instance), but do not involve the genetic code directly. We will focus now on the latter.

The basic mechanisms involved in any particular step of gene expression, are shared by all genes within a genome, and probably in other species as well. Indeed, using standard technics of molecular biology, the various elements of a given gene, like promoter, coding sequence, ribosome assembly line at the 5' end of the mRNA, transcription termination sequence etc..., can be often successfully replaced by similar elements borrowed from other genes of the species, or even (especially for coding sequence) from other species. In most cases, the protein specified by the coding sequence will be produced, possibly with an efficiency specific of the particular construction made. This is so because in most instances, the signals needed for the expression of a gene are embedded in the nucleic acid, at or near the place where a particular step of the expression occurs. The enzymatic machineries in charge of expression (RNAP, ribosomes,...) "recognize", then sense these signals, which in turn provide them with the instructions they need to perform their task within precise limits.

This raises a series of fundamental questions, dealing with the exact nature of a particular signal, and how it triggers the enzyme action: the molecular physics of gene expression.

Signal elements.

Control and/or regulation signals are usually encoded on the nucleic acid at, or close to, the locus where they are sensed by the enzymes. The base sequence bearing the signal can be at present easily established. As an example, we mentionned earlier that over thousand promoters of E.coli have been sequenced. The comparison of their sequences, all different, is frustrating, since only very tiny, insuffcent and unreliable homologies are observed. The same holds for other signals, like for instance the transcription termination signals and the ribosome assembly sequences located on the 5' end of mRNAs. Obviously, even in a given species, a specific signal is encoded in different genes by a bewildering variety of synonymous base sequences. All we know is that each of the synonymous sequences must express a common set of basic, physical properties, capable of getting hold of the enzyme and triggering its activity. In addition, this activity must take place within precise limits, in keeping with the control and regulation characteristics of individual genes. It is the superposition of these basic and regulatory or tuning instructions which is probably responsible for the diversity of the signal sequences, as seen for promoters for instance.

Signal essentials.

What could be the "basic physical properties" characterizing a given signal? Since signals can be encoded by diverse base sequences, the precise chemical identity of individual bases or base pairs seems not directly involved. Rather, short or long range collective and/or cooperative properties, shared by sequences of bases or base pairs, would better suite these constraints (no clear-cut border will be set here between collective and cooperative properties).

A basic collective property of a DNA sequence is of course the precise local DNA structure. As discussed earlier, the conformation, and conformation flexibility, of the double helix are controlled by the base sequence and its environment. DNA segments with diverse sequences can be designed, sharing under identical or close physiological environments, closely similar conformations and/or other structural characteristics. The same holds for the steric positioning, in both DNA grooves, of immobilized water molecules, or of proton donors and acceptors exposed by the bases. Other local helix parameters, like those characterizing its torsion or bending properties, are linked to local constraints in base or base pair distributions, which can be matched by a variety of sequences also.

Cooperative properties between bases or base pairs are involved in gene expression as well. Constraints in the local distributions of AT and GC pairs for instance, give rise to well-characterized cooperative domains (see "Cooperativity" section), which can be encoded by diverse sequences, provided that they share the AT /GC constraint and distribution. Each domain has a specific "stability", as it unwinds at a characteristic value of the denaturing parameter.

The biological relevance of DNA unwinding is obvious, because the genetic information is invariably derived from a single- strand (see "Transcription" section, the same holds for DNA replication). At the promoter or following a pause for instance, the helix must be locally opened for transcription to proceed. One would therefore expect that the "open" segment would be a cooperative domain of defined stability. Common, specific constraints in AT and GC distributions are indeed observed in promoter sequences.

A variety of sequences were synthesized, matching the cooperative stability domains shared by a set of E.coli promoters (DNA and RNA, up to 100 bases long, can be automatically synthesized quite easely). Probed in vivo (23-25), these synthetic sequences direct transcription of a gene attached downstream, confirming that part at least of the promoter signal is with the promoter- specific cooperativity pattern. However, in this experiment, the efficiencies among the synthetic promoters, measured by the amount of mRNA produced per unit time, varied tenfold, and transcription started at several, unexpected places. The signals setting these important parameters were obviously missed in the selected AT/GC constraints.

Signal coordination.
A genuine promoter sequence bears actually a whole list of signals, enabling it to successively bind, then complex RNAP; force RNAP to select the

template strand, to open the promoter at a given rate, and to start transcription at a selected base; set the rates of promoter closure and clearance, of entrance of the next RNAP (promoter efficiency) etc... Each of the individual signal elements may be optimally encoded by a dedicated base sequence, of the order of a helix turn (10 bases) or more. Furthermore, almost all promoters are recognized and processed by one and the same RNAP. Therefore, a particular signal is most probably located at a similar place in all promoter sequences, where it can be sensed by the corresponding receptor of the enzyme subunit in contact, and trigger its activity. Given the small space (70-80 base pairs) covered by RNAP, it is likely that the many signal sequences overlap. Instead of being encoded each by its optimal sequence, the superposed signals might content with degraded, yet functional synonymous sequences of the signals. The compromises may be hierarchized, as more or less essential signals are encoded more or less optimally. The actual promoter sequence would then be the result of the condensations of the ensemble of promoter signals. Different promoter sequences would then reflect different compromises

It is amazing to observe how many hierarchized informations are squeezed into a single linear promoter sequence of, say, less than 100 base pairs. Moreover, this can be achieved in thousands of ways, since the sequences of the many E.coli promoters differ, but are nevertheless recognized and processed by the same RNAP. Rephrased in terms of information science, we have a set of several thousand sentences, each made up of a linear, ordered sequence of less than 100 characters of four symbols (ATGC). The sentences have a quite similar meaning, "promoter", supported by a series of words (the signals). The number of words is unknown, but is expected to be of the order of ten at least. The length of a word is unknown also, but should be of the order of 10 characters on the average. The dictionary of the words is unknown, with a few exceptions (AT/GC constraint for instance), but most words are believed to be characterized by specific constraints in the distributions of the four symbols. The words may overlap, as part or all characters of a word may be included in another word. The position of a given word is the same, or almost the same, in all the sentences. Would this be enough information for back- engineering, aimed at uncovering the words and completing their dictionary?

Signal tracks.
Situations similar to that observed at promoters are met at other places of the gene where the expression is in need of control and regulation, perhaps with lesser signal density. In addition to the basic information defining the aminoacid sequence of the protein, the gene sequence carries at least three distinct layers of instructions, one directing transcription (promoter, pausing, termination), one for translation (ribosome assembly, translation kinetics, ribosome traffic), and one for DNA replication (not discussed in this contribution). In prokaryotes, transcription and translation are tightly coupled (see fig. 10), meaning that the layers for transcription and translation must be carefully adjusted, to coordinate the kinetics of both processes. In particular, pausing of RNAP should correlate with translation of a row of "slow" codons, as is indeed observed in a few cases (our unpublished observations). Do we know of another example of a linear, uninterrupted full score, written with four symbols only and devoid of any punctuation (the gene sequence), performed (expressed) in perfect concert by several musicians (the enzymes in charge of gene expression and replication), each of them deriving all instructions needed for his instrument from a dedicated pattern of symbols, which may change from place to place? Questions for information scientists: assume the size of the

pattern is limited to some reasonable value (say 10 symbols); does a score of given size (say 1000 symbols) have a maximal instruction content, can this maximum be denumbered, how does it depend on the pattern size? How does the information density of the score compare to that of informations encoded conventionnally by present-day methods?

Signal transduction.

We did so far deal with the signals encoded in the nucleic acid. How the signals are sensed by the enzyme, and trigger the process it has to perform, is at present poorly understood.

Some structural data exist. The aminoacids of a protein, in close contact with bases of the nucleic acid, can be identified biochemically. The "footprint" method determines the portion of the nucleic acid protected by the complexed protein against enzymatic digestion. Crosslinking, including photocrosslinking, reveals the aminoacids of the protein and the nucleic acid constituants in close contact. In a few cases, a specific protein (or a fragment of it) complexed to its nucleic acid target could be crystallized and the structure of the complex established by X-ray diffraction. Recent technical progresses in NMR enable to study in solution the structure of nucleoprotein complexes of fair molecular weight. Other spectroscopy techniques, in particular vibration spectroscopies, were developed to meet the demands of molecular biologists for precise informations on local structural and dynamic properties of nucleic acids, of specific proteins and of their complexes.

During gene expression, the links between nucleic acid and protein are of several types, but are never covalent. Electrostatic interactions are the most common. In uncomplexed nucleic acids, the phosphate groups are screened by counterions. In monovalent salt buffers with concentrations ranging from 10^{-4} to .5M, each phosphate of a double helix is permanently screened by an average of .88 monovalent counterion. A protein can bind to the phosphates via the two positively charged aminoacids (lysine and arginine), which may compete with the couterions. However, displacing the screening counterions is difficult, as they remain tightly associated with the phosphate groups even at ionic strength as low as 10^{-4} M.

Hydrophobic interaction may take place between domains of the protein, enriched in a selection of the seven hydrophobic aminoacids, and the nucleic acid bases. More or less strongly bound water molecules, in particular those located in both grooves of the double helix, may be removed during the complexation step. The positively charged aminoacids of the protein may then come close enough to the phosphates, and compete and replace the screening counterions.

Complexes of specific proteins and their nucleic acid targets, in particular those tight enough to allow the identification by biochemical methods of the elements in contact, may involve a mixe of electrostatic and hydrophobic links. The promoter- RNAP complex is probably of this type. It is indeed observed that some 12-15 monovalent screening counterions are removed from the promoter DNA as it is complexed by RNAP (16). This desorption could be responsible for the bending of the promoter DNA helix. If it is assumed that the counterions are removed from the phosphates located on one side only of the double helix, the repulsion between the unscreened, or partially screened phosphates, could then be strong enough to force the helix to bend, as is indeed observed.

The structures of the few nucleproteins crystallized so far reveal also water bridges, where one or two water molecules link a base or sugar ring to polar or charged aminoacids. Hydrogen bonds, between proton donors and acceptors located on the bases and the aminoacids, are observed also, but appear to play a less crucial role than was thought earlier.

While the structures of nucleoproteins can at the present time be explored experimentally and modelized to some extent, little is known about how the signals trigger the enzymatic mechanisms. The physics of the signal transduction at the molecular level is an almost unexplored field.

IV. The protein folding problem.

A given protein is in charge of a dedicated function. It can perform its task only once it has adopted a unique three- dimensional structure (27, 28), selected among an astronomic number of possible structures, as each aminoacid can adopt a number of rotational conformations (fig. 12).

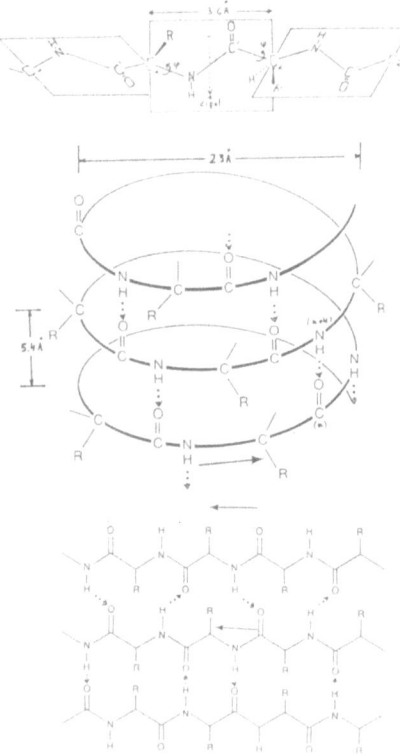

Figure 12. Basics of protein structure. Top. The peptide unit, which includes the C_α and the C'=O of an aminoacid, and the NH and C_α of the next aminoacid, is planar. Each unit has two degrees of freedom, as it can rotate around N-C_α (ϕ) and C_αC' (ψ). Due to hindrance resulting from interactions of adjacent residues, the values of ϕ and ψ are in most cases found in the "core" intervals (-170°, -45°; 180°, 100°), and (-120°, -30°; 30°, -60°). The first interval corresponds to αhelices with 3.6 residues per turn (middle of the figure), and βstructures (bottom), either parallel (upper) or antiparallel (lower), the two main structures found in proteins, linked by more or less random coils. Notice the H-bonds (dotted lines) linking non nearest neighbour aminoacids in these structures.

The knowledge of the protein structure (see fig. 12 for basic elements of protein structure) is a prerequisit for the understanding of the mechanisms of enzymatic reactions, or the reasons why an altered protein cannot perform its function. The protein structure is also important in many applications. In pharmacology for instance, knowing the protein structure allows to design drugs fitting closely the protein surface, and to position the active part of the drug at the right spot of the target protein.

To date, the experimental determination of the stucture of a protein takes of the order of a year, is expensive (X-ray diffraction equipment, synchrotron radiation facility, complex computations), and is not always feasable, as certain proteins could not be crystallized, despite years of efforts...

Protein refolding.

A protein can be "denatured", meaning that its function, and the associated native structure, have been lost, although the aminoacid chain keeps its native sequence and remains unbroken. Denaturation can be brought about in many ways, by heating or mechanical shaking of the protein solution, or by chemicals, like urea, guanidinum chloride, etc...(29).

It has been known for long that a particular protein, once denatured, could recover most of its function, some time after it had been brought back to physiological conditions. It was therefore concluded that the information needed to achieve the native structure, and hence the function of the protein, is carried in full by its aminoacid sequence. Folding and refolding would be equivalent.

Refolding and structure prediction.

This statement was the start for many studies aimed at finding the rules linking the aminoacid sequence of a protein to its structure. With the help of the genetic code, it is indeed simple to derive the aminoacid sequence from the base sequence of the gene. An algorithm building the native structure from the aminoacid sequence would then avoid the burden of the experimental structure determination.

The general procedure ("global" approach) consists in writing down the complete aminoacid sequence of the protein, set the initial coordinates of all atoms, write the (empirical) interaction potential between all atoms not linked directly by a covalent bond and wiyhin a given distance, write the total energy (sum of covalent and non covalent interactions) and minimize it. This should yield the stable conformation of the protein. However, many local minima with very close energies are usually obtained, corresponding each to quite different conformations. Despite a wealth of often brilliant attempts, no algorithm has been found so far predicting the native conformation of a protein. The best performing ones predict about 80% only of the protein structure established experimentally. Obviously, in addition to flaws due to the empirical potential functions and to neglecting the solvent, some crucial informations, needed for proper protein folding, are missing in the aminoacid sequence, but are provided in the cell.

A series of experimental observations challenge the general validity of the equivalence of folding and refolding. The proteins able to refold, usually taken of small size, need tens of minutes, or even hours, to recover their activity once brought back to physiological conditions. For larger proteins (say about 150 aminoacids long), renaturation may take days, and still larger ones (>200 aminoacids) may just permanently fail to renature. In the cell however, all proteins (average lenth 300 aminoacids) are active by the time, or a few seconds after, their synthesis is completed. Obviously, in the cell a folding pathway must exist for each protein, which speeds up by orders of magnitude the completion of its folding.

Protein folding in vivo.

Let us consider protein synthesis and folding as it occurs in the cell. Protein synthesis is vectorial, as the aminoacids are fused to the nascent protein in the (5' to 3') order of the corresponding codons on the mRNA. The fusion takes place at a particular site of the large subunit of the ribosome. A kind of tunnel links this site to the ribosome surface, through which the nascent protein is extracted. The tunnel is some 100A long and accomodates about 27 aminoacids of the nascent protein. Since in the most extended conformation of the protein, an aminoacid measures 3.2A, the part of the protein chain in the tunnel must be in the most extended conformation. The chain is free to fold only once it has emerged at the ribosome surface.

Folding vs. refolding.

This has three kinds of consequences on protein folding:

a) the part of the nascent protein, which has already left the ribosome, remains anchored at the tunnel exit, and the volume occupied by the ribosome is not available for folding;

b) the aminoacid at the tunnel exit is immobilized to some degree. It will be subjected to interactions from behalf of its local environment: the ribosome (see a)), the solvent (water), the nascent protein which has already left the ribosome (specially the aminoacids just preceding). The conformation adopted by the aminoacid at the tunnel exit will be determined by interactions specified by the type of the aminoacid (hydrophobic, polar, charged), the type of the aminoacids of the nascent protein in close vicinity, and possibly the type of the aminoacids or proteins of the ribosome, which form and suround the tunnel exit. As this aminoacid leaves the tunnel exit, some of the constraints will progressively soften and vanish.

c) the aminoacid at the tunnel exit will remain there for the time it takes to fuse the aminoacid to the nascent end of the protein, 27 aminoacids down in the tunnel. Therefore, the time left to the aminoacid at the tunnel exit to find its energy minimum at this place, is set by the rate of translation of the codon that the ribosome is about to process. The latter is located on the mRNA 27 codons downstream (3') the codon which corresponds to the aminoacid present at the tunnel exit.

These features have important consequences on folding, but are ignored, or appear in part only, in the aminoacid sequence, to which the refolding process restricts.

Concerning point a), let us assume that, at the tunnel exit, the ribosome is equivalent to an unlimited plane, and that the aminoacids of the protein are restricted to fold along the edges of a cubic lattice. We are interested in counting the number of different conformations which can be taken by a protein of N aminoacids. This amounts then to counting the number $P(N)$ of different, self-avoiding pathes which can be made in N steps in the cubic lattice. Denumbering was made by computer simulation (N<30, (30)).

If the limitations mentionned in a) are ignored (global approach), each configuration would be a self-avoiding random walk in an unlimited cubic lattice; then $P(N)=4.87P(N-1)$. If the limitations are accepted, we consider two cases: a walk, where the growing point is with the walker, and a growth, where it is at the start site (the exit of the ribosome tunnel). The latter mimics best translation in vivo.

For the self-avoiding random <u>walk</u> in a cubic lattice, limited by a solid plane (ribosome), all walks starting at the same origin in the plane, P(N)=4.65P(N-1). For the self-avoiding random <u>growth</u> in a cubic lattice, limited by a solid plane (ribosome), each growth starting at the same origin in the plane, P(N)=4.15P(N-1). The difference in the number of pathes between walk and growth comes from the fact that once the walker has left the plane, he may have the option to move towards the plane again, which is excluded during growth, since once the first step has left the plane, it can never come back to it. The ratio of the number of pathes in the first and the third case is $(1.175)^N$. Despite the oversimplification of the model, it is clear that neglecting the constraints mentionned in a) amounts to consider many conformations which cannot form in the cell and may contribute to the minima degeneracy.

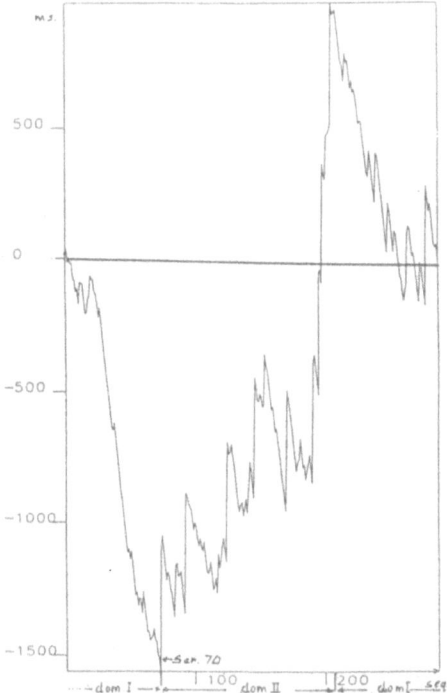

Figure 13. Translation acceleration chart, drawn for the protein β-lactamase of E. coli. Abscissa: sequence in aminoacids. Ordinate: the time taken to translate the protein to aminoacid (i), minus i times the average time taken to translate an aminoacid of the protein, in milliseconds. Negative (or positive) figures indicate that the actual translation is in advance (or retardates), compared to translation proceeding at the constant, average rate (horizontal line at ordinate 0). The diagram shows that translation is very fast (codons with abundant cognate tRNAs) up to serine 70 (the active center of the protein), where it is 1500 ms in advance on the average, then begins to lose slowly its advance (less frequent codons), before translation of a row of about 20 codons with rare tRNAs (aminoacids 190 to 212), which cause a retardation of 1500 ms, the speeds up again. The protein has been crystallized. It is formed by two domains, domain I largely helical, domain II mainly in conformation. The domain boundaries are indicated below the abscissa scale. It is stricking to observe that domain II fits precisely the aminoacid sequence 70 to 212, corresponding to the slow translation step.

The limitations in b) show that the computations made in the global approach are probably unrealistic in several aspects. For instance, the global approach considers interactions for a given aminoacid (like that at the tunnel exit mentionned in b)) with all the others in the total chain, although those of the part of the chain which remains to be synthesized cannot contribute. The observation that a nascent protein can adopt its final, native conformation 20-30 aminoacids away from the tunnel exit of the ribosome, indicates that the definitive conformation of an aminoacid can be reached soon after leaving the ribosome. Also, due to the temporary immobilization of the aminoacid at the tunnel exit, the coordinates of its atoms remain fixed in a first approximation, during the energy minimization procedure. The minima obtained will most probably differ from those obtained by the global method. The role of the ribosome, as a possible catalyst for protein folding, cannot be ruled out (auxiliary proteins, the chaperonins, are present in the cell and may help in the catalysis the protein folding).

The time restrictions mentioned in c) may prevent the aminoacid at the tunnel exit to find its conformation minimum, before it is displaced by the following aminoacid. As the aminoacid moves away from the ribosome, the constraints mentioned in b) relax, its conformation could be permanently trapped in the metastable state it had attained at or close to the tunel exit. This scenario would be consistent with the observation that sometimes, the native conformation of a protein, or of part of it, is not that corresponding to the energy minimum.

We mentioned in the "translation" section that the translation of certain codons is fast, compared to that of others. We examined the translation speed chart (19) of some 2000 genes of E.coli and observe that quite often, fast as well as slow codons come in clusters of tens of codons. This means that the corresponding segments of the protein are translated quickly or slowly, hence the time left to segments 25 to 30 aminoacids ahead, for searching their stable conformation, is short or long, respectively. In a few cases, the fast and slow segments overlap rather accurately with the native structural domains of the protein (see fig. 13). This infers that folding could indeed be linked to the translation kinetics, written not in the aminoacid, but in the codon sequence of the gene. Studies are in progress to see whether this observation is just coincidental, applies to a class or proteins only, or is of general validity.

A data base including several hundred protein structures at resolution better than 3A is available, together with their aminoacid sequences. It could be a good training set for testing new approaches to the folding pathway problem.

Conclusion.

Obviously, the true molecular investigation of biology, as understood in "hard" sciences, is in its very preliminary stage, at best. There is probably no need for new concepts to understand the life processes. The main difficulty comes from the high structural complexities of the actors of the processes, and the extremely sophisticated interactions and feed-backs they experience. These complexities built up in the course of evolution. It can be envisioned that the primitve processes were rather basic. Sophistication, or adaptation, occured progressively, adding many layers of processes which enabled to improve the fitness of living beings. Today, we face the resulting complexity and have to back- engineer the evolutionary pathes, to try to grasp the basic steps.

The superb tools provided by the biotechnologies will be of great value in this quest. The few examples mentioned above illustrates an important point: with reasonable investments in time and material, present- day biotechnologies enable to subject most theoretical models to experimental (in vitro or in vivo) molecular probing, opening the way to constructive, multidisciplinary cooperations. It is hoped that the present contribution may convince that multidisciplinarity is the only way to progress in the quest for "What is Life?".

Acknowledgement. This contribution was supported in part by EEC grant SC1*-CT91-0705. I am indept to my present and former collegues, especially J. Gabarro, R. Ehrlich, M. A Jacquet, G. Duval-Valentin, B. Schmitt F.J. Wang, A. Deana, T. Lesnik and J. Solomovici for their valuable and enlightening contributions.

References.
1. W. Saenger. Principles of Nucleic Acid Structure. Springer Verlag, 1984.
2. C.R. Calladine & H. Drew. Understanding DNA. Academic Press, 1992.
3. V.A. Bloomfield, D.M. Crothers & I. Tinoco. Physical Chemistry of Nucleic Acids. Harper & Row 1974.
4. M.Y. Azbel. Rhys. Rev. Lett. 31, p. 589-92, 1973.
5. J. Gabarro-Arpa & F. Michel. Biochimie 64, p.99-112, 1982.
6. H. Marcaud, J. Gabarro-Arpa, R. Ehrlich & C.Reiss. Nucleic Acids Res. 14, p. 551-8, 1986.
7. F. Michel, J. Gbarro-Arpa & B. Dujon. Biochimie 64 p. 113-126, 1982.
8. J. Gabarro-Arpa, P. Tougard & C. Reiss. Nature 280, p. 515-7, 1979.
9. J. Freifelder. Molecular Bology. Jones & Bartlett, 1988.
10. P.H. von Hippel, D.G. Bear, W.D. Morgan & J.A. McSwiggen. Ann. Rev. Biochem. 53, 389-446, 1984.
11. B. Schmitt & C. Reiss. Biochem. J. 277, p. 435-443, 1991.
12. G. Duval-Valentin & R. Ehrlich. Nucleic Acids Res. 15, p. 575-594, 1987.
13. H. Heumann, M. Ricchetti & W. Werel. EMBO J. 7, p. 4379-81, 1988.
14. F.J. Wang. Contrôle et régulation de la Transcripton par l'ARN polymérase d'E.coli: étude de l'élongaton. Thesis University Paris VI, 1991.
15. F. J. Wang & C. Reiss. Bichem. & Mol. Biol. Internat. 30, P. 984-91, 1993.
16. B. Schmitt & C. Reiss. Biochem. J. 306, p. 12_8, 1995.
17. V. Emilsson, A.K. Näslund &C.G. Kurland. J. Mol. Biol. 230, p. 483-91, 1993.
18. S. Pedersen. EMBO J. 12, p. 2895-8, 1984.
19. T Lesnik. Exploitation d'une base de données d'E. coli: Analyse de la composition et de la traduction des gènes. Thesis University Paris XI, 1994.
20. A. Deana. Etude du rôle des codons synonymes dans l'expression du gène ompA d'E. coli. Thesis, University Paris VII, 1993.
21. A. Deana & C. Reiss. C.R. Acad. Sci. Paris 316, p. 628-32, 1993.
22. T. Ikemura. Mol. Biol. Evol. 2, p.13-34, 1985.
23. M.A. Jacquet, R. Ehrlich & C.Reiss. Nucleic Acids Res. 17, p. 2933-45, 1989.
24. M.A. Jacquet & C. Reiss. Nucleic Acids Res. 18, p.1137-43, 1990.
25. M.A. Jacquet & C. Reiss. Molec. microbiol. 6, p.1681-91, 1992.
26. T.M. Lohman, P.L. Dehaseth & M.T. Record Jr. Biophys. Chem. 8, p.281-94, 1978.
27. T. E. Creighton. Proteins. Freeman 1993.
28. C. Branden & J.Tooze. Introduction to Protein Structure. Garland Pub. Inc., 1991.
29. T.E. Creighton. Protein Folding. Freeman &Co, 1992.
30 B.T. Nall &K.A. Dill. Conformations and Forces in Protein Folding. AAAS 1991.
31. R. Corme & C. Reiss, manuscript in preparation.

Modelling the DNA double helix:
techniques and results

R. Lavery

Laboratoire de Biochimie Théorique,
Institut de Biologie Physico-Chimique,
13 rue Pierre et Marie Curie,
75005 Paris, France

1. INTRODUCTION

Biological systems confront us with an extremely high level of complexity and organisation. If we wish to describe even a small part of. such systems at an atomic level, analytical models must. generally be replaced by numerical models and quantum mechanics by classical mechanics. The size of macromolecules, their number of degrees of freedom and the extent of the conformational space they can occupy pushes us to and beyond the limits of current computational possibilities. In addition, it must not be forgotten that the biological world is one of perpetual movement, movement that is a fundamental characteristic of life itself. The dynamics of macromolecules is thus a fundamental component of their behaviour. This dynamics covers a daunting range of time scales, from the femtosecond (10^{-15} sec) vibrations of certain chemical bonds to the seconds or minutes necessary to refold a protein or replicate a fragment of the genome. These time scales are also coupled to ranges of movement which can vary from a fraction of an angstrom (10^{-10} m) to tens or hundreds of angstroms.

Despite these difficulties, molecular modelling already plays an important role in refining and studying the conformations and interactions of biological macromolecules, often addressing questions which cannot be directly studied by experiment. The concepts developed by such studies have already contributed to our understanding of the nature of biological systems in a number of fields. Continued improvements both in the quality of force fields and in simulation algorithms, coupled with assured improvements in computer power, should allow extensions of the range of systems which can be effectively treated and thus enable modelling to usefully guide the experimentalist both in the interpretation of his results and in the design of new experiments.

In the specific case of DNA, understanding the processes by which specific binding sites are recognised by proteins, drugs, mutagens and other molecules would represent a fundamental step towards understanding its biological activity. It is clear that such understanding must pass by a detailed knowledge of the structure and dynamics of DNA as a function of its base sequence. Their is ample experimental evidence today to show that base sequence indeed has an important impact on DNA structure and thus on the behaviour of DNA within the cell. X-ray and NMR studies of oligonucleotides point to important conformational heterogeneities as a function of sequence. Nucleosome positioning analysis, gel migration and cyclisation studies have also brought to light specific sequences which lead to static curvature or influence the overall flexibility of DNA [1,2]. In addition, DNA-protein interaction studies have shown that so-called indirect read out of information can play an important role in recognition processes [3,4]. Such effects rely on local variations in the structure and dynamics of DNA and complement direct interactions between amino acid side chains and the nucleic acid bases.

Despite this progress, detailed knowledge of DNA structure is accumulating relatively slowly, on one hand due to difficulties in crystallising DNA oligomers and, on the other hand, due to the difficulties in obtaining full conformational data using NMR spectroscopy. Since learning to predict the structure and the dynamics of DNA for an arbitrary base sequence will undoubtedly require data on a very large number of test sequences, it becomes important to consider whether molecular modelling can help in creating such a structural database. This article will attempt to briefly summarise the progress that has been made over the last few years both in the algorithms necessary to simulate the behaviour of DNA fragments and in understanding the mechanics of this very flexible macromolecule.

2. MODELLING TECHNIQUES

Virtually all modelling of macromolecular systems is based on classical Newtonian mechanics. Macromolecules are represented as complex mechanical systems composed of atomic masses interacting via an empirically defined force field [5,6]. This simplified picture, already embodied in the term "molecular mechanics" (MM), clearly has its limitations, amongst which is the impossibility of studying processes strongly influenced by quantum effects, such as bond breaking and bond making or electron or proton transfer. However, its great strength is to provide a reasonably detailed model of conformational changes and intermolecular interactions for systems containing thousands, or even tens of thousands, of atoms. With an appropriately formulated and parameterized force field it is possible to search for the stable states of the system under study and to follow the evolution of this system with time, under controlled physico-chemical conditions.

The empirical potentials used to determine the conformational energy of macromolecules are built up from two types of terms which model respectively, interactions between chemically bound atoms (deformations of bond lengths, valence angles and dihedral angles) and interactions between non-bounded atoms (van der Waals interactions, electrostatics, etc.). The parameters which enter into empirical force fields can be obtained directly from experiment, indirectly from experiment by attempting to fit certain experimental observables or from accurate quantum mechanical calculations on small molecules. Despite the difficulties in obtaining a coherent and complete set of experimental data on a wide variety of molecules, the empirical nature of the terms composing the force field generally results in the former approach being more successful than an *ab initio* technique. It should be remarked that the simplicity of terms constituting empirical potentials often leads to inter-term compensations and couplings. This implies that different elements of different force fields cannot easily by combined with one another and also that detailed interpretations of the role of different terms are somewhat suspect. Today the number of major force fields is roughly 5 or 6. Although a large number of more specific force fields exist, the effort required to parameterize a general-purpose force field has led to a tendency for collaborative efforts - to the advantage of all users of such methods.

The precision of any modelling exercise is directly related to the quality of the force field employed, however computational limitations often impose their own barriers. Force fields are thus generally a compromise between quality and the computational cost of carrying out the modelling. Three typical examples of such economic constraints are: (a) the representation of molecular charge distributions using simple atomic monopoles; (b) the absence of terms describing atomic polarizability; (c) the introduction of cut-off distances beyond which non-bonded atomic interactions are ignored. As computer speed improves it will be relatively easy to correct these weaknesses and thus to make an important step towards more accurate calculations.

Having defined a force field, it is generally necessary to obtain a starting conformation for the system under study. This can come from experimental data, such as the structures generated by X-ray diffraction. In the case of proteins, and certain RNA's, one can attempt to build starting conformations by homology with known structures using structural databases and the tools of interactive graphics. For less structured systems, such as denatured biopolymers or lipid membranes the situation is much less well understood and still constitutes an area of active research. Once a starting conformation exists, a wide variety of algorithms can be used to study the macromolecule or the molecular complex in question. Searching for most stable conformations can be carried out using energy minimisation, however the complexity of the conformational space to be searched may render such algorithms ineffective. Most macromolecules have energy landscapes covered with local minima, separated from one another by energy barriers of widely varying heights; landscapes worthy of the world's best mountain ranges transposed from a space of three dimensions to a space of tens of thousands of dimensions. In most cases, we are therefore prevented from finding the global minimum and may not even approach it if the starting point is poor.

To overcome these problems, one answer is to introduce temperature into the systems and to move to a dynamic model. The technique termed "molecular dynamics" (MD) allows us to obtain a time trajectory for a molecular system by the numerical integration of Newton's equations of motion. Alternatively we can sample conformational space in a statistical manner using a Monte Carlo (MC) algorithm. Both techniques, given sufficient time, can help us to approach more stable states of the systems and will, moreover, teach us something of its conformational behaviour. Over the last few years, molecular dynamics in particular has undergone a rapid evolution and has greatly contributed to improving our knowledge of the role of movement within biological macromolecules [5-7]. Nevertheless, even with today's most powerful computers, dynamic simulations rarely exceed 1 nanosecond (10^{-9} sec) in duration, due both to the number of atoms in the systems and the very short integration steps (typically 1 femtosecond, 10^{-15} sec), imposed by rapid bond vibrations. Such simulations still represent only a brief instant on biological time scales.

A further difficulty with macromolecular simulations is the need to take into account the environment within which such molecules function, an environment consisting of both water and a variety of salts. The most rigorous way to include such effects is to study a macromolecule contained in a droplet consisting of both water molecules and ions. For one turn of DNA, or for a medium-sized protein, such a droplet must contain several thousand water molecules and this will naturally increase the difficulty of studying the system, not only because of its increased size, but also because of the time which must be spent to reach thermal equilibrium, before extracting a reliable dynamic trajectory. Attempts are already being made to replace at least a part of such explicit models by continuum representations of the solvent, which could considerably help in lightening the computation load [8].

In addition to studying macromolecular dynamics, both MD and MC techniques can be used to search for stable conformations using simulated annealing algorithms. This approach consists of beginning a simulation at high temperature and then slowly allowing the system to cool, leaving sufficient time for re-equilibration to occur as the temperature drops. Many other sophisticated search techniques are still the subject of active research [9]. Some methods attempt to simplify part of the system under study using stochastic models (e.g. Brownian dynamics, BD), thus allowing much longer simulations to be made. Others attempt to modify or combine the variables representing the flexibility of the macromolecules in order to simplify the conformational space to be searched and to allow larger search steps to be taken. Still others, attempt to artificially modify the energy surface to make local minima disappear and thus force the macromolecule towards its optimal conformation.

Having reached the area of conformational space where a stable state of the system exists, it becomes possible to extract a wide variety of information concerning the factors responsible for its conformation and its dynamics. One can also study interactions between the macromolecule and the surrounding solvent or with other species. Until recently, the thermodynamic aspects of such analyses were limited to estimates of enthalpies since the conformational sampling necessary to calculate entropies was out of reach. Modern advances have however partly overcome these limitations by allowing at least the calculation of free energy differences resulting from restricted changes to the systems under study [7,10]. Thus it is now possible to study how the chemical modification of a drug will influence its binding to DNA or how the modification of an active site amino acid may affect the enzymatic properties of a protein. Such calculations require multiple MD or MC simulations which differ by modifications in a perturbational parameter representing the mutation of the system between its two states. To take the example of drug binding, since it is not possible to simulate the binding process itself, changes in binding free energy following drug modifications are obtained "alchemically" by comparing the stability of the DNA-drug complexes and of the free solvated drugs as a function of the desired structural modification. Completing a thermodynamical cycle then allows the necessary information to be extracted. In the best cases, such modelling can lead to results which approach the quality of experimental measurements.

2.1 Modelling nucleic acids using adapted coordinate systems

Several years ago our interest in nucleic acid behaviour led us to look for ways of simplifying the modelling of helical DNA and RNA. Since we were primarily interested in the effects of base sequence it was clear that we would need to perform many hundreds of simulations of different nucleic acid fragments and this, in turn, would require very rapid computations. Energy minimisation seemed a possible route, since when extended to the calculation of energy surfaces, it can be a very powerful tool for looking at structural deformations. However, the usefulness of minimisation in macromolecular systems is generally hindered by severe problems connected with local minima. The complexity of the problem can easily be appreciated when we note that a single turn of double helical DNA (~600 atoms) represents ~ 2000 Cartesian variables. We thus set out to adapt our representation of DNA to reduce these problems, firstly, by reducing the number of independent variables and, secondly, by choosing variables adapted to representing the movements occurring within nucleic acids.

The choice of variables used to model macromolecular structures has considerable consequences for the performance of the resulting algorithm. In the case of the nucleic acids, helicoidal variables are obviously of great interest. We thus developed and refined a method ("Jumna", JUnction Minimisation of Nucleic Acids, [11,12]) which enables such variables to be exploited in conjunction with an internal variable model of nucleotide flexibility (torsional angles plus valence angles within the flexible sugar rings). This approach allows both regular and irregular nucleic acid segments to be constructed easily, including unusual structures such as bends, junctions, loops and triple helices. Secondly, chosen conformational features may be "frozen" while all other variables are energy relaxed. This, in turn, opens the way for calculating optimal energy pathways involved in a range of structural deformations including, for example, transitions between the different allomorphic forms of DNA. Lastly, a considerable reduction can be achieved in the number of variables necessary to model a given segment of nucleic acid, roughly 10 times less than with conventional Cartesian coordinate molecular mechanics. Introducing helical symmetry can further simplify the model, typically reducing the number of independent variables by a further factor of 10. It should also be remarked that these variables are better adapted to describing the nature of the flexibility of helical nucleic acids, allowing energy optimisation to be faster and avoiding many local minima.

Jumna begins by breaking a DNA oligomer down into 3'-monophosphate nucleotide units (with the exception of the 3'-termini which are simple nucleosides). A ring break is also introduced into each sugar moiety between the atoms C4' and O1'. These units are positioned in space with respect to a common helical axis system using a set of 6 helicoidal variables. The internal flexibility of each nucleotide is represented by single bond rotations at the glycosidic (base-sugar) link, within the phosphodiester backbone and the sugar ring. Valence angles variations are also taken into account for the sugar ring. All other valence angles and all bond lengths are taken to be fixed. The independent variables associated with each nucleotide are consequently, 3 translations and 3 rotations which position the nucleotide with respect to the helical axis system, the glycosidic dihedral, 3 valence angles (O1'-C1'-C2', C1'-C2'-C3' and C2'-C3'-C4') and 2 dihedrals (O1'-C1'-C2'-C3' and C1'-C2'-C3'-C4') within the sugar moiety and 2 backbone dihedrals ε (C4'-C3'-O3'-P) and ζ (C3'-O3'-P-O5'). Other sugar and backbone dihedrals are dependent and are determined by the "closure" conditions which involve the C4'-O1' bond length within each sugar ring, the inter-nucleotide O5'-C5' bond and the valence angles P-O5'-C5' and O5'-C5'-C4'. These constraints are imposed via harmonic energy penalty terms.

Jumna contains a wide variety of options for modelling special situations. One of the most extensive choices that can be made involves helical symmetry. By regrouping sets of symmetry equivalent variables it is possible to impose symmetries with unit cells containing any number of nucleotides, with or without symmetry relationships between different strands in the complex. It is also remarked that when symmetry is imposed on a DNA oligomer, it is possible to avoid end-effects and to gain time by calculating only the energy of a single unit cell interacting with a sufficient number of neighbouring cells. Other constraints which can be imposed using Jumna include inter-atomic distance and dihedral angle constraints (used to fit conformations to NMR data), axis bending and supercoiling. Other options allow helicoidal variables to be controlled, for example, to provoke conformational transitions. It is also possible to directly control sugar phase and amplitude which, as we shall see shortly, is of great interest for investigating DNA fine structure.

The force field we use to model nucleic acids has a rather standard form [13-15], consisting of pairwise additive non-bonded terms (electrostatic interactions, repulsion and dispersion) complemented by harmonic bond length and angle penalties and sinusoidal dihedral barriers. The sugar-phosphate backbone requires some special consideration especially through the use of anomeric (or 'gauche') dihedral terms which appear to be necessary for correctly reproducing both phosphodiester and sugar puckering conformations (see, for example, [16-18]). The only notable points in our formulation are explicit terms for the angular dependence of hydrogen bonding interactions and the use of a distance dependent dielectric function [19]. We have reformulated this function as shown below so that it is possible to vary both the plateau value of the dielectric reached at long distance (D) and the slope of the sigmoidal segment of the function (S).

$$\varepsilon(R) = D - (D-1)/2 \ [\ (RS)^2 + 2RS + 2 \ ] \ e^{(-RS)}$$

The values used presently are D=78 and S=0.16, this slope having been found most appropriate for modelling B-DNA in aqueous solution [12]. This damping, combined with a reduction of phosphate net charges to -0.5e, represents a very crude, but relatively effective model of solvent and counterion effects on helical DNA.

2.2 Conformational analysis

Increasing interest in the fine details of macromolecular conformation has led to a search for quantitative methods of describing such structures. Such analysis is one of the major challenges facing the simulation of increasingly complex biological systems and this task will undoubtedly require the development of tools at least as sophisticated as those used for the simulation itself [20,21]. This need has been very clearly felt in the case of nucleic acids, notably since sequence dependent curvature has been shown to play an important biological role. However the problem of correctly describing irregular helical conformations is much more complicated than it may seem at first sight. Several solutions have been proposed [22-27] including our own approach termed "Curves" [28,29] which is based on the calculation of a unique curvilinear helical axis describing the nucleic acid fragment under study. This algorithm redistributes departures from perfect helical symmetry between deformation of the helical axis and irregularities in the position of successive bases with respect to this axis, using a least squares optimisation procedure. Once the optimal helical axis has been determined, a full set of independent helical parameters can be calculated, whose definitions are in accord with those laid out in the so-called Cambridge convention for nucleic acids [30]. Recent versions of the Curves program have been extended to treat 3- and 4-stranded complexes and to measure the width and depth of helical grooves [31]. Curves has also been extended for the analysis of dynamic trajectories as part of the "Dials and Windows" package developed by the group of David Beveridge [21]. This approach allows both helicoidal and backbone parameters to be represented graphically in a compact way and also offers interesting possibilities for studying differential parameter changes which can greatly help in locating and characterising conformational transitions. Axial variations can also be used for calculating the persistence length of irregular DNA fragments [32].

3. DNA FINE STRUCTURE

In order to understand base sequence effects on DNA it is necessary to take a systematic approach, studying a broad range of base sequences. Our studies have all been carried out on infinite, regular polymers so that the complicating end-effects of oligonucleotides could be eliminated (see section 2.1). The first step [33] involved a set of six regular polymers having base sequences containing all the possible dinucleotide steps which can be built from the four standard DNA bases. These polymers are shown below, followed by the dinucleotide steps that they contain. Note that although 16 dinucleotide sequences can be formed from the 4 bases, there are only 10 unique double stranded dinucleotide steps, since the dinucleotides indicated by stars are simply the paired strands of the unstarred dinucleotides on the same line.

$(AA)_n$	AA	TT*		
$(GG)_n$	GG	CC*		
$(CG)_n$	CG	GC		
$(TA)_n$	TA	AT		
$(CA)_n$	CA	AC	TG*	GT*
$(GA)_n$	GA	AG	TC*	CT*

Following the base sequences, these polymers were constrained to obey mononucleotide (AA..., GG...) or dinucleotide (CG..., TA..., TG..., GA...) symmetry. In addition, when appropriate (CG..., TA...) interstrand dyad symmetry was imposed (homonomous constraints). Early calculations were made taking into account the interactions of 7 neighbouring nucleotide pairs with the central unit cell, but more recently this number was increased to 10 pairs. Looking for the stable minima of these polymers involved a large number of minimisations, using a variety of starting points including both canonical B-DNA fibre coordinates [34,35] and the relaxed conformations subsequently found for each base sequence. A second technique was also used to verify the stability of the minima found. This involved using the constraint possibilities available in Jumna (notably, overwinding-underwinding and stretching-compression) to perturb the optimised conformations and thus to show that each lay within a clearly defined energy well. These calculations also enabled several new minima to be located. It was also noted that the perturbations imposed of the double helix could, in certain cases, change the relative stabilities of the known minima and could also provoke transitions between different minima [36]. Recently it has been possible to automate the search for stable conformational states using techniques described below. These improvements as well, as the minor modifications introduced since our earlier work (in particular, reparameterization of the thymine methyl rotation and an increase in the number of neighbours interacting with the central unit cell), have led to some changes in the number of minima detected and in the detailed conformation of certain minima (Tab. I). We also discovered several new minima for the AA and GG homopolymers [37], once dinucleotide symmetry was allowed to develop.

Table I. Energies per unit cell (Kcal/mol) for the stable sub-state conformations of the 6 mono- and dinucleotide polymers studied. (Note: Conformations containing purine nucleotides with O1'-endo sugar puckers are indicated by the symbol #. Homopolymer conformations which in fact display dinucleotide symmetry are indicated by the symbol *.

Sequence	Sub-states					
$(AA)_n$	-82.5*	-82.4*	-81.9*			
$(GG)_n$	-118.1	-115.2	-115.1*	-114.2*		
$(CG)_n$	-119.4	-118.7	-118.0	-114.0#		
$(TA)_n$	-81.6	-80.8	-80.2	-80.0#		
$(CA)_n$	-100.9	-100.8	-100.6	-100.4	-100.3	-100.2
	-99.6	-99.6	-99.2	-97.2#	-97.1#	-95.2#
	-95.0#					
$(GA)_n$	-99.5	-98.4#	-98.3	-97.6	-97.0	-96.7#

The main result of this study was that each of the sequences investigated had a number of distinct minima with similar energies, but very different conformations. The range of conformations is shown in Table II which lists the range of values adopted by each of the helical and backbone parameters. These results agree with recent crystallographic and NMR studies and confirm that the term "B-DNA" actually represents a family of structures occupying a rather large volume of conformational space. It will be noted that a number of backbone parameters vary by 20°-50° and that the sugar phase angle covers a remarkable 100°

range. Equally, several helical parameters are very variable, twist covering a 17° interval and propeller and buckle roughly 30°, while the translational parameters rise and Xdisp vary respectively by 1.2Å and 3.3Å. Despite these variations, the mean values resulting from our simulations and from experimental data are very similar.

Table II. Range of backbone and helical parameters for the stable sub-states of the regular DNA polymers studied. For comparison both fibre and crystallographic data on B-DNA are shown.

Jumna :	Minimum	Maximum	Mean	Experiment :	Fibre	Crystals
α	-72	-58	-65		-41	-64
β	159	188	175		135	166
γ	50	67	58		37	50
δ	91	150	133		139	132
ε	-177	-161	-172		-134	-179
ζ	-145	-90	-110		-102	-94
χ	-149	-94	-118		-102	-104
Phase	74	184	150		154	150
Amplitude	28	46	38			39
Xdisp	-3.5	-0.2	-1.6		0.0	0.5
Inclin	-18.0	12.0	-0.3		1.5	-1.2
Propeller	-22.0	4.0	-6.4		-13.3	-11.0
Buckle	-8.0	21.0	-1.8		0.0	0.9
Rise	2.8	4.0	3.3		3.4	3.4
Twist	28.0	45.0	35.5		36.0	36.2

It should also be stressed that important changes in conformation occur not only between different polymers, but also between the conformational sub-states of a given polymer. In fact, a given regular base sequence is only rarely associated with distinct values of either helical or backbone parameters if all its possible energy minima are taken into account.

Table III. Sub-states classified according to their sugar puckering (S: C2'-endo, low amplitude, X: C2'-endo, high amplitude, E: O1'-endo). Sugars are given in the 5'→3' sense and follow the base order in the first column (Note: # indicates purines with O1'-endo sugars; * indicates homopolymers showing dinucleotide symmetry)

Sequence	1	2	3	4	5	6
(AA)$_n$	SX:SS*	SX:ES*	SX:SE*			
(GG)$_n$	XX:XX	SS:XX	XS:SX*	SX:ES*		
(CG)$_n$	SX:SX	XS:XS	ES:ES	SE:SE#		
(TA)$_n$	SX:SX	XS:XS	ES:ES	XE:XE#		
(CA)$_n$	ES:XS XS:ES SE:ES#	XS:SX SX:SX	XS:XS SX:ES	ES:SX SE:XS#	SX:XS SE:XX#	ES:ES SE:SE#
(GA)$_n$	XS:SX	ES:SX#	SX:ES	XS:SE	SX:SE	ES:SE#

In attempting to rationalise these results, we noted that the sub-states of each polymer always differed in terms of their sugar puckers [for a definition of sugar puckering see, 5]. The sugars were found to lie in three groups: C2'-endo with low amplitude (termed 'S'), C2'-endo with high amplitude (termed 'X') and O1'-endo (termed 'E'). It was found that each sub-state could be uniquely identified by a simple label describing its sugar pucker groups (Tab. III). In consequence, it appears that the sugars play a central role in translating base sequence into the local conformation of the double helix.

3.1 Energy surface mapping

Since sugar puckers appear to be sufficient for characterising the conformational sub-states we located, it seemed worthwhile to investigate the energy surfaces defined by these variables [37,38]. Jumna was consequently modified to make 1D and 2D energy maps in terms of sugar phase angles, all other variables being relaxed by an energy minimisation at each fixed value of phase. Since both the mononucleotide conformations of the homopolymers poly(dA).poly(dT) and poly(dG).poly(dC) and the dinucleotide conformations of the alternating polymers with dyad symmetry poly(dCG).poly(dCG) and poly(dTA).poly(dTA) each have only two symmetry distinct sugars, it was possible to search all the sugar conformations of these polymers with 2D maps. One such map is shown in Figure 1, covering both the A- and B-family conformations of the CG alternating polymer. The results of this study led to several important conclusions. Firstly, the maps obtained were independent of the starting conformation used and maps built up from several independently calculated zones could be fitted together smoothly (as in the case of Fig. 1). Secondly, all the energy minima which had been found for the polymers studied were also present within the sugar maps. Both these findings confirm that, for a given base sequence, once sugar puckers are fixed, the conformation of the double helix is also fixed.

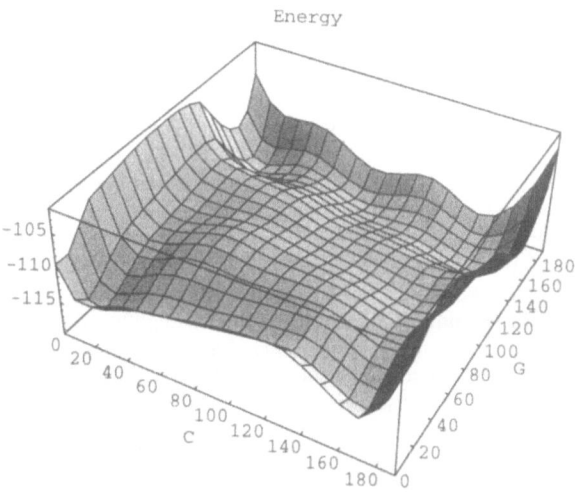

Figure 1. Energetic surface of the $(CG)_n$ polymer as a function of its sugar phase angles (both dinucleotide and dyadic symmetry are imposed on the polymer, leading to the presence of only two symmetry distinct sugars)

These maps also enabled calculations of the energy barriers separating the various sub-states of each polymer. In general, it was found that the S and X puckering domain minima are separated by very small barriers amounting to only a few tenths of a Kcal/mol. In contrast, sub-states containing E (O1'-endo) sugars are generally isolated by larger barriers of the order of 1-2 Kcal/mol. Similar barriers also oppose the passage from the E to the N domain (C3'-endo) as seen in Figure 1, which also illustrates the optimal pathway for passing between the B (top left) and A (bottom right) forms of the double helix.

3.2 Combinatorial searches

The map shown in Figure 1 already contains the results of roughly 600 energy minimisations. Consequently, making sugar maps in more than 2D would become prohibitively expensive. It is therefore generally impossible to use this technique for polymers with more than two symmetry distinct sugars, although extensions to 4-sugar cases are possible in special cases [38]. However, the information gained by this technique has enabled us to move on to the study of more complex sequences.

We recall that all the sub-state conformations of DNA detected so far have sugars belonging to the S, X, E or N families and can be uniquely characterised by their sugars. It therefore seems reasonable to suppose that if all possible combinations of sugar conformations are used as starting points for energy minimisation, it will be possible to locate all of the stable sub-states for any given sequence. This idea was put into practice by modifying Jumna to read a chosen set of possible puckers for each nucleotide within a given polymer and then to cycle automatically through full combinatorial set of puckers, in each case performing a constrained minimisation followed by a free minimisation. Tests of this procedure for the dinucleotide repeat polymers discussed above confirmed that it is a reliable way of finding sub-states.

Such automated searches have enabled us to build up a library which now contains the stable sub-states of the 136 unique tetranucleotide sequences. These sequences, which can be obtained via studies of 39 regular repeating polymers (Tab. IV), represent a much more detailed guide to sequence effects since they can be used to study the effect of all possible neighbouring sequences on any given dinucleotide step within the double helix. This represents a move from the nearest neighbour model of our early studies (and all other current attempts to explain sequence effects on DNA [39-41 and references therein]) to a next nearest neighbour model [42]. A step which, due to the scarcity of experimental data mentioned in the introduction, is currently possible only through molecular modelling.

The combinatorial study of the tetranucleotide repeat polymers has led to several interesting conclusions. Firstly, each polymer, and therefore the tetranucleotide sequences it contains, again exhibits a number of stable sub-states (20 on average) with quite widely differing conformations. Secondly, it is still possible to characterise these minima by their sugar puckers (S, X, E, N), although these classes are less tightly defined than in the case of the dinucleotide sequences. Thirdly, mono- and dinucleotide repeat sequences and also sequences with dyad symmetry commonly exhibit stable energy minima with lower symmetry than that implied by their base sequences. The symmetry reduction effect found with poly(dA).poly(dT) [43] is thus a general phenomenon, although it should be added that the most stable sub-state of each tetranucleotide does, in general, reflect the fundamental symmetry of the corresponding base sequence. This finding nevertheless suggests that conformational asymmetries induced by neighbouring sequences, by thermal excitation, or by interactions with other molecules are probably very common.

Table IV. 39·regular polymer sequences containing the 136 unique tetranucleotides

$(GGGG)_n$	GGGG		
$(AAAA)_n$	AAAA		
$(CGCG)_n$	CGCG	GCGC	
$(TATA)_n$	TATA	ATAT	
$(GTGT)_n$	GTGT	TGTG	
$(GAGA)_n$	GAGA	AGAG	
$(CCGG)_n$	CCGG	CGGC	GGCC
$(TTAA)_n$	TTAA	TAAT	AATT
$(TCGA)_n$	TCGA	CGAT	GATC
$(TGCA)_n$	TGCA	ATGC	CATG·
$(ACGT)_n$	ACGT	CGTA	GTAC
$(AGCT)_n$	AGCT	TAGC	CTAG
$(AGGT)_n$	AGGT	GGTA	GTAG	TAGG
$(AGGA)_n$	AGGA	GGAA	GAAG	AAGG
$(TGGT)_n$	TGGT	GGTT	CAAC	TTGG
$(GCGG)_n$	GCGG	CGGG	GGGC	GGCG
$(ACGA)_n$	ACGA	CGAA	GAAC	CGTT
$(AGCA)_n$	AGCA	TTGC	CAAG	AAGC
$(AGCG)_n$	AGCG	GCGA	CGAG	GAGC
$(ATAA)_n$	ATAA	TAAA	AAAT	AATA
$(ATAC)_n$	ATAC	TGTA	ATGT	CATA
$(GATG)_n$	GATG	ATGG	TGGA	GGAT
$(GATA)_n$	GATA	ATAG	TAGA	AGAT
$(GGAG)_n$	GGAG	GAGG	AGGG	GGGA
$(AGAA)_n$	AGAA	GAAA	AAAG	AAGA
$(CGAC)_n$	CGAC	GGTC	CGGT	CCGA
$(TGAT)_n$	TGAT	AATC	CAAT	TTGA
$(CAGC)_n$	CAGC	AGCC	TGGC	CTGG
$(TAGT)_n$	TAGT	AGTT	TAAC	CTAA
$(CAGA)_n$	CAGA	AGAC	TGTC	CTGT
$(CAGT)_n$	CAGT	AGTC	TGAC	CTGA
$(GGTG)_n$	GGTG	GTGG	TGGG	GGGT
$(AGTA)_n$	AGTA	GTAA	TAAG	AAGT
$(CGTC)_n$	CGTC	GGAC	CGGA	ACGG
$(TGTT)_n$	TGTT	AAAC	CAAA	TTGT
$(CTGC)_n$	CTGC	GGCA	AGGC	CAGG
$(ATGA)_n$	ATGA	TGAA	GAAT	AATG
$(GTGC)_n$	GTGC	CGCA	ACGC	CGTG
$(GTGA)_n$	GTGA	TGAG	GAGT	AGTG

It should be added that the number of sub-states found for each polymer is very much less than the total number of combinations which may be formed from the sugar pucker classes. If we consider that purines most commonly adopt either S or X puckers, while pyrimidines adopt S, X or E puckers, then the eight symmetry distinct sugars in a tetranucleotide can lead to a total of $2^4 \times 3^4 = 1296$ different combinations. This number is roughly 60 times bigger than the number of sub-states typically found for each tetranucleotide repeat. Each base sequence therefore has a strong discriminatory role in selecting only a certain number of pucker combinations, which, in turn, translate this sequence into the specific three dimensional properties of the corresponding double helix. It must nevertheless be noted that with 20 or more sub-states of similar energies for most tetranucleotide repeats, it is difficult to understand how any sharply defined sequence dependent features can arise. The sub-states conformations of a given tetranucleotide generally differ quite strikingly from one another, as evidenced by the backbone RMS values of roughly 10° (between sub-states containing only S and X sugars) and roughly 20° (in the case of any E sugars). Since the energy range is small (see Tab. V for an example, the ACGT repeat) many of these sub-states should be occupied at room temperature, leading to averaging both helical and backbone conformations over a rather wide ranges, not, if fact, much narrower than the variation of these parameters over the full tetranucleotide database.

Table V. Sub-states of the (ACGT)n polymer (energies in Kcal/mol). (Only the 19 symmetry unique minima are shown, with symmetry equivalent copies there are a total of 45 minima. The last two columns indicate minima with the inversion symmetry of the sequence and minima with unusual $\alpha\gamma$ backbone conformations).

No.	ACGTTGCA	Energy	Sym	$\alpha\gamma$
1	XSXSSXSX	-199.542	*	.
2	XXSSSSXX	-199.515	*	.
3	XSXSSSXX	-199.470	.	.
4	XESSSSXX	-199.232	.	.
5	XXSSSXSES	-199.221	.	.
6	SESXXSXS	-199.054	.	.
7	SESXXSES	-199.037	*	.
8	XSXSXSXS	-198.961	.	.
9	XESSSSEX	-198.832	*	.
10	XXSSESXX	-198.678	.	.
11	SESXSSEX	-198.663	.	.
12	XXSSSXSXS	-198.663	.	.
13	SXSXXSXS	-198.419	*	.
14	XSXSXXSS	-198.294	.	.
15	XXSEESXX	-198.154	*	.
16	XXSSXXSS	-198.043	.	*
17	SESESSXX	-197.762	.	.
18	SESEXSES	-197.529	.	.
19	SESEXSXS	-197.359	.	.
20	SXSEXSXS	-197.356	.	.
21	XESESXSX	-197.213	.	.
22	XXXSXSSS	-197.104	.	*
23	SESESSEX	-196.949	.	.
24	SESEEES	-196.516	*	.
25	SSXXESES	-196.352	.	.
26	XSXSXXSS	-196.201	.	*

In constrast with this situation, earlier studies of DNA oligomers with more complex sequences have shown that we can generally locate only one or two stable minima, which, moreover, often differ by only minor structural rearrangements. Such behaviour was observed in the particular case of two dodecamers, CATGACGTCATG (which contains the CRE site

Figure 2. Superposition technique for building long sequences from the tetranucleotide database.

(a) Tetranucleotides composing the Cre sequence CATGACGTCATG

C	A	T	G	A	C	G	T	C	A	T	G
C	A	T	G								
	A	T	G	A							
		T	G	A	C						
			G	A	C	G					
				A	C	G	T				
					C	G	T	C			
						G	T	C	A		
							T	C	A	T	
								C	A	T	G

(b) Variables taken into account during the superposition tests (indicated in the schematic diagram by bold lines). After successful superposition the two nucleotide indicated by the black circles are added to the polymer under construction.

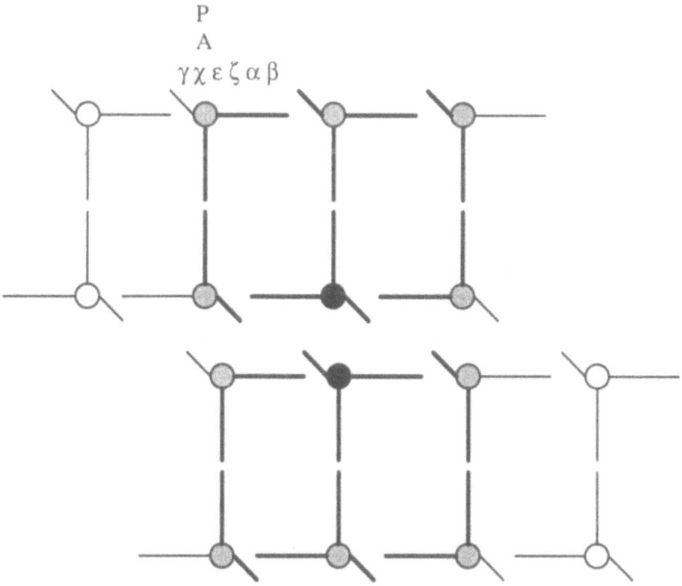

[44,45] recognised by various dimers of b-ZIP proteins including those of JUN and FOS) and GTACTGCAGTAC (the inverse of the preceding sequence). These oligomers were studied with the help of ^{1}H and ^{31}P NMR data [46]. The first of these sequences gave rise to two minima which differed mainly within the central ACGT segment, one of these minima was in very close agreement with the NMR data and constraining to these data only induced minor structural changes. The second oligomer only gave rise to one minimum which was also quite close to agreement with the experimental information. Both sequences exhibited sequence dependent variations from the canonical B-DNA form, although these variations were much more marked in the case of the former, biologically active sequence.

Figure 3. Successive sub-state elimination during the construction of the Cre sequence CATGACGTCATG (the subscript i indicates that two sub-states n and ni are related by inversion symmetry).

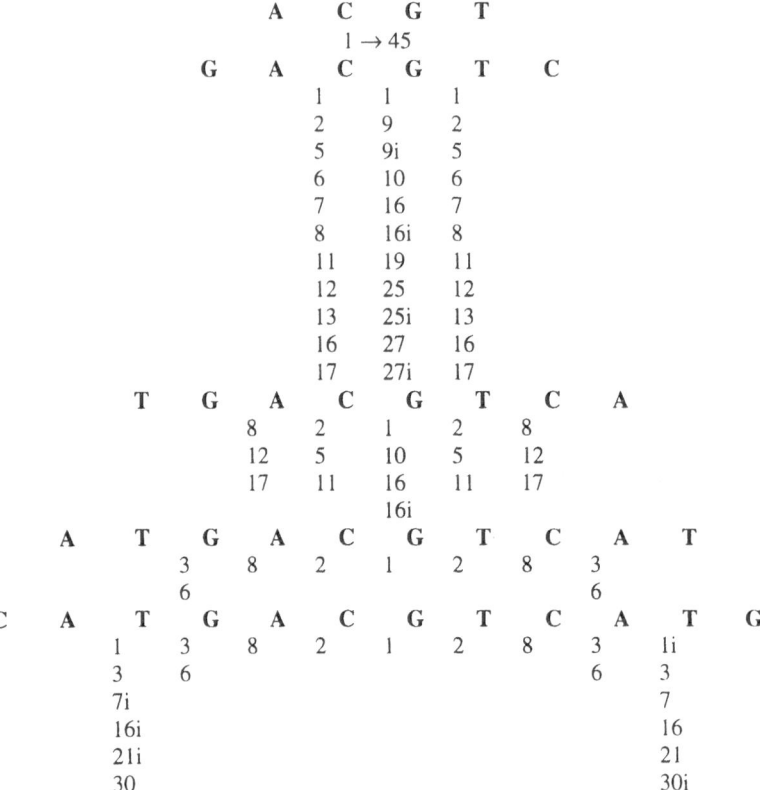

3.3 Building long sequences by superposition

In an attempt to reconcile these two results, we have considered whether we could also approach the construction of DNA fragments with complex sequences using the data contained within the tetranucleotide database. If we assume that next nearest neighbour interactions will generally be sufficient to describe sequence effects, then our present database must contain the information needed to assemble any chosen sequence. The technique we have developed for this construction is illustrated schematically in Figure 2 for the Cre oligomer CATGACGTCATG discussed above. Firstly, we decompose the sequence into its constituent tetranucleotides which consequently overlap by trinucleotide sequences. Next we search the database for tetranucleotide sub-states which have the same structure as their neighbours for the six overlapping nucleotides. As indicated in Figure 2, each overlap is tested using an angular RMS value which takes into account the phosphodiester backbone angles (α, β, γ, ϵ, ζ), the sugar conformation (phase P and amplitude A) and the glycosidic angle (χ) of each nucleotide. Excluding backbone linkages outside the trinucleotide overlapping fragment leads to a total of 38 parameters which contribute to the RMS value calculated. It might be suspected that since neighbouring tetranucleotides within an irregular sequence are generally derived from different DNA polymers (amongst those shown in Tab. IV) that there would be little chance of obtaining a good superposition. However, the tests we have carried out so far suggest that this is not the case and successful superpositions can generally be made with a remarkable quality of around $2° \rightarrow 4°$ RMS for the 38 parameters compared.

The second surprising finding concerns the number of conformations which can be constructed for a given sequence. In the case of the Cre sequence, the number of sub-states in the database for each of its constituent tetranucleotides enables us to calculate that their are roughly 3×10^{12} possible structural combinations which could be made. However, once superposition is taken into account a very different picture emerges. As illustrated in Figure 3, fitting together tetranucleotides within an irregular sequence leads to a dramatic decrease in structural multiplicity and, in general, it is found that the final DNA fragment is compatible with only a small number of global conformations. For Cre, the central tetranucleotide ACGT begins with 45 stable sub-states. Adding the two neighbouring tetranucleotides (GACG on the 5' side and CGTC on the 3' side) and searching for fits of better than 4° RMS leads to elimination of all but 11 of the ACGT sub-states and, simultaneously, to elimination of half of the 22 sub-states of the adjoining tetranucleotides. If we now continue the construction by adding further tetranucleotides at either end of the central fragment, the number of acceptable ACGT sub-states is reduced successively to 11 possibilities, 4 possibilities and, finally, to a single sub-state. At the end of the construction all of the central part of the oligomer has a unique structure and only its ends maintain a number of possible conformations. It should also be noted that although we only use tetranucleotide fragments for this construction, this does not exclude the appearance of sequence effects well beyond the next nearest neighbour range. As shown in the case discussed, the conformation of the central nucleotides of the Cre sequence is finally chosen by fixing the base sequence 5 nucleotides away from the centre of the oligomer.

Having built a conformation for our oligomer, we must now determine whether this conformation corresponds to that obtained by direct minimisation of the fragment. The answer is contained in Table VI which shows that, with the exception of the terminal nucleotides, the database constructed fragment coincides almost perfectly with one of the two conformations found for the Cre dodecamer in our earlier study [46]. (Note that this comparison was made with an arbitrary choice for the terminal nucleotides of the fragment which, as shown in Fig. 3, have a number of possible conformations. It should be remarked that, in any case, one should not expect a perfect agreement at the ends of an oligomer, since all the database conformations belong to infinite regular polymeric sequences. Consequently, the explicit end-effects, present in the minimised oligomer, certainly cannot be reproduced).

Table VI. RMS fit between the structure of the Cre dodecamer CATGACGTCATG obtained by minimisation (conformation Ib in [46]) and that obtained by construction from the tetranucleotide database (values are in degrees, 8 angular variables are compared for each nucleotide).

Res	γ	ε	ζ	α	β	χ	Pha	Amp	Rms
C1	---	4.1	39.8	2.4	7.7	32.5	77.3	6.0	35.3
A2	3.3	0.3	0.1	1.5	2.3	6.9	1.5	0.3	2.93
T3	1.3	0.7	0.2	1.0	1.2	2.8	0.1	0.9	1.29
G4	1.6	0.9	2.6	1.2	1.1	5.8	8.2	1.4	3.82
A5	1.4	1.7	3.4	1.1	0.7	4.4	0.7	0.1	2.16
C6	0.3	0.7	2.8	1.3	0.7	1.4	2.5	0.1	1.54
G7	0.2	0.9	1.8	0.9	3.7	3.5	3.5	1.0	2.35
T8	1.4	0.1	7.4	2.0	0.2	1.9	0.3	1.5	2.88
C9	1.5	0.1	2.0	0.1	2.4	0.9	6.8	2.1	2.81
A10	0.1	0.0	1.7	0.0	2.7	3.5	2.3	0.7	1.87
T11	1.0	0.2	0.6	0.9	6.3	1.0	7.4	1.3	3.52
G12	2.5	---	---	---	---	8.8	12.0	5.8	8.08
G13	2.6	---	---	---	---	8.7	12.0	5.7	8.08
T14	1.0	0.2	0.7	0.8	6.2	1.0	7.3	1.3	3.49
A15	0.2	0.0	1.7	0.1	2.6	3.6	2.2	0.7	1.87
C16	1.5	0.1	2.0	0.1	2.5	0.8	6.9	2.1	2.84
T17	1.4	0.1	7.4	2.1	0.1	2.0	0.4	1.5	2.90
G18	0.2	0.9	1.8	0.9	3.8	3.5	3.6	1.1	2.38
C19	0.4	0.7	2.8	1.2	0.6	1.4	2.5	0.1	1.53
A20	1.4	1.7	3.5	1.1	0.6	4.3	0.7	0.1	2.17
G21	1.5	0.9	2.7	1.3	1.1	5.8	8.3	1.4	3.83
T22	1.3	0.7	0.2	1.0	1.2	2.8	0.1	0.9	1.28
A23	3.2	0.2	0.1	1.4	2.4	7.0	1.5	0.3	2.94
C24	---	4.0	39.8	2.3	7.7	32.5	77.3	6.0	35.3

These results suggest that it should indeed be possible to construct long fragments of DNA using the tetranucleotide database. This would enable computational barriers which would normally hinder the study of such fragments to be overcome. One important application in view is the problem of sequence dependent DNA curvature which is known to play an important role in protein-DNA interactions.

4. FLEXIBILITY AND STRUCTURAL TRANSITIONS

4.1 Base pair opening

Having discussed the stable conformational states of DNA, we will now pass to the various deformations which the double helix can undergo. One specific, but very important type of local DNA deformation, involves the opening of base pairs. This process is a necessary part of both DNA transcription and replication. It has been extensively studied by hydrogen exchange [47] which follows reactions involving the exchange of labile DNA protons (belonging to imino and amino groups) with the surrounding solvent. Such exchange, within double stranded nucleic acids, requires opening one or more base pairs in order to expose protons which are otherwise sterically hindered by their participation in hydrogen bonds. Despite the accumulation of a considerable amount of data concerning proton exchange kinetics, little is known about the structure of the transitory open state. Moreover, since this exchange is a rather slow process typically occurring in the millisecond range [48,49], it is well beyond the scope of normal molecular dynamics simulations.

In order to study the energetics of base pair opening, we began by forcing the opening of the central pair within a B-DNA oligomer $(dA)_5.(dT)_5$, using distance constraints between pairs of atoms belonging to the Watson-Crick hydrogen bonds (see also [50]). These studies showed that the bases open mainly by a rotational movement and that a mean axis of rotation is perpendicular to the base plane and passes through a point close to the centre of the sugar ring attached to the opening base. This information was put to use in a preliminary study of opening with a simplified model of B-DNA consisting of five stacked A-T base pairs in the standard B-conformation [35], without phosphodiester backbones. Both breaking the base pair hydrogen bonds and decreasing stacking interactions with the neighbouring base pairs were seen to contribute to the cost of the opening process. It was also found that a variety of open states could be formed with roughly similar stabilities [51,52]. Interactions between the rotating bases showed that opening occurs most easily towards the major groove. Adenine and thymine can rotate into this groove either independently or in a concerted fashion. Rotating either base towards the minor groove led to severe steric hindrance, unless its partner was first rotated some way into the major groove. The various pathways studied and the deformation energies necessary to rotate a base by 50° are listed in Table VII.

Table VII. Base pair opening energies (Kcal/mol) as a function of the radius of curvature (R) of DNA. The open state corresponds to a 50° base rotation (or a 50° path length in the case of the concerted opening routes 6 and 7). The second column gives the energy deduced from the simplified base only model. Paths defined in reference [52].

Path	Base only	R=∞	R=30Å	R=15Å
1	15.5	28.4	24.7	21.8
2	23.5	37.4	33.2	33.8
3	16.4	30.0	26.5	30.2
4	23.5	39.2	34.4	35.7
5	23.5	37.5	33.2	34.1
6	10.2	29.3	24.8	32.7
7	13.5	25.0	20.2	26.5

Using the possibilities of Jumna, we were able to extend these studies to a full DNA oligomer, regenerating the pathways studied with the simple model, but now starting from an energy minimised oligomer conformation and including the hindrance and inter-nucleotide coupling due to the sugar-phosphate backbones [51,52]. The results of this study are shown in the 3rd column of Table VII. Opening to 50° within the full oligomer is seen to be roughly 13 Kcal/mol more difficult than within the base only model, but the relative energies associated with different pathways are hardly changed with respect to our simple model. It should also be remarked that opening one or both bases belonging to the central pair of the oligomer had very little effect on the rest of the molecule. The base pairs on either side of the opening pair showed only minor deformations and most of the backbone deformation was absorbed in the phosphodiester linkages on either side of the opening nucleotides.

An analysis of the deformation energy associated with opening showed that, for small opening angles, the major energy component came from base-base interactions. At roughly 30° rotation, the base pair hydrogen bonds are more or less destroyed, but base stacking interactions are still present and continue to exert a restoring force on the opening base. As the opening angle increases further, the limit of the flexibility of the sugar-phosphate backbones is reached and backbone deformation becomes the principal component of the deformation energy.

These studies with Jumna brought to light a further feature of the opening process. It was found that when thymine was rotated into the major groove, the stability of the open state could be improved by bending the oligomer towards the minor groove. Similar results were subsequently observed for most of the alternative opening pathways (see Tab. VII). This effect was found to be due to strain within the backbones which builds up as DNA is bent. Opening a base pair releases this strain by locally increasing the flexibility of the double helix [53]. Opening and bending are thus closely linked. Bending DNA makes opening easier, opening a base pair makes bending DNA easier. The coupling of these deformations is summarised for the case of path 1 by the thermodynamic cycle shown in Figure 4. A further implication of this coupling is that bending the double helix could serve as a mechanism to concentrate strain at a given point and triggering base pair opening.

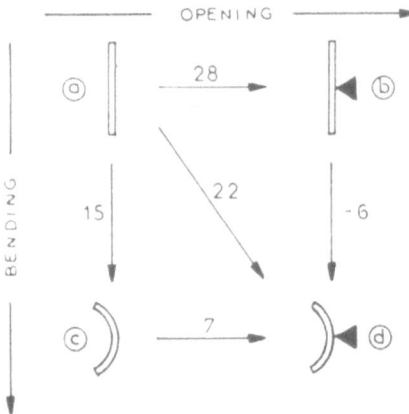

Figure 4. Thermodynamic cycle for DNA bending and thymine opening: bending favours opening and vice versa. (Deformation energies correspond to bending to a radius of curvature of 15Å and opening thymine towards the major groove by 50° rotation).

A subsequent study [54] used the deformation energy curve generated for thymine rotation as the basis for a Brownian dynamics study of the opening process. Such dynamics involves numerically integrating a Langevin equation which takes into account the equilibrium restoring force on the base (coming from our studies), a frictional drag and a random torque due to base-solvent interactions. This approach enables much longer simulations than conventional Newtonian dynamics. Studies covering 0.08 ms were made and showed a proportionality between the inverse log of the time interval separating two fluctuations and the opening angle achieved. Extrapolation to opening angles corresponding to breaking the base pair hydrogen bonds led to base pair lifetimes of 15 ms at room temperature and an activation energy of roughly 20 Kcal/mol. Both these values are in good qualitative agreement with experimental measurements [55]. The physical interpretation of this study is that opening can occur by a stochastic random-walk process and its slow kinetics are associated with the large scale base rotations which have to be made.

4.2 Backbone transitions

A particularly interesting case of local backbone deformation within B-DNA involves the B_I and B_{II} conformations which have been observed crystallographically [56]. These conformations influence the position of the phosphate group with respect to the grooves of the double helix. The more common B_I state places the phosphate in a roughly symmetric position with respect to both grooves, while the B_{II} state swings the phosphate around towards the minor groove. This transition involves coupled changes of two dihedral angles ε (C4'-C3'-O3'-P) and ζ (C3'-O3'-P-O5') which pass from (tg-) in B_I to (g-t) in B_{II}. The conformation of the phosphate within a given dinucleotide junction can therefore be characterised by the difference ε–ζ, which passes from roughly -90° in B_I to roughly +90° in B_{II}.

We have recently carried out a study of B_I-B_{II} transitions within the Cre oligomer mentioned previously [58]. ^{31}P data on this dodecamer show three phosphates resonating at relatively low field (T3pG4, C6pG7, C9pA10), which could suggest the presence of B_{II} conformations [59,60], but, in the fine structure we proposed [46], there is only a tendency towards the B_{II} state (with ε-ζ values of roughly -30° compared to +90° in B_{II}). Starting from this conformation, we therefore induced controlled B_I-B_{II} transitions within the dodecamer in order to study both the energetic and conformational aspects of such deformations. Transitions were made by imposing chosen values for the ε dihedral of the desired dinucleotide junction via harmonic constraints. Note that since the Cre oligomer has a sequence showing inversion symmetry, two symmetrically equivalent phosphate conformations should change together. This implies that a transition at the central C6pG7 junction will result in two B_{II} sites facing one another, while, in all other cases, the B_{II} sites will be staggered.

Table VIII. Energy change and barriers for B_I-B_{II} transitions within the dodecamer CATGACGTCATG (Kcal/mol).

Junction	$\Delta E(B_{II}-B_I)$	Barrier
C6pG7	1.9	4.6
T3pG4	7.1	7.6
C9pA10	8.1	8.6
T3pG4+C9pA10	-3.6	15.0
G4pA5	5.1	6.4
A5pC6	3.4	11.0
G7pT8	10.0	15.0
T8pC9	2.3	7.6

We began by looking at the phosphates showing a tendency towards the B_{II} conformation, T3pG4, C6pG7 and C9pA10. The deformation energy curves for the transitions [58] show that, in all cases, an energy barrier occurs at an $\varepsilon-\zeta$ value of roughly 70°. We can thus conclude that the $\varepsilon-\zeta$ values of roughly -30° observed for these steps in the optimal Cre oligomer conformation clearly belong to the B_I family. It is also found that the B_{II} minima are narrower than the B_I states. The barrier heights and transition energies are given in Table VIII. The central CpG dinucleotide is found to transit very easily to the B_{II} state, losing less than 2 Kcal/mol after crossing a barrier of less than 5 Kcal/mol. The other two junctions yield less stable B_{II} conformations and also have higher transition barriers. A coupled transition of T3pG4 and C9pA10 leads to the most stable state, but this requires passing a very high energy barrier of 15 Kcal/mol and also leads to major structural changes. The remaining junctions we have studied (G4pC5, A5pC6, G7pT8, T8pC9), although they show no tendency towards B_{II} in the optimal oligomer conformation, in fact can transit roughly as easily as the other dinucleotide junctions. However, they do create much stronger structural perturbations, a fact which was subsequently linked to sugar puckers. Junctions showing a tendency towards B_{II} were all found to have sugars of the type XpS (see section 3.1), while other junctions were SpX. After transition to the B_{II} state, all junctions became XpS or XpX. The amplitude of the sugar at the 5'-side of the junction is thus closely coupled to the value of $\varepsilon-\zeta$. It was also found that the B_{II} junctions could bend DNA towards the minor groove (with a large negative roll angle at the modified junction), but only when two B_{II} phosphates face one another within the duplex, as for transitions of the central CpG junction of the Cre oligomer (Fig. 5).

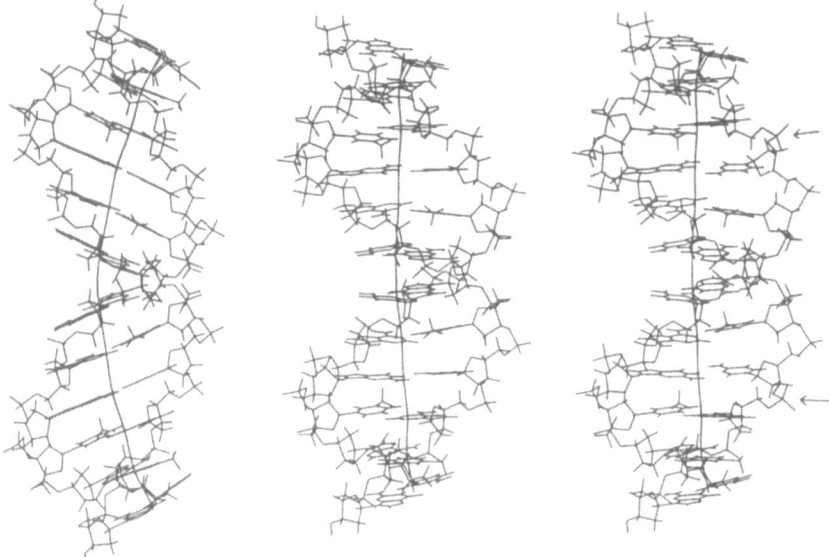

Figure 5. Conformations of the CRE oligomer with and without B_{II} junctions. Centre: optimal conformation with only B_I junctions, Left: a central B_{II} junction in both strands at C6pG7 causing kinking of the duplex, Right: two symmetry related single strand B_{II} junctions at C9pA10 (indicated by arrows). The optimal helical axis, calculated by Curves [29], is shown in each case.

It should be added that crystallographic oligomers which exhibit B_{II} junctions also exhibit negative roll but no overall curvature is seen since the neighbouring steps adopt positive rolls [61]. This observation has been interpreted as crystal packing effect, which would seem to agree with our observations. Modelling studies using molecular dynamics have also observed rapid, but infrequent, B_I-B_{II} transitions [62-64]. In two different investigations, both simulating DNA surrounded by water and counterions, a single transition occurred during respectively 60ps and 140ps, within a homopolymeric AT sequence and at a GpA junction.

4.3 Transitions between allomorphic forms

We will lastly consider the nature of transitions between allomorphic forms of the DNA double helix. Certain transitions of this type can occur smoothly with only small energy barriers, such as the B-A transition already mentioned in connection with the calculation of energy surfaces defined by sugar puckering (see Fig. 1 in section 3.1). Other transitions involve more substantial energy barriers and major conformational changes. This is the case for B-Z transitions which we will now consider.

We began our studies by constructing junctions between DNA fragments in the right-handed B and left-handed Z conformations using a GC alternating base sequence (which can be experimentally induced to adopt the Z-form by high salt concentrations or negative supercoiling stress). As the base sequence chosen has a dinucleotide repeat, it is possible to build two distinct junctions, the interface between the two allomorphic forms occurring either at a 5'-CpG-3' step (termed type I) or at a 5'-GpC-3' step (termed type II). Each of these junctions was roughly constructed on a graphics system by simply juxtaposing fragments of optimised Z-DNA and B-DNA. It is worth noting that this procedure avoided any temptation to maintain a common helical axis across the junction, the juxtaposition primarily being aimed at rejoining the phosphodiester backbones. The resulting structures were then analysed to obtain the "kink" parameters describing the axis disruption at the interface. With this data it was possible to reconstruct a junction containing fragment within Jumna and carry out a new energy minimisations.

The results obtained showed that both B-Z junctions were indeed principally characterised by an important dislocation of the helical axis (roughly 10Å) and also by an important axis bend of 30°-40°. Apart from these features, very little change occurred in the conformations of the B and Z fragments on either side of the junctions. The type I junction resulted in axis bending into the major groove of the B-fragment, while the type II junction led to bending towards the minor groove. This can be explained by the fact that the B to Z transition occurs by right-handed rotation of the base pairs around the G → C long axis [65]. During the transition, guanosine changes its glycosidic angle from anti to syn, thus hardly distorting the backbone linkages. In contrast, cytidine hardly changes its glycosidic angle. This would result in breaking the phosphodiester linkage on the 3'-side of the nucleotide unless some compensation is made. One way to achieve this is simply to incline the Z-fragment in the directions noted above - towards the major groove when the first Z base pair is G-C and the minor groove when the first pair is C-G. This observation also has an effect on the relative stability of the two junctions since, for type II junctions, the Z-fragment fits neatly into the B-DNA minor groove allowing considerable stacking interactions to be conserved at the interface. For type I junctions, steric hindrance between the B and Z bases at the interface forces a much greater rise and consequently inhibits stacking. The result is that the type I junction costs almost twice as much to form as the type II junction (19 Kcal/mol versus 10 Kcal/mol).

The alternating directions of bending for the two types of B-Z junction led us to suggest a possible transition pathway between these allomorphs. Since the base pairs must rotate 180° in passing from the B to Z conformation it is necessary to facilitate this process by opening up the equivalent of an intercalation site on one side of the transiting pair. The location of this site should be above G-C pairs and below C-G pairs due to the right-handed sense of rotation around the G → C axis. As a given base pair undergoes its rotation, the junction changes its type. For rotating C-G pairs, the junction changes from type I to type II. This means that the Z-fragment changes its bending direction from the major groove to the minor groove, or, in other words, rotates anti-clockwise around the G → C axis of the rotating pair. The consequence of these contrary rotations is that stacking energy at the junction is recovered before the rotating pair has completed its full 180° turn. For C-G pairs the same is true, the base pair and the junction bend again rotate in opposite directions. We concluded that the B-Z interface can progress along the double helix by a flip-flop movement coupling base pair rotation and junction bending. Some preliminary calculations of this pathway (illustrated in Figure 6) were carried out by constraining the tip angle of the bases at the B-Z interface. The results suggested that the energy barrier should not exceed roughly 25 Kcal/mol. The most difficult point along the path was found roughly at the half-way stage when the rotating base pair was vertical and entirely contained within the double helix. At this point, the phosphodiester strands are forced apart and are under strain. In one simulation, this strain resulted in breaking the transiting pair, whose component bases then comfortably stacked on one another within the helix cavity. Further rotation pulled the bases apart and reformed the usual Watson-Crick base pair (similar results have recently been found in a simulation carried out by Ron Elber using the Charmm program [66]).

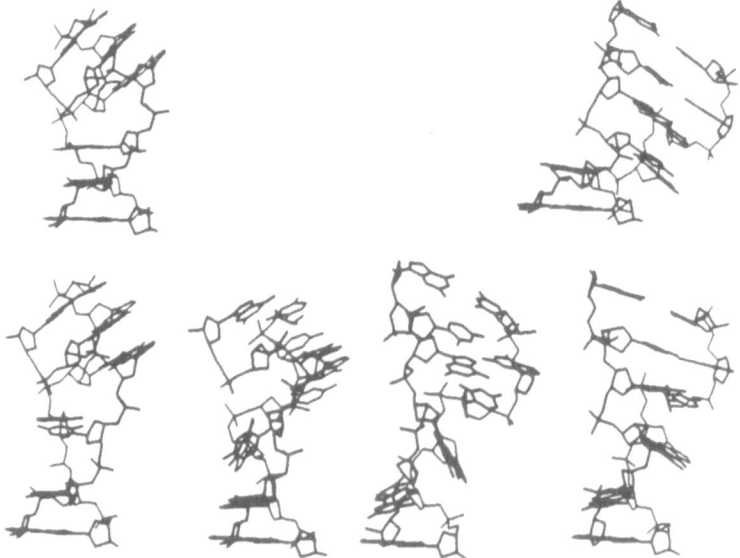

Figure 6. Steps along the B-Z transition pathway. The top left hand image shows a type I B-Z junction with a B-DNA segment below and a Z-DNA segment above. Starting top left and rotating anti-clockwise to top right

corresponds to the passage of one nucleotide pair from the B to the Z conformation and the creation of a type II junction.

The fact that the B-Z junction appears to be localised at a single dinucleotide junction within the double helix has also recently found experimental support [67]. It will be noted that our model pathway concentrates on rotation of the base pairs around their long axis, rather than the passage from a right-handed helix to a left-handed helix, which is often thought of as the major difference between B- and Z-DNA. In fact, this change can occur very easily and has little influence on the B-Z interface which, according to our model, has almost no twist and is moreover disrupted by the transitory unstacking necessary to rotate intact base pairs.

5. DNA DYNAMICS AND LARGE SCALE SIMULATIONS

Molecular dynamics [68,69] and Monte Carlo [70,71] simulations have been applied to nucleic acids both *in vacuo* and with a surrounding shell of explicit water molecules, these simulations are, however, far from being routine. Interesting results have been obtained concerning the behaviour of different force fields [63,72], the positioning of hydrating water molecules and counterions [68,73,74], hydrogen bonding [75] and the role of dynamics underlying nuclear Overhauser signals in NMR spectra [62,76,77]. Molecular dynamics calculations in water are the most refined approach to the problem of modelling nucleic acids in their natural environment, but such calculations remain very expensive and the time period simulated rarely exceeds a fraction of a nanosecond. This period is nevertheless sufficient for oligomeric DNA double helices to be destroyed, unless very careful equilibration is performed. In certain cases it is also necessary to impose artificial constraints on hydrogen bonding to prevent such denaturation [64,68]. Even in properly stabilised systems, other problems exist, such as the very small number of counterions employed, generally limited to one cation per phosphate group to ensure electroneutrality, and the low mobility of these ions during the simulation. One must also note that nucleic acid force field parameters have generally received much less attention than those used for studying proteins. Dynamics nevertheless plays an increasingly important role in refining DNA oligomer structures in conjunction with both X-ray and NMR experimental data.

A final area of study, which seems very promising, involves the development of simplified models of the nucleic acids which allow very large scale systems to be simulated. The largest systems to have been treated in atomic detail are probably dynamic studies of tRNA[Phe] [78], of a 4-stranded junction [79] and of a DNA nodule [80]. These systems involve up to roughly 100 nucleotides. In order to study the behaviour of still larger systems, of which DNA plasmids are a good example, some detail has to be sacrificed. Work in this field began with phenomenological models of sequence induced DNA curvature (for a recent review, see [81]). More detailed models have allowed the mechanics and dynamics of plasmid supercoiling to be simulated [82,83] and recent work from Tamar Schlick and Wilma Olson has made very elegant progress in this field [84].

6. CONCLUSIONS

Molecular modelling has already made considerable contributions to the study of nucleic acids, firstly, as an indispensable tool for refining the structures of oligomers studied by X-ray diffraction or NMR spectroscopy and, secondly, as a means for studying the mechanics and energetics of the double helix at an atomic scale. Modelling studies have helped in understanding both the static and dynamic aspects of DNA structure and the mechanisms of DNA interactions with other molecules.

In our own work, the use of a combination of helicoidal and internal coordinates for modelling nucleic acids has proved useful. Although our understanding of base sequence effects is far from complete, recent work on regular sequences has emphasised the central role of sugar puckering in determining the structural features of the nucleic acid helices. We are also beginning to understand something of the mechanisms of DNA deformation and the local and global conformational transitions that the double helix can undergo. These results should contribute to understanding the remarkably precise way in which proteins are capable of recognising their DNA target sequences.

It should be mentioned that, while the present article has concentrated on helical DNA, the second family of nucleic acids, RNA's, also represent a fascinating and broad field of research. RNA is capable of folding into a wide variety of compact structures which are only partially helical. Many biologically active RNA's involve many hundreds of nucleotides and although the range of basic building blocks in RNA (guanosine, adenosine, cytidine and uridine nucleotides) is smaller than the 20 common amino acids found in proteins, the existence of numerous modified bases and the variety of internucleotide interactions that can occur makes these macromolecules equally complex. Indeed, certain RNA's (ribozymes) share with proteins the possibility of specifically binding molecules and of catalysing chemical reactions. Despite the relatively small number of known RNA structures, attempts at modelling, made in conjunction with experimental and phylogenetic data, are very promising and have already had an impact on this field [85,86].

Molecular modelling clearly has important limitations. Force fields are far from exact, computational limitations restrict both the size, the quality and the duration of simulations and simulation algorithms can certainly be improved. Many processes are still far beyond the scope of all-atom models. Nevertheless, one can hope that the results obtained by modelling small systems, over short time scales, will be able to contribute to large scale studies by suggesting which detailed features need to be conserved and in helping to parameterize such models.

References

[1] Trifonov E.N. *CRC Crit. Rev. Biochem.* **19** (1989) 89-106 86-106.

[2] Trifonov E.N. in Theoretical biochemistry and molecular biophysics Vol.1 DNA (Adenine Press New York 1991) pp. 377-388.

[3] Suck D. in Structural tools for the analysis of protein-nucleic acid complexes Eds. Lilley D.M.J. Heumann H. and Suck D. (Birkhäuser, Basel 1992) pp. 127-141.

[4] Travers A.A. *Curr. Opinion in Struct. Biol.* **2** (1992) 71-77.

[5] McCammon J.A. and Harvey S.C. Dynamics of proteins and nucleic acids. (Cambridge University Press, U.K. 1987).

[6] Brooks C.L. III, Karplus M. and Pettitt B.M. Proteins: A theoretical perspective of dynamics, structure and thermodynamics *Advances in Chem. Physics* Vol. 71 (Wiley, NewYork 1988).

[7] van Gunsteren W.F. and Berendsen H.J.C. (1990) *Angew. Chem. Int. Ed.* **29** (1990) 992-1023.

[8] Harvey S.C. *Proteins: Structure, function and Genetics* **5** (1989) 78-92.

[9] Elber R. *Curr. Op. Struct. Biol.* **3** (1993) 260-264.

[10] McCammon J.A. *Curr. Op. Struct. Biol.* **1** (1991) 196-200.

[11] Lavery R. in "Unusual DNA structures" Eds. Wells R.D. and Harvey S.C. (Springer-Verlag 1988) pp. 189-206.

[12] Lavery R. in "Structure and expression. Vol.3, DNA bending and curvature" Eds. Olson W.K., Sarma R.H., Sarma M.H. and Sundaralingam M. (Adenine Press, New York, 1988) pp. 191-211.

[13] Lavery R., Sklenar H., Zakrzewska K. and Pullman B. *J. Biomol. Struct. Dyn.* **3** (1986) 989-1014. ·

[14] Lavery R., Parker I. and Kendrick J. *J. Biomol. Struct. Dyn.* **4** (1986) 443-461.

[15] Lavery R., Zakrzewska K. and Pullman A. *J. Comp. Chem.* **5** (1984) 363-373.

[16] Schlick T., Peskin C., Broyde S. and Overton M. *J. Comp. Chem.* **8** (1987) 1199-1224.

[17] Olson W.K. and Sussman J.L. *J. Am. Chem. Soc.* **104** (1982) 270-278.

[18] Hayes D.M., Kollman P.A. and Rothenberg S. *J. Am. Chem. Soc.* **99** (1977) 2150-2154.

[19] Hingerty B., Richie R.H., Ferrel T.L. and Turner T.E. *Biopolymers* **24** (1985) 427-439.

[20] Lavery R. and Sklenar H. in Structure and methods. Vol.2, DNA Protein complexes and proteins Eds. Sarma R.H. and Sarma M.H. (Adenine Press, New York 1990) pp. 215-235.

[21] Ravishankar G., Swaminathan S., Beveridge D.L., Lavery R. and Sklenar H. *J. Biomol. Struct. Dyn.* **6** (1989) 669-699.

[22] Fratini A.V., Kopka M.L., Drew H.R. and Dickerson R.E. *J. Biol. Chem.* **257** (1982) 14686-14707.

[23] von Kitzing E. and Diekmann S. *Eur. J. Biophys.* **15** (1987) 13-26.

[24] Soumpasis D.M. and Tung C.-S. *J. Biomol. Struct. Dyn.* **6** (1988) 397-420.

[25] Battacharya D. and Bansal M. *J. Biomol. Struct. Dyn.* **6** (1988) 93-104.

[26] Babcock M.S., Pednault P.D. and Olson W.K. *J. Mol. Biol.* **237** (1994) 125-156.

[27] Soumpasis D.M., Tung C.S. and Garcia A.E. *J. Biomol. Struct. Dyn.* **8** (1991) 867-888.

[28] Lavery R. and Sklenar H. *J. Biomol. Struct. Dyn.* **6** (1988) 63-91.

[29] Lavery R. and Sklenar H. *J. Biomol. Struct. Dynam.* **6** (1989) 655-667.

[30] Dickerson R.E., Bansal M., Calladine C.R., Diekmann S., Hunter W.N., Kennard O., Lavery R., Nelson H.C.M., Olson W.K., Saenger W., Shakked Z., Sklenar H., Soumpasis D.M., Tung C.-S., von Kitzing E., Wang A.H.-J. and Zhurkin V.B. *J. Mol. Biol.* **205** (1989) 787-791.

[31] Stofer E. and Lavery R. *Biopolymers* **34** (1993) 337-346.

[32] Prévost C., Louise-May S., Ravishankar G., Lavery R. and Beveridge D.L. *Biopolymers* **33** (1993) 335-350.

[33] Poncin M., Hartmann B. and Lavery R. *J. Mol. Biol.* **226** (1992) 775-794.

[34] Arnott S. and Hukins D.W.L. *J. Mol. Biol.* **81** (1973) 93-105.

[35] Arnott S., Chandrasekaran R., Birdsall D.L., Leslie A.G.W. and Ratliff R.L. *Nature* **283** (1980) 743-745 and coordinates communicated to our laboratory by S. Arnott.

[36] Poncin M., Piazzola D. and Lavery R. *Biopolymers* **32** (1992) 1077-1103.

[37] Lavery R. in Advances in computational biology Vol. 1 Ed. Villar H.O. (JAI Press, Connecticut 1994) pp. 69-145.

[38] Lavery R. and Hartmann B. *Biophys. Chem.* **50** (1994) 33-45.

[39] Bolshoy A., McNamara P., Harrington R.E. and Trifonov E.N., *Proc. Natl. Acad. Sci. (USA)* **88** (1991) 2312-2316.

[40] Cacchione S., De Santis P., Foti D., Palleschi A. and Savino M. *Biochemistry* **28** (1989) 8706-8713.

[41] Tan R.K.-Z. and Harvey S.C. *J. Biomol. Struct. Dynam.* **5** (1987) 497-511.

[42] Yanagi K., Privé G.G. and Dickerson R.E. *J. Mol. Biol.* **217** (1991) 201-214.

[43] Zakrzewska K., Poltev V.I., Oguey C. and Lavery R. *J. Mol. Struct. (Theochem)* **286** (1993) 219-230.

[44] Montminy M.R., Sevarino K.A., Wagner J.A., Mandel G., Goodman R.H. *Proc. Natl. Acad. Sci. (USA)* **83** (1988) 6682-6686.

[45] Ziff E.B. *Trends in Genetics* **6** (1990) 69-72.

[46] Mauffret O., Hartmann B., Convert O., Lavery R. and Fermandjian S. *J. Mol. Biol.* **227** (1992) 852-875.

[47] Englander S.W. and Kallenbach N.R. *Quart. Rev. Biophys.* **16** (1983) 521-655.

[48] Guéron M., Kochoyan M. and Leroy J.L. *Nature* **328** (1987) 89-92.

[49] Hartmann B., Leng M. and Ramstein J. *Biochemistry* **25** (1986) 3073-3077.

[50] Keepers J., Kollman P.A. and James T.L. *Biopolymers* **23** (1984) 2499-2511.

[51] Ramstein J. and Lavery R. *Proc. Natl. Acad. Sci. (USA)* **85** (1988) 7231-7235.

[52] Ramstein J. and Lavery R. *J. Biomol. Struct. Dyn.* **7** (1990) 915-933.

[53] Manning G.S. *Biopolymers* **22** (1983) 689-729.

[54] Briki F., Ramstein J., Lavery R. and Genest D. *J. Am. Chem. Soc.* **113** (1991) 2490-2493.

[55] Leroy J.L., Broseta D. and Guéron M. *J. Mol. Biol.* **184** (1985) 165-178.

[56] Privé G.G., Heinemann U., Chandrasekharan S., Kan L.S., Kopta M.L. and Dickerson R.E. *Science* **238** (1987) 498-504.

[58] Hartmann B., Piazzola D. and Lavery R., *Nucleic Acids Res.* **21** (1993) 561-568.

[59] Nikonowicz E.P. and Gorenstein D.G. *Biochemistry* **29** (1990) 8845-8858.

[60] Roongta V.A., Jones C.R. and Gorenstein D.G. *Biochemistry* **29** (1990) 5245-5258.

[61] Heinemann U. and Hahn M. *J. Biol. Chem.* **267** (1992) 7332-7341.

[62] Withka J.M., Swaminathan S., Beveridge D.L. and Bolton P.H. *J. Am. Chem. Soc.* **113** (1991) 5041-5049.

[63] Swaminathan S., Ravishanker G. and Beveridge D.L. *J. Am. Chem. Soc.* **113** (1991) 5027-5040.

[64] Brahms S., Fritsch V., Brahms J.G. and Westhof E. *J. Mol. Biol.* **223** (1992) 455-476.

[65] Harvey S.C. *Nucleic Acids Res.* **11** (1983) 4867-4878.

[66] Elber RE in Modelling biomolecular structures and mechanisms. Proceedings of the 27th Jerusalem Symposium (Kluwer Adademinc Press, Doortrecht 1994) in press.

[67] Dai Z., Dauchez M., Thomas G. and Peticolas W.L. *J. Biomol. Struct. Dynam.* **9** (1992) 1155-1183.

[68] Beveridge D.L. and Ravishankar G. Curr. Opin. Struct. Biol. 4 (1994) 246-255.

[69] Goodfellow J.M. and Williams M.A. *Curr. Op. in Struct. Biol.* **2** (1992) 211-216.

[70] Zhurkin V.B., Ulyanov N.B., Gorin A.A. and Jernigan R.L. *Proc. Natl. Acad. Sci. (USA)* **88** (1991) 7046-7050.

[71] Olson W.K., Marky N.L., Jernigan R.L. and Zhurkin V.B. *J. Mol. Biol.* **232** (1993) 530-554.

[72] Srinivasan J., Withka J.M. and Beveridge D.L. *Biophys J.* **58** (1990) 523-547.

[73] Subramanian P.S., Ravishankar G. and Beveridge D.L. *Proc. Natl. Acad. Sci. (USA)* **85** (1988) 1836-1840.

[74] Jayaram B., Swaminathan S., Beveridge D.L., Sharp K. and Honig B. *Macromolecules* **23** (1990) 3156-3165.

[75] Fritsch V. and Westhof E. in "Modeling of molecular structures and properties" Ed. Rivail, J.L., Ed. (Elsevier, Amsterdam 1990) pp. 627-634.

[76] Withka J.M., Swaminathan S. and Bolton P.H. *J. Magn. Res.* **89** (1990) 386-390.

[77] Withka J.M., Swaminathan S., Beveridge D.L. and Bolton P.H. *Science* **255** (1992) 597-599.

[78] Harvey S.C., Prabhakaran M. and McCammon J.C. *Biopolymers* **24** (1985) 1169-1188.

[79] von Kitzing E., Lilley D.M. and Diekmann S. *Nucleic Acids Res.* **18** (1990) 2671-2683.

[80] Sprous D. and Harvey S.C. *J. Biol. Chem.* **267** (1992) 5502-5512.

[81] Trifonov E.N. *Trends in Biochem. Sci.* **16** (1991) 487-490.

[82] Tan R.K.Z. and Harvey S.C. *J. Mol. Biol.* **205** (1989) 573-591.

[83] Tan R.K.Z. and Harvey S.C. in "Theoretical biochemistry and molecular biophysics" Eds. Beveridge D.L. and Lavery R. (Adenine Press, New York 1990) pp.125-137.

[84] Schlick T. and Olson W.K. *J. Mol. Biol.* **223** (1992) 1089-1119.

[85] Michel F. and Westhof E. *J. Mol. Biol.* **216** (1990) 585-610.

[86] Westhof E. and Altman S. *Proc. Natl. Acad. Sci (USA)* **91** (1994) 5133-5137.

Potential-of-mean-force description of ionic interactions and structural hydration in biomolecular systems

G. Hummer, D.M. Soumpasis* and A.E. García

*Theoretical Biology and Biophysics Group T-10,
MS K710, Los Alamos National Laboratory,
Los Alamos, NM 87545, U.S.A.*
** Biocomputation Group, Max-Planck-Institute
for Biophysical Chemistry, P.O. Box 2841,
37018 Göttingen, Germany*

1. INTRODUCTION

To understand the functioning of living organisms on a molecular level, it is crucial to dissect the intricate interplay of the immense number of biological molecules. However, large biological macromolecules are not the only players in the field. Most of the biochemical processes in cells occur in a liquid environment formed mainly by water and ions. This solvent environment plays an important role in biological systems [1]. It mediates biochemical reactions and has a strong influence on the structural equilibrium of certain molecules. Therefore, the development and application of theoretical descriptions of solute-solvent interactions provides relevant insight into biological processes.

The potential-of-mean-force (PMF) formalism attempts to describe quantitatively the interactions of the solvent with biological macromolecules on the basis of an approximate statistical-mechanical representation. At its current status of development, it deals with ionic effects on the biomolecular structure and with the structural hydration of biomolecules. The underlying idea of the PMF formalism is to identify the dominant sources of interactions and incorporate these interactions into the theoretical formalism using PMF's (or particle correlation functions) extracted from bulk-liquid systems.

In the following, we shall briefly outline the statistical-mechanical foundation of the PMF formalism and introduce the PMF expansion formalism, which is intimately linked to superposition approximations for higher-order particle correlation functions. We shall then sketch applications, which describe the effects of the ionic environment on nucleic-acid structure. Finally, we shall present the more recent extension of the PMF idea to describe quantitatively the structural hydration of biomolecules. Results for the interface of ice and water and for the hydration of deoxyribonucleic acid (DNA) will be discussed.

2. THE STATISTICAL-MECHANICAL FOUNDATION OF THE POTEN-TIAL-OF-MEAN-FORCE FORMALISM

The structural equilibrium properties of fluid systems are best described in terms of a hierarchy of particle correlation functions. Within the framework of statistical mechanics applied on classical-mechanical systems, the particle correlation functions can be expressed as integrals over the configuration space of the system. In the following, we consider a monoatomic liquid in a canonical (N, V, T) ensemble. The N particles are subject to interactions described by a potential U depending on the positions $r_1, \ldots, r_N$ of the N particles. Then, the n-particle correlation functions are defined as

$$g^{(n)}(\mathbf{r}_1, \ldots, \mathbf{r}_n) \; = \; \rho_0^{-n} \, \frac{N!}{(N-n)!} \, \frac{\int d\mathbf{r}_{n+1} \cdots d\mathbf{r}_N \, \exp[-\beta \, U(\mathbf{r}_1, \ldots, \mathbf{r}_N)]}{\int d\mathbf{r}_1 \cdots d\mathbf{r}_N \, \exp[-\beta \, U(\mathbf{r}_1, \ldots, \mathbf{r}_N)]} \, . \tag{1}$$

Here, $\beta = 1/k_B T$ is the inverse temperature; and $\rho_0 = N/V$ is the bulk density of the solvent. No external field is present; and the 1-particle correlation function is independent of the position: $g^{(1)}(\mathbf{r}) \equiv 1$.

To introduce the concept of potentials of mean force, we fix n of the N particles at positions $\mathbf{r}_1, \ldots, \mathbf{r}_n$. In the canonical ensemble, the mean force acting on particle i $(1 \leq i \leq n)$ is calculated as

$$\mathbf{F}_i(\mathbf{r}_1, \ldots, \mathbf{r}_n) \; = \; \frac{\int d\mathbf{r}_{n+1} \cdots d\mathbf{r}_N \, (-\partial U/\partial \mathbf{r}_i) \, \exp[-\beta \, U(\mathbf{r}_1, \ldots, \mathbf{r}_N)]}{\int d\mathbf{r}_1 \cdots d\mathbf{r}_N \, \exp[-\beta \, U(\mathbf{r}_1, \ldots, \mathbf{r}_N)]} \, . \tag{2}$$

Changing the order of differentiation and integration yields a potential $W^{(n)}$ for the mean force,

$$\mathbf{F}_i(\mathbf{r}_1, \ldots, \mathbf{r}_n) \; = \; -\frac{\partial}{\partial \mathbf{r}_i} \, W^{(n)}(\mathbf{r}_1, \ldots, \mathbf{r}_n) \, , \tag{3}$$

where the n-particle potential of mean force (PMF) is obtained as

$$W^{(n)}(\mathbf{r}_1, \ldots, \mathbf{r}_n) \; = \; -k_B T \, \ln g^{(n)}(\mathbf{r}_1, \ldots, \mathbf{r}_n) \, . \tag{4}$$

This provides an important link between the microscopic structure of the fluid and the thermodynamic properties of inhomogeneous systems. The mean force $\mathbf{F}_i$ integrated over a path $\mathbf{r}_i(\tau)$ connecting $\mathbf{r}_i^1$ and $\mathbf{r}_i^2$ gives the reversible work of bringing particle i from $\mathbf{r}_i^1$ to $\mathbf{r}_i^2$ in the presence of $n-1$ particles constrained to $\mathbf{r}_1, \ldots, \mathbf{r}_{i-1}$ and $\mathbf{r}_{i+1}, \ldots, \mathbf{r}_n$. Correspondingly, if all particles j are moved from $\mathbf{r}_j^1$ to $\mathbf{r}_j^2$, the free energy (reversible work!) required for this operation is given by $W^{(n)}(\mathbf{r}_1^2, \ldots, \mathbf{r}_n^2) - W^{(n)}(\mathbf{r}_1^1, \ldots, \mathbf{r}_n^1)$.

To achieve practical relevance for these formulae, one has to devise ways to calculate $W^{(n)}$ (or, correspondingly, $g^{(n)}$) or find at least approximate representations. However, already the accurate calculation of $g^{(2)}$ for simple (pair additive, spherically symmetric) models for the interaction potential U proves to be a nontrivial task [2]. This is even more so for higher-order correlations $(n \geq 3)$. Nevertheless, by means of computer simulations or integral-equation techniques, two- and three-particle correlations can be calculated with reasonable accuracy. This guides the development toward expansion of the higher-order PMF's in terms of expressions of lower-order. This goal is accomplished by the so-called PMF expansion [3]. We attempt to rewrite our function $W^{(n)}(\mathbf{r}_1, \ldots, \mathbf{r}_n)$ (which is symmetric in its n arguments) as a sum of functions $w^{(m)}$ of m arguments ($1 \le m \le n$),

$$W^{(n)}(\mathbf{r}_1, \ldots, \mathbf{r}_n) = \sum_i w^{(1)}(\mathbf{r}_i) + \sum_{\substack{i,j \\ n \ge i > j \ge 1}} w^{(2)}(\mathbf{r}_i, \mathbf{r}_j) + \sum_{\substack{i,j,k \\ n \ge i > j > k \ge 1}} w^{(3)}(\mathbf{r}_i, \mathbf{r}_j, \mathbf{r}_k) + \cdots .$$

(5)

Inversion of Eq. 5 yields

$$\begin{aligned}
w^{(1)}(\mathbf{r}) &= W^{(1)}(\mathbf{r}) \\
w^{(2)}(\mathbf{r}_1, \mathbf{r}_2) &= W^{(2)}(\mathbf{r}_1, \mathbf{r}_2) - W^{(1)}(\mathbf{r}_1) - W^{(1)}(\mathbf{r}_2) \\
w^{(3)}(\mathbf{r}_1, \mathbf{r}_2, \mathbf{r}_3) &= W^{(3)}(\mathbf{r}_1, \mathbf{r}_2, \mathbf{r}_3) - W^{(2)}(\mathbf{r}_1, \mathbf{r}_2) - W^{(2)}(\mathbf{r}_2, \mathbf{r}_3) - W^{(2)}(\mathbf{r}_3, \mathbf{r}_1) \\
&\quad + W^{(1)}(\mathbf{r}_1) + W^{(1)}(\mathbf{r}_2) + W^{(1)}(\mathbf{r}_3) .
\end{aligned}$$

(6)

For a homogeneous system, $g^{(1)} \equiv 1$ and $W^{(1)} \equiv 0$. From Eqs. 4, 5, and 6 we obtain an expansion for $W^{(n)}$ in terms of particle correlation functions,

$$-\beta W^{(n)}(\mathbf{r}_1, \ldots, \mathbf{r}_n) = \sum_{\substack{i,j \\ n \ge i > j \ge 1}} \ln g^{(2)}(\mathbf{r}_i, \mathbf{r}_j)$$

$$+ \sum_{\substack{i,j,k \\ n \ge i > j > k \ge 1}} \ln \frac{g^{(3)}(\mathbf{r}_i, \mathbf{r}_j, \mathbf{r}_k)}{g^{(2)}(\mathbf{r}_i, \mathbf{r}_j)\, g^{(2)}(\mathbf{r}_j, \mathbf{r}_k)\, g^{(2)}(\mathbf{r}_k, \mathbf{r}_i)} + \cdots .$$

(7)

Eqs. 4 and 7 combined yield a product expansion for n-particle correlation functions,

$$g^{(n)}(\mathbf{r}_1, \ldots, \mathbf{r}_n) = \left[\prod_{\substack{i,j \\ n \ge i > j \ge 1}} g^{(2)}(\mathbf{r}_i, \mathbf{r}_j) \right]$$

$$\times \left[\prod_{\substack{i,j,k \\ n \ge i > j > k \ge 1}} \frac{g^{(3)}(\mathbf{r}_i, \mathbf{r}_j, \mathbf{r}_k)}{g^{(2)}(\mathbf{r}_i, \mathbf{r}_j)\, g^{(2)}(\mathbf{r}_j, \mathbf{r}_k)\, g^{(2)}(\mathbf{r}_k, \mathbf{r}_i)} \right] \cdots .$$

(8)

Generalized superposition approximations for the n-particle correlation functions are obtained by setting $w^{(n)} \equiv 0$ for $n \ge n_0$. In particular, the Kirkwood superposition approximation (KSA) [4] is obtained for $n_0 = 3$,

$$g_{\mathrm{KSA}}^{(3)}(\mathbf{r}_1, \mathbf{r}_2, \mathbf{r}_3) = \prod_{\substack{i,j \\ 3 \ge i > j \ge 1}} g^{(2)}(\mathbf{r}_i, \mathbf{r}_j) .$$

(9)

The Fisher-Kopeliovich superposition approximation [5] corresponds to $n_0 = 4$,

$$g_{\mathrm{FKSA}}^{(4)}(\mathbf{r}_1, \mathbf{r}_2, \mathbf{r}_3, \mathbf{r}_4) = \frac{\prod_{\substack{i,j,k \\ 4 \ge i > j > k \ge 1}} g^{(3)}(\mathbf{r}_i, \mathbf{r}_j, \mathbf{r}_k)}{\prod_{\substack{i,j \\ 4 \ge i > j \ge 1}} g^{(2)}(\mathbf{r}_i, \mathbf{r}_j)} .$$

(10)

These two approximations form a cornerstone of the PMF formalism to describe biomolecular systems in solution.

3.   INTERACTIONS OF NUCLEIC ACIDS WITH IONIC SOLUTIONS

DNA forms an important component of biological systems as the carrier of genetic information. DNA molecules show a remarkable variability in their structure depending on base composition and solution environment [6]. With respect to salt effects on the structure of DNA, the transition of some base sequences from a right-handed B form (low salt) to a left-handed Z form has been particularly well characterized [7]. This structural transition offers an important test ground for any theory attempting to describe ionic interactions with biomolecules.

The PMF formalism can indeed successfully account for the experimentally observed behavior, i.e., the free-energy difference between the two DNA structures depending on ion types and concentrations [8–10]. In the following, we shall briefly outline the general ideas underlying the PMF description of DNA interacting with ionic solutions. More general presentations of the PMF method and its applications on DNA-ion interactions can be found in Refs. [11, 12].

Nucleic acid molecules form negatively-charged polyelectrolytes. The sugar-phosphate backbone connecting the bases carries one negative charge per phosphate group (PO_4^-). As mentioned in the introduction, the key idea of the PMF formalism is to identify the dominant interactions in the system. With respect to the structure dependence on the ionic environment, the charge-charge interaction is expected to be of foremost importance. In low-concentration electrolytes, the negatively-charged phosphate groups are exerting a strong charge repulsion on each other. However, at higher ionic concentrations this picture changes. It is well known from the theory of liquids that at high ionic concentrations the effective interactions, i.e., the PMF's, are of oscillatory behavior reflecting attractive interactions over some distance regions. This qualitative change in the ion-ion PMF's results in a shifted structural equilibrium for certain DNA sequences.

The simplest model for the DNA in the PMF formalism is that of a group of negatively-charged anions centered at the positions of the phosphate atoms. A structural transition of the DNA then corresponds to a rearrangement of these anions. The anions representing the PO_4^- groups are described as restricted-primitive-model (RPM) ions. This pair-interaction model contains the essential features that characterize ions in water, i.e., a hard-sphere interaction modeling the short-range repulsion of hydrated ions and a dielectrically screened long-range Coulomb interaction.

The formal development of the previous section provides us with a tool to find approximate expressions for the free-energy difference between two DNA structures owing to ionic contributions. Other contributions, in particular energetic [13] and vibrational components [14], have to be accounted for differently. For a given solution environment (temperature, ion types and concentrations) the difference in free energy ΔF between two conformations 1 and 2 is calculated as

$$\Delta F(1 \to 2) \;=\; W^{(n)}(\mathbf{r}_1^2, \ldots, \mathbf{r}_n^2) - W^{(n)}(\mathbf{r}_1^1, \ldots, \mathbf{r}_n^1) \;, \tag{11}$$

where n is the number of phosphate groups (modeled as anions) and $\mathbf{r}_i^1$, $\mathbf{r}_i^2$ are their initial and final positions, respectively. Employing the expansion Eq. 7 of $W^{(n)}$ in terms of particle correlation functions of anions in RPM electrolytes, we obtain an approximate expression

$$\Delta F(1 \to 2) = -k_{\mathrm{B}}T \sum_{\substack{i,j \\ n \geq i > j \geq 1}} \ln \frac{g_{--}^{(2)}(\mathbf{r}_i^2, \mathbf{r}_j^2)}{g_{--}^{(2)}(\mathbf{r}_i^1, \mathbf{r}_j^1)} - k_{\mathrm{B}}T \sum_{\substack{i,j,k \\ n \geq i > j > k \geq 1}} \ln \frac{\Gamma_{---}^{(3)}(\mathbf{r}_i^2, \mathbf{r}_j^2, \mathbf{r}_k^2)}{\Gamma_{---}^{(3)}(\mathbf{r}_i^1, \mathbf{r}_j^1, \mathbf{r}_k^1)} + \cdots,$$

$$(12)$$

where

$$\Gamma_{---}^{(3)}(\mathbf{r}, \mathbf{s}, \mathbf{t}) = \frac{g_{---}^{(3)}(\mathbf{r}, \mathbf{s}, \mathbf{t})}{g_{--}^{(2)}(\mathbf{r}, \mathbf{s})\, g_{--}^{(2)}(\mathbf{s}, \mathbf{t})\, g_{--}^{(2)}(\mathbf{t}, \mathbf{r})}. \tag{13}$$

In the case of 1:1 RPM electrolytes in the concentration regime up to about 4 mol/l, substantial contributions to the free energy come mainly from the pair PMF's. This is a consequence of the reasonable quality of the KSA (Eq. 9) for the RPM system [15]. The three-particle PMF's add only minor corrections to ΔF, as shown in the case of the B- to Z-DNA transition in NaCl [16]. The required particle correlation functions can be calculated accurately by means of integral equation methods such as the hypernetted chain (HNC) or exponential mean-spherical approximation (EXP-MSA).

The approximate expression for the free energy difference in terms of pair PMF's of RPM anions has been used to describe the salt dependent transitions of right-handed B-DNA to left-handed Z-DNA in sodium chloride solution [8] and in solutions containing other alkali cations [10]. The equilibrium between different DNA structures (A, B, C, and Z) has been studied in Ref. [9]. The PMF formalism has been compared with other theoretical approaches (Poisson-Boltzmann equation, counterion-condensation theory) in Ref. [17]. The salt-concentration dependent equilibrium between hairpin, mismatched duplex, and single-stranded DNA has been discussed in Ref. [18]. A similar approach has been used to calculate the salt-dependent melting temperature of DNA (helix-coil transition) within the PMF method [11]. The PMF interaction between charged atoms has been incorporated into force-field descriptions of biomolecules [13]. This allows an approximate representation of the effective, water-averaged Coulomb interactions emerging from a McMillan-Mayer picture of ionic systems.

A similar method has been used to study the harmonic vibrations of left- and right-handed DNA depending on the salt concentration [14]. The effective Coulomb interaction between different phosphate groups has been described by PMF's between anion pairs. In this study of the normal modes of B- and Z-DNA it has been observed that the lowest-frequency eigenmode of B-DNA shows a significant frequency decrease when increasing the salt concentration. This indicates that a soft-mode mechanism might play an important role in the B-Z transition.

4. A STATISTICAL-MECHANICAL THEORY OF BIOMOLECULAR HYDRATION

4.1 Introductory remarks

In the original PMF formalism dealing with ionic effects on DNA, water is described as a dielectric continuum in which the (hydrated) ions are immersed. This simplified picture works well as long as *structural* water does not play an important role in the effects studied.

In the case of DNA, the experimental results (reviewed, for instance, in Refs. [6, 19–21]) show a sequence- and conformation-dependent network of water molecules close to the DNA surface caused by short-range hydrophilic interactions with polar DNA atoms (oxygen, nitrogen). The most prominent feature in the case of B-DNA is the so-called *spine of hydration* [22, 23] covering tracts of A·T base pairs (1) in the minor groove, with water molecules bridging adjacent base pairs and themselves being bridged by other water molecules. Such highly localized water molecules at the DNA surface are thought to affect the structural stability of the DNA and to interfere with the binding of drug molecules to DNA [24].

It becomes evident that water must be seen as an integral part of biomolecular systems. Changes in the biomolecular structure or binding reactions (enzyme catalysis, antibody-antigen binding, drug binding, binding to carrier molecules, etc.) are accompanied by a reorganization of water molecules at the molecular surface. Water mediates interactions between different groups on biomolecules and between different molecules. Our understanding of these biologically important processes is therefore intimately coupled to having insight into the role of water in these systems.

In the following, we shall describe a method to computing quantitative measures of the biomolecular hydration based on the ideas of the PMF formalism. A similar approach has been used to compute ion-density distributions near biological macromolecules [25, 26].

4.2   Density expansion in inhomogeneous liquid systems

We shall study a system comprising a biological macromolecule (or a biomolecular assembly) and water molecules. The biomolecule (solute) is treated as a source of inhomogeneity affecting the structural organization of water in its vicinity. The central quantity to describe the structural equilibrium properties of the water phase is the position-dependent one-particle density of water $\rho(\mathbf{r})$. The primary goal of the theoretical development is to devise approximate expressions for $\rho(\mathbf{r})$ (in terms of known quantities) in the presence of a biological macromolecule (or other solid-like sources of inhomogeneity).

The biomolecule is represented by a set of positional coordinates $\{\mathbf{r}_{i_\alpha}\}$ of N_α atoms of type α and M different atom types α. We shall assume a rigid equilibrium structure of the molecule, as known from X-ray or neutron diffraction crystallography, NMR, or theoretical modeling. This assumption can be relaxed by performing an additional average over representative ensembles of structures, as, for instance, obtained from NMR spectroscopy.

To keep the notation simple, we shall develop the theory for an atomic solvent. The generalization to molecular solvents (water) is straightforward, either by introducing additional angular coordinates or by using an atom-site description. We consider a canonical ensemble of N solvent particles at positions $\{\mathbf{r}_i\}$ within the framework of statistical mechanics of classical-mechanical systems. The total potential energy of the system com-

---

(1) The following abbreviations for the DNA bases are used: Adenine (A), thymine (T), guanine (G), and cytosine (C). The deoxyribose in the sugar-phosphate backbone of DNA is denoted by $d(\cdots)$. The strand polarity of the backbone is represented by writing the base sequence in the 5' to 3' direction. Dinucleotide steps XY are written as 5'-d(XpY), where p represents the sugar-phosphate linkage. A dot denotes Watson-Crick base pairing (i.e., A·T and G·C). The DNA molecules studied in this work are all double stranded forming Watson-Crick base pairs.

prising solute and solvent molecules is denoted by U. For pair-additive interactions, it can be split into two parts such that the solute appears as an external field acting on the solvent molecules. The conditional solvent density at a point r_1 can be expressed as an integral over the configuration space of the solvent molecules,

$$\rho(\mathbf{r}_1|\{\mathbf{r}_{i_\alpha}\}) = N \frac{\int d\mathbf{r}_2 \cdots d\mathbf{r}_N \exp[-\beta U(\{\mathbf{r}_i\}, \{\mathbf{r}_{i_\alpha}\})]}{\int d\{\mathbf{r}_i\} \exp[-\beta U(\{\mathbf{r}_i\}, \{\mathbf{r}_{i_\alpha}\})]} . \tag{14}$$

We now introduce higher-order particle correlation functions describing the correlations of sets of atoms $\{N_\alpha\}$ and water, as generalized for multiparticle systems from Eq. 1,

$$\rho(\mathbf{r}|\{\mathbf{r}_{i_\alpha}\}) = \rho_0 \frac{g^{(1;\{N_\alpha\})}(\mathbf{r}, \{\mathbf{r}_{i_\alpha}\})}{g^{(\{N_\alpha\})}(\{\mathbf{r}_{i_\alpha}\})} , \tag{15}$$

where $\rho_0 = N/V$ is the bulk density of the solvent and

$$g^{(\{N_\alpha\})}(\{\mathbf{r}_{i_\alpha}\}) = \left(\prod_{\alpha=1}^{M} N_\alpha! \rho_\alpha^{-N_\alpha}\right) \frac{\int d\{\mathbf{r}_i\} \exp[-\beta U(\{\mathbf{r}_i\}, \{\mathbf{r}_{i_\alpha}\})]}{\int d\{\mathbf{r}_{i_\alpha}\} d\{\mathbf{r}_i\} \exp[-\beta U(\{\mathbf{r}_i\}, \{\mathbf{r}_{i_\alpha}\})]} , \tag{16}$$

$$g^{(1;\{N_\alpha\})}(\mathbf{r}_1, \{\mathbf{r}_{i_\alpha}\}) = N\rho_0^{-1} \left(\prod_{\alpha=1}^{M} N_\alpha! \rho_\alpha^{-N_\alpha}\right) \frac{\int d\mathbf{r}_2 \cdots d\mathbf{r}_N \exp[-\beta U(\{\mathbf{r}_i\}, \{\mathbf{r}_{i_\alpha}\})]}{\int d\{\mathbf{r}_{i_\alpha}\} d\{\mathbf{r}_i\} \exp[-\beta U(\{\mathbf{r}_i\}, \{\mathbf{r}_{i_\alpha}\})]} , \tag{17}$$

with $\rho_\alpha = N_\alpha/V$. Eq. 15 is analogous to the well-known formula for conditional probabilities of events A and B,

$$P(A|B) = \frac{P(A \cap B)}{P(B)} . \tag{18}$$

The exact calculation of the higher-order correlation functions $g^{(\{N_\alpha\})}$ and $g^{(1;\{N_\alpha\})}$ is not possible with currently available means. However, the PMF expansion Eq. 8 provides the tool to expand them in terms of lower-order correlations. We obtain a product expansion,

$$\rho(\mathbf{r}|\{\mathbf{r}_{i_\alpha}\}) = \rho_0 \left[\prod_{\alpha=1}^{M} \prod_{i_\alpha=1}^{N_\alpha} g^{(1;\alpha)}(\mathbf{r}, \mathbf{r}_{i_\alpha})\right]$$
$$\times \left[\prod_{\alpha=1}^{M} \prod_{i_\alpha=1}^{N_\alpha} \prod_{\beta=1}^{M} \prod_{i_\beta=1+\delta_{\alpha\beta}i_\alpha}^{N_\beta} \frac{g^{(1;\alpha,\beta)}(\mathbf{r}, \mathbf{r}_{i_\alpha}, \mathbf{r}_{i_\beta})}{g^{(1;\alpha)}(\mathbf{r}, \mathbf{r}_{i_\alpha})g^{(\alpha,\beta)}(\mathbf{r}_{i_\alpha}, \mathbf{r}_{i_\beta})g^{(\beta;1)}(\mathbf{r}_{i_\beta}, \mathbf{r})}\right] \cdots , \tag{19}$$

where $\delta_{\alpha\beta}$ is the Kronecker symbol. This expression relates the conditional solvent density to products of correlation functions involving one solvent particle and one, two, etc., solute atoms. The first term in the expansion Eq. 19 is the bulk-density of the solvent, the next order yields a product over pair correlations involving individual solute atoms; and the triplet term involves a product over all distinct pairs of solute atoms.

4.3 The water-density profile at the interface of ice and water

We shall first illustrate the applicability of the expansion formula Eq. 19 in the context of a planar solid-liquid interface formed by ice and water. A comparison with reference data from computer simulations of the same system allows us to assess the quality of the density

ice   water   ice

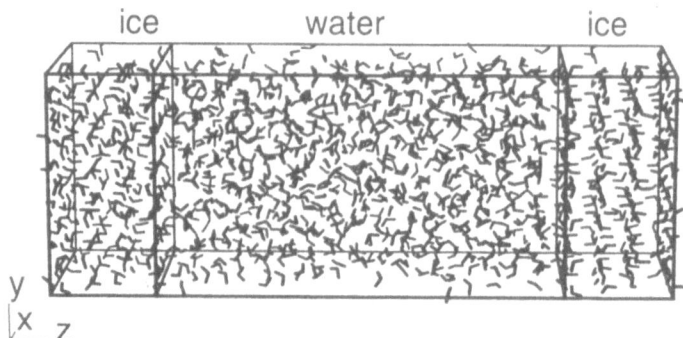

Figure 1: A snapshot of an MC simulation of the ice-water interface using 1000 water molecules. The simulation box comprises one ice and one liquid water layer. The second ice layer arises from the use of periodic boundary conditions.

expansion method under well-defined conditions. In particular, we are able to investigate the reliability of the truncation of the expansion after the two- and three-particle level in the case of water. Results for the ice-water interface have been presented in Ref. [27]. Here, we shall review this study to give an instructive example for applications of the density expansion methods.

To model the interface, we fix water oxygens at ideal hexagonal-ice Ih lattice positions $r_1, \ldots, r_n$, but do not specify molecular orientations. The thermal disorder in the interfacial region of the solid phase is therefore limited to reorientation of the water molecules. This ice layer is embedded into liquid water, giving rise to a structured density distribution in the liquid phase. This distribution has been calculated both using the PMF expansion formalism and extensive Monte-Carlo (MC) computer simulations.

The application of the general expression Eq. 19 is simplified since only water molecules are present. We obtain an expression for the water-oxygen and hydrogen density in the liquid phase in terms of two- and three-particle correlation functions of bulk water,

$$\rho_X(\mathbf{r}|\mathbf{r}_1,\ldots,\mathbf{r}_n) \;\approx\; \rho_X \prod_{i=1}^{n} g_{XO}^{(2)}(\mathbf{r},\mathbf{r}_i) \prod_{j=1}^{n-1} \prod_{k=j+1}^{n} \frac{g_{XOO}^{(3)}(\mathbf{r},\mathbf{r}_j,\mathbf{r}_k)}{g_{XO}^{(2)}(\mathbf{r},\mathbf{r}_j)g_{OO}^{(2)}(\mathbf{r}_j,\mathbf{r}_k)g_{OX}^{(2)}(\mathbf{r}_k,\mathbf{r})} \;, \quad (20)$$

where X=O or H. Since only the pair but not the triplet correlations of fluid phases are accessible to an experimental determination using currently available techniques, we performed computer simulations of bulk water to obtain the correlation functions. $g^{(2)}$ and $g^{(3)}$ were calculated for the simple-point-charge (SPC) model of water [28], as described in Ref. [27].

To provide reliable density data for the interface of ice Ih and water that allow a careful testing of the PMF expansion method, a series of MC simulations of the interfacial system has been performed. We position 240 water molecules in an ideal ice-Ih lattice. The oxygens are constrained to lattice sites; but the molecules are free to rotate. A rectangular box is used with sides $L_x = 2.245$ nm and $L_y = 2.333$ nm. The two basal planes of the ice layer are oriented perpendicular to the z axis. The volume above the plane is filled with mobile water molecules. The temperature is kept at 298 K. System sizes of 600 and 1000 water molecules in total at an average density of $\rho_0 = 33.33$ nm^{-3} are

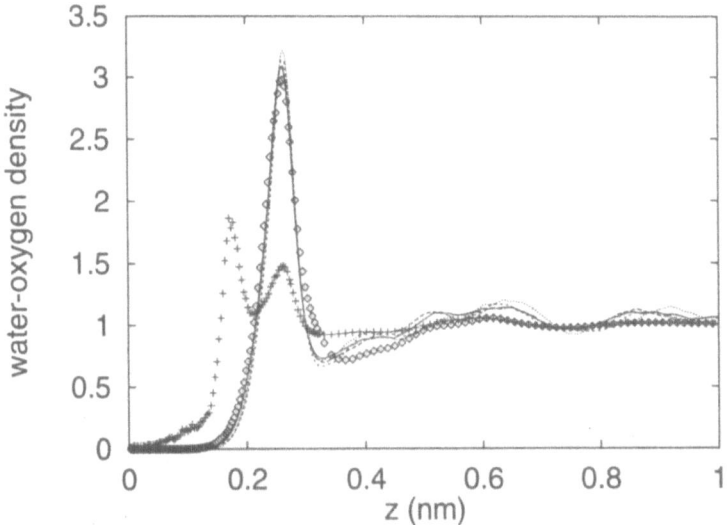

Figure 2: Water-oxygen density at the ice-water interface in units of the bulk-water density ρ_0. z measures the distance from the closest layer of oxygen atoms in the ice phase. ($+$), PMF expansion including only pair correlations; ($\diamond$), PMF expansion including pair and triplet correlations. MC simulations: (—), GRF, $N = 600$, $r_c = 0.9$ nm; (– –), GRF, $N = 600$, $r_c = 0.8$ nm; (- - -), GRF, $N = 1000$, $r_c = 0.9$ nm; ($\cdots$), Ewald, $N = 600$.

studied, resulting in $L_z = 3.436$ and 5.726 nm, respectively. Periodic boundary conditions are applied in all three directions, such that the system effectively consists of an ice layer (1.192 nm thick) in contact with liquid water on both sides. Fig. 1 shows a snapshot of an MC simulation with the liquid phase enclosed by two ice layers. The charge-charge interaction has been treated with Ewald summation [29] and a generalized reaction field (GRF) method [27, 30] using different cutoff radii r_c.

The z-dependent oxygen and hydrogen density in layers parallel to the interface (basal plane) obtained from four simulations with different system sizes and charge interactions is shown in Figs. 2 and 3. Also shown are the corresponding results obtained from PMF calculations, truncating Eq. 20 at the pair and triplet level.

Not surprisingly, if only the pair correlations are included in the PMF expansion, we observe serious disagreement between simulations and PMF calculations. In particular, the oxygen density shows two distinct peaks: The first one is too close and the second one is too weak. The PMF hydrogen density calculated with pairs alone shows somewhat better agreement with the simulations.

However, if the three-particle corrections are applied in the PMF expansion, the agreement drastically improves. The first peak in the oxygen density is quantitatively reproduced (position and height). Also, the strong second peak in the hydrogen density is found in quantitative agreement with the simulations. The first peak of the hydrogen density is somewhat too small. We ascribe this discrepancy to problems with the discretization of the $g_{OOH}^{(3)}$ correlations. These are known with a resolution of 0.02 nm, which results

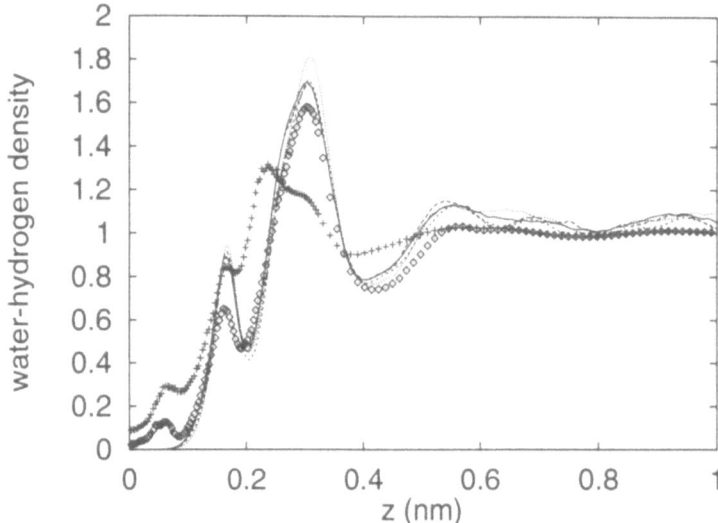

Figure 3: Water-hydrogen density at the ice-water interface in units of $2\,\rho_0$. Details as in Fig. 2.

in large jumps at small OH pair distances only poorly described by linear interpolation. For larger distances ($z > 0.4$ nm), the agreement between the PMF expansion and the Monte-Carlo simulation is somewhat less satisfactory. However, qualitative agreement is observed. In particular, the computer simulations consistently show minima in the oxygen and hydrogen densities at $z = 0.75$ and 0.8 nm, respectively. These minima also appear in the PMF calculations at the right positions but more shallow.

It becomes evident that it is essential to retain the water-triplet correlations in the expansion Eq. 20. They contain the information about the preferentially tetrahedral, ice-like organization of water molecules caused by hydrogen bonding. Three water molecules in close contact typically form isosceles triangles with edges of 0.275, 0.275, and 0.45 nm length rather than the KSA prediction of equilateral triangles with edges of 0.275 nm length. This KSA prediction results in the erroneous peak of the oxygen density at small z values and the underestimation of the contact peak at $z = 0.26$ nm. The inclusion of the water-triplet correlations on the other hand results in agreement of phase and amplitude of the density distribution. The quality of the expansion results is particularly high in the interfacial region, where the strongest variations in the density profile occur. This observation is of special relevance for further studies of biomolecular hydration where the major emphasis lies on the identification of highly localized water.

4.4 Hydration of biological macromolecules: Results for DNA

A prohibitively large number of correlation functions would be required to describe the hydration of solvated macromolecules such as proteins and nucleic acids using Eq. 19 directly. As a consequence, it is important to simplify the theoretical description with regard to the number of different atom types α used to model the macromolecules, recalling that the main interactions between water and a solvated macromolecule are hydrogen bonds.

In the case of nucleic acids, we classify the atoms as polar (oxygen, nitrogen) and nonpolar (carbon). The hydrogen bonds involving oxygen and nitrogen atoms are within narrow bands concerning energy, bond length, and bond angle; and they are similar to those formed between water molecules. In the following we shall equate all nitrogen and oxygen atoms to water oxygen with respect to their interactions with water. The nonpolar atoms (carbon) are represented as excluded volume (unit step function for $g^{(2)}$, no triplet correction). This treatment of the nonpolar atoms might require some refinement in the case of large hydrophobic surface regions, e.g., by using correlation functions of methane in water. However, in the case of nucleic acids studied here, the molecular surface is essentially hydrophilic justifying the simplification. The negative net charge of the phosphate groups PO_4^- on the DNA is expected to be well represented by its four oxygens, each carrying a negative partial charge. As a consequence, we use Eq. 20 (disregarding excluded volume effects of the nonpolar atoms) to compute the water density, where $r_1, \ldots, r_n$ are the positions of the electronegative atoms oxygen and nitrogen on DNA.

Some results for the hydration of DNA oligomers are described in Ref. [31]. Preliminary results for the hydration of proteins have been presented [32]. A comparison with published simulation results for canonical B-DNA oligomers by Forester and McDonald [33] has been submitted for publication.

In this study we use water pair and triplet correlations computed separately in collaboration with P. Procacci and G. Corongiu at IBM, Kingston [34], from an MD simulation of the highly refined *ab initio* Niesar-Corongiu-Clementi model of water, which includes n-body polarization effects and yields very good results for thermodynamic, structural, spectroscopic, and transport properties of the liquid [35]. The extensive pair and triplet correlational database thus obtained at a discretization of 0.02 nm step size is used in Eq. 20 along with multilinear interpolation subroutines to obtain the results described next.

In this work we shall report results for DNA duplexes in the B-family of conformations. Water-oxygen densities are computed at vertices of a three-dimensional cylindrical volume with a height of 4.2 nm and a radius of 1.2 nm. Global helix axes are obtained from diagonalization of the molecules' tensor of inertia as discussed in Ref. [36]. Grid spacings are used of $\Delta r = 0.015$ nm in radial, $\Delta z = 0.03$ nm in axial, and $\Delta \phi = 2\pi/100$ in angular direction. The atomic coordinates have been taken from the Brookhaven database for biomolecular structures. The carbon atoms are surrounded by a sphere of radius 0.25 nm, from which water oxygens are excluded. The triplet correction in Eq. 20 is applied to triangles with all edges between 0.23 nm and 0.79 nm, which is the range considered in the water triplet simulations [34].

In Fig. 4, we present plots of the main features of DNA structural hydration obtained for the Dickerson B-dodecamer d(CGCGAATTCGCG)$_2$ [22] re-refined by Westhof [37] and two relatives, d(CGCATATATGCG)$_2$ [38] and d(CGCAAAAATGCG)$_2$ [39], of which the so-called "up" structure is analyzed. The water-oxygen densities are depicted on cylindrical surfaces with different radii around the DNA molecules. Black stands for low, white for high water densities. The water densities are averaged over 0.2 nm thick cylindrical shells.

In all cases, we find a marked *spine of hydration* in the narrow minor groove containing the A·T base pairs, which splits into two less pronounced side-by-side water *ribbons* in

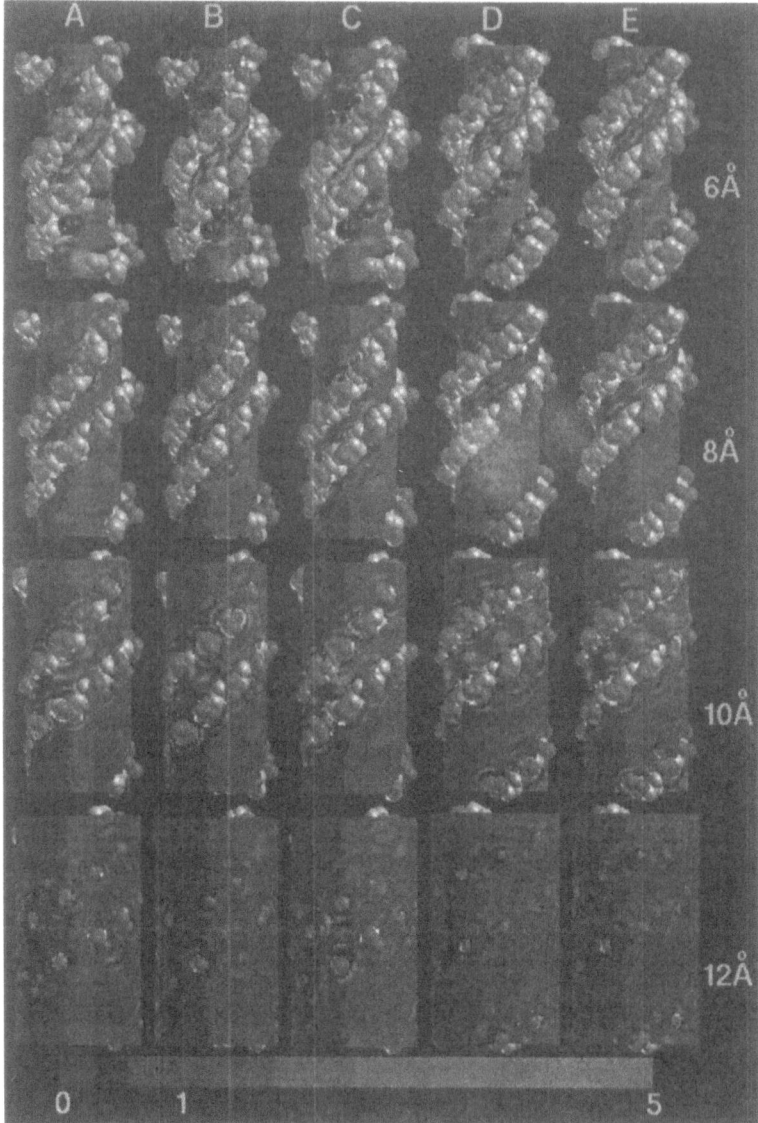

Figure 4: Water-oxygen distributions around B-DNA structures on cylindrical surfaces with radii 6, 8, 10, and 12 Å. The densities are averaged over cylindrical shells with radii 4–6, 6–8, 8–10, and 10–12 Å, respectively. Black stands for low, white for high densities, as indicated by the scale. Crystal structures: (A) d(CGCGAATTCGCG)$_2$, (B) d(CGCATATATGCG)$_2$, and (C) d(CGCAAAAATGCG)$_2$; Canonical B-DNA: (D) d[(AATT)$_3$]$_2$ and (E) d[(CCGG)$_3$]$_2$.

the wider minor groove at the two ends of the structures, where the G·C base pairs are situated. By and large, the major grooves do not contain strongly localized water, and localization along the sugar-phosphate backbone is less pronounced except for high probability regions at the phosphates. O-3' and O-5' neighborhoods as well as the sugars are free of strongly localized water. All these findings are in complete agreement with the experimental data.

Another interesting result of the computation concerns the sequence dependence of the *spine of hydration* in the A+T-rich central regions of the molecules. Reflecting the different geometrical arrangement of the electronegative atoms, the 5'-d(TpA) dinucleotide organizes the water in its minor groove less than the 5'-d(ApT) dinucleotide, with the latter being a strong promoter of water localization. As can be seen particularly for d(CG-CATATATGCG)$_2$, the continuity of the high probability (*spine*) region is interrupted at the 5'-d(TpA) steps. Disruption of the *spine of hydration* at the 5'-d(TpA) step of another oligomer, namely d(CGATTAATCG)$_2$, has experimentally been found [40] and the weak tendency of 5'-d(TpA) to support a *spine* formation has also been observed in simple energy calculations [41]. However, it must be stressed that energetic considerations alone do not suffice in dense many-body systems where the central quantities are free energies. (PMF's are free energies *by definition* and often dominated by entropy contributions).

In Fig. 4 also, data are shown from computations with canonical (Arnott-Hukins) B-form structures [42] for d[(AATT)$_3$]$_2$ and d[(CCGG)$_3$]$_2$. Unlike crystallographic B-DNA, canonical B-DNA does not exhibit variations in the local helical parameters. In particular, the minor groove width is constant while crystallographic structures show a distinct sequence dependence with the minor groove in general being narrow at A·T and wide at G·C base pairs. Many features of DNA structural hydration described above are also found with the canonical structures. A·T sequences again exhibit the most pronounced localization (well-defined minor-groove *spines*); whereas, the minor groove of C·G base pairs is covered with a *double ribbon* of water molecules.

What we consider to be a particularly interesting result of our computations is the consistently observed unfavorable effect of the 5'-d(TpA) step on the hydration *spine* formation. It is present both in the case of crystal and canonical B structures, although less distinct for the latter. As has already been mentioned, it agrees both with X-ray studies [40] and simple (*in vacuo*) energy calculations [41]; and it could play a key role in many puzzling DNA structural phenomena. We mention here two examples.

Leroy *et al.* [43] have found that base-pair lifetimes in the T$_3$A$_3$ center of a DNA oligomer are much smaller than for the inverted sequence A$_3$T$_3$. Burkhoff and Tullius [44] report that A$_4$T$_4$ is better protected from cleavage by hydroxyl radicals than T$_4$A$_4$. We believe that this behavior is related to the absence of strongly localized water molecules, which as a byproduct results in poor stacking. If stacking were the principal cause, the 5'-d(TpA) *dinucleotide* in solution should stack as well as 5'-d(ApT) since these mini helices are much more free to optimize their geometry; but as has been found by NMR [45] 5'-d(TpA) shows less geometric overlap than 5'-d(ApT). However, the presence of a strongly localized water molecule in the minor groove at 5'-d(ApT) steps could explain the longer base-pair lifetimes. For the bases to open, this localized water molecule must break its *hydrogen bonds* to the bases, resulting in a higher energetic barrier to be overcome.

5. CONCLUSIONS

We have reviewed the formal basis of the PMF method and described applications on ionic effects in biomolecular systems. Expressing the Coulomb interactions between the negatively-charged phosphate groups on DNA as solvent-averaged PMF's, we achieve a conceptually clear and technically simple description of complex systems involving DNA molecules in electrolytic solution. This description captures the essential features, as established by comparison of experimental results and theoretical calculations for various salt-induced structural transitions of DNA [11]. Moreover, incorporation of the PMF's into a force field description of biomolecules allows efficient ways to study salt concentration dependent harmonic and anharmonic motions [14].

However, many important properties of biomolecular systems are strongly related to hydration phenomena. We have also presented a method to study the structural hydration of biomolecules. It approximates the water-density distribution near a macromolecule in solution (or, generally, at a solid-liquid interface) as a product expansion in terms of particle correlation functions. A calculation of the water-density profile at the interface of ice and water illustrates that by inclusion of three-particle corrections quantitative agreement with results of extensive computer simulations is achieved. In particular, position and magnitude of the high-density region are predicted correctly by the expansion formula.

This observation is of particular relevance for the study of biomolecular hydration, where a major interest lies in the identification of regions of strong water localization at the biomolecular surface. In this work, we have described some results for the hydration of B-DNA. We have made observations in agreement with the experimental studies. The narrow minor groove of B-DNA exhibits the most pronounced localization of water. In the case of A·T base-pair regions it shows a peaked, *spine*-like hydration, which is replaced by two double *ribbons* of high water density in C·G regions. Interestingly, 5'-d(TpA) steps are found to organize water less than 5'-d(ApT) steps.

We would like to point out two important aspects of the PMF hydration method. First, the required particle correlation functions need to be calculated only once (for given ρ and T). Then they can be used for various geometries and systems, minimizing the computational effort necessary. Second, the expansion method is local in space, i.e., to compute the density at a given position no explicit information about the density at any other point is required. Put together, this results in a computationally efficient method which allows to study large numbers of molecules (e.g., different mutants of a protein), large molecules or molecular assemblies (tRNA, complexes between proteins and nucleic acids), and localized regions in space (e.g., reaction centers). Various studies of this kind are planned for the near future.

Acknowledgments

We wish to thank Drs. T. M. Jovin, M. Neumann, R. Klement, P. Procacci, G. Corongiu, and E. Clementi for collaborations and discussions that greatly contributed to the realization of the project. This work has been funded by the Department of Energy (U.S.), the Bundesministerium für Forschung und Technologie, and the Max-Planck-Society (both F.R.G).

References

[1] Rupley J. A. and Careri G., *Adv. Protein Chem.* **41** (1991) 37–172.

[2] Hansen J.-P. and McDonald I. R., Theory of Simple Liquids (Academic Press, London, UK, 1986).

[3] Münster A., Statistical Thermodynamics (Springer, Berlin, 1969), Vol. 1, p. 338.

[4] Kirkwood J. G., *J. Chem. Phys.* **3** (1935) 300–313.

[5] Fisher I. Z. and Kopeliovich B. L., *Dokl. Akad. Nauk. SSSR* **133** (1960) 81–83 [*Sov. Phys.-Dokl.* **5** (1960) 761–763].

[6] Saenger W., Principles of Nucleic Acid Structure (Springer, Berlin, 1984).

[7] Jovin T. M., Soumpasis D. M., and McIntosh L. P., *Ann. Rev. Phys. Chem.* **38** (1987) 521–560.

[8] Soumpasis D. M., *Proc. Natl. Acad. Sci. USA* **81** (1984) 5116–5120.

[9] Soumpasis D. M., Wiechen J., and Jovin T. M., *J. Biomol. Struct. Dyn.* **4** (1987) 535–552.

[10] Soumpasis D. M., Robert-Nicoud M., and Jovin T. M., *FEBS Lett.* **213** (1987) 341–344.

[11] Soumpasis D. M., Garcia A., Klement R., and Jovin T., "The potentials of mean force (PMF) approach for treating ionic effects on biomolecular structures in solution", Theoretical Biochemistry and Molecular Biophysics, Beveridge D. L. and Lavery R., Eds. (Adenine Press, Schenectady, NY, 1990), pp. 343–360.

[12] Soumpasis D. M., "Formal aspects of the potentials of mean force approach", Computation of Biomolecular Structures, Soumpasis D. M. and Jovin T. M., Eds. (Springer, Berlin, 1993), pp. 223–239.

[13] Klement R., Soumpasis D. M., von Kitzing E., and Jovin T. M., *Biopol.* **29** (1990) 1089–1103.

[14] García A. E. and Soumpasis D. M., *Proc. Natl. Acad. Sci. USA* **86** (1989) 3160–3164.

[15] Hummer G. and Soumpasis D. M., *J. Chem. Phys.* **98** (1993) 581–591.

[16] Hummer G., Ph.D. thesis (University of Vienna, Austria, 1992).

[17] Soumpasis D. M., *J. Biomol. Struct. Dyn.* **6** (1988) 563–574.

[18] Garcia A. E., Gupta G., Soumpasis D. M., and Tung C. S., *J. Biomol. Struct. Dyn.* **8** (1990) 173–186.

[19] Texter J., *Prog. Biophys. Mol. Biol.* **33** (1978) 83–97.

[20] Westhof E. and Beveridge D. L., Water Science Reviews, Franks F., Ed. (Cambridge Univ. Press, Cambridge, UK, 1990), Vol. 5, pp. 24–136.

[21] Berman H. M., *Current Opinion Struct. Biol.* **4** (1994) 345–350.

[22] Drew H. R. and Dickerson R. E., *J. Mol. Biol.* **151** (1981) 535–556.

[23] Kopka M. L., Fratini A. V., Drew H. R., and Dickerson R. E., *J. Mol. Biol.* **163** (1983) 129–146.

[24] Marky L. A. and Kupke D. W., *Biochem.* **28** (1989) 9982–9988.

[25] Klement R., Soumpasis D. M., and Jovin T. M., *Proc. Natl. Acad. Sci. USA* **88** (1991) 4631–4635.

[26] Klement R., "Computation of ionic distributions around charged biomolecular structures using the PMF approach", Computation of Biomolecular Structures, Soumpasis D. M. and Jovin T. M., Eds. (Springer, Berlin, 1993), pp. 207–222.

[27] Hummer G. and Soumpasis D. M., *Phys. Rev. E* **49** (1994) 591–596.

[28] Berendsen H. J. C., Postma J. P. M., van Gunsteren W. F., and Hermans J., "Interaction models for water in relation to protein hydration", Intermolecular Forces: Proceedings of the 14th Jerusalem Symposium on Quantum Chemistry and Biochemistry, Pullman B., Ed. (Reidel, Dordrecht, Holland, 1981), pp. 331–342.

[29] Ewald P., *Ann. Phys.* **64** (1921) 253–287.

[30] Hummer G., Soumpasis D. M., and Neumann M., *J. Phys.: Condens. Matt.* **23A** (1994) A141–A144.

[31] Hummer G. and Soumpasis D. M., "A new approach to calculate the hydration of DNA molecules", Structural Biology: The State of the Art; Proceedings of the Eighth Conversations in the Discipline Biomolecular Stereodynamics, Sarma R. H. and Sarma M. H., Eds. (Adenine Press, Schenectady, NY, 1994), Vol. 2, pp. 273–278.

[32] García A. E., Hummer G., and Soumpasis D. M., *Biophys. J.* **66** (1994) A130.

[33] Forester T. R. and McDonald I. R., *Mol. Phys.* **72** (1991) 643–660.

[34] Soumpasis D. M., Procacci P., and Corongiu G., IBM DSD report, October 1991, IBM.

[35] Niesar U., Corongiu G., Clementi E., Kneller G. R., and Bhattacharya D. K., *J. Phys. Chem.* **94** (1990) 7949–7956.

[36] Soumpasis D. M., Tung C.-S., and García A. E., *J. Biomol. Struct. Dyn.* **8** (1991) 867–888.

[37] Westhof E., *J. Biomol. Struct. Dyn.* **5** (1987) 581–600.

[38] Yoon C., Privé G. G., Goodsell D. S., and Dickerson R. E., *Proc. Natl. Acad. USA* **85** (1988) 6332–6336.

[39] DiGabriele A. D., Sanderson M. R., and Steitz T. A., *Proc. Natl. Acad. Sci. USA* **86** (1989) 1816–1820.

[40] Quintana J. R., Grzeskowiak K., Yanagi K., and Dickerson R. E., *J. Mol. Biol.* **225** (1992) 379–395.

[41] Chuprina V. P., *Nucleic Acids Res.* **15** (1987) 293–311.

[42] Arnott S. and Hukins D. W. L., *Biochem. Biophys. Res. Commun.* **47** (1972) 1504–1509.

[43] Leroy J.-L., Charretier E., Kochoyan M., and Guéron M., *Biochem.* **27** (1988) 8894–8898.

[44] Burkhoff A. M. and Tullius T. D., *Nature* **331** (1988) 455–457.

[45] Hosur R. V., Govil G., Hosur M. V., and Viswamitra M. A., *J. Mol. Struct.* **72** (1981) 261–267.

Inelastic neutron scattering studies of oriented DNA

H. Grimm and A. Rupprecht*

Institut für Festkörperforschung,
Forschungszentrum Jülich GmbH,
P.O. Box 1913, 52425 Jülich, Germany
* Arrhenius Laboratory, University of Stockholm,
10691 Stockholm, Sweden

Abstract

The spectrum of self-correlation of DNA hydrogens has been measured by thermal neutron scattering for two orientations of the helical axis relative to the momentum transfer. Spectra typical for glass formers are observed for energy transfers below 6 meV in the temperature range $200 \leq T \leq 300$ K. Analysis in terms of coupled oscillatory (phonons) and jump (pseudo-spins) motions results in a spin-phonon coupling energy of $\simeq 8$ meV and an average jump frequency rising from 10 to 140 μeV in this temperature range. High resolution measurements for $270 K \leq T \leq 350$ K reveal in addition a much narrower quasielastic component of $\simeq 20$ μeV which is related to jump distances of $\simeq 3.4Å$.

The "glass spectra" are found to be essentially isotropic. Anisotropic contributions to the scattering are identified as being due to compressional waves along the helix and the low lying optical band observed in Raman scattering. Both modes show strong coupling to relaxational modes at the zone center (10-th layer line). It is argued that the observability of the optical band at this position is due to an antiparallel displacement of the base pair centers along the helix direction.

1. INTRODUCTION

The investigation of low frequency spectra of DNA - and thus of its thermodynamical stability - is a fascinating problem by itself because of the central role of this molecule for life. On the one hand one may consider DNA just as another charged polymer - then unnecessarily complex - with the corresponding expectations for its linear response. On the other hand, DNA is a highly "engineered" molecule due to evolution and one cannot *a priory* exclude a "special tuning" of geometry and forces which might allow for non-linear excitations as a means for energy localization and transport.

More specifically, detailed experimental information on the low frequency dynamics of DNA is relevant for questions concerning soft mode driven transition between conformational states [1] glass-like phenomena within conformational substates of biomolecules [2,3] and non-linear excitations driving base pair opening below the denaturation temperature [4-6].

Thermal neutron scattering from oriented DNA fibers [7] may contribute significantly to those questions. Methodically, this probe offers the advantage of a dominant incoherent cross section for hydrogens and the simultaneous exploration of a wide range of frequency (ω) and momentum ($\vec{Q}$) space. Self- and distinct correlation as well as scattering contributions from DNA and its water hull (D_2O versus H_2O humidification) may thus be separated.

A major problem is the availability of large enough oriented samples of DNA in order to exploit the discrimination of atomic displacements by means of their projection on $\vec{Q}$. Existing thermal neutron sources require about 1 cm^3 DNA in order to reach the level of statistical significance for signals due to collective excitations within the practical limits of days measuring time. This problem has been partly overcome by the development and perfection of the wet-spinning method [8]. This method allows the controlled production of highly oriented (within a few degrees) thin films (1 to 100 μm) by winding up DNA fibres which are continuously stretched during precipitation into an aqueous alcohol solution. Films up to 45×275 mm^2 have been obtained in this way. A sample for neutron scattering suitable for *in situ* variation of humidity or D_2O/H_2O-exchange is shown in Figure 1.

Fig.1 - DNA sample for *in situ* humidification. The total area is 35 mm×28 mm×21 films of oriented DNA. The film thickness is $\simeq$ 50 μm. The films are kept under slight tension by the four spring loaded cylinders which connect the two supporting grids.

The amount of needed material and the high molecular weight of $\simeq 10^7$ necessary for good orientation limits those samples to native DNA. However, smaller samples consisting of arrays of single pulled fibres from native and synthetic DNA have been used successfully for structural work, i.e. the location of bound water by means of the strong D_2O/H_2O-contrast for neutrons [9]. Attempts to use such fibre arrays made from fully deuterated DNA for dynamical studies as well are under way [10]. In view of the reduction of the incoherent "background" by deuteration ($\sigma_{inc}(H)\simeq80$ barns, $\sigma_{inc}(D)\simeq2$ barns), much smaller samples might prove sufficient for the observation of collective excitations.

2. SELF CORRELATION

2.1 Time-of-flight spectra

On the other hand, the large incoherent cross section of protons provides the means to measure rather directly the density of states in DNA since most of the hydrogens are tightly bound. Humidifying, in addition, the DNA with D_2O rather than with H_2O allows for the suppression of the water contribution to this signal.

A time-of-flight instrument is a suitable tool to achieve an overview of the spectral density since a large surface in $(\vec{Q},\omega)$-space is monitored simultaneously by many detectors. Another important advantage is the orthogonality of ω- and $\vec{Q}$-resolution which allows for favouring the former at the expense of the latter.

The spectrometer MIBEMOL of the Laboratoire Léon Brillouin at CEN-Saclay was used for this purpose. In this case, the fixed variable is the incident energy of the neutrons E_i ($\simeq 2\ k_i^2$ meVÅ^2, k_i denotes the incident wavevector). A value of $E_i = 3.27$ meV was chosen since it combines a good energy resolution of $\delta\omega \simeq 75\ \mu$eV (HWHM = half width at half maximum) close to the "elastic" line together with the accessibility of the interesting Q-range of $2\pi/d\simeq 1.9\text{Å}^{-1}$ where d denotes the base pair stacking distance of about 3.4Å. The interest in this range results from the possibility to observe single mode contributions as will be discussed below. Since those contributions turn out to be indeed observable and since their location in $\vec{Q}$-space is not straightforward for a time-of-flight instrument an illustration of the relevant scattering surface seems appropriate.

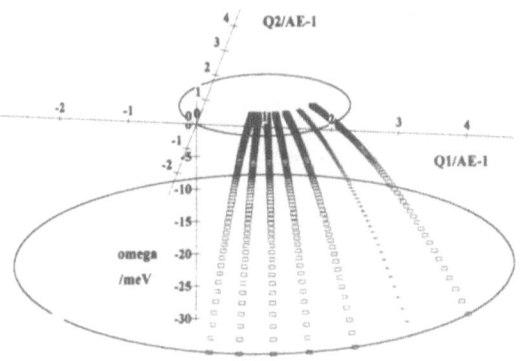

Fig.2 - Average paths in $\vec{Q},\omega$-space of the seven used detector groups of the time-of-flight spectrometer MIBEMOL. The symbols represent every third time channel and their distance indicates the energy resolution. Elastic circle (Ewald sphere) and the corresponding path at ω=-25meV are indicated. The component Q_1 is parallel to the helical axis for the sample orientation "helix in scattering plane". The path of the 90°-detector group (crosses) intersects the elastic plane close to this axis.

The sum of energy $(\omega/2 = \vec{Q}(\vec{k_i} + \vec{k_f}))$ and momentum $(Q^2 = \vec{Q}(\vec{k_i} - \vec{k_f}))$ conservation for initial and final wave vectors defines this surface for a $k_i = const$ instrument, i.e.

$$(\vec{Q} - \vec{k_i})^2 = -(\omega - E_i)/2$$

The $\omega = 0$ section denotes the usual Ewald sphere. Relevant in the present case is the paraboloid of the $Q_3 = 0$ section where Q_3 is the component of $\vec{Q}$ perpendicular to the scattering plane of the instrument.

Fig. 2 shows the average paths in $\vec{Q}, \omega$-space for the seven groups of detectors arranged at scattering angles Φ between 40 and 140 degrees used in this experiment. The 'visibility' of a mode or its 'strength' is governed by the dynamical structure factor $F(\vec{Q})$ which represents essentially the dot product of the momentum transfer $\vec{Q}$ and the atomic displacements in this mode. In order to exploit the discrimination of spectra by this weight factor, the helical axis of the sample was oriented either perpendicular ("perp"-configuration) to the scattering plane (Q_3) or along Q_1, i.e. the elastic momentum transfer of the detector group around $\Phi \simeq 90^o$ ("para"-configuration).

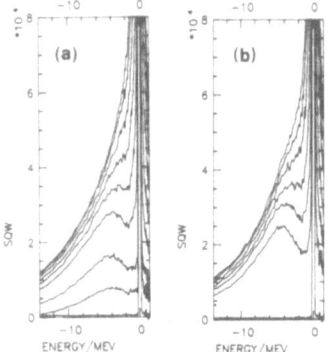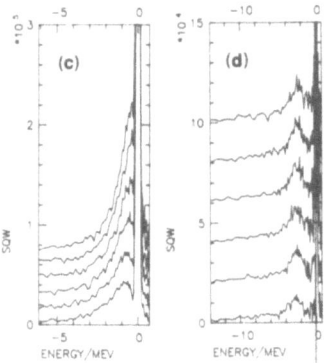

Fig.3 - Time-of-flight spectra from oriented DNA (B-conformation, humidified with D_2O, 75% r.h.). The intensity scale SQW refers to Φ, ω-space. (a) Average over all detectors. Orientation of the helical axis within the scattering plane ("within", see text). ¿From bottom to top: Va-scan, DNA at T = 77, 134, 208, 235, 254, 270, 283, 297 K. (b) same as (a) but with orientation of the helical axis perpendicular ("perp") to the scattering plane and without spectra for T = 75 and 125 K. (c) Detector group around $\Phi = 90^o$. Intensity difference. "within" minus "perp"-orientation. ¿From bottom to top: T = 208, 235, 254, 270, 283, 297 K. The intensity scale applies for the T = 208 K curve. Consecutive curves are vertically shifted by 0.15×10^5 units of SQW. (d) same as (c) but for the detector group around $\Phi = 74^o$ and vertical shift of 2×10^4.

Figs. 3(a,b) shows the sum over all spectra obtained for those two configurations in the range from T=77 K to ambient temperature. Summation seems justified because of a simple Q^2-dependence of the inelastic signal. Two qualitative informations are obvious, (i) a rather dramatic softening (anharmonicity) of all modes up to 6 meV for temperatures above 200 K accompanied be strong quasielastic scattering and - *cum grano salis* - (ii) a spectral similarity for motions of the DNA-hydrogens parallel and perpendicular to the helix direction. Small but significant differences as shown in Figs.3(c,d) are due to distinct correlations and will be discussed below.

The data are very similar to those obtained recently for myoglobin [3]. On the basis of length scale considerations, a decoupling of vibrational and relaxational response was assumed in this case. Since such a length scale is difficult to specify from the Q-dependence of the present data, such a decoupling will not be assumed *a priori* here. A restricted ansatz shall be made which includes, however, the level of harmonic oscillators with exponentially decaying memory in order both to extract quantitative information from the data and to probe the limit of linear response.

2.2 Scattering Function

The restrictions are (i) discarding diffusional motion on the basis of the simple Q^2 dependence of the data (at least in the observed range) (ii) and going to the extreme case of anharmonicity, i.e. jump motion (pseudo spins). The advantage of the latter restriction is the quadratic Hamiltonian both in the phonon and spin variables.

Temperature scaling of the data also shows the dominance of single excitations in the spectra which allows for a corresponding restriction in the ansatz, i.e.

$$S(\vec{Q},\omega) = 2/\beta * g(\beta\omega) * |F(\vec{Q})|^2 * S(\omega)$$

where g(x) = x/(1-exp(-x)) denotes the average energy of a mode with frequency ω in units of temperature (= $1/\beta$). The reduced scattering function $S(\omega)$ is then even in ω and related to the system properties by linear response via

$$S(s) = a_0/(s + \alpha_1 + a_1/(s + \alpha_2 + a_2/(s + \alpha_3 + a_3/...)))$$

and

$$S(\omega) = 1/\pi \ Re(S(s))|_{s=-i*\omega}$$

where the system eigenvalues are given by $a_0 = S_0$, $a_1 = S_2/S_0$, $a_2 = S_4/S_2 - a_1$, $a_3 = (S_6/S_2 - (a_1 + a_2)^2)/a_2$ etc. and S_k denotes the k-th moment of $S(\omega)$, i.e.

$$S_k = \int d\omega\omega^k S(\omega)$$

The α_k represent dissipative terms due to coupling to random forces [11].

The question then arises at which moment one may truncate or approximate the continued fraction representation of $S(\omega)$ in view of the probable information content of the experimental data. The data show clearly a three peak structure (peaks at $\omega \neq 0$ and one at $\omega = 0$) which could mean either an independent superposition of relaxational (truncation at S_2) and oscillatory (truncation at S_4) response or alternatively a coupled spin-phonon response (truncation at S_6). Because of the rather continuous shift with temperature from oscillatory to relaxational response the latter type of response is assumed as minimum requirement.

Tunneling (or kinetic energy) of the spins is neglected as well as a significant contribution from spin-phonon correlation due to phase incoherence. Thus the underlying Hamiltonian contains three parameters only, the phonon eigenvalue λ, the spin-spin interaction J and the spin-phonon coupling W [12].

Introducing the effective jump frequency $\Omega_J = \Omega(1-\beta J)$ of the spins and the shift due to the spin-phonon coupling by $\Omega_W = \Omega\beta W$, the phonon-phonon scattering contribution $S_{phon}(s)$ is determined by

$$a_1 = \lambda(1 - \Omega_W/\Omega_J) = 1/a_0;\ a_2 = \lambda - a_1$$

with the non-zero dissipative terms $\alpha_2 = \Gamma$ representing a possible phonon damping due to phonon-phonon interaction and $\alpha_3 = \Omega_J$.

The dissipative terms for the spin-spin contribution to the scattering $S_{spin}(s)$ are given by

$$\alpha_1 = \Omega_J - \Omega_W;\ \alpha_2 = \Omega_W;\ \alpha_3 = \Gamma$$

and the eigenvalues by

$$a_0 = \beta\Omega/\alpha_1;\ a_1 = -\Omega_W * \alpha_1;\ a_2 = \lambda$$

All three parameters of the model, the phonon eigenvalue λ, the jump frequency Ω of the uncoupled pseudo spins and the spin-phonon interaction W have distributions. Comparison to the data from MIBEMOL indicates that consideration of the average values of W and Ω is sufficient for their description.

2.3 Eigenvalue density

The distribution function for the phonon eigenvalues $p(\lambda)$ is given by the measurement at temperatures low enough to consider the spins frozen on the time scale set by the experimental resolution. The data obtained at the two lowest temperatures (Fig.3(a), T=77 and 134 K) are therefore taken as a guide for a coarse modeling of the eigenvalue density. Below $|\omega| \simeq 2$ meV, they exhibit the constant intensity level for acoustic phonons which means that the DNA-hydrogens occupy single equilibrium positions at temperatures below 200 K. A density peak of low lying optic phonons around 4 meV is followed by a $\simeq 1/\omega$-decay above $|\omega| \simeq 5$ meV which extends up to $\simeq 25$ meV, where a distinct change of slope indicates the onset of multiple excitations. The high frequency information results from all spectra scaled to a common temperature.

A $1/\omega$-decay in the selfcorrelation function corresponds to a constant level in eigen-value space $\lambda = \omega^2$ and the Debye type acoustic density of states corresponds to a $\sqrt{\lambda}$ dependence. A coarse modeling of $p(\lambda)$ by these two types of densities in connection with a phonon-phonon interaction $\Gamma = 1.5$ meV for the optic range and 0.4 meV for the acoustic range was found sufficient to represent the low temperature data. They determined range and weight for four frequency ranges in the interval up to $\omega = 25$ meV as shown in Fig.4.

A qualitative understanding of this ansatz results from a next neighbor interaction model for a very simplified double helix [13]. It also helps for a better understanding of the coherent scattering discussed in the next section.

This "DNA" is build from the center of gravity of average base, sugar, and phospate (308 amu placed on the x-axis at a distance of 4.1 Å from the helical or z-axis) with the three symmetry operations S_{10z}, C_{2x} and C_{2z}. The twofold axis C_{2z} decouples the sixty modes of this model into two non-interacting groups representing in (ip) and out of phase (oop) motions of the two strands of the double helix. The tenfold screw axis S_{10z} suggests the presentation in the tenfold extended zone scheme shown in Fig.5. Note that this symmetry element causes the finite and "reflected" slope of the dispersion curves at the zone boundary which means no additional symmetry determined van Hove singularities in the density of states resulting from the interior of the extended zone. The three ip-branches represent with increasing restoring force phase related motions along the axes (ϕ,z,r) of a cylindrical coordinate system, i.e. 'libration (LIB)', 'longitudinal acoustic (LA)' and 'breathing'.

As a consequence of the screw axis, the oop-counterparts of LA and LIB need a phase shift of $36°$ in order to reach the vanishing restoring force for the 'transverse acoustic (TA-x,y)' modes.

Frequency range and stability of the helix narrow the range of choice for the four force constants of the model appreciably. The qualitative features of an acoustic region followed by a higher density of low lying optic modes in the range of 4 meV concluded by the breathing modes up to 25 meV ($f_0 \simeq 1.9$ mdyne/Å) are persistent.

2.4 Analysis of Self Correlation Spectra

Integration of $S_{phon}(s)$ and $S_{spin}(s)$ over the above simplified eigenvalue density $p(\lambda)$ can be done analytically. The ratio w of spin to phonon contribution is given by the structure factors. The simplest assumption for the inelastic structure factor of the pseudo spins (1-EISF) is the two state model, i.e. assuming a time averaged density of

$$\rho(\vec{r}) = p * \delta(\vec{r}) + (1-p) * \delta(\vec{r} + \vec{R})$$

where p is the population at one of the two sites separated by the distance R. EISF$(\vec{Q})$ denotes the absolute square of the Fourier transform of $\rho(\vec{r})$. Taking the isotropic average, and neglecting Q^4 terms, the ratio w becomes

$$w = p * (1 - p) * (R^2/3)/u^2$$

where u^2 stands for the mean squared vibrational amplitude. This ansatz is folded numerically with the resolution determined by the Va-measurement and compared to the incoherent spectra (Fig.3(a,b)).

The response of S_{phon}(s) and S_{spin}(s) is very similar in the quasielastic region ($\omega \simeq \Omega_J$). Thus spectral data alone cannot give precise information on the weight w.

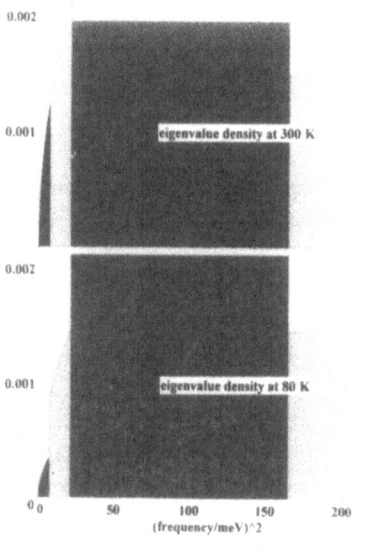

Fig.4 - Density of states in eigenvalue space $\lambda = \omega^2$ as obtained from the incoherent spectra under the restriction of either a $\sqrt{\lambda}$ or no dependence on λ. The upper range extends up to 625 (meV)^2. The phonon-phonon interaction Γ is set to 1.5 meV apart from a value of 0.4 meV for the acoustic range (lowest range).

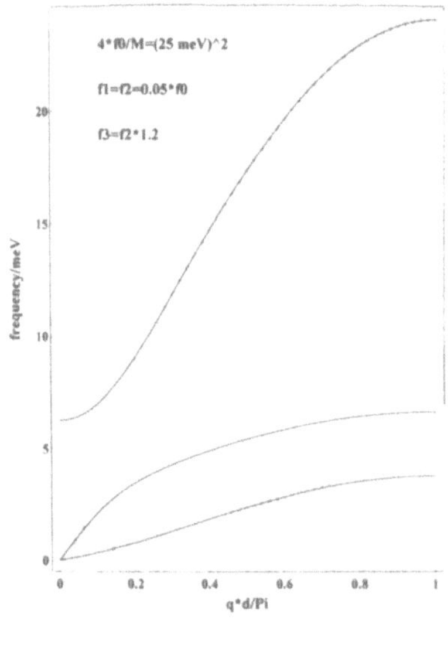

Fig.5 - Dispersion curves for a double helix of identical mass points with next neighbor interaction [13]. The force constants are determined by a maximum frequency of 25 meV and sound velocities of $\simeq$ 5.4, 1.4, and 1.0 km/sec for the LA-, TA-, and LIB-mode, respectively. The abscissa of the extended zone representation covers ten Brillouin zones. Full lines correspond to in-phase motion, open circles to out-of-phase motion of the two strands.

Because of this uncertainty w was set to zero in the present fit. It was verified, however, that this restriction has only a minor influence on the determination of effective jump frequency Ω_J and the spin-phonon interaction W as one may expect from the limita-

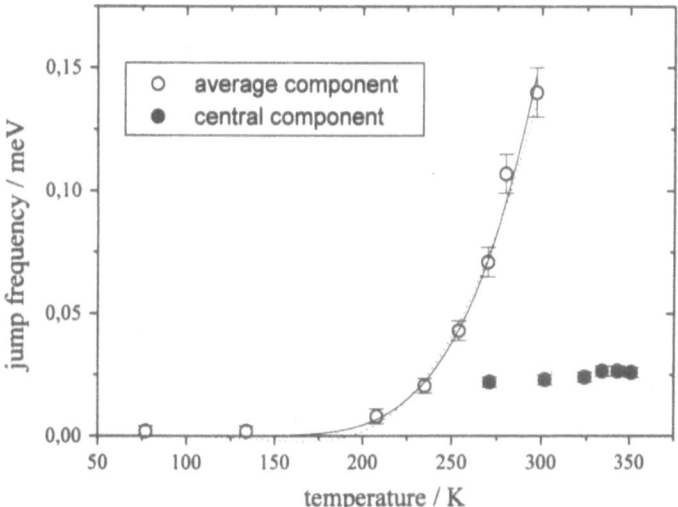

Fig.6 - Temperature dependence of the jump frequency Ω_J (open circles) and the central component Ω_c (filled circles). The solid line corresponds to (198 meV)*exp(-2139 K/T), the dashed line to (25 meV)*(1-193 K/T)*exp(-1237 K/T) with T=temperature.

tion of $S_{spin}(s)$ to the quasielastic region. The same is valid for the small quasielastic contribution from acoustic phonon being overdamped due to Γ=0.4meV.

The fit to the spectra delivers a rather strong spin-phonon coupling W$\simeq$8meV together with an effective jump frequency which increases rapidly above T=200 K and reaches a value of $\simeq$ 0.14 meV at ambient temperature (Fig.6). The quality of the data description is shown in Fig.7 together with the limit of no spin-phonon interaction showing the renormalization of the phonon spectrum.

Describing the temperature dependence of Ω_J in terms of an Arrhenius ansatz

$$\Omega_J = \Omega_{max} * exp(-\beta B)$$

leads to an unreasonable prefactor Ω_{max}. Allowing for a finite spin-spin interaction J and fixing Ω_{max} to 25 meV results in J $\simeq$ 190 K and B $\simeq$ 1250 K (Fig.6). Of physical relevance should be the order of magnitude for the barrier, only, since both descriptions ignore the effect of random fields which should become important approaching the "glass transition" temperature of $\simeq$ 200 K. These random fields express the fact that the DNA is not translationally invariant and thus an attempted phase transition at T = J is "smeared out" due to local freezing.

3. DISTINCT CORRELATION

The increase of jump frequency together with the strong coupling should soften the frequencies within the range of W. Therefore, an appreciable redistribution of modes within the chosen three lowest frequency ranges takes place between $200 \text{ K} \leq \text{T} \leq 300$ K as shown in Fig.4(b). The acoustic range gains by about a factor of 3 which means

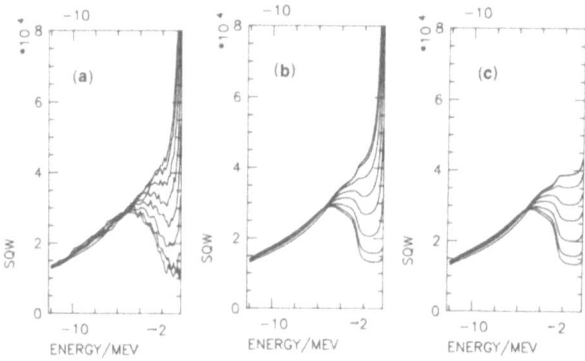

Fig.7 - (a) Time-of-flight spectra as in Fig.3(a) but without the contributions from the detector groups at $\Phi = 74°$ and $90°$. The data are scaled to the temperature T=297 K by multiplication with (297 K/T)/g(energy/T). (b) S_{phon}(energy) with spin-phonon coupling W = 8 meV and other parameters as described in the text. (c) as (b) but W=0.

that the density averaged sound velocity has decreased by a factor of $3^{1/3}$ or about 30% in this temperature interval.

This is indeed recognizable in the difference intensity for the $90°$ detector group as shown in Fig.3(c). For a better understanding of this signal the paths in $(\vec{Q}, \omega)$-space of the seven detectors of this group are illustrated in Fig.8 as they intersect an acoustic dispersion surface for a speed of 2 km/sec. The zone center $\vec{G}$ at $Q_1 = 1.87 \text{Å}^{-1}$ corresponds to the base pair stacking distance of 3.36Å along the helix direction. A decrease in the sound velocity by 30% would mean a doubling of peak intensity which seems to be an underestimate of the observed temperature dependence. Mean values are however misleading here due to the strong $1/\omega^2$ dependence for the acoustic mode intensity.

A three-axis-spectrometer is a suitable instrument for a very controlled scanning of $(\vec{Q}, \omega)$-space because of the flexible selection of $\vec{k_i}$ and $\vec{k_f}$. Scanning ω precisely at $\vec{G}$ results in the difference intensity shown in Fig.9(a). The facility H7 at Brookhaven National Laboratory was used for those experiments.

In previous experiments - with coarser resolution - a monotonous variation of the intensity was observed at ambient temperature [13]. In the present case, with $\delta\omega$ tightened by

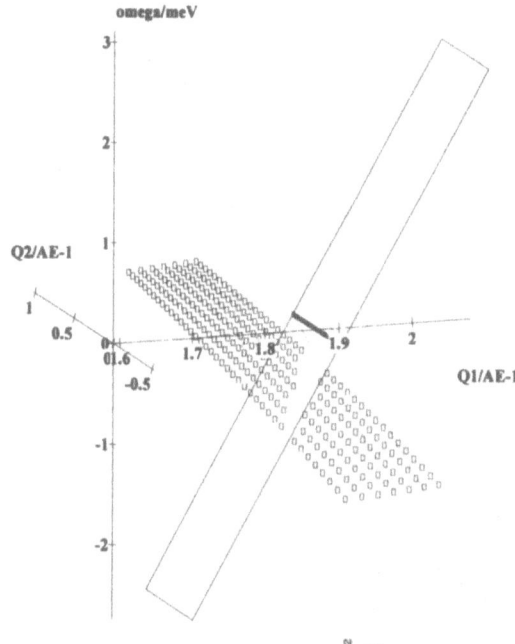

Fig.8 - Paths in $\vec{Q}, \omega$-space for the seven detectors of the 90°-group contributing to the intensity shown in Fig.3(c). The slope of the indicated acoustic dispersion surface corresponds to a speed of 2 km/sec.

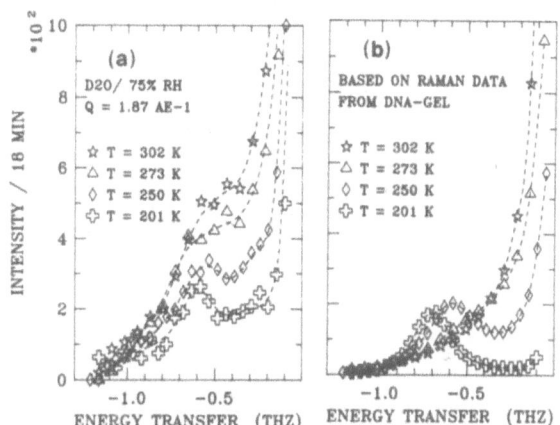

Fig.9 - (a) Temperature dependence of the coherent part of the scattering at the center of the Brillouin zone as obtained from const-Q-scans at H7. The dashed lines are guides to the eye, only. (b) Calculated scattering for oscillator coupled to relaxator. Parameters are those obtained for Raman scattering from DNA-gel by Tominaga et al. [16], interpolated to the temperatures of the neutron data. The structure factor of the pseudospin is set to zero as in [16].

more than a factor of four, a saddle point in the intensity curve is observed which reveals an optic band renormalizing to $\omega_o \simeq 0.6$ THz at low temperatures. This optic band is also recognizable in the difference spectrum for the 74°-group (Fig.3(d)). The average path of this group intersects this band at momentum transfers $Q = (1.85 \pm 0.18) \text{Å}^{-1}$ making an angle of $(20 \pm 5)^o$ with the helical axis. It is remarkable that no softening is observed at this position.

The low frequency optic band (LOB) at $\omega_o \simeq 2.4$ meV or 20 cm^{-1} is well known from Raman scattering experiments. Its nature, i.e. eigenvector, is a question of intensive experimental and theoretical effort ([1],[15-17]). The fact that this mode has been also observed for hydration levels much higer than in the present case (Fig.9(b), [16]) demonstrates its essentially intrahelical nature - interacting, of course, with water, counterions and surrounding DNA molecules, but without phase preservation.

The additional information resulting from the neutron data is best seen by considering again the simplified "DNA" model [13]. The eigenvector basis for LOB modes is built from 3 components, only, i.e. the oop-displacement of the "base pair" along z together with the ip- and oop-rotations around z. Thus - in this simplified model - no LOB mode should be visible for neutron scattering with $\vec{Q}$ parallel to the helical axis, since the visibility is determined by the sum of displacements projected on $\vec{Q}$ times the positional phase factor. Inclination of the base pair (loss of C_{2z}) switches on, both the ip-oop interaction and the visibility of LOB modes having oop-components along z. The difference in the z-component of Adenin-Thymin and Guanin-Cytosin pairs is about 0.58Å for B-DNA and 0.82Å for C-DNA [18] giving rise to an intensity ratio at the Γ-point of LOB/LA$\simeq$1/3 and 1/1, respectively. ¿From Figs.3(c,d) follows e.g. a ratio of $\simeq$0.4 for T=193 K in reasonable agreement with the C-DNA admixture evident from elastic scans. Similar ratios follow from the H7 data obtained for $\vec{Q}$ in the neighborhood of the Γ-point. Thus the conclusion is that the eigenvector of the soft optic mode shown in Fig.9(a) contains the out-of-phase translation of the base centers along the helical axis besides the libration [15] and that it strongly interacts with the LA-mode.

4. CENTRAL COMPONENT

The jump frequency Ω_J represents an average based on the response of DNA for frequencies above the resolution of $\delta\omega \simeq 75 \ \mu eV$. This does not exclude much slower relaxation processes within this resolution window.

This frequency range has been explored by using the IRIS spectrometer at Rutherford Appleton Laboratory. The fixed variable is in this case $k_f = 0.938 \text{Å}^{-1}$ which is determined by using near backscattering Bragg reflection, whereas time-of-flight together with a suitable distance from the pulsed source determines k_i. A resolution of $\delta\omega \simeq 8 \ \mu eV$ (HWHM) is thus achieved in combination with a relatively large Q-range and still good flux.

Data for the configuration "helix perpendicular to scattering plane" have been obtained up to now. They have been analyzed in terms of multiple Lorentzian contributions. Persistent is a central component with $\Omega_c \simeq 20 \mu eV$ and a possible second component in the order of 100μeV which - by the information from the MIBEMOL data - is identified as Ω_J. Because of the much narrower resolution, the quasielastic scattering shown in Figs.3(a,b) has the appearance of a temperature dependent "background" within the covered ω-range and Ω_J can hardly be determined. The results for Ω_c are shown in Fig.6.

The dynamic structure factor of this central component (1-EISF) exhibits a clear deviation from the Q^2 dependence within the explored range of momentum transfer (Fig.10). This shows that Ω_c is connected with much larger jump distances than Ω_J. Analysis with the two state model results in a distance R$\simeq 3.4\overset{\circ}{A}$ and an increase of the minority population from $\simeq 7\%$ to 18% between T = 271 K and 351 K. The inspection of the sample after this experiment showed irreversible denaturation.

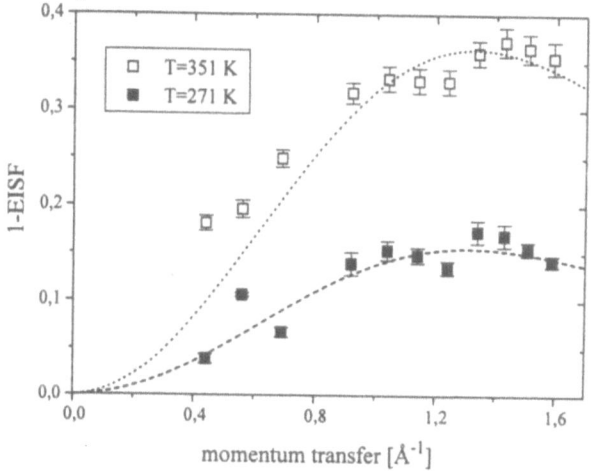

Fig.10 - Structure factor of the central component at T = 271 K (filled) and 351 K (open symbols). The lines correspond to $2p(1-p)(1-\sin(QR)/(QR))$ with the jump distance R=3.4$\overset{\circ}{A}$, Q=momentum transfer and p=0.07 (dashed) and p=0.18 (dotted).

5. CONCLUSIONS

Apart from the discussed quantitative experimental information being important for model parameters, the present interest focuses on possible spectral indications for non-linear excitations leading to partial base pair opening prior to complete denaturation. Observations for the self correlation are that the central component (Fig. 6) is beyond the description within linear response up to S_6. The distinct correlation has not yet been followed up to the denaturation temperature. However, the present data show that the soft mode signal at $\vec{G}$ involves both the parallel and antiparallel component of the strand motion along the helix. On the basis of the simplified "DNA"-model one would expect that both components mix with the ip- and oop-libration around the helix. However, the intrinsic effects of force and mass disorder for native DNA might provide a sufficient basis for a model description of the quasielastic scattering (Fig.9(a)) and the apparent gap in the LA-response [14]. Thus, further experimental attempts to disentangle linear from non-linear response seem both necessary and not hopeless. Incorporation of relevant parts of force- and mass disorder in model calculations should improve the prospects of this task.

Acknowledgements

We thank P.M.Gehring, S.M.Shapiro, R.Kahn, H.D.Middendorf and A.V.Belushkin for their collaboration in the various experiments summarized here. We also appreciate the advice of G.Coddens and C.Carlile in the use of the facilities MIBEMOL and IRIS. It is a pleasure to thank M.Peyrard for stimulating the denaturation experiment and illuminating discussions. A.R. acknowledges the support by the Swedish Medical Research Council. H.G. gratefully acknowledges the hospitality experienced at the Physics Department of Brookhaven National Laboratory, at the Laboratoire Léon Brillouin, CEN-Saclay and at ISIS, Rutherford Appleton Laboratories. Work at BNL was supported by the Division of Material Sciences, U.S. Department of Energy under Contract No. DE-AC02-76CH000016.

References

[1] Lindsay S.M.,Lee S.A.,Powell J.W.,Weidlich T.,Demarco C., Lewen G.D.,
     Tao N.J. and Rupprecht A., Biopolymers **27** (1988) 1015-1043.

[2] Frauenfelder H.,Petsko G.A. and Tsernoglou D., Nature **280** (1979) 558-563.

[3] Doster W., Cusack S. and Petry W.,
     Nature **337** (1989) 754-756 and Phys.Rev.Lett.**65** (1990) 1080-1083.

[4] Englander S.W.,Kallenbach N.R.,Heeger A.J.,Krumhansl J.A. and Litwin S.,
     Proc.Natl.Acad.Sci.USA **77** (1980) 7222-7226.

[5] Dauxois T., Peyrard M. and Bishop A.R., Phys.Rev.**E47** (1993) R44-47; Flach S. and
     Willis C.R., this issue.

[6] Chen Y.Z. and Prohofsky E.W., Phys.Rev.**E49** (1994) 873-881.

[7] Grimm H. and Rupprecht A., Physica**B174** (1991) 291-299.

[8] Rupprecht A., Acta Chem.Scand.**20** (1966) 494-504.

[9] Langan P., Forsyth V.T., Mahendrasingam A., Alexeev D., Fuller W. and Mason S.A.,
     Physica**B180** (1992) 759-761.

[10] Fuller W., Forsyth V.T., private communication.

[11] e.g. Hansen J.P. and McDonald I.R., Theory of simple liquids (Academic Press, 1986)
     pp.303-310.

[12] Yamada Y., Takatera H., and Huber D.L., J.Phys.Soc.Japan**36** (1974) 641-648.

[13] Capellmann H., Biem W., Z.f.Phys.**209** (1968) 276-288.

[14] H.Grimm,H.Stiller,C.F.Majkrzak,A.Rupprecht and U.Dahlborg,
     Phys.Rev.Lett.**59** (1987) 1780-1783.

[15] Urabe H., Sugawara Y., Tsukakoshi M., and Kasuya T.,
J.Chem.Phys.**95** (1991) 5519-5523.

[16] Tominaga Y.,Shida M., Kubota K., Urabe H., Nishimura Y., and Tsuboi M.,
J.Chem.Phys.**83** (1985) 5972-5975.

[17] Prabhu V.V., Schroll W.K., van Zandt L.L., and Prohofsky E.W.,
Phys.Rev.Lett.**60** (1988) 1587-1588.

[18] Arnott S., and Hukins D.W.L., Biochem. Biophys. Res. Comm. **47** (1972) 1504-1509;
Arnott S., and Selsing E., J.Mol.Biol. **98** (1975) 265-269.

Model simulations of base pair motion in B-DNA

M.A. Collins and F. Zhang

Research School of Chemistry,
Australian National University,
Canberra. ACT. 0200, Australia

1. INTRODUCTION

Why study base-pair motion in DNA? Both replication and transcription of DNA involve the recognition of base sequences by other molecules. The transcription of a gene begins with the recognition by RNA Polymerase of a segment of DNA called a "promoter site" [1]. Apparently, RNA Polymerase can recognise and bind to these characteristic sequences of about 50 base pairs. This recognition is a dynamical process. The genetic code is statically inscribed in the base-pair sequences, but proteins must interact with a moving DNA molecule.

The largest amplitude motions of a segment of DNA about 50 base pairs in length must involve bending, twisting, unwinding and stretching motions. Of these, stretching is likely to be the stiffest, lowest amplitude motion. We might reasonably ask whether the bending, twisting and unwinding motions play any special role in this recognition process, or generally in either transcription or replication. Unwinding of DNA is certainly an important element of the overall kinetics of replication. Moreover, as described by Prof. Reiss in this volume, base-pair opening occurs at least locally in transcription as well as in replication.

We propose to ask simple questions about the general characteristics of DNA base-pair motion. Can one understand the dynamical consequences of differing base-pair sequences? Letting our imagination run free for the moment, could one hope to detect any special characteristics of the motion of the sequences in a typical promoter site, adjacent to the site of m-RNA transcription. A generic form for such a promoter site in E. coli, for example, is shown below. [2]

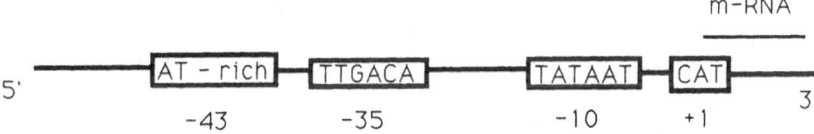

Figure 1: m-RNA transcription begins at base sequences following the CAT sequence. The sequence of about 50 base-pairs preceding the CAT sequence makes up the promoter site, including the highly conserved sequences at the positions shown.

The TATAAT sequence at about -10, known as the "Pribnow Box", is highly conserved (appears unchanged in many promoter sites in many related organisms). This sequence, and the AT rich region near -43 are expected to be more amenable to base pair opening (an AT base-pair, with two hydrogen bonds, unpairs more easily than a GC pair which has three hydrogen bonds). Similarly, in eukaryotic cells an analogue to the Pribnow box (a sequence such as TATAAATA) typically appears about 23 base pairs upstream from the start of transcription. [3]

Can we understand the dynamical consequences of such base-pair sequences? Or at least, can we understand in some detail the characteristic motion of sequences which are more or less rich in AT pairs? There is, after all, a well-known experimental correlation between the temperature at which DNA melts and the percentage of AT content. [4]

2. WHAT SORT OF MODEL?

How should we study DNA motion? There seem to be two extremes: "simple" mathematical models and all atom molecular dynamics (MD) [5-17]. MD would appear to be the method of choice by virtue of the fact that it represents the exact classical solution, if the molecular potential energy surface were known. Unfortunately, MD is particularly difficult to apply to DNA for a number of reasons. Coulomb effects are important as DNA is charged. This gives rise to the need for long-range forces, as well as proper account of the solvent and counterions. These requirements lead to unreliable results based on short simulation times, small system size, and possibly poor H_2O models. Moreover, there are no crystal structures for large segments of native DNA which can be used to initiate an MD simulation.

At the other extreme, simple mathematical models, based on few degrees of freedom, may be a little premature, since we do not yet really know what the important degrees of freedom are.

We consider approximate model dynamics, hoping to put in only enough detail to mimic reality, but preserve maximum simplicity. The initial model described here proceeds in the spirit of an MD simulation, but discards as many degrees of freedom as possible. Indeed, these same simulations may only serve to show that some discarded freedoms must be retrieved. The validity of the model is examined by comparing the observed dynamical averages with those measured in all atom MD simulations of short base-pair sequences. The model is then used to study longer sequences over longer times than are possible for all atom MD. Perhaps then simple analytically tractable models may suggest themselves.

3. THE MODEL

Figure 2 depicts a short segment of the B-DNA double helix structure, where only the heavy atoms are included for clarity.

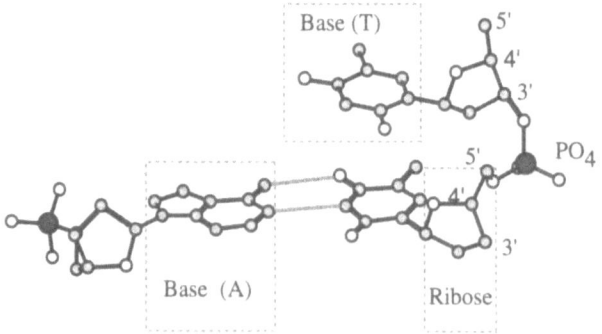

Figure 2: Schematic of a segment of B-DNA, showing only the heavy atoms of an AT base pair with ribose rings and phosphate groups, connected on one strand only (for clarity) to an adjacent thymine base. Atoms and groups which constitute the strand backbone are identified.

Each strand is composed of a sequence of purine (Adenine and Guanine) and pyrimidine (Thymine and Cytosine) bases or nucleotides, which are all flat molecules composed of one (T and C) or two (A and G) rings of nitrogen and carbon atoms [4]. These molecules are quite stiff with respect to all molecular distortions, so that we (and most other workers) treat them as rigid bodies. Each base is attached to a five membered deoxyribose ring. This ring is not flat, and has a number of preferred conformations. Different types of DNA structure (e.g. those called A, B and Z DNA) feature different structures for these ribose rings. We sometimes emphasis that the bases are attached to deoxyribose rings by the notation dA, dT, etc. The ribose ring is connected to two phosphate (PO_4) groups; directly at a ring position, PO_4 - (3') , and via an intermediate CH_2 group (denoted the 5' site), (4') - CH_2(5') - PO_4. The backbone thus connects ribose rings in order (4') - CH_2(5') - O - PO_2 - O - (3') - (4') ... This backbone is quite flexible by comparison with the ribose ring itself. Thus, in order to discard as many freedoms as possible, we ignore distortions of the ribose ring (including "puckering" or conformational motion) and fix it rigidly to each base in the standard conformation it takes in idealised B-DNA [18]. This is clearly a serious approximation, taken in the interests of simplicity. Finally then, we treat the flexible backbone itself as a structureless flexible rod. Each strand of this B-DNA model is then a sequence of rigid base-ribose bodies connected by flexible rods.

The bases of each strand are hydrogen bonded to the complementary bases on the other strand; A to T and G to C. In the standard equilibrium B-DNA structure, the bases in a pair are nearly coplanar. In this first simplest model, we will only allow these rigid base-ribose bodies to move in this XY plane. The motion of each rigid body can then be completely described in terms of the (X,Y) coordinates of the body's centre of mass, and an angle, φ, which determines the orientation of the body with respect to say the X axis. This model is depicted schematically in Figure 3.

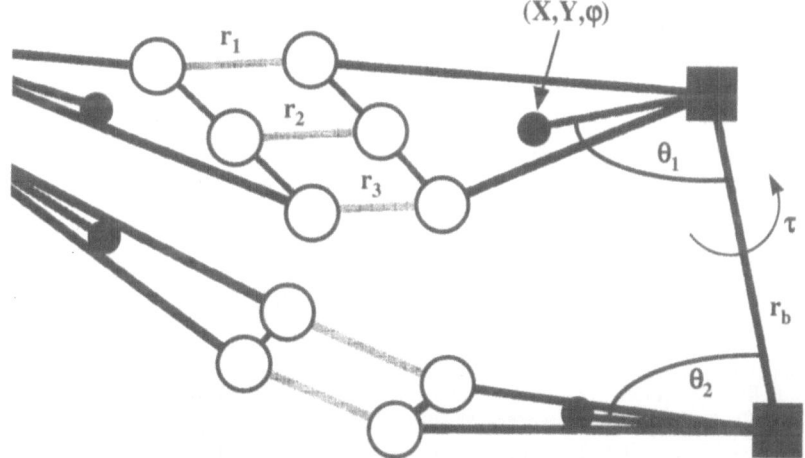

Figure 3: Schematic representation of the B-DNA double helix of rigid base-ribose bodies, showing the centre of mass (o), hydrogen bonding sites (O), and backbone connections (■), with bond lengths, angles and torsion angles defined.

All hydrogens are discarded from the model, so that the forces due to hydrogen bonding are taken to act directly at the oxygen and nitrogen atoms involved (see Figures 2 and 3). These forces then decompose into forces on the centres of mass of the bodies and torques about those centres (preserving the asymmetry of the major and minor grooves). The flexible rods of the backbone are connected to the rigid bodies at the 3' and 4' positions of the ribose rings, producing additional centre of mass forces and torques. The 3' and 4' positions project onto virtually the same position in the base plane; so we connect the rods at the average position in this plane. Forces arise from changes in the rod length, r_b, from bending the angles, θ_1 and θ_2, which the rods make with the rigid bodies, and from twisting the rod to allow changes in the rigid body orientation.

3.1 Model potential surface

The equilibrium geometry of the double helix is taken to be that for B-DNA [18]. The model potential energy, V, is composed of the following simple terms (see Figure 3 for definitions of lengths and angles):

$$V \quad = \quad V_b + V_\theta + V_\tau + V_{LJ} + V_H \tag{1}$$

where we take quadratic backbone stretching and bending potentials:

$$V_b \quad = \quad \frac{1}{2} k_b \ (r_b - r_b^{eq})^2 ; \tag{2}$$

$$V_\theta \quad = \quad \frac{1}{2} k_\theta \left[\cos(\theta_i) - \cos(\theta_i^{eq}) \right]^2 \quad , i = 1, 2 ; \tag{3}$$

and a periodic torsion potential,

$$V_\tau \quad = \quad k_\tau \left[1 - \cos(\tau - \tau_{eq}) \right] \quad ; \tag{4}$$

with non-bonded interactions between bases to avoid unphysical overlap,

$$V_{LJ} \quad = \quad 4\varepsilon \left[\left(\frac{\sigma}{R_{cm} - R_{cm}^{eq} + 2^{\frac{1}{6}}\sigma} \right)^{12} - \left(\frac{\sigma}{R_{cm} - R_{cm}^{eq} + 2^{\frac{1}{6}}\sigma} \right)^{6} \right] \quad ; \tag{5}$$

and Morse potentials to describe the stretching and contraction of each hydrogen bond,

$$V_H \quad = \quad V_o \left\{ \exp\left[-a(r - r_o) \right] - 1 \right\}^2 \quad ; \tag{6}$$

In Eq. (5), R_{cm} denotes the distance between the centres of mass of the rigid bodies in a pair, while in Eq. (6), r refers to the hydrogen bond lengths r_1, r_2 and (for GC pairs) r_3 of Figure 3. The magnitudes of the parameters are: k_b = 1.0 kcal mol^{-1} Å^{-2} ; k_θ = 10 kcal mol^{-1} rad^{-2}; and k_τ = 40 kcal mol^{-1}. The values of k_b and k_τ are close to the values used by Zan and Harvey [19] in a molecular mechanics study of the B-DNA structure. The value of k_θ is taken to be a typical value for single bond angle bending. All three parameters in combination are consistent with the observed elastic modulus for twisting of DNA [20, 21]. The hydrogen bond parameters are those of Chen and Prohofsky [22].

All the dynamical variables in V are expressed in terms of the centre of mass coordinates [$X_1(n)$, $Y_1(n)$] and [$X_2(n), Y_2(n)$] and orientation angles $\varphi_1(n)$ and $\varphi_2(n)$ of the two rigid bodies in each (nth) pair. The total Hamiltonian for this model is then

$$H \quad = \quad \sum_n \frac{1}{2M_1(n)}\Big[P_{X1}(n)^2 + P_{Y1}(n)^2\Big] + \frac{1}{2M_2(n)}\Big[P_{X2}(n)^2 + P_{Y2}(n)^2\Big]$$

$$+ \frac{P_{\varphi1}(n)^2}{2I_1(n)} + \frac{P_{\varphi2}(n)^2}{2I_2(n)} + V. \tag{7}$$

Here, P represent the momentum conjugate to the variable indicated, $M_i(n)$ denotes the total mass of the nth base-ribose rigid body in the ith strand, $I_i(n)$ denotes the corresponding moment of inertia about the base-ribose centre of mass, and the total potential is a sum of V in Eq. (1) for each base pair.

4. SOME DYNAMICAL AVERAGES OF THIS MODEL DNA MOTION

The dynamics of Eq. (7) have been evaluated using standard microcanonical molecular dynamics methods for chains of various lengths and base-pair compositions. Unless stated otherwise, results reported here are obtained from averaging over time intervals of 200 psec. Dynamical averages are often shown below as a function of temperature. The temperature shown is simply proportional to the average kinetic energy during a microcanonical simulation.

The first feature of note in the dynamics of this model, is that the double helical structure is basically stable. The "helix repeat number" is the average number of base pairs contained in a 360° twist. For standard B-DNA (the starting geometry at zero energy), the helix repeat number is exactly 10. A value higher than 10 indicates that the helix is "unwound" by comparison with the ideal structure. Figure 4 shows the variation of the helix repeat number for DNA segments of 100 base pairs.

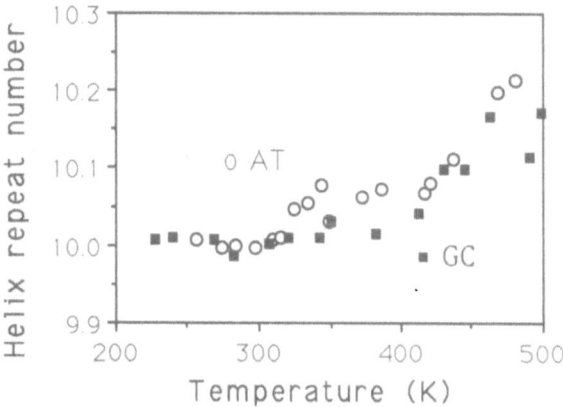

Figure 4: Helix repeat number versus temperature for helices of Poly(dA).Poly(dT) and Poly(dG).Poly(dC).

It is clear that both helices are stable at low temperature, and that both unwind gradually as the temperature rises. There are significant uncertainty in these averages, so that we could only infer that Poly(dA).Poly(dT) (all A in one strand, all T in the other) unwinds to about the same extent as Poly(dG).Poly(dC). We recall that the hydrogen bonding is stronger in GC pairs.

A measure of the fluctuations in the unwinding of DNA is given by the rms average of the angle between successive interbase centre of mass vectors, $\mathbf{R}_{cm}(n)$ and $\mathbf{R}_{cm}(n+1)$. This is displayed in Figure 5.

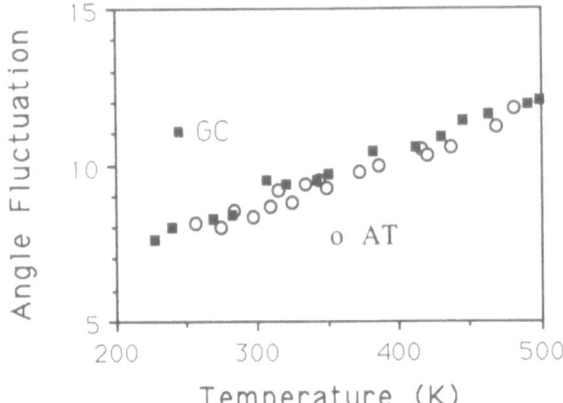

Figure 5. The average rms angle fluctuations versus temperature for helices of Poly(dA).Poly(dT) and Poly(dG).Poly(dC).

These angle fluctuations are comparable to but slightly larger than values observed in simulations of d(CGCGAATTCGCG)$_2$ by Rao and Kollman [11], and comparable to values observed for d(CGCAACGC)/d(GCGTTGCG) over 80 psec by Van Gunsteren et. al. [10]. There is no significant effect of base pair composition apparent in these angular fluctuations.

4.1 Hydrogen bond breaking

The effect of base pair composition becomes evident however when we measure the extent of hydrogen bond breaking or "opening". Following Chen and Prohofsky [22], we say that a hydrogen bond is "broken" or "open" if its length exceeds its equilibrium value by a few tenths of an Angstrom. Figure 6 shows that, as expected, there is a much greater tendency for the strands of Poly(dA).Poly(dT) to pull apart than is the case for Poly(dG).Poly(dC).

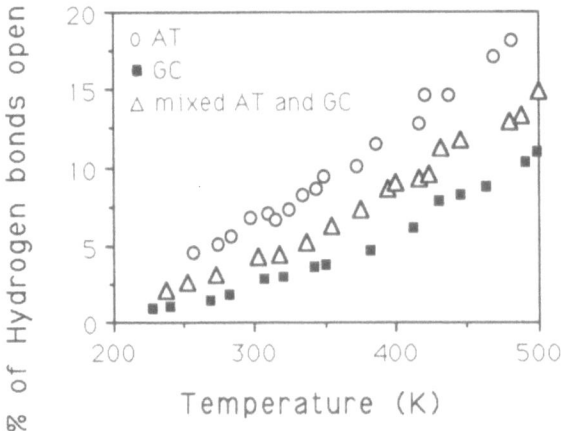

Figure 6: The average percentage of hydrogen bonds open [22] at any time versus temperature for 100 base pair sequences of Poly(dA).Poly(dT), Poly(dG).Poly(dC), and a mixed sequence of AT and GC pairs.

4.2 RMS "atomic" displacements

An important qualitative aspect of the motion of the DNA double helix can be deduced from calculations of the rms deviations of the atoms of DNA from their equilibrium positions. These rms atomic displacements have been reported in a number of MD simulations of DNA [7, 10] and RNA [23]. Here, the corresponding observation is the rms displacements of various positions on the rigid bodies. Figure 7 shows, for a 200 psec simulation of d(CGCGAATTCGCG)$_2$, the average rms displacements of the positions where hydrogen bonding occurs and where the backbone rods are attached.

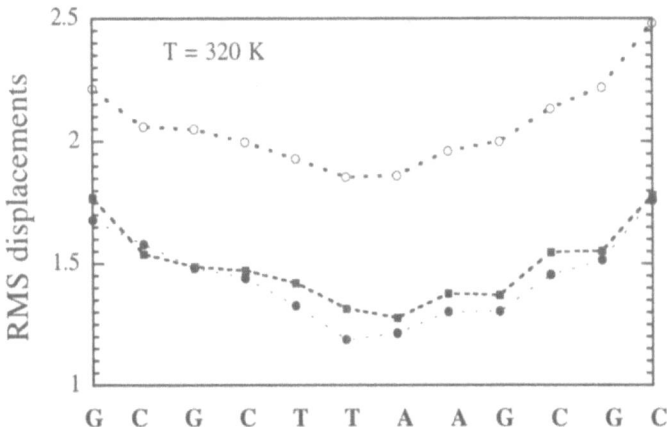

Figure 7: The rms deviations (Å) from equilibrium of the hydrogen bonding sites, (■) nearest the major groove and (●) next nearest, and the backbone connection site (o), the model applied to d(CGCGAATTCGCG)$_2$.

These results are qualitatively similar to those reported in MD simulations (though somewhat larger, indicating a slightly too floppy molecule). Firstly, we note that the outer region of the double helix moves most. Secondly, the hydrogen bonding site(s) nearest the major groove of B-DNA move more than the site nearer the minor groove. We have then a picture of the base-pair motion in which the base-ribose bodies move in and away from their partners while undergoing hindered rotation. The overall picture is **not** one in which the base and ribose swing about a hinge supplied by the backbone; rather the reverse, where the base, ribose and backbone are hinged (flexibly) about the hydrogen bonds.

It appears that this model does reproduce some basic characteristics of the base pair motion in B-DNA, both simple averages and average fluctuations; at least, this model is in good qualitative and semi-quantitative agreement with all-atom simulations of short double helices. It remains to be seen, as the model is further refined, whether more subtle features of the dynamics can be seen in a model with so few degrees of freedom.

To go beyond these observations of simple averages to look at cooperative dynamical events (of possibly low probability) requires much more extensive simulations, over longer times and larger helices. At this early stage in our study, we only report such measurements as suggestive, and subject to considerable uncertainty. With these reservations clearly in mind, two final observations are worth mentioning.

4.3 Large amplitude excitations

We commonly observe infrequent large amplitude excursions in H-bond lengths. Figure 8 shows a schematic representation of such a large amplitude excursion.

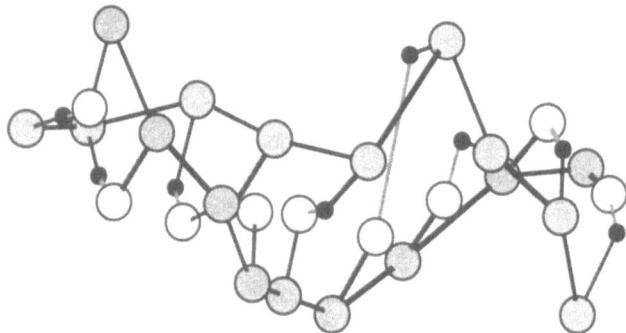

Figure 8. Schematic of a segment of model B-DNA, showing only the centres of mass, $\bigcirc$, of each strand, and the hydrogen bonding sites adjacent to the major groove on one strand, $\bullet$, and on the other, $\bigcirc$. A broken hydrogen bond is apparent four pairs from the right.

This event occurs during a simulation of a B-DNA fragment of 100 base pairs of mixed AT and GC content. We have introduced sequences of four or five contiguous AT and TA base pairs in this B-DNA segment (analogous to similar sequences in promoter sites); sequences expected to be more than usually prone to hydrogen bond breaking. Indeed, we only see such large amplitude hydrogen bond breaking in such regions of the DNA. The break sketched in Figure 8 persists for several psec. It involves a swing of both bases into the major groove, and the "pull-back" of only one base from the helix axis. There is some suggestion that such excitations repeat at nearby base-pairs; smaller "breaks" or large amplitude excursions are seen for the AT base pairs neighbouring that which undergoes the massive displacement (occurring just before and just after the major break).

4.4 Correlations

Correlation functions, $< (r_i(t) - < r_i >)(r_j(0) - < r_j >) >$, have been calculated for hydrogen bond length variations in a GC base-pair in a 100 base-pair segment of mixed AT and GC content (accumulated using standard methods over 2000 psec. The results provide some further evidence for a picture of base-pair motion in which both linear separation of the hydrogen bonds and hindered rotation of the bases are significant: Thus, r_2 is positively correlated at time $t = 0$ with both r_1 and r_3 (stretching in unison dominates), but the combination of stretching and base rotation cancel to give no net correlation between r_1 and r_3 at the same time. These correlations decay within 1 psec, and possibly on a longer timescale (but the statistical errors are very large here!). We see no correlations between the hydrogen bond lengths of adjacent base pairs, within the errors apparent in these early, limited, simulations.

5. CONCLUDING REMARKS

This simple model reproduces a number of the qualitative features of all-atom MD simulations of B-DNA segments. With the crudely approximate potential parameters used, the overall flexibility of the model DNA appears to be somewhat exaggerated by comparison with the MD results. It is interesting that the approximation of rigid base-ribose bodies does not preclude these characteristic motions or substantially understate the flexibility of B-DNA.

Several improvements to this model ought to be tested before one can reliably judge which degrees of freedom are essential to the description of unwinding or base pair "melting". Explicit account of nonbonded interactions between the bases (the "stacking" energy [4]) and allowance of motion in the Z direction are just two examples of necessary improvements. Nevertheless, there appears to be some grounds for optimism in pursuing simplified models of the base pair motion in DNA.

REFERENCES

[1] Kornberg R., DNA Replication (Freeman, San Francisco, 1974) p. 242.
[2] Bujard H., in DNA makes RNA makes Protein, T. Hunt, S. Prentis and J. Tooze Eds (Elsevier Biomedical, Amsterdam, 1983) pp. 10-17.
[3] Kornberg R., DNA Replication (Freeman, San Francisco, 1974) p. 254.
[4] Saenger W., Principles of Nucleic Acid Structure (Springer-Verlag, New York, 1984).
[5] Levitt M., Computer Simulation of DNA Double-helix Dynamics. Cold Spring Harbor Symposia on Quantitative Biology, 47 (1983) 251-262.
[6] Tibor B., Irikura K. K., Brooks B. R., and Karplus M., J. Biomol. Struct. Dyn., 1 1983) 231.
[7] Singh U. C., Weiner S. J., Kollman P., Proc. Natl. Acad. Sci. USA, 82 (1985) 755-759.
[8] Seibel G. L., Singh U. C., and Kollman P., Proc. Natl. Acad. Sci. USA, 82 (1985) 6537.
[9] Prabhakaran M. and Harvey J. A., J. Phys. Chem., 89 (1985) 5767.
[10] van Gunsteren W.F., Berendsen H.J.C., Geurtsen R.G., and Zwinderman H.R.J., Annals of the New York Academy of Sciences 482 (1986) 287-303.
[11] Rao S.N. and Kollman P., Biopolymers 29 (1990) 517-532.
[12] Swaminathan S., Ravishanker G., and Beveridge D.L., J. Am. Chem. Soc. 113 (1991) 5027-5040.
[13] Prévost C., Louise-May S., Ravishanker G., Lavery R., and Beveridge D.L., Biopolymers 33 (1993) 335-350.
[14] Falsafi S. and Reich N. O., Biopolymers 33 (1993) 459.
[15] Fritsch V., Ravishanker G., Beveridge D.L., and Westhof E., Biopolymers, 33 (1993) 1537-1552.
[16] Ferentz A.E., Wiorkiewicz-Kuczera J., Karplus M., and Verdine G.L., J. Am. Chem. Soc. 115 (1993) 7569-7583.
[17] Briki F. and Genest D., J. Biomol. Struct. Dyn. 11 (1993) 43-56.
[18] Arnott S. and Hulins D.W.L., Biochem. Biophys. Res. Comm. 47 (1972) 1504-1509.
[19] Tan R.K.Z. and Harvey S.C., J. Mol. Biol. 205 (1989) 573-591.
[20] Barkley M.D. and Zimm B.H., J. Chem. Phys. 70 (1979) 2991-3007.
[21] Hagerman P.J., Ann. Rev. Biophys. Biophys. Chem. 17 (1988) 265-286.
[22] Chen Y.Z. and Prohofsky E.W., Biopolymers 33 (1993) 797-812.
[23] Harvey S.C., Prabhakaran M., Mao B., and McCammon J.A., Science 223 (1984) 1189-1190.

A nonlinear model for DNA melting

T. Dauxois and M. Peyrard

Laboratoire de Physique,
Ecole Normale Supérieure de Lyon,
URA 1325 du CNRS, 46 allée d'Italie,
69007 Lyon, France

1. INTRODUCTION

Understanding the mechanism of the transcription of DNA is a very important challenge. Knowing the map of the genome is indeed important, but not sufficient if we do not understand how the genome is expressed. One fundamental aspect is the regulation of the expression, in particular how many times a gene is read, or how fast it is expressed. This is a difficult and interdisciplinary, but fascinating problem which will require a cooperation between biologists and physicists.

We address one aspect of this regulation problem by investigating the initiation of the transcription, namely the formation of the transcription "bubble". This bubble is a local opening of the double helix which is due to the action of the RNA polymerase. As long as the transcription has not yet started no chemical energy is brought to the DNA molecule. However the formation of the bubble requires the breaking of many hydrogen bonds. We propose a mechanism according to which energy is collected on the DNA molecule through nonlinear effects. A test of this idea is provided by the investigation of the thermal denaturation of DNA which starts also by the formation of a local bubble and is more amenable to physical treatment because it does not involve the enzyme.

At temperature well below the denaturation temperature, DNA shows also large amplitude motions known as "fluctuational openings" in which base pairs open for a very short time and then close again. These motions are important because, when the base pairs close again can trap some external molecules causing a defect in the sequence. Moreover, the fluctuational openings can be considered as intrinsic precursors to the denaturation.

The denaturation or "melting" transition is the separation of the two complementary strands. It can be induced by heating or by changing the ionicity of the solvent. This is a well defined physical problem, which can be seen as a phase transition from the native state at low temperature to the denatured state in the high temperature range as schematically shown in figure (1a).

In addition, this transition is associated with an energy localization where the localized excitations excited at the beginning of the process (see figure 1b) are growing when the temperature is increasing. These precursor events of the transition, probably generated by thermal fluctuations, combine with each other into larger bubbles until the complete separation of the strand is reached.

This localization phenomenon could have two different origins.

(i) The disorder of the DNA sequence, corresponding to the genetic code, could induce Anderson localization [1] which is possible in a *linear* system.

(ii) In an homogeneous but *nonlinear* media, the energy evenly distributed can concentrate itself spontaneously into spatially localized nonlinear excitations. This phenomenon corresponds to the self-focusing in optics or plasma.

One important question is to determine the dominant mechanism in a nonlinear inhomogeneous system. Could both effects localize the energy in the same way, or the two effects will compete and the presence of nonlinearity and disorder will not enhance the creation of localized excitations ?

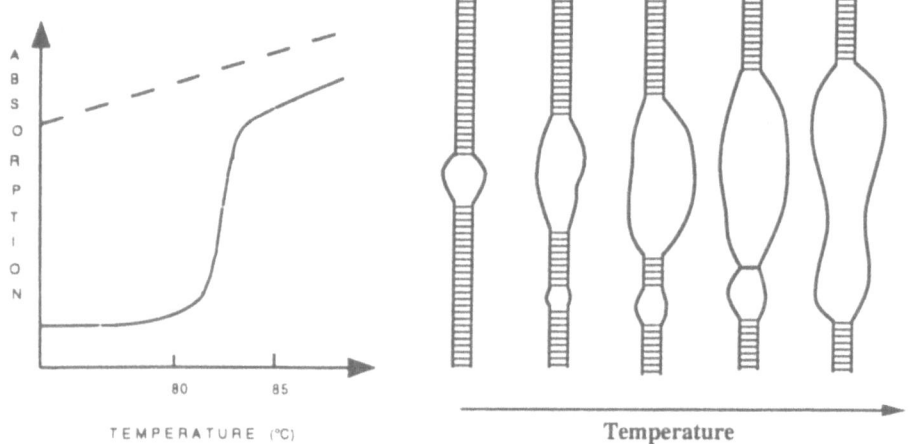

FIG. 1. DNA denaturation : (a) UV Absorbance of a DNA sample versus temperature. (b) Schematic evolution of DNA for increasing temperature, showing the growing of localized bubbles.

The denaturation process has been extensively investigated experimentally and models have been proposed to explain the complicated denaturation curves found in the experiments. Wartell and Benight have written an extensive review that summarized much theoretical work and compared the results with experimental observations [2]. However these models are essentially Ising-like where a base pair is considered as a two-state system which is either closed or open: such an approach cannot reproduce the full dynamics of the denaturation.

Our aim is to design a dynamical model which could extend these Ising models to include the intrinsic nonlinear dynamics of the process but preserving the simplicity of the model by considering only one or two degrees of freedom per base pair. The idea is to select the relevant degrees of freedom in the molecule to allow for the simulation of long sequences. Moreover nonlinear interactions must be included due to the large amplitude motions.

With this approach, we expect to understand the origin of the energy localization, and maybe to determine the possible extensions of the model which can be proposed to go beyond denaturation and study the transcription. The second expected output is to describe DNA melting from "basic parameters", like the potential parameters, while the Ising models use phenomenological parameters, such as the probability of base-pair breaking, or the cooperativity parameters,...

In summary we would like to define a model which can exhibit a thermal behavior in qualitative agreement with the experiments. Preserving the simplicity of the model, we should be able to do the statistical mechanics of this melting transition. In addition, we would like to compute relevant data of the dynamics which could be compared with experiments. The simplicity of the model prevents us from being able to reproduce quantitatively experimental results. This is certainly a weakness but we also believe that keeping the model as simple as possible is a strength because if it can reproduce to a reasonable accuracy some observed phenomena this indicates that this phenomena are in fact, controlled by very few degrees of freedom which have been identified. Only the fine details are the due to all the forgotten degrees of freedom.

Maybe the last point to notice is to recall, as H. Frauenfelder did during the workshop, the following citation by Stan Ulam :

"Ask not what physicist can do for biology, ask what biology can do for physicist."

Trying to answer to the first question, we obtain some answers to the second one ! The biological problem we tried to solve, lead us to the study of the localized modes in a thermalized system, and we find the new mechanism of localization by collision presented in section IV. Moreover, trying to define the improved model presented in section V, we found a one dimensional system, with nearest neighbor interaction, extremely close to a first order phase transition. We will also discuss the biological consequences of our work.

2. THE NONLINEAR DYNAMICAL MODEL

In the simplest model we have considered, a base-pair is characterized by only one degree of freedom, the stretching y_n of the hydrogen bond connecting the bases. These variables are coupled between neighboring sites by a *nonlinear* potential $W(y_n, y_{n-1})$. The interaction between the two strands of DNA appear in such a model as an on-site potential $V(y_n)$ which tends to maintain y_n around its equilibrium value. The hamiltonian of the model is

$$H = \sum_n \frac{1}{2} m \dot{y}_n^2 + V(y_n) + W(y_n, y_{n-1}) \tag{1}$$

The first term corresponds to the kinetic energy term for bases of mass m. The on-site Morse potential

$$V(y_n) = D(e^{-a y_n} - 1)^2 \tag{2}$$

represents not only the H-bonds connecting two bases belonging to opposite strands, but also the repulsive interactions of the phosphates, and the surrounding solvent effects. As we want to study the denaturation process which involves large stretching, and consequently nonlinear excitations, we can not limit the study to linear deviations around the mean position ; the intrinsic nonlinearity of the potential is crucial. However, the analytical choice of this potential is not crucial. It was chosen according to the following requirements :
(i) The plateau of the potential for large stretching represents the possibility of the breaking of the transversal bond.
(ii) The strong repulsive part for negative stretching reproduces the impossibility for bases to become too close.

The interaction between two neighboring base pairs (also called stacking energy) is described in the simplest model by the following harmonic potential chosen because of analytical convenience. As discussed in the last section, $W(y_n, y_{n-1})$ has to be improved to give realistic results, but this simple form can be used to get some analytical understanding.

$$W(y_n, y_{n-1}) = \frac{K}{2}(y_n - y_{n-1})^2 \quad .\tag{3}$$

Therefore the model corresponds to the motion of an harmonic chain in a Morse potential (Fig. 2b). With this simplified view, one can easily distinguish the two different states.

(i) In the low temperature regime the chain will be in the bottom of the well and this state corresponds to the native state of DNA, since the dynamics of the different nucleotides are small in comparison with the distance between nucleotides of different strands.

(ii) In the high temperature regime, the chain is on the Morse plateau, and the system corresponds to a chain on a flat potential. This state corresponds to the denaturated state where the two strands are decoupled and can move rather freely over large distances.

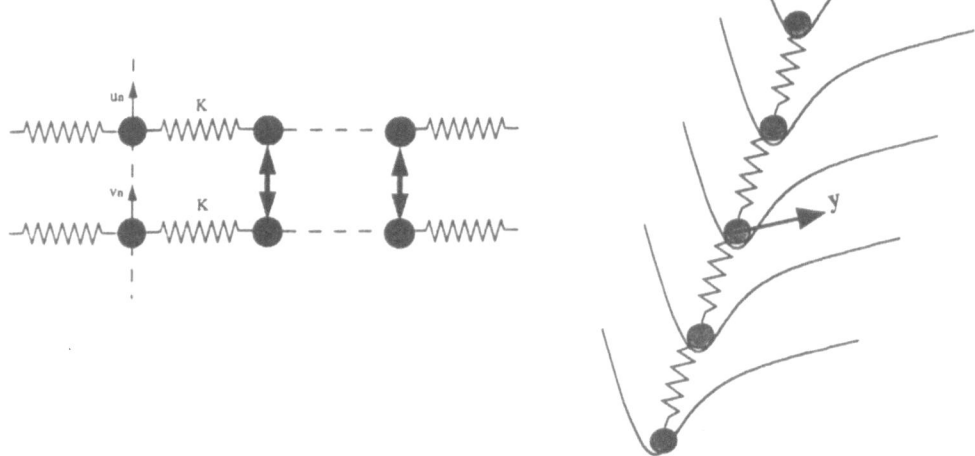

FIG. 2. Schematic views of the model : (a) analogous view with longitudinal and transversal springs (b) simplified view of the native state.

Once the model is defined, we first have to check that its behavior is qualitatively equivalent to the experimental observations and particularly to the dynamical effects introduced by temperature fluctuations.

3. DYNAMICS OF THE MELTING

We have investigated the dynamics of the model in contact with a thermal bath by molecular dynamics simulation with the Nosé-Hoover method [3,4]. This methods couples the physical system of interest to two additional non local variables s and p_s according to the equations

$$\frac{dy_n}{dt} = \frac{p_n}{m} \quad \text{and} \quad \frac{dp_n}{dt} = -\frac{\partial V}{\partial y_n} - \frac{p_n p_s}{Q}$$

$$\frac{dp_s}{dt} = \sum_n \frac{p_n^2}{m} - Nk_BT \quad \text{and} \quad \frac{ds}{dt} = \frac{sp_s}{Q}$$

where p_n is the conjugate momentum of the coordinate y_n, V the potential part of the hamiltonian, N is the number of base pairs in the chain, k_B is the Boltzmann constant and Q a parameter that sets the timescale of the thermostat. Although the variable s is not directly coupled to the coordinates and momenta, it is convenient to follow its evolution because the extended hamiltonian $H' = H + p_s^2/2Q + Nk_BT\ln(s)$ is a conserved quantity which can be tested to check the numerical accuracy.

(a)   (b)   (c)

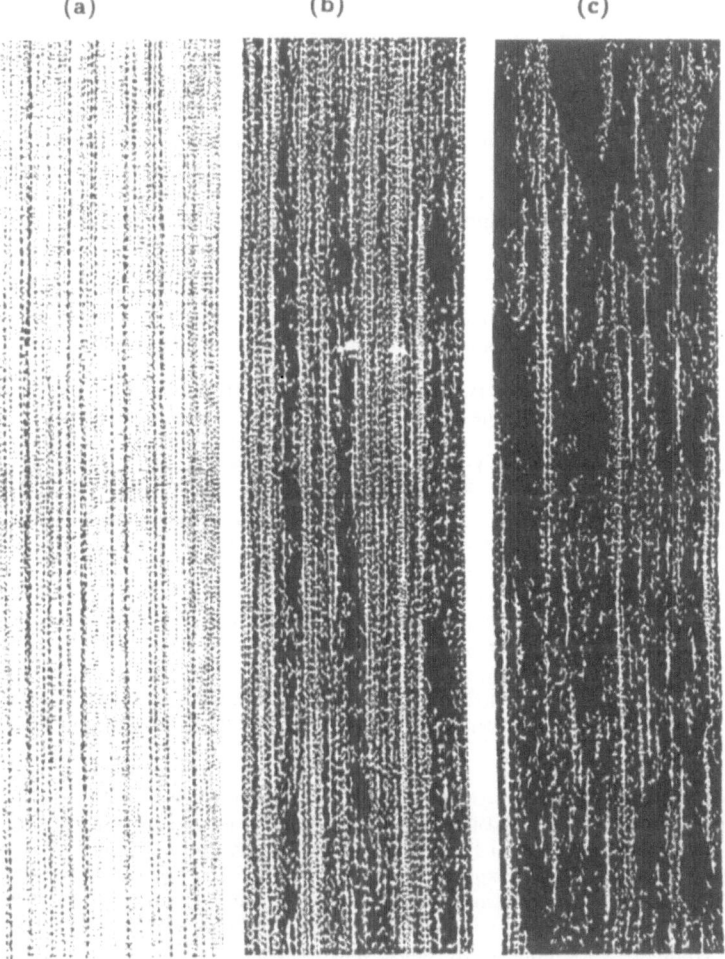

FIG. 3. Evolution versus time of the stretching for three equilibrium temperatures : (a) $T = 150\ K$, (b) $T = 340\ K$, (c) $T = 450\ K$. The horizontal axis indicates the position along the 256 cells of the molecule and the vertical axis corresponds to the time. The grey scale goes from $y \leq -0,1\overset{\circ}{A}$ (white) to $y \geq 1\ \overset{\circ}{A}$ (black).

Figures (3a), (3b) and (3c) show the time evolution of a short chain (256 base pairs) for different temperatures. Looking at these figures, one notices immediately different features for the three temperatures. The dark regions which correspond to large stretching are organized into lines which attest that some excitations are long lived in the system but the size of the bubbles and their behavior depend on the temperature.

(a) Following one of these lines, on the first figure one can notice that it is interrupted regularly and looks like a dotted line. This is due to the internal breathing of the localized excitations that oscillate between a large amplitude (black dots in the figure) and a small amplitude state (light dots) in a regular manner. This excitations are very reminiscent of the small fluctuational openings that are observed experimentally [5,6].

(b) When the temperature increases, as one moves on horizontal direction, i.e. along the molecule for a given time, one notices that the amplitude of the stretching varies very much from site to site. This shows that there is no equipartition of energy in this nonlinear system on a time-scale which is very long with respect to typical periods of the molecular motions, but on the contrary this points out a tendency for the energy to localize at some points. This tendency becomes more and more pronounced as temperature increases. However, the "hot-spots" due to nonlinear energy localization are dynamical entities. They, move, appear and die, and, although they are extremely long lived compared to typical time-scales of the system, on a macroscopic time-scale one can consider that the average energy of all the sites is the same.

(c) The figure shows large black regions which correspond to denaturated regions of the molecule. These black areas are the denaturation bubbles observed experimentally.

From these three figures, it is possible to emphasize the following qualitative points, which are in accordance with the biological behavior : the first small localized and oscillating excitations are the precursors events of the bigger bubbles. The dynamics of the system is therefore determined by the evolution of the *fluctuational openings* which are growing with temperature to give the *denaturation bubbles*. As many physical phenomena, this system clearly involves a *localization of energy* in space that we will discuss in the following section.

In figure (4), we plot the frequency of the phonons as a function of the temperature. This figure has been obtained by calculating the structure factor $S(q,\omega)$ for $q = \pi/4$ after a convolution with a gaussian in order to smooth the data. The errors bars are defined as the width at half-height of the peak of the spectra for the different temperatures: they are large at intermediate temperature, because the phonon peak has considerably broadened indicating that the phonon lifetime is shorter. But the decrease of the frequency as the temperature increases is clear.

Indeed at very low temperature, we have a well defined peak around $\omega = 0.75$ corresponding to the low temperature phase phonon peak, $\omega^2 = (2a^2D + 4K \sin^2(q/2))/m$. At high temperature ($T = 450$ K), a well defined phonon peak is again present around $\omega = 0.02$. The frequency corresponds to the high temperature phonon's peak. Recalling that, at this temperature, most of the particles are on the plateau of the Morse potential, it is clear that this peak results from the phonons on the plateau: $\omega^2 = 4K/m \sin^2(q/2)$. This is a complete softenning, in the sense that the frequency of the lower band edge has reached zero: the potential being flat on the plateau, there is no frequency gap.

This softenning of the phonons is qualitatively in good agreement with Raman experiment on DNA gels which were performed by H. Urabe and Y. Tominaga [7]. Hans Grimm has also reported the observation of an optic mode of similar behavior in neutron scattering and was able to confirm these results by measurements using time-of-flight machine (see article in this proceeding). These measurements represent a possible experimental test of the model, and could allow an accurate choice of the parameters of the model, which are only crudely chosen. Work along this line is in progress.

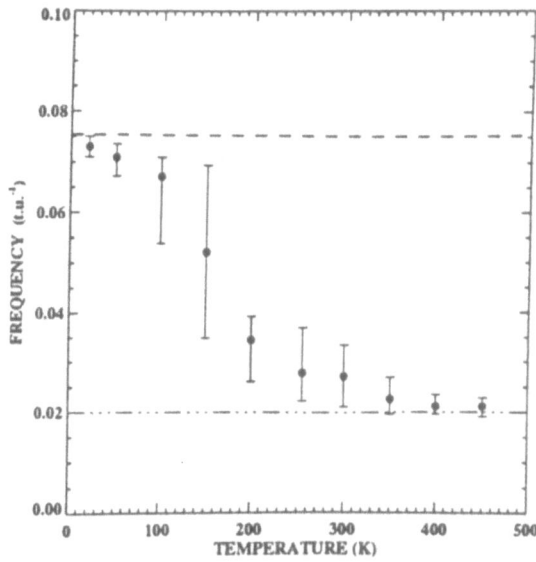

FIG. 4. Frequency of the phonons versus temperature for $q = \pi/2$. The error bars indicates an interval where the frequency lies at those temperatures for which a single mode cannot be identified. The horizontal lines corresponds to the phonons frequency at the bottom of the well (dashed) and on the Morse plateau (dashed-dot-dot-dot).

4. FORMATION OF DENATURATION BUBBLES

If the previous figure could help very much in the choice of the parameter, it didn't explain the creation of big denaturation bubbles. If the phonons modes are seen as first events as expected, we have to understand the connection between these intrinsically linear excitations and the localized and large excited states that we can see on the figure (3c).

Some of our recent works [8,9] in nonlinear physics could give accurate explanations of the creation of such nonlinear excitations in a homogeneous system. Nonlinear energy localization in continuous media has been extensively investigated since Benjamin and Feir discovered the modulational instability of Stokes waves in fluids, but very little has been done in lattices although it would be of wide interest for solids or macromolecules. In these studies, we pointed out that, in a discrete lattice, nonlinear energy localization is very different from its counterpart in a continuum medium. In particular, we showed that, besides the familiar mechanism of modulational instability, which is itself strongly modified by discreteness effects, there is an additional channel for energy concentration which appears as a very general process leading to localization of energy in a lattice.

The first step toward the creation of localized excitations can be achieved through modulational instability which exists in a lattice as well as in a continuum medium, although discreteness can drastically change the conditions for instability. Indeed, these systems exhibit an instability that leads to a self-induced modulation of the steady state as a result of an interplay between the nonlinear and dispersive effects. However the maximum energy of the breathers created by modulational instability is bounded because each breather collects the energy of the initial wave over the modulation length λ so that its energy cannot exceed $E_{max} = \lambda e$ where e is the energy density of the plane wave. Consequently, although modulational instability can lead to a strong increase in energy *density* in some parts of the system, it cannot create breathers with a *total* energy exceeding E_{max}. In our physical problem, that would say that the system could create small breathers, but that the large amplitude breathers could not be created.

For a given initial energy density, one can however go beyond this limit if one excitation can collect the energy of several breathers created by modulational instability. Such a mechanism is not observed in a continuum medium because there the breathers generated by modulational instability are well approximated by solitons of the Nonlinear Schrödinger (NLS) equation which can pass through each other without exchanging energy. On the contrary, when discreteness effects are present, the energy of each excitation is not conserved in collisions, and, the important point is that the exchange tends to favor the growth of the larger excitation.

This mechanism of discreteness-induced energy localization by collisions that we have described in a general context [8] appears also in the nonlinear model for DNA dynamics that we discuss here. This "new" mechanism which is specific to lattices, where some excitations grow at the expense of the others is perhaps at the basis of the creation of the denaturation bubbles.

5. STATISTICAL MECHANICS

5.1 Melting transition

Since we are interested in the thermal denaturation transition of the molecule, the classical approach is the statistical mechanics. Due to the one-dimensional character of the system, and because the interactions are restricted to nearest neighbor interactions, it can be treated exactly, including fully the nonlinearities, with the transfer operator method [10]. In the continuum limit approximation, the transfer integral method can be solved exactly and one finds a melting temperature $T_d = 2\sqrt{2KD}/ak_B$. However experiments on proton exchange in DNA [11] show some evidence of exchange limited to a single base pair which suggests that discreteness effects can be extremely large in DNA. Some numerical simulations (see article in this proceeding) have also confirmed the motion of isolated base pairs, proof of high discreteness effects. Therefore the calculation must be completed by an investigation able to describe the discrete case and therefore we have solved numerically [12] the transfer operator.

The results are shown on the figure (5a) where one sees that the strong increase of the mean value of the stretching corresponds in the discrete case to temperatures higher than the continuum denaturation temperature T_d. If the transition between the native state at low temperature to the denaturated state is clear on the figure, we can easily notice that the denaturation is very smooth, while the experimental observations show a very sharp transition. Moreover, introducing the disorder does not change this feature : although this nonlinear dynamical model exhibited a thermal behavior in qualitative agreement with the experiments, an essential feature is still missing.

5.2 Entropy driven DNA denaturation

If the model is clearly very simple, it lacks an essential point; the stacking energy is not a property of *individual* bases but a character of base *pairs* themselves. That's why we introduced a very small change with large consequences [15] by replacing the harmonic coupling by a *nonlinear* coupling:

$$W(y_n, y_{n-1}) = \frac{K}{2}\left(1 + \rho e^{-\alpha(y_n+y_{n-1})}\right)(y_n - y_{n-1})^2 .$$

(4)

When the hydrogen bonds connecting the bases break, the electronic distribution on bases are modified, causing the stacking interaction with adjacent bases to decrease. In Eq. (4), this effect is enforced by the prefactor of the usual quadratic term $(y_n - y_{n-1})^2$. This prefactor depends on the *sum* of the stretchings of the two interacting base pairs and decreases from $\frac{1}{2}K(1+\rho)$ to $\frac{1}{2}K$ when either one (or both) base pair is stretched. This coupling potential is in agreement with the properties of chemical bonds in DNA and also provide the cooperativity effects that were introduced phenomenologically in the Ising models. A base pair that is in the vicinity of an open site has lower vibrational frequencies, which reduces its contribution to the free energy. Simultaneously a lower coupling along the strands gives the bases more freedom to move independently from each other, causing an entropy increase which drives a sharp transition.

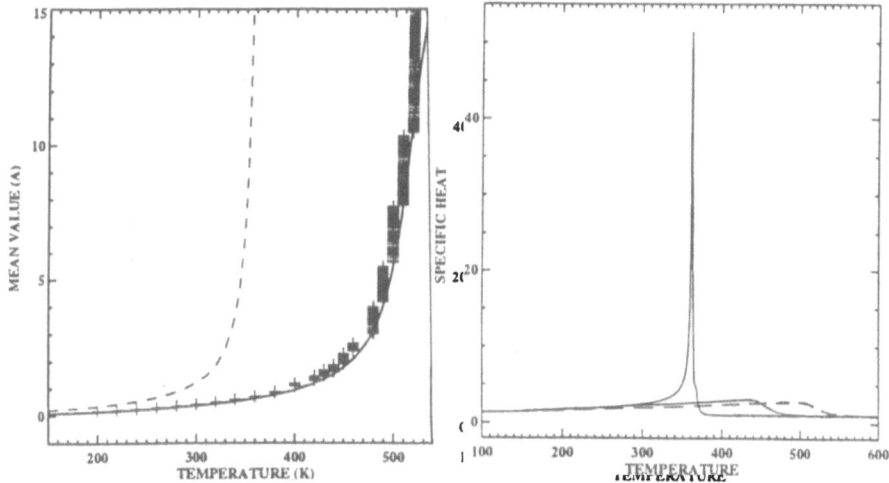

FIG. 5. (a) Variation of $\langle y \rangle$ versus temperature: the solid line corresponds to the exact numerical calculation, the dotted line to the results obtained with the continuum approximation and the plus signs to molecular dynamics simulations. (b) Variation of the specific heat versus temperature. The very narrow peak corresponds to the anharmonic case ($\alpha = 0.35, \rho = 0.5$), the dotted curve and the solid broad peak to harmonic coupling ($K' = 1.5K$ and $K'' = K$, respectively).

The effect of this entropy driven transition is shown on the figure (5b) where the specific heat of the model is plotted versus temperature for the anharmonic and the harmonic coupling potential. The extremely sharp peak in the first case is extremely close to a first order phase transition [13] and it must be stressed that the qualitative change of behavior introduced by the anharmonic coupling is not due to special values of the parameters (one can find a discussion if whether or not the melting transition is of the first order type in the work by Chen and Prohofsky [14] and in our work published elsewhere [15]). The model is now a good qualitative description of DNA melting and its dynamics.

6. CONCLUSION

Our work shows that the nonlinearity is more efficient than disorder to localize energy and that the breathers (or fluctuational openings) are important precursors of the melting. The second point which has to be emphasized is that the nonlinear coupling along the strands is essential and is related to cooperativity effects in the opening of the bubbles. But some properties of the DNA dynamics are not described at this stage of the work. The denaturation process involves maybe (the biological data are not extremely clear) an intermediate state, where the cooperativity effects are not strong enough to prevent the opening of only one base pair. The cooperativity effects would only be more important for the end of the denaturation process, ie for higher temperatures.

Of course many extensions have to be done in order to reach the original goal: understand DNA transcription. Some of them, such as the comparison of the evolution of the frequency versus temperature with experiments to define better model parameters, or the study of the interplay of disorder and nonlinearity are currently in progress. The open question to introduce in some way the helicoïdal geometry is also important, because it will introduce a coupling at large distance, which could for example enhance a sharper transition as longer range forces do in classical statistical mechanics problem. A second extension would be to take into account that the bending of the macromolecule is an efficient way to break one base pair. It means that we have to understand the interplay between the acoustic and the stretching modes.

The more important question is maybe to validate such a simple approach. Indeed, the extreme simplicity of the model is of course its main weakness, but it could emphasize the main features of the process and help to construct a more serious model. This way, to construct step by step a simple model in order to describe a complex system is maybe as successful as to start from the complex system and try to extract the relevant data from a huge quantity of informations. The right way is probably to combine the two ideas, keeping in mind the main goal, which is to be as close as possible to the biological evidences. But, anyway, the right answer will be given in the long term.

Acknowledgements
Part of this work has been supported by the CEC grant SC1-CT91-0705. Most of the computations have been performed with the facilities of the Advanced Computer Laboratory of the Los Alamos National Laboratory which is gratefully acknowledged.

References

[1] P. W. Anderson, *Phys. Rev.* **109** (1958) 1492.
[2] R. M. Wartell and A. S. Benight, *Physics Reports* **126** (1985) 67-107.
[3] S. Nosé, *J. Chem. Phys.* **81** (1984) 511-519.
[4] W. G. Hoover, *Phys. Rev. A* **31** (1985) 1965-1967.
[5] H. Teitelbaum and E. Englander, *J. Mol. Biol.* **92** (1975) 55.
[6] E. W. Prohofsky, et al, *Plys. Lett. A* **70** (1979) 492-494.
[7] Y. Tominaga et al, *J. Chem. Phys.* **83** (1985) 5972-5975.
[8] T. Dauxois and M. Peyrard, *Phys. Rev. Lett.* **25** (1993) 3935-3938.
[9] T. Dauxois, M. Peyrard and C.R. Willis, *Phys. Rev. E* **48** (1993) 4768-4778.
[10] M. Peyrard and A. R. Bishop, *Phys. Rev. Lett.* **62** (1989) 2755-2758.
[11] J.L.Leroy *et al*, *Mol. Biol.* **200** (1988) 223.
[12] T. Dauxois, M. Peyrard and A. R. Bishop, *Phys. Rev. E* **47** (1993) 684-695.
[13] M. Peyrard and T. Dauxois, "DNA melting : a phase transition in one dimension ?", Lingby 1-4 August 1994, Christiansen and Mosekilde Eds. (IMACS, 1994), pp. 52-62.
[14] Y. Z. Chen and E. W. Prohofsky, *Biophysical Journal* **66** (1994) 202-206.
[15] T. Dauxois, M. Peyrard and A. R. Bishop, *Phys. Rev. E* **47** (1993) R44-R47.

Dynamics of conformational excitations in the DNA macromolecule

A.M. Kosevich and S.N. Volkov*

*Institute for Low Temperature Physics and Engineering,
National Academy of Sciences,
Kharkov 310086, Ukraine
* Bogolyubov Institute for Theoretical Physics,
National Academy of Sciences,
Kiev 252143, Ukraine*

1. INTRODUCTION

The effective functioning of the genetic apparatus of biological systems is, to a great extent, due to the specific conformational abilities of the DNA macromolecule. The comparatively low energy barriers for the transitions between the stable states and small free-energy difference of double helix forms allow the macromolecule to change, within broad ranges, the mutual position of its structural elements (nucleic bases, sugar rings and phosphate groups) [1].

In the structural dynamics of macromolecules an important role is played by the conformational excitations representing the collective motions of structural elements of the system. For DNA these excitations are comparatively low-energy and can participate directly in the biological functioning of the macromolecule. In the case of small amplitudes of the structural elements displacements the conformational excitations of the system prove to be the normal conformational vibrations. For large amplitudes - the conformational excitations may be manifested as a nonlinear waves.

The problem of the collective structural excitations description for DNA is stated in a general form since 1983 [2], but the progress has been achieved only in some narrow directions in spite of a great interest to the studies in this field [3].

In this paper we present a consistent approach to the description of the collective conformational excitations in DNA macromolecule, which we used under investigations of its linear and nonlinear dynamics [4-12]. The main attention is paid to a possible identification of the structural elements motions, which are weakly sensitive to nucleotide heterogeneity. In the linear approximation the agreement with experiment for DNA low-frequency Raman spectra is demonstrated. For the nonlinear case the developed theory allowed us to explain the mechanism of the known data on the long-range transmission in DNA.

2. THE MODEL CONSTRUCTING

Proceeding from the results of the macromolecule conformational mechanics study [1] the DNA structural elements mobility may be classified into three types of motions. This is, first, the nucleic bases and sugar rings (nucleosides) torsion motions with respect to backbone chains - the analogue of librations in molecular crystals. These motions take place because of the correlated changes in the torsional angles of the backbone. The second one is the nucleic bases motions under the changes in nucleoside conformation (the geometry of the sugar ring and the glycoside bond angle), that is the intranucleoside mobility. In a double-stranded DNA the conformation of nucleoside changes in accordance with the pseudorotation coordinate and can be regarded as a unitary motion. And the third one is the motion of the nucleotide as a whole (nucleic base, sugar and backbone group). The movements described can be accompanied with hydrogen bonds stretching in the base pairs and backbone bending and torsion.

To consider the above mentioned types of mobility of the structural elements in the DNA macromolecule the four-mass model for the monomer link can be used: two nucleoside masses and two phosphate group masses of the backbone. The position of masses in a monomer link of double helix is determined by the parameters which represent the reduced length of the nucleoside regarded as the physical pendulum (l_i) and the angle of nucleoside suspension to the backbone (Θ_i) (Fig. 1). Here the index i = 1,2 numerates the double helix chains. The backbone masses (m_0) are homogeneous along the chain of macromlolecule but the nucleoside masses (m_i) are varied as they include the different base masses.

The expression for the energy of conformational excitations of the system within the four-mass model is:

$$ E = \frac{1}{2} \sum_n \sum_{i=1}^{2} \left\{ m_0 \, \dot{\mathbf{R}}_i^2(n) + m_i \, \dot{r}_i^2(n) + U[r_i; \mathbf{R}_i] \right\} . \tag{1} $$

Here the summation in n is carried out over all monomers of the double chain. The radius-vectors r_i and $\mathbf{R}_i$ describe the displacements of mass centers of the n-th nucleoside and backbone group, respectively.

It is natural to consider that the potential energy of the system depends on the relative displacements of the structural elements in the double helix. Taking into account the structural organization of a macromolecule the potential energy U can be presented as the sum of five terms:

$$ U = U_1[\delta(n)] + U_2[\rho(n)] + U_3[r(n, n-1)] + U_4[R(n, n-1)] + $$
$$ + U_5^{(1)}[\rho(n); R(n+1, n)] + U_5^{(2)}[\rho(n); R(n, n-1)] . \tag{2} $$

The first of them describes the contribution from the hydrogen bond stretching in the base pairs and depends on displacements of the nucleosides inside the pair $\delta(n) = |r_1(n) - r_2(n)|$. The term U_2 is the energy of intranucleoside mobility associated with the displacement of the mass center of the nucleoside with respect to the point of its suspension to the backbone: $\rho_i(n) = |r_i(n) - \mathbf{R}_i(n)|$. The next two terms ($U_3$ and U_4) describe the energies of the nearest-neighbour interactions along the macromolecule chain (base stacking and sugar-phosphate backbone). Here the following notations are accepted: $r(n, n-1) = |r(n) - r(n-1)|$ and $R(n, n-1) = |R(n) - R(n-1)|$. Last terms $(U_s^{(1)} + U_s^{(2)})$ are the energies of the interaction between the intramonomer displacements and the displacements of a monomer link as a whole. Numerous data of the conformational analysis of double-stranded DNA and polynucleotides [1] testify that variations in the position and conformation of the structural elements inside the

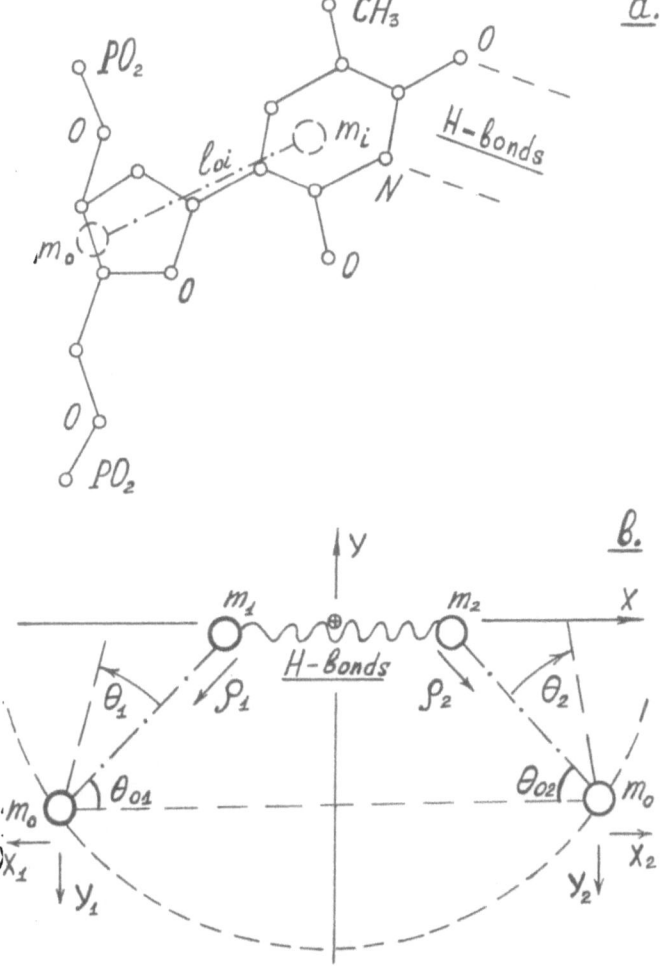

Figure 1: The four-mass model for the DNA conformational mobility study: a. - the mass centres of the nucleoside (m_i) and backbone (m_0) for the fragment of a polynucleotide chain; b - the position and displacements (showed by arrows) of the masses in the base-pair plane.

monomer link are accompanied by the changes in the position of the link itself in the polynucleotide chain. The energy U_5 could be neglected only under studying of small displacements of the structural elements within certain form of the double helix.

The given model is a generalization of the one used to study the DNA conformational vibrations [4, 6, 10] and the model for the study of DNA conformational transition dynamics [5, 7-9].

An important property of DNA macromolecules, in particular for biology, is the heterogeneity of their nucleotides content. Thus, the geometric parameters and masses of nucleosides essentially depend on the kind of nucleic base included (Tab. I).

Table I: Parameters for the model of DNA conformational mobility (as calculated for the B-form).

Nucleoside	m_i (a.u.m.)	l_i (Å)	Θ_i (deg.)	M (a.u.m.)	μ (a.u.m.)	μ_p (a.u.m.)	J (10^{-37} g cm^2)
Ade	203	5.1	25	615	140.7	99.2	33.7
Thy	194	4.6	35				
Gua	219	5.4	23	616	140.8	98.5	34.5
Cyt	179	4.5	30				

The energy of conformational excitations of the system must be also sensitive to the DNA heterogeneity. For the kinetic energy this sensitivity becomes obvious if we go over to the "natural" coordinates for the displacements in the four-mass model (see [12]). The heterogeneity of the potential energy is mostly manifested in the term U_1 since the A•T- and G•C-base pairs are coupled by the different number of hydrogen bonds. For small displacement amplitude the differences in force constants of H-bond stretching constitute no more than 10%, but for the large amplitudes of displacements the difference can exceed 40%. These estimations are obtained by us on the basis of conformational calculations of the hydrogen bond stretching energy in complementary pairs [13].

There are two possible ways to resolve the problem. The first one is to obtain the solutions of the equations of motion for the averaged parameters of the model (1), (2) and then take into accout the heterogeneity effect as the perturbation in calculations. This variant was used by us in studying of the DNA small vibrations. Another way is associated with the separation of the structural elements whose mobility are insentitive to DNA heterogeneity. Such approach can be applied to studies of the large-amplitude excitation dynamics.

3. THE CONFORMATIONAL VIBRATIONS

In the linear approximation we can correlate our approach with the data on DNA low-frequency Raman spectra (10-200 cm^{-1}). We have analyzed the normal vibrations of the DNA macromolecule within the four-mass model regarding that monomers had the averaged over nucleotide content parameters $\left(l_0 = \langle l_{0i} \rangle, \Theta_0 = \langle \Theta_{0i} \rangle, m = \langle m_i \rangle \right)$. The expression for the energy of small vibrations written in the continuum approximation is following [10]:

$$E = \sum_{i=1}^{2} \int \frac{dz}{2h} \{ M_0(\dot{X}_i^2 + \dot{Y}_i^2) + m(l_0^2 \dot{\Theta}_i^2 + \dot{\rho}_i^2 + 2l_a \dot{\Theta}_i \dot{X}_i + 2l_b \dot{\Theta}_i \dot{Y}_i + 2b\dot{\rho}_i \dot{X}_i -$$

$$- 2a\dot{\rho}_i \dot{Y}_i + \alpha(\delta)^2 + \beta(\Theta_i)^2 + \varepsilon(\rho_i)^2 + fh^2[(\Theta_i')^2 + (\rho_i')^2] + gh^2[(X_i')^2 + (Y_i')^2] \} .$$

(3)

Here h is the double helix step, $M_0 = m + m_0$, $a = \sin \Theta_0$, $b = \cos \Theta_0$, $l_0 = al_0$, $l_b = bl_0$, X_i and Y_i are the coordinates of the transverse displasements of backbone masses, Θ_i is the nucleoside torsion motion coordinate and ρ_i is the intranucleoside mobility coordinate.

In expr. (3) the potentail energy of the system is written in the harmonic form that corresponds to studying of small vibrations. The force constants describe the interaction in the hydrogen-bonded pairs (α), nucleoside torsion motion (β), intranucleoside mobility (ε) and the interactions along the macromolecular chain (f and g).

We consider only the transverse displacements to make our model simpler and because the transverse mobility of DNA macromolecules have received the most study (see [1]). Then we assume everywhere that the motions under the study have corresponding longitudinal components, which are not considered explicitly in expr. (3).

Using the energy (3) the normal vibrations of the system are obtained: five branches of optical vibrations and three branches of acoustic ones. The expressions for frequencies of limitedly long-wave optic vibrations have the view:

$$\omega_{1,2}^2 = \left[\varepsilon_0 + \alpha_0 \pm \sqrt{(\varepsilon_0 + \alpha_0)^2 - 4q\varepsilon_0\alpha_0} \right] \frac{M_0}{2m_0} ;$$

$$\omega_3^2 \approx \omega_5^2 = \frac{\beta_0 M_0}{m_0} ;$$

(4)

$$\omega_4^2 = \frac{\varepsilon_0 M_0}{m_0} ,$$

where $\alpha_0 = \frac{2\alpha m_0}{M_0 m}$, $\beta_0 = \frac{\beta}{ml_0^2}$ and $\varepsilon_0 = \frac{\varepsilon}{m}$.

The numerical analysis of the mutual position of the ω_1, ω_2 and ω_4 branches showed that at all possible ratios of the force constants α_0 and ε_0 the inequality is valid [11]: $\omega_1 > \omega_4 > \omega_2$. Since, considering that for such macromolecules as DNA it is fulfilled: $\alpha_0 \sim \varepsilon_0$ and $\alpha_0, \varepsilon_0 \gg \beta_0$, we obtain the hierarchy of the vibration branches in the long-wave spectrum:

$$\omega_1 > \omega_4 > \omega_2 > \omega_5 \approx \omega_3.$$

(5)

The analysis of exprs. (4), (5) makes it possible to conclude that the wide band observed in the Raman spectra of the DNA with the maximum about 85 cm^{-1} [14] consists of three branches of vibrations: ω_1, ω_2 and ω_4. The lower-frequency peak (~16 cm^{-1}) is created by the vibrations ω_3 and ω_5 which are degenerated in the long-wave limit. The vibrations ω_1 and ω_2 occur with H-bond stretching and are the most sensitive to the DNA heterogeneity. The vibrations ω_3, ω_4 and ω_5 are dependent on the nucleic bases composition to a much less extent.

To compare the results of the theory with the experiment on low-frequency Raman spectra we calculated the frequency values of optical branches in long-wave limit for DNA in the B-form. In our calculatios we used the model parameters and the force constants averaged over the nucleotide content [6, 10]. The results obtained show a good agreement whith the experiment (Tab. II).

Table II: Comparison of the experimental values of vibration frequencies with the theoretical results for B-DNA (cm^{-1}).

Raman spectra	Band			Peak
H. Urabe, et al [14] (fiber)	85 ± 30			16
Om.P.L.Lamba, et al. [15](crystal)	120	96	68	18
T.Weidlich, et al. [16] (film)	98	76	41	15
Our results	ω_1	ω_4	ω_2	ω_5, ω_3
	114	85	62	16, 17
Estimated deviations	±5	–	±5	±1

Notice, that our estimations of frequency value deviations caused by the nucleotide content also show that three of the five optical modes ω_3, ω_4 and ω_5 are practically independent on the nucleotide heterogeneity (Tab. II).

The approach developed in the linear aproximation allowed us to interpret experimental data so far, in particular, the DNA Raman spectra dependences on the double-helix conformation, the nucleotide content and temperature [6, 10-12].

4. THE CONFORMATIONAL SOLITONS

In this section we study the dynamics of the macromolecule with large amplitudes of dispplacements of the structural elements. To describe such motions it is necessary to make use of nonlinear approach taking into account the anharmonic terms in the potential energy. The view of this terms determine the form of the nonlinear excitations in the system. Among possible excitations the stationary solitonic waves seem to be the most interesting for different applications. The solitons remain stable under the collision with other excitations and can propagate along the system on a large distance. For solitons the asymptotic principle of superposition is valid: if in some period of time the solitonic excitations exist they will exist during a long time as well [17].

But not all nonlinear excitations possible in the DNA macromolecule can propagate along the chain [9]. It is clear that the excitations, which are less sensitive to the DNA heterogeneity, have the advantage for their existence in the macromolecule. Furthermore, the conformational motions created the excitation have to be cooperative to transmit excitation along the macromolecule.

Considering the nonlinear excitations in DNA we first of all study the possibility of singling out the motions weakly sensitive to the nucleotide content. Proceeding from the expr. (1), (2) we transform our model by introducing to the reference system the joint and relative motions of the structural elements, which are typical to conformational transition dynamics. We introduce the displacements for the centre of mass (c.m.) of the both nucleosides in the pair and c.m. of the backbone groups: r_p and r_q. Then we take into account the displacements of the DNA monomer link c.m. (**R**), relative displacements of the nucleosides with respect to the backbone (**r**) and to the each other position in the link (δ):

$$R = \frac{m_p r_p + 2 m_0 r_q}{M}; \quad r = r_p - r_q; \quad \delta = r_1 - r_2, \tag{6}$$

were $m_p = m_1 + m_2$, $M = m_p + 2 m_0$ are the base-pair mass and the total mass of the monomer link, respectively.

Using notations (6) and taking into account the smallness of the terms concerning the relative displacements of the backbone groups of the different chains in the monomer link the expression for energy of the system can be written as

$$E = \frac{1}{2h} \int dz \left[M \dot{R}^2 + \mu \dot{r}^2 + \mu_p \dot{\delta}^2 + \Phi(\delta) + \Phi(r) + \Phi(R', r', \delta') + F(r, \delta, R') \right]. \tag{7}$$

Here we used the continuum approximation. In expr. (7) $\mu = \frac{m_p \cdot 2 m_0}{M}$ and $\mu_p = \frac{m_1 \cdot m_2}{m_p}$ are the reduced masses. The potential energy in expr. (7) represents the sum of terms, which describe, respectively, the energy of the H-bond stretching, intranucleoside joint motions, interaction along the chain and the energy of the interrelation between the intra- and intermonomer degrees of freedom.

It follows from our estimations (Tab. I) that the mass coefficients in the kinetic energy in expr. (7) are independent on the nucleotide content. Furthermore, if conformational excitations occur without essential change in the distance between the nucleosides (therefore $\Phi(\delta) \approx 0$) the energy of the system becomes practically insensitive to heterogeneity of the DNA. The motions of structural elements under the condition $\delta \approx 0$ may take place in real DNA macromolecules. It is known [1] that for such conformational transitions as B-A in DNA the geometry of the base pairs remains unchanged during the transitions. The existence of the motions with $\delta \approx 0$ in the dynamics of macromolecule is evident also from our normal modes studies [4, 6, 10].

Using the approach developed we study the dynamics of B-A transition in the DNA chain. Much evidence has recently been accumulated to show the important role of DNA B-A transitions in the transcription regulation. There is indication that A-shaped section appears in B-DNA when it interacts with RNA-polymerase and some proteins [18, 19]. In 1988 the hypotehsis on the mechanism of the long-range effect transmission in DNA was proposed [5]. According to [5] the attachment of regulatory protein to DNA triggers a B-A transition in the double helix and creates a propagation of local B-A transitions along the macromolecule. Under physiobiological conditions DNA is in the B-form of a double helix and only local transitions to the metastable pairs (as the A-form) occur. Moving along the DNA double chain local B-A transitions can transfer energy, which is sufficient to the long-range activation processes observed for DNA [20, 21, 22].

Let us consider the possibility for the local B-A transition to propagate along the DNA double helix. Under B-A transitions in DNA the motions of the pairs occur along the diade axis and are connected with torsion and bending of the macromolecule [1]. Taking into account that the conformational energy of the double-helix is usually considered as a function of the coordinates of torsion and pair displacement along the diade axis, it is expedient to choose these components for a description of B-A excitations dynamics. With $\delta \approx 0$ the excitation energy (7) for the torsion (φ) and the pair displacement (u) components can be represented in the form:

$$E = \int \frac{dz}{2h}\left[\mu\left(\dot{u}^2 + s_1^2\,u'^2\right) + J\left(\dot{\varphi}^2 + s_2^2\,\varphi'^2\right) + \Phi(u) + \kappa\,F(u)\varphi'\right].\tag{8}$$

In expr. (8) J is the moment of inertia of the monomer link, $s_{1,2}$ are the elastic constants of the interaction along the chain.

The potential function $\Phi(u)$ represents a nonsymmetric double well corresponding to the transition from the ground state (B-form) to the metastable one (A-) (see Fig. 2). In this form of the potential the real DNA properties under physiological conditions are taken into account. The term with the coefficient κ describes the interrelation between the base-pair displacement and the DNA torsion. The function $F(u)$ was choosen in the form of a symmetric barrier on the interval between the stable states of the B- and A-form (showed by the dashed line on Fig. 2).

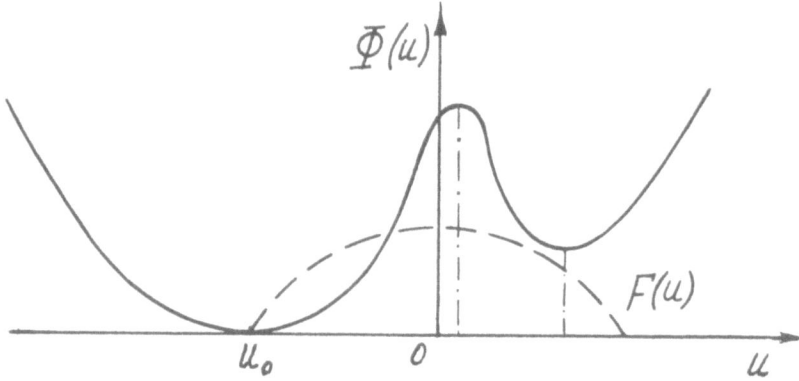

Figure 2: The view of the potential functions used for the description of the local B-A transition dynamics.

The equations of motion for the components u and φ have the form:

$$\begin{cases}\ddot{u} = s_1^2 u'' - \dfrac{1}{2\mu}\dfrac{\partial\Phi}{\partial u} - k_1\dfrac{\partial F}{\partial u}\varphi'; & (9)\\[3mm]\ddot{\varphi} = s_2^2\varphi'' + k_2\dfrac{\partial F}{\partial u}u', & (10)\end{cases}$$

where $k_1 = \dfrac{\kappa}{2\mu}$; $k_2 = \dfrac{\kappa}{2J}$.

We seak for the soliton-like solutions of eqs. (9, 10) for the base-state asimptotic ($u = u_0$, $\Phi(u_0) = F(u_0) = 0$) of the bistable system (8). After the wave substitution ($\xi = z - vt$) and one time integration we obtain the following expression for the energy conservation equation of a nonlinear oscillator

$$u_\xi^2 + Q(u) = 0 \ . \tag{11}$$

Here $Q(u) = \dfrac{\Phi(u) + \gamma \left[F(u)\right]^2 - C}{\mu \left(v^2 - s_1^2\right)}$, $\gamma = \dfrac{k_2^2 J}{\left(v^2 - s_2^2\right)}$ and C is the integration constant.

To derive the eq. (8) we take into account that $\varphi_\xi = \dfrac{k_2 F(u)}{\left(v^2 - s_2^2\right)}$ and $\varphi_\xi = 0$ at $u = u_0$.

The examination of dynamic trajectories of the oscillator (11) shows that three types of soliton-like solutions for the base-state assimptotic exist in the system (8) [23]. But only the one of them has a physical sence for the reference system. The nonlinear excitation for the u-component has a pulse form and for the φ-component - the step one [5, 7, 8, 23]. The two-component soliton occurs under the condition: $s_2 > s_1$. The form and amplitude of the nonlinear wave found depend on the soliton velocity, which obeys an inequality: $s_1 < s_p < v < s_2$. Here s_p is the velocity of the transverse "sound" for the two-component system (7). In the velocity interval of existance the amplitude (width) of the solitary wave changes from zero (infinity) at $v \to s_p$ up to the maximum value (zero) at $v \to s_2$. We note that the solutions obtained behave as excitations of the impact wave and from general classification are the dynamical solitons [24]. Our numerical study showed that the excitations found may be stable in the discrete chain under the collisions with other excitations and under the external friction [25]. When the wells in the potential $\Phi(u)$ become symmetric the obtained solutions transform into two two-component kinks.

From our estimations it follows that the soliton excitation can propagate in DNA as a local B-A transition only under the transformation of macromolecule to the elastically stressed state. In such a state the elastic constant s_2 must become more then s_1 and the condtions of cooperativity of excitation are fulfilled [5, 7, 8].

The results obtained allow us to explain the mechanism of long-range action known for DNA [20-22]. As observed experimentally the action transmission in DNA is always accompanied by the additional stresses in the chain. From our point of view these stress (in our model $s_2 > s_1$) is necessary for soliton propagation in the DNA chain. A local B-A transition moving along the double helix can transfer an energy in the order of the value of B-A-boundary energy, which is sufficient to fulfil the action.

Acknowledgement

Authors are grateful to the State Committee of Science and Technology of Ukraine for the partial support of this work, Grant # 2.3/135.

References

[1] Saenger W., Principles of Nucleic Acid Strcture (Springer-Verlag, Berlin, 1984) ch.9, 11.

[2] Krumhansl J.A., Alexander D.M., in: Structure and Dynamics: Nucleic Acids and Proteins, E.Clementi and R.H.Sarma Eds. (Adenine Press, 1983) pp.61-80.

[3] Krumhansl J.A., *Physica D* **68** (1993) 97-103.

[4] Volkov S.N., Kosevich A.M., *Mol.Biol.*, Moscow **21** (1987) 797-806.

[5] Volkov S.N., *Doklady Ac.Sci.Ukr.SSR* **A** (1988) 46-49.

[6] Volkov S.N., Kosevich A.M., G.E.Weinreb, *Biopolymery i Kletka*, Kiev **5** (1989) 32-39.

[7] Volkov S.N., *Phys.Lett.A* **136** (1989) 41-44.

[8] Volkov S.N., *J.Theor.Biol.* **143** (1990) 485-496.

[10] Volkov S.N., Kosevich A.M., *J.Biomolec.Struct.Dynamics* **8** (1991) 1069-1083.

[11] Volkov S.N., *Biopolymery i Kletka*, Kiev **7** (1991) 40-49.

[12] Volkov S.N., *Mol.Biol.*, Moscow **26** (1992) 835-846.

[13] Poltev V.I., Shulyupina N.V., *Mol.Biol.*, Moscow **18** (1984) 1549-1561.

[14] Urabe H., et al., *J.Chem.Phys.* **82** (1985) 531-535.

[15] Lamba Om.P., Wang A.H.-J., Thomas G.J.Jr., *Biopolymers* **28** (1989) 667-678.

[16] Weidlich T., et al., *J.Biomolec.Struct.Dynamics* **8** (1990) 139-171.

[17] Kosevich A.M., Theory of the crystal lattice (Vyshcha Shkola, Kharkov, 1988) pp. 117-128.

[18] Ivanov V.I., *Biopolymary i Kletka*, Kiev **1** (1985) 5-13.

[19] Klug A., *Nature* **365** (1993) 486-487.

[20] Crothers D.M., Fried M., *Cold Spring Harb.Symp.Quant.Biol.* **47** (1983) 263-269.

[21] Luchnik A.N., *Bio Essays* **3** (1985) 249-252.

[22] Lazurkin Yu.S., *Biopolimery i Kletka*, Kiev **2** (1986)   283-292.

[23] Volkov S.N., Savin A.V., *Ukr.Fiz.Zhurn.*, Kiev **37** (1992) 498-504.

[24] Kosevich A.M., Kovalev A.S., Introduction in the Nonlinear Physical Mechanics (Naukova Dumka, Kiev, 1989) pp. 150-169.

[25] Savin A.V., Volkov S.N., *Mathematic Modelling*, Moscow **4** (1992) 36-49.

Nonlinear dynamics of plasmid *pBR*322 promoters

M. Salerno

Department of Theoretical Physics,
University of Salerno,
84100 Salerno, Italy

Abstract

We investigate the role played by DNA promoters of *pBR*322 plasmid as dynamical activators of the transport process of the RNA-polymerase along the molecule. This is done in terms of discrete model of DNA which includes informations about specific base sequences. The results of direct numerical integrations of the model are compared with an effective potential analysis for a sine Gordon kink on a discrete nonuniform background.

1. INTRODUCTION

The gene expression of a cell in a living organism is regulated in a very precise manner by certain enzymes (RNA-polymerase) which selectively bind to specific DNA regions, called promoters, which control both the expression and the transcription frequency of the gene. In procaryotic cells promoters and genes are almost contiguous, while in eucaryotic cells the promoters are usually located far away from the genes they regulate, so that a mechanism which bring the action of the regulatory protein from the promoter to the gene must be involved. This action can be induced in two different ways. The first involves specific protein-protein interactions to fold the double-helix in order to bring the promoter close to the gene. The second may be due to solitary waves which could be created as local changes in DNA conformation in the binding process of the RNA polymerase to the DNA. These excitations could then travel along DNA carrying the regulatory protein from the promoter region to the gene. It is known that the binding of the RNA-polymerase to the DNA in the promoter region involves several steps. First, one has the formation of the so called "closed complex" in which the RNA polymerase and the DNA promoter are bound, but the hydrogen bonds between the bases along the double helix are closed. Then, there is the formation of the so called "open complex", i.e. a state in which the RNA polymerase is bonded to DNA with the hydrogen bonds along DNA open. The interacting region between the enzyme and the promoter extends over 50 base pairs (bp) while the open complex (open "bubble") is only 20 bp long. The existence of an open "bubble" inside the DNA-RNA-polymerase interaction region, strongly suggest the idea of a solitary wave. In previous papers [1, 2], assuming the formation of solitary waves in the RNA-DNA binding process, we showed that the specificity of the base sequences in the promoter regions may induce dynamical effects on this wave that, in turn, can influence the transcription and the regulation process of the genes. In particular, for promoters of the $T7$ family the existence of a dynamically "active" regions inside the promoter was shown [1, 2].

The aim of the present paper is to further extend these studies by considering the dynamical properties of the promoters of a circular DNA molecule known as plasmid $pBR322$. Plasmids occur naturally in E.coli and other bacteria and are often used as cloning vectors. They contain a replicon (i.e. a DNA sequence which allow the host cell to replicate them), several restriction-enzyme clavage sites (which allow foreign-DNA inserts) and DNA sequences corresponding to antibiotic-resistance genes. Plasmid $pBR322$ is a ring of 4361 bases which contains two antibiotic resistance genes: the gene "bla" which encode for an ampicillin resistance protein (beta-lactamase) and the gene "tet" which encode for a tetracycline resistance protein (the ring sequence of this plasmid can be readily obtained from the Genbank library). It is known that the gene "bla" has two promoters, called P_3 and P_1, located upstream to the gene [3]. Here we investigate the dynamical properties of the promoters of the gene "bla" by considering a fragment of the ring consisting of 1600 bases which includes both the gene and the promoters.

As a result we show the existence of two dynamically active regions, one located in the P_3 region, the other in the P_1 region, in which a static solitary wave acquires a finite velocity to travel downstream toward the ampicillin-resistance gene. This is in contrast with what happens in other regions, in which the solitary wave oscillates around the initial position or remains static. Furthermore, by inserting promoter P_3 downstream in reversed order (to simulate an antisense transcription) we find that the direction of the wave is effectively reversed. These results are in good agreement with biochemical studies on plasmid $pBR322$ [3, 4] and can be analytically understood in terms of an effective potential for a kink of a discrete sine-Gordon equation in a non uniform background.

The paper is organized as follows. In section 2 we discuss the model. In section 3 we present numerical studies on plasmid $pBR322$ and derive an effective potential analysis for the nonlinear wave which is responsible of the nonlinearity-supported transport mechanism. We show that this analysis gives results in good agreement with our numerical investigations. In the last section we summarize the main results of the paper.

2. THE MODEL

We consider the B-form of the DNA and concentrate the attention on the degrees of freedom characterizing base rotations in the plane perpendicular to the helical axis around the back-bone structure. This dynamics plays an important role for DNA functioning since, under certain circumstances, it can open the hydrogen bonds between conjugated pairs, exposing the unpaired bases to the action of external ligands. We assume that each base of a strand is coupled with the next neighbor bases by an uniform elastic stacking force, and with the complementary base in the opposite strand by a cosine potential modeling the hydrogen bond. This phenomenological model was introduced long ago by Englander et al. [5] and was further improved by many authors [6, 7, 8, 9].

Our modifications to the model consists in taking into account the specificity of the base sequence of real DNA, using the fact that the hydrogen bond involved in the pairings is double for Adenine-Thymine (A-T) and triple for Guanine-Cytosine (G-C). This leads to a chain of nonuniform pendula which corresponds to a specific DNA sequence by fixing the ratio between the pendula masses corresponding to A-T and G-C pairs to be 2/3. The Hamiltonian of the model is

$$H = \sum_{n=1}^{N} \{\frac{1}{2}I_n(\dot{\psi}_n^2 + \dot{\theta}_n^2) + \frac{1}{2}k_n(\psi_{n+1} - \psi_n)^2 + \frac{1}{2}\bar{k}_n(\theta_{n+1} - \theta_n)^2$$

$$+\eta_n[1 - \cos(\psi_n - \theta_n)]\} \qquad (1)$$

where ψ_n and θ_n are the deflection angles which two complementary bases form with the line passing between the attaching points of the bases to the backbone double helix. The parameters k_n and $\bar{k}_n$ denote the spring constants due to the stacking of the bases along the two helixs, I_n is the moment of inertia of individual bases, N is the number of base pairs in the chain, and η_n is a nonlinear parameter modeling the strength of hydrogen bonds between complementary bases. As mentioned before, we choose the coefficients β_n in Eq.(1) according to the rule: $\delta_n = \lambda_n\delta$ with $\lambda_n = 2$ if it refers to $A - T$ or $T - A$ pairs, $\lambda_n = 3$ otherwise, with δ free parameter to be fixed later. For simplicity in the following we consider only uniform stacking forces and uniform moments of inertia along the two strands of DNA, so we fix $k_n = \bar{k}_m = K$, $I_n = I$, $n, m = 1, \ldots N$. The equations of motion obtained from Eq.(1) are then

$$I\ddot{\psi}_n = K(\psi_{n+1} - 2\psi_n + \psi_{n-1}) - \frac{\delta_n}{2}\sin(\psi_n - \theta_n),$$

$$I\ddot{\theta}_n = K(\theta_{n+1} - 2\theta_n + \theta_{n-1}) - \frac{\delta_n}{2}\sin(\theta_n - \psi_n), \qquad (2)$$

which can be cast into an equation for the angle difference $u \equiv \psi - \theta$ between complementary bases

$$\ddot{u}_n = (u_{n+1} - 2u_n + u_{n-1}) - \beta_n\sin(u_n) \qquad (3)$$

with $\beta_n = \delta_n/K$. In this equation time has been rescaled according to $t \to (I/k)^{1/2}t$ so to leave as parameter in the equation just the ratio between anharmonicity and dispersion. Note that β_n in Eq. (3) takes only two values (say β_1 for A-T pairs and β_2 for G-C pairs). Denoting with β the average value

$$\beta \equiv < \beta_n > = \sigma\beta_1 + (1 - \sigma)\beta_2 \qquad (4)$$

with σ characterizing how often we find $A - T$ pairs in the sequence, and using the estimate of δ and K as in Ref. [10], we have a value of β of the order of 10^{-3}. This value is found to be consistent with the requirement of existence and stability of solitary wave solutions for equation (3) as shown in Ref. [1, 2]. We see that Eq.(3) in the continuum limit ($\beta << 1$) and homogeneous case ($\beta_n \to \beta$), reduces to the sine-Gordon equation with the well known kink solution

$$u(x,t) = 4\tan^{-1}[\exp(\gamma(x - vt - x_0))], \qquad (5)$$

where γ stands for the Lorentz contraction factor. In the next section Eq. (5) will be used as initial condition to integrate Eq.(3) with the $\beta's$ corresponding to a $pBR322$ fragment containing the ampicillin resistant promoters.

3. NUMERICAL EXPERIMENT AND ANALYSIS

In order to investigate dynamical effects due to the specificity of the bases, we integrate Eq. (3) with β_n values corresponding to the bases sequence of the $pBR322$ plasmid. The

whole sequence is 4361 base pairs (bp) long, but here we concentrate only on the region
relative to the ampicillin resistance gene which includes the promoters P_3 and P_1. In
the numeration given by the Genbank (in which the 0 is in the middle of the EcoRI site
and increases by moving first through the tetracycline resistant gene and then through
the ampicillin gene) this region goes from bp 3200 to bp 438 for a total of 1600 bp. To
simplify the notation, we numerate the above fragment simply from 1 to 1600 (with 1
corresponding to bp 3200 in the Genbank numeration).

From biochemical studies it is known that the RNA polymerase binds to the P_3 promoter in the region going from bp 970 to bp 1160, so we expect this region to be dynamically active. To evidence differences in soliton dynamics we have performed several
integrations of Eq.(3) with the initial position of the static kink varied along the $pBR322$
plasmid (we have taken the size of the solitary wave to vary between 30 and 40 bp, this
corresponding to a choice of β_n in Eq.(3) of the order of 10^{-3}).

In Figure 1 the position of the center of the waves versus time is reported for kinks
placed in the plasmid region going from bp 500 to bp 1300. The dotted lines refer to
trajectories starting respectively (from the left to the right) at bp $500, 690, 914, 1114, 1273$.
These trajectories represent typical behaviors of the waves along the ring fragment i.e.
they stay static or oscillate around the initial positions. Exceptions to this behavior
occur at bp 1056 and bp 1281 (i.e. respectively, at bp 4256 and bp 119 of the Genbank
numeration) in which the wave acquires a finite velocity to travel downstream toward the
ampicillin resistance gene. This is clearly seen in Figure 1 by the circled and triangled
trajectories which start, respectively, at bp 1056 and at bp 1281. It is worth to remark
that these initial conditions are just in the flanking regions of promoters P_3 (bp 1056)
and P_1 (bp 1281), i.e. just where the RNA polymerase binds to initiate the transcription
process. We also note that the direction of the waves correlates with the position of the
gene "bla" which is on the left of the promoter (i.e from about bp 1 to bp 853).

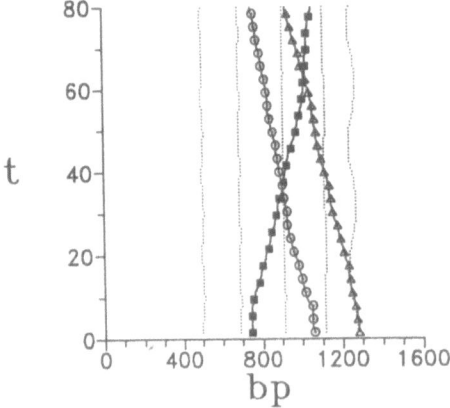

Figure 1: Space-time diagrams for the center of a nonlinear wave of Eq.3. The dotted lines
(from left to right) refer to initial conditions respectively at bp $500, 690, 914, 1114, 1273$ while
the circled and triangled trajectories refer to initial conditions respectively at bp 1056 and bp
1281. The black squared trajectory refer to a solitary wave placed in the antisense P_3 promoter
at bp 744.

A further confirm of the existence in the promoters of a dynamically active region is obtained by simulating an antisense transcription. More precisely, we construct a new *pBR322* fragment in which the region going from bp 966 to bp 1156 (containing the P_3 promoter plus flanking regions), is replaced with the region going from bp 833 to bp 966. The removed promoter is then inserted in reversed order (i.e. going from bp 1156 to bp 966) in the original place of the sequence 833 − 966. Having reversed the position of the bases in the P_3 promoter, we expect the promoter to initiates the transcription of the opposite DNA strand (antisense transcription). This implies that the wave should move in opposite direction with respect to the original *pBR322* sequence considered before. In Figure 1. the black squared trajectory refers to a wave initially placed in the antisense P_3 promoter at bp 744, from which we see that the direction of the wave is indeed reversed. This result correlates with recent biochemical studies on the P_3 antisense promoter of the *pBR322* plasmid [4].

These dynamical behaviors can be understood in terms of a chain of pendula with lighter and heavier masses, respectively, corresponding to $A-T$ and to $G-C$ base pairs. A soliton placed on the top of the heavier pendula region has a bigger rest mass than the one placed on the lighter one. On almost uniform background the solitary wave remains static or oscillates around some equilibrium position while in the transition region where the wave partially overlaps the "heavier" part of the chain there is an effective potential which puts the wave in motion. The kinetic energy acquired by the wave can be approximated by the difference between its initial and final rest mass. This analysis was performed in ref.[1] in terms of a continuous parametrically perturbed sine Gordon equation.

A more detailed description of these phenomena can be performed in terms of an effective potential for the collective variable of a kink of the discrete sine-Gordon equation on a non uniform background [11]. To this end we note that the total energy

$$E = \sum_{n=1}^{N} \frac{1}{2} \dot{u}_n^2 + \frac{1}{2}(u_{n+1} - u_n)^2 + \beta_n [1 - \cos(u_n)] \tag{6}$$

is conserved for Eq. (3). Since in our model β is very small, we may consider a solution of the discrete inhomogeneous model in the form close to (5) but with slowly varying parameters. Thus we look for (approximate) kinks in the model (3) as

$$u_n \simeq 4 \tan^{-1} \left[\exp(\sqrt{\beta}n - X) \right], \tag{7}$$

where the kink's coordinate X has a meaning of a collective variable. We can then find an effective evolution equation for this collective coordinate. In the present case this equation follows directly from the conservation law, provided radiation effects are neglected. Substituting (7) in (6) and using the relations

$$1 - \cos(u_n) = \frac{2}{\cosh^2(z_n)}, \quad \dot{u}_n = -\frac{2\dot{X}}{\cosh(z_n)}$$

$$(u_{n+1} - u_n) \simeq \frac{2\sqrt{\beta}}{\cosh(z_n)} \tag{8}$$

with $z_n = \sqrt{\beta}n - X$, we can express the energy of the kink as

$$E = \frac{1}{2}\dot{X}^2 M(X) + U(X) \tag{9}$$

where

$$M(X) = \sum_n \frac{4}{\cosh^2(z_n)}, \quad U = 2\sum_n \frac{(\beta_n + \beta)}{\cosh^2(z_n)},$$

E being a conserved quantity. To introduce an effective potential for the kink dynamics we consider as initial condition, a kink being initially at rest at $X(0) = X_0$ i.e. $\dot{X}(0) = 0$. For such a kink we have $E = U(X_0)$ and therefore Eq. (9) can be rewritten as

$$\frac{1}{2}\dot{X}^2 + W(X, X_0) = 0, \tag{10}$$

where

$$W(X, X_0) = \frac{\sum_n(\beta_n + \beta)[\mathrm{sech}^2(z_n) - \mathrm{sech}^2(z_n^{(0)})]}{2\sum_n \mathrm{sech}^2(z_n)} \tag{11}$$

and $z_n^{(0)} \equiv \sqrt{\beta}n - X_0$. Equation (10) can be considered as an equation for an-unit-mass particle moving in the effective potential $W(X, X_0)$. This potential can be used to predict the direction in which the kink will move due the nonuniformity of the chain.

To check this result we have considered Eq. (3) with λ_n values corresponding to the above 1600 bases fragment of $pBR322$.

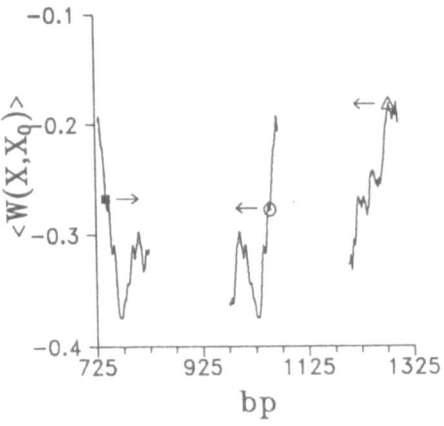

Figure 2: Effective potentials for a kink corresponding to the marked trajectories (black squares, circles and triangles) of Figure 1.

In Figure 2 the effective potentials of a kink placed in the active regions of both promoter P_3, P_1 of the original $pBR322$ sequence, and in the antisense P_3 promoter described above, are reported. The second and third curves from the left of this figure, refer respectively to the P_3 and P_1 promoters, with the initial kink placed respectively at bp 1056 and bp 1281, while the first curve on the left refers to the antisense promoter with the initial kink placed at bp 744. We see that there is a good agreement between these results and the numerical integrations reported in Figure 1.

4. CONCLUSIONS

We have used a discrete model of DNA which includes informations about the specificity of the base sequences along DNA double helices, to show the existence of two active regions for the ampicillin resistant gene in the $pBR322$ plasmid. These regions are located at bp 1054 (in promoter P_3) and bp 1281 in promoter P_1. The direction of the wave correlates with the effective direction of transcription experimentally observed in this plasmid. Furthermore we showed that for a P_3 antisense promoter, the direction of the solitary wave is reversed. This is in agreement with recent experimental results on antisense transcription in plasmid $pBR322$ [4]. In conclusion we have showed that the specificity of base sequences of DNA promoters can have dynamical effects on a solitary wave which can be important for the understanding of the transcription and regulation of the genes.

Acknowledgments

I am grateful to prof. C.Reiss and Dr. P.Lio for useful discussions and informations about plasmid $pBR322$. I also wish to thank Prof. J.A.Krumhansl for discussions and encouragements. Financial support from the Istituto Nazionale di Fisica della Materia (INFM), Italy, is also acknowledged.

References

[1] M. Salerno, Phys. Rev. A **44** (1991) 5292.

[2] M. Salerno, Phys. Lett. A **167** (1992) 43.

[3] J. Brosins, R.L.Cate, A.P.Perlmutter, J.Biol.Chem. **257** (1982) 9205.

[4] C.Reiss, private communications.

[5] S.W. Englander, N. Kallenback, A. Heeger, J.A. Krumhansl, S. Litwin, Proc. Nat. Acad. Sci. (USA), **77** (1980) 7222.

[6] S. Yomosa, Phys. Rev. A **27** (1983) 2120; **30** (1984) 474.

[7] S. Homma and S. Takeno, Prog. Theor. Phys. **72** (1984) 679.

[8] Chung-Ting Zhang, Phys. Rev. A **35** (1987) 886.

[9] L.V. Yakushevich, Phys.Lett. A **136** (1989) 413.

[10] V.K. Fedyanin and L.V. Yakushevich, Stud. Biophys. **103** (1984) 171.

[11] M.Salerno, Yu.S. Kivshar, "DNA promoters and nonlinear dynamics" ,Phys.Lett.A (to appear).

Helical geometry and DNA models

G. Gaeta

Centre de Physique Théorique, Ecole Polytechnique,
91128 Palaiseau, France
Departamento de Fisica Teorica II,
Universidad Complutense,
28040 Madrid, Spain

1. Introduction

We have several reasons to think that nonlinear excitations, and in particular solitons, could play a functional role in DNA, as it was first proposed by Englander, Krumhansl and others in 1980 [1]. In particular, these excitations could play a role in the process of DNA transcription: here the RNA-Polymerase has to open the double helix in order to read the base sequence (see e.g. the discussion by Reiss in these proceedings or [2]), so that a number of H-bonds has to be broken. This number is typically of the order of thousand H-bonds per second; the presence of a solitonic excitation travelling along the DNA double helix, i.e. of a moving region in which the double helix is unwinded, would then permit to the RNAP to read the sequence without having to provide or concentrate energy to break the H-bonds; sometimes the pictorial image of "RNAP surfing a topological soliton" is used to describe this scenario. Another process in which nonlinear excitations of a different nature (breathers) could play a functional role is that of DNA melting, as discussed e.g. in the contribution by Dauxois to this conference.

Here we will not discuss the question of the *role* of nonlinear excitations in DNA in general, but focus on some special aspect of *modelling* nonlinear excitations in DNA (having in mind mainly the problem of transcription).

Several models have been proposed to study DNA nonlinear dynamics in a quantitative way; a short review of these is given e.g. in [2]. A longer review, including a discussion of comparison with experimental data, is given by Yakushevich in [3]; see also her recent paper [4] in which the useful concept of a *hierarchy* of DNA models is discussed, and her contribution to the present volume. Particularly successfull models were proposed by Peyrard and Bishop [5] (aiming at a model for DNA denaturation) on the one side, and by Yakushevich [6] (aiming at a model for DNA transcription) on the other; these models will be referred to, respectively, as the PB and the Y model.

Such models select *one* relevant degree of freedom out of the many degrees of freedom of each unit (base plus nucleoside) of the DNA double chain, and relevant interactions between degrees of freedom of different units.

This is quite a bold reduction, and has been addressed in two different ways: on the one side, we can formulate phenomenological models, in which the one degree of freedom should actually be seen as an order parameter (in the sense of Landau theory), i.e. as providing a phenomenological description of the state of the DNA unit, adequate to the phenomenon we want to study and not aiming at a complete description of the molecule. On the other side, it has been suggested that some relevant characteristics of the state of the DNA molecule for specified processes can be described in terms of a small number of degrees of freedom.

To give a concrete example, when a base pair opens, the base are essentially rotating rigidly around a point, and the rotation takes place in a plane (see the contribution by Lavery to this conference); if we are interested in base-pair openings, it is therefore natural to attempt a description of the state of the DNA molecule in terms of a rotation angle for each base.

It should be stressed that we have no reason to think *a priori* such a simple description will work: e.g., it is known that the base opening is correlated with bending of the double chain, so that if we only consider rotational degrees of freedom we are probably missing some important ingredient. On the other side, a description in simple terms can permit to isolate which features are dominant in a specified process; after all, DNA is a very complicated molecule with very complicate structure and with very complicate tasks: maybe not all of its features are relevant to all of its tasks! Simple models of DNA dynamics bets on an economic choice of nature, i.e. suppose that for a single process only a small number of degrees of freedom are relevant. This tells nothing of how nature can make so different exigences cohexist in the same molecule, but maybe we should not be too ambitious, and try to understand only one little piece of the mosaic at a time. I will not discuss the general issue of how sensible it is to attempt a description of DNA dynamics in terms of simple models, but just consider these simple models, and actually a specific feature of them.

Our main goal here is to discuss the relevance of considering, beyond nearest-neighbour (in the DNA sequence) interactions, also other interactions, connecting bases which are half-pitch of the helix away in the DNA sequence, but near in space due to the helical geometry of DNA; such interactions are therefore also called *helicoidal*. I will discuss in which respects the introduction of these gives a qualitatively different behaviour of the model, and the possible relevance in the context of modelling real DNA processes.

2. Selection of relevant degrees of freedom

The DNA molecule is a very complex one; each unit of the molecule is made up of a nucleoside and a base; we have, *a priori*, about one hundred degrees of freedom per unit, just considering the positions of different atoms. Such a number - to be multiplied by the number of units in the DNA sequence - is a formidable obstacle to any attempt of computation of DNA dynamics.

However, we are mainly interested in the behaviour of DNA at body temperature; it happens that at such temperatures (say, of the order of $300^0 K$) most of the degrees of freedom are actually frozen, just on the basis of quantum mechanics. We can therefore consider the DNA unit as a near- rigid complex, with only a limited number of motion being actually possible at the energies at which DNA assolves its biological function. Such motions include bending of the molecule as a whole, and motions internal to each unit; the latter are essentially of two types (see the discussion in [2]): elongation of the H bonds joining pairing bases, and rotations (torsion) of the bases in a fixed plane.

There is therefore a good physical ground for considering simplified models of DNA, in which only *two* degrees of freedom per base are present; it should be stressed that there is no *a priori* reason which ensures we should be able to consider each one of these two degrees of freedom separately. It is a fact, however, that models considering only one of these degrees of freedom (elongation for the PB model, torsions for the Y model) had a good success[1] in describing the processed they are aimed at modelling, so that we are justified *a posteriori* in considering separately the two degrees of freedom.

It is remarkable - and again justified only *a posteriori* by the good success of these models - that we can make the approximation that *all the bases are equal*, both for what concerns their dynamical properties and the strenghts of their mutual interactions, and still obtain a satisfactory description of DNA dynamics[2].

3. Description of the interactions

The models of DNA dynamics mentioned above can be most simply described in terms of an Hamiltonian generating the dynamics. We will denote the degree of freedom associated to the base at site n of the chain i (here $n \in \mathbf{Z}$, $i = \pm 1$) by $x_n^{(i)}$; for the PB model $x_n^{(i)} \in \mathbf{R}$, while for the Y model $x_n^{(i)}$ is an angular coordinate, so that $x_n^{(i)} \in [0, 2\pi)$.

The Hamiltonian will be written as the sum of kinetic (T) and potential (V) energy parts

$$H[x, \dot{x}] = T + V \tag{1}$$

The kinetic energy part T, which will just be the sum of the kinetic energies associated to all the bases

$$T = \sum_i \sum_n \frac{1}{2} \mu \dot{x}_n^{(i)} \tag{2}$$

where μ is the mass (PB model) or the moment of inertia around the N-C bond (Y model) of the concerned base.

---

[1] Quite obviously, these models do not solve the problem of DNA transcription and/or melting; they are successful in that they give some right order of magnitude, and can be a good starting point in the understanding of the involved processes; i.e. they point out what could be the relevant features of DNA for the specified processes. It is remarkable that we get sensible predictions by such rough models: this suggests that indeed they succeed to capture relevant features of DNA.

[2] Actually, recent work [7] shows that attempts to refine these models by taking into account the differences among different bases result in poorer agreement with experimental observations. This surprising feature is conjecturedly [7] due to the fact that nonlinear dynamics of DNA involve collective behaviour of groups of bases, so that the dynamics itself provides a sort of self- averaging over the characteristics of the bases.

Moreover, we will have interactions among the different degrees of freedom, which will be modelled by a potential energy term V. In the DNA molecule, the main interaction terms are due to the H bonds bridging pairing bases (*transversal* interactions) and to the stacking energy among bases at consecutive sites (*stacking* interactions); such interactions will be modelled by terms V_T and V_S respectively, which will be of the form

$$V_T = \sum_n f(x_n^{(1)}, x_n^{(-1)}) \tag{3}$$

$$V_S = \sum_i \sum_n g(x_n^{(i)}, x_{n-1}^{(i)}) \tag{4}$$

Notice that, as mentioned above, we have considered an *homogeneous* DNA chain, i.e. we are considering all the bases to be equal, both for what concerns their dynamical properties (embodied in μ) and their interaction properties (embodied in the f, g).

Thus, the potential energy part of our Hamiltonian will be of the form

$$V = V_T + V_S \tag{5}$$

These models are also called *planar*; the reason for this name is that the DNA molecule - or rather the simplified lattice of degrees of freedom involved by the modellization - can be depicted as in figure 1.

As mentioned before, such a modellization does not take into account a peculiar feature of the DNA molecule, i.e. its helical geometry. Due to this, bases which are half-pitch of the helix away in the DNA sequence, are actually nearby in three-dimensional space, as the two bases pointed out in figure 2. Therefore, we expect that such bases are also interacting; in real DNA, such interactions exist and are mediated by water filaments.

These interactions, which in this context will be called *helicoidal*, are weaker than transversal and stacking ones; nevertheless, their presence introduce some *qualitative* - and not only quantitative ! - changes in the dynamics [8,9]. We are indeed going to discuss in the following how these weak terms change the "experimental predictions" of the models.

Once we introduce helicoidal interactions in the Hamiltonian, the potential energy term will have an helicoidal part V_H as well, and we write

$$V = V_T + V_S + V_H \tag{6}$$

In this case, we will say we have a *helicoidal* model, as opposed to the planar models considered above. If the pitch of the helix is 2ℓ in base units, the helicoidal term will be of the form

$$V_H = \sum_i \sum_n h(x_n^{(i)}, x_{n+\ell}^{(-i)}) \tag{7}$$

It should be mentioned that such an hamiltonian description implies there is no dissipation in DNA dynamics. This assumption has been questioned by several authors, see e.g. [10]. For what concerns travelling solitons, it should be mentioned that the mathematical theory describing solitons for dissipative systems is different from the one describing solitons for hamiltonian ones; solitons for models of the kind considered here but in the presence of dissipation have been studied by [10]; see also [11].

4. DNA hamiltonians

In order to discuss the differences induced by the helicoidal term, we should write down the actual PB and Y hamiltonians. This will also be the occasion to discuss a little further the general form of DNA models.

It should be realized that we have a well-defined hierarchy of interactions: indeed, stacking forces are quite stronger than pairing ones (a phenomenological estimate suggests a factor three for their ratio [9,12]); therefore, the system will remain near to the bottom of the stacking potential, and experience greater deviations from the equilibrium situation in what concerns the transverse potential. This means that we can keep the harmonic approximation for V_S, while we should consider anharmonic terms in V_T.

As for V_H, we have said that in real DNA this is mediated by water filaments; this means that elongations of this will result in the elongation of a number of H bonds; moreover, due to the distance among the involved bases, the movements of the bases will result in quite minor variations in their mutual distance. Thus, we expect that for V_H as well we remain in the region in which the harmonic approximation is satisfactory.

This means that we have for g and h quadratic functions:

$$g(x_n^{(i)}, x_{n-1}^{(i)}) = \frac{1}{2} K_s \left(x_n^{(i)} - x_{n-1}^{(i)} \right)^2 \tag{8}$$

$$h(x_n^{(i)}, x_{n+\ell}^{(-i)}) = \frac{1}{2} K_h \left(x_n^{(i)} - x_{n+\ell}^{(-i)} \right)^2 \tag{9}$$

where K_s and K_h are coupling constants.

As for f, this will go beyond the harmonic approximation; we will not discuss here the reasons which led Peyrard and Bishop on the one side and Yakushevich on the other one to their proposal for f, but just mention that in the PB model we have

$$f(x_n^{(i)}, x_n^{(-i)}) = D \left[e^{-\alpha(x_n^{(1)} + x_n^{(-1)})/2} - 1 \right]^2 \tag{10}$$

with D and α constant parameters (as K_T below), while in the Y model we have

$$f(x_n^{(i)}, x_n^{(-i)}) = \frac{1}{8} K_T \left[\left(4 - \cos x_n^{(1)} - \cos x_n^{(-1)} \right)^2 + \left(\sin x_n^{(1)} + \sin x_n^{(-1)} \right)^2 \right] \tag{11}$$

Notice that the harmonic terms in these are just the same; this fact will lead to having the same small-amplitude (linearized) dynamics.

5. DNA dynamics

From the Hamiltonians considered above we can easily derive the equations of motion. It is somewhat convenient, both for notational semplicity and for the sake of analytical computation, to pass to the continuum approximation[3]. This means that the arrays $x_n^{(\pm 1)}(t)$ of coordinates are substituted by two fields $\phi^\pm(x,t)$, the correspondence being given by

$$\phi^\pm(n\delta, t) = x_n^{(\pm 1)}(t) \tag{12}$$

where δ is the x distance between successive sites; this approximation makes sense only if we obtain solutions with scale lengths λ large compared to δ, as it will be the case. Actually, the soliton solutions we are interested in should have a size, by experimental observations, of the order of tens of site units, so that such an approximation is reasonable.

In discussing the dynamics, it is also convenient to pass to the fields

$$\psi^\pm(x,t) = (1/2)[\phi^+(x,t) \pm \phi^-(x,t)] \tag{13}$$

With these, the field equations for (the continuum version of) our Hamiltonian are

$$\psi_{tt}^\pm = \kappa_s \triangle \psi^\pm \pm \kappa_h \mathcal{W}_\mp(\psi^\pm) - \kappa_t b^\pm(\psi^+, \psi^-) \tag{14}$$

with $\kappa_s, \kappa_h, \kappa_t$ parameters which are functions of the $\mu, K_T, K_S, K_H, \delta$ [14]; the $\mathcal{W}_\pm$ are linear nonlocal operators given by

$$[\mathcal{W}_\pm(\varphi)](z,t) = \varphi(z+w,t) \mp 2\varphi(z,t) + \varphi(z-w,t) \qquad w \equiv \ell\delta \tag{15}$$

and for the Y model we have $b^+ = \sin\psi^+ \cos\psi^-$; $b^- = \sin\psi^-(\cos\psi^+ - \cos\psi^-)$.

We are mainly interested in two features of these equations: 1) their dispersion relations (i.e. the small amplitude dynamics); 2) the soliton solutions.

6. Dispersion relations, and breather solutions

The small amplitude dynamics is governed by (14) once we linearize the nonlinear terms $b^\pm$ (as mentioned above, we obtain the same linearization for the Y and for the PB models). The dispersion relations are obtained by looking for solutions of the form of Fourier waves $A_{q,\omega}|q, \omega>$, where $|q, \omega>= \exp[i(qx + \omega t)]$

Now we get an explicit formula for the action of $\mathcal{W}_\pm$ on $|q, \omega>$, and we have that solutions of this form are possible provided the following dispersion relations are satisfied (the subscript $\pm$ refers to the $\psi^\pm$ fields):

$$\begin{aligned}
\omega_+^2(q) &= (1/\mu)[(\kappa_s \delta^2)q^2 + 2\kappa_t + 4\kappa_h \sin^2(qh\delta/2)] \\
\omega_-^2(q) &= (1/\mu)[(\kappa_s \delta^2)q^2 + 4\kappa_h \cos^2(qh\delta/2)]
\end{aligned} \tag{16}$$

---

[3] It has been remarked that in many case the discreteness can be very relevant to the behaviour of the models and to the processes in DNA, see e.g. the contributions by Dauxois, Peyrard and Salerno to these proceedings, and references therein; see also [13]. The discussion presented in the following for the continuum approximation could be repeated keeping the discrete description, giving similar results; in this framework the continuous setting requires simpler notation and computations.

The curves corresponding to these dispersion relations are plotted (for physically significant values of the parameters [2,9,14]) in figure 3 for the planar Hamiltonian (i.e. for K_H and therefore κ_h equal to zero), and in figure 4 for the helicoidal Hamiltonian. There is a clear qualitative difference among the two cases, despite the small ratio $\kappa_h/\kappa_s \simeq 0.1$.

A first clear difference consists in the fact that in the planar case we have an *acoustic branch*: i.e., for $q \to 0$, we have $\omega_- \to 0$. Thus, we have phonon excitations, and long wavelength modes can be excited with very small energy. In the helicoidal case, instead, this phonon mode disappear[4], as the V_H term favours excitations with wavelength $\ell\delta$ (indeed, the minimum for the $\omega(q)$ is reached for $q \simeq \ell\delta$) [14].

It should be recalled that the nonlinear modes building up from these equations will have length scale near to the wavelengths corresponding to the minima of the dispersion relations, so that these minima also give us an information about the length scale of the nonlinear excitations present in the model. By the same argument, other physical quantities relative to modes q corresponding to minima of the dispersion relation curves, give informations about the observable quantities.

Predictions of these for the helicoidal models are reported in [2,14]; these should be compared in particular with experimental data, according to which the length scale of the transcription bubble (or the denaturation promoters, i.e. breathers) is of the order of tens of bases, and the oscillation times for breathers is of the order of picoseconds. The data corresponding to minima of the dispersion relations for the helicoidal PB model (obtained for an apprpriate choice of parameters [9]), show good agreement with these experimental figures [2,14]. Further numerical investigations, taking into account the full nonlinear dynamics [13], show that the agreement goes beyond these semiqualitative considerations, and indeed helicoidal models reproduce quite satisfactorily[5] the experimentally observable features of DNA dynamics.

In the case of the helicoidal Y model, the results from linear dynamics apply less directly to the nonlinear excitations of interest, i.e. travelling soliton; but we notice however that the length scale corresponding to the minima of dispersion relations [14] does again fit the length scale of the transcription bubble. It should be stressed that here the parameters are chosen following independent computations [12], and are *not* considered as phenomenological parameters to be fitted in order to meet experimental observations [14].

7. Soliton solutions

I would like to stress that the question of helicoidal interactions is not directly related to the issue of travelling solitons, although it was first raised [8] in the context of a model [6] aiming at the study of such objects.

However, we are also interested in (travelling) soliton solutions for the Y model; these are solutions of the form

$$\psi^{\pm}(x,t) \equiv \psi^{\pm}(x - vt) \tag{17}$$

---

[4] Obviously, this concerns only the (rotational) degrees of freedom described by the model: phonon modes corresponding to the overall configuration of the DNA molecule (bending) are not described by such models, and cannot disappear - neither appear - in them.

[5] Again, satisfactorily in the sense of footnote 1 above; the agreement is even more remarkable if we consider all the approximations involved in the formulation of the models.

which should moreover, to be physically meaningful, have *finite energy*. Inserting the ansatz (17) into the field equations for the Y model, we obtain the equation for Y solitons [15]. The finite energy condition requires that for $x \to \pm\infty$ we have $\nabla\psi^\pm(x,t) \to 0$ (to control the kinetic energy term); and that $\psi^+(x,t) \to 2\pi n_\pm^+$; $\psi^-(x,t) \to 2\pi n_\pm^-$ for integers n (to control the potential energy contribution). The integers $\nu^\pm = (n_+^\pm - n_-^\pm)$ control how many time the fields $\psi^\pm$ wind around the circle, and also provide a topological classification [15,16] of the soliton solutions; i.e., for any pair of integers (ν^+, ν^-) we have a soliton solution.

In the case of simplest soliton solutions, i.e. those involving only one field, we can obtain exact solutions for the planar Y model; these are given respectively by

$$\psi^+ = \arccos\left[\frac{\sinh^2(Az) - 1}{\sinh^2(Az) + 1}\right] \;\;;\;\; \psi^- = 0 \text{ and } \psi^- = \arccos\left[\frac{(Bz)^2 - 1}{(Bz)^2 + 1}\right] \;\;;\;\; \psi^+ = 0 \;\; (18)$$

for the (1,0) and the (0,1) soliton; here $z \equiv (x - vt)$, and we have taken the solution to be centered in zero. Notice that v is here a free parameter, only subject to the condition [15] $v^2 \le v_{max}^2 \equiv (K_S\delta^2/\mu)$. For the physical values of parameters [2,9,14], v_{max} is of the order of 10^6 bases/second, i.e. several orders of magnitude above[6] the transcription speed v_0, which is of the order of 10^3 b/s. The energy of the soliton depends on v (with $\gamma = v/v_{max}$ and A, B numerical parameters, we have $E(\gamma) = A/(B - \gamma^2)$), but this dependence is very weak at $v \simeq v_0$ [15]. In the case of helicoidal models, we can proceed numerically; it appears [15] that the presence of helicoidal terms does not affect significantly neither the size neither the energy dependence (on speed) of soliton solutions.

8. Ising model approach

Apart from the "mechanical" models considered so far, another possible, and quite natural, modellization of DNA is in terms of an Ising model; the two states of each "spin", i.e. of each base pair, correspond then to the state, "open" or "closed", of the H bond bridging the base pair. Such a model is also particularly convenient as the most accessible experimental information is indeed the average number of open base pairs (through measures of UV absorbance [12,17]). In this case, the helicoidal terms correspond to introducing an interaction between spins n and $n + \ell$. In the planar case, such a model can be dealt with in the context of transfer matrix method; this can be generalized to study the helicoidal case too, by considering transfer matrices of higher dimension. This kind of approach is discussed in [18]; here we just mention that, as already observed for the mechanical models, the introduction of helicoidal terms lead to a strongest cooperative behaviour, i.e. to a steepest melting transition; at this stage I am not able to make quantitative statements about the relevance of this increase in steepness.

---

[6] It should be mentioned, anyway, that v_{max} is curiously coinciding with the "speed" of random search for identification of promoters in the phase preceding the starting of transcription, as mentioned by Reiss in his talk to the conference.

9. Conclusions

We have considered nonlinear models of DNA dynamics, such as those proposed by Peyrard and Bishop [5] or by Yakushevich [6]. After briefly discussing the schematization of DNA which leads to such models, we focused on the interaction terms which should be introduced in the models [8,9] to take into account the helicoidal geometry of DNA. Numerical computations [15] show that the soliton solution of the Y model are very little sensitive to the presence of such interactions. On the other side, the dispersion relations, and therefore the small-amplitude dynamics (identical for the Y and the PB model), are modified - not only quantitatively but also qualitatively - by the introduction of such terms. The Y and PB models, with the introduction of "helicoidal terms", give predictions for observable quantities [14] (length scale of structures, time scale and amplitude of oscillations) of the same order of magnitude as observed in experiments, suggesting that they consider and capture relevant features of DNA dynamics.

References

[1] S.W. Englander, N.R. Kallenbach, A.J. Heeger, J.A. Krumhansl and S. Litwin, *Proc. Nat. Acad. Sci. USA* **77** (1980), 7222

[2] G. Gaeta, C. Reiss, M. Peyrard and Th. Dauxois: "Simple models of nonlinear DNA dynamics", *Riv. N. Cim.* **17** (1994), issue n.4

[3] L.V. Yakushevich: "Nonlinear dynamics of biopolymers: theoretical models, experimental data", *Quart. Rev. Biophys.* **26** (1993), 201

[4] L.V. Yakushevich: "Nonlinear dynamics of DNA: hierarchy of the models", to appear in *Physica D* (1994)

[5] M. Peyrard and A. Bishop, *Phys. Rev. Lett.* **62** (1989), 2755

[6] L.V. Yakushevich, *Phys. Lett. A* **136** (1989), 413

[7] G. Gaeta, "A realistic Y model for DNA dynamics; and selection of soliton speed", to appear in *Phys. Lett. A* (1994)

[8] G. Gaeta, *Phys. Lett. A* **143** (1990), 227

[9] Th. Dauxois, *Phys. Lett. A* **159** (1991), 390

[10] V. Lisy and V.K. Fedyanin, *J. Biol. Phys.* **18** (1991), 127

[11] L.V. Yakushevich, *Studia Biophys.* **121** (1987), 201

[12] W. Saenger, *"Principles of Nucleic acid Structure"*; Springer (Berlin), 1988

[13] Th. Dauxois, M. Peyrard and C.R. Willis, *Physica D* **57** (1992), 267; Th. Dauxois, M. Peyrard and A.R. Bishop, *Phys. Rev. E* **47** (1993), R44 and 684; Th. Dauxois and M. Peyrard, *Phys. Rev. Lett.* **70** (1993), 3935; M. Peyrard, Th. Dauxois, H. Hoyet and C.R. Willis, *Physica D* **68** (1993), 104

[14] Gaeta, *Phys. Lett. A* **172** (1993), 365

[15] G. Gaeta, *Phys. Lett. A* **168** (1992), 383

[16] A. Dubrovin, S.P. Novikov and A. Fomenko, *"Modern Geometry"*; Springer (Berlin), 1984

[17] D. Pörschke, *Biopolymers* **10** (1971), 1989

[18] G. Gaeta, "Ising model on a helix, and the statistical mechanics of DNA melting"; in preparation

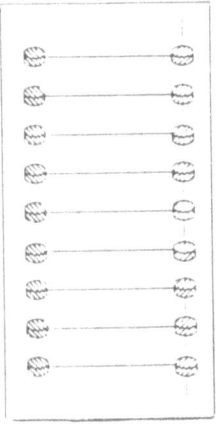

Figure 1

Schematic view of planar
DNA models.

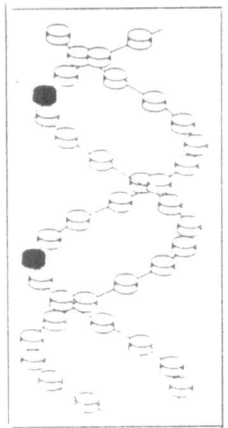

Figure 2

Helicoidal DNA models;
a pair of bases interacting
via the helicoidal term is
pointed out.

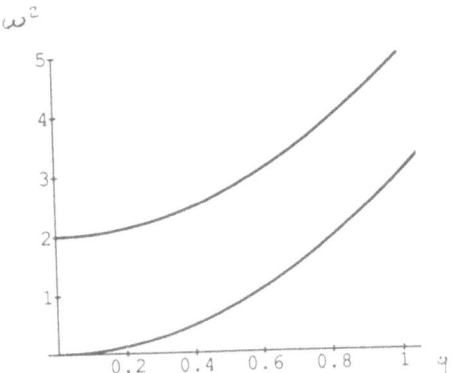

Figure 3

Dispersion relations for the planar
Y model; see text.

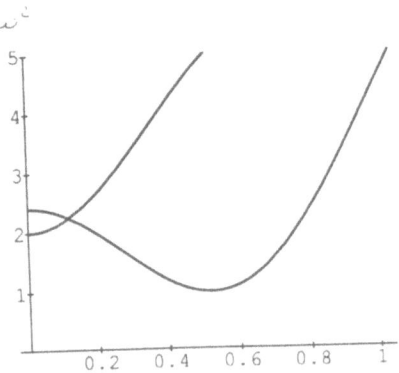

Figure 4

Dispersion relations for the
helicoidal Y model; see text.

Nonlinear localized excitations and the dynamics of H-bonds in DNA

S. Flach[(1)] and C.R. Willis

Department of Physics, Boston University,
590 Commonwealth Avenue,
Boston, Massachusetts 02215, U.S.A.

1. INTRODUCTION

The functioning of biomolecules has been a field of very extensive research in modern biology. One of the interesting questions e.g. is the understanding of transcription and copying of DNA molecules. In this contribution we apply recent results we obtained for vibrational energy localization in (arbitrary) nonlinear Hamiltonian lattices to models which might reflect parts of the dynamics of H-bonds in DNA molecules. We show using an abstract level of analysis that there exist generic features of whole classes of models with respect to the existence of (very) localized vibrational excitations. By analyzing the properties of energy (heat) flows in the presence of these localized vibrations we show the localized vibrations to control the heat flow, and thus can define functioning on the level of these physical models. Let us however at the beginning spend a few paragraphs in order to express our views on the advantages and limitations of theoretical physics approaches to the initial problem of functioning in biomolecules.

The formulation of the problem of functioning of biomolecules (let us for clarity constrain ourselves to DNA transcription) in terms of theoretical physics amounts to describing specific nonequilibrium processes of a complex system of many constituents ('particles'). These processes might correspond to certain (nonlinear) dynamic processes within a choosen physical model. Still measurements of the biological system might reveal certain equilibrium properties as well (whatever their connection to the functioning is). A good example is denaturation of DNA, which reminds us of a first order phase transition. Thus we see two approaches in which abstract theory of many-particle systems might help to understand DNA functioning. Both of them start with certain model classes known to be good abstractions of different realizations of condensed matter (here the crucial point is the choice of 'reductionism' of a certain biological constituent to a 'particle', e.g. assuming that a base of a DNA-strand is a rigid body, connected to its neighbours by certain bonds). The first approach leads us to the definition of necessary conditions our models have to have in order to say reveal certain equilibrium properties of DNA. A good example is the work of M. Peyrard and coworkers [1,2], who have studied the denaturation

[(1)] *present address*: Max-Planck-Institut für Physik Komplexer Systeme, Bayreuther Str. 40 H.16, 01187 Dresden, Germany

of DNA. After the answer is found (with respect to the necessary conditions) we will have some restrictions on the choice of models and parameters.

The second approach starts again with the choosen model classes (or it might also start from the subclass of models obtained from the analysis using the first approach) and poses the question: 'what are the typical nonlinear excitations my class of models posess, and how can the presence of these excitations in the original DNA molecule contribute to functioning?'. Let us emphasize that the two approaches are not alternative - in fact after obtaining the results from the first approach one has to proceed with the second approach anyway. In other words, the first approach yields certain constraints (with respect to the choice of a model) which have to be obeyed in carrying out the second approach. In order to complete the second approach (assuming that we found some typical or generic nonlinear excitations or processes) a nontrivial feedback with experimental knowledge of functioning of DNA is needed.

In fact there exists a third way, which is solely computer simulation of models closer and closer to the original DNA. Besides the always present computer-assisted limitations, this approach is fundamentally different from the first two described ways, because it does not pretend to understand a physical phenomenon and then to find its trace in the biological system (this will be simply impossible because of the complexity of the biomolecule and its functioning). Still this approach might be as (or even more) successful in answering questions about functioning of say DNA than the first two ways.

In the following we will present our results following the second approach, and of course we will have to stop at the end because of lack of the mentioned feedback. Still we hope that our results and thoughts will help to interprete experimental data and to focus and direct future experimental investigations.

2. H-BOND DYNAMICS

2.1 Model classes

As indicated above, we will focus on the dynamics of H-bonds between two bases of a basepair in DNA. The simplest approximation of this problem might be to consider the bases as rigid masses, connected with their second partner of a basepair through an interaction potential $V(z)$ which provides bound states, and connected to their neighbour bases along the strand through a second interaction potential $\Phi(z)$. We assume discrete translational invariance of the system (which will not be crucial, but demonstrates the novelty of the results in the best way), i.e. we exclude from our considerations differentiation between different bases and/or disorder. For the moment we consider nearest neighbour interaction along the strand and assign to every rigid mass one degree of freedom $Q_{\pm 1,l}$ where ± 1 indentifies the two different strands and l counts the bases along the strand (Q could be a rotation angle or just a scalar displacement). The interaction potentials are assumed to be given by differences of the involved degrees of freedom $(Q_{+1,l} - Q_{-1,l})$ and $(Q_{\pm 1,l} - Q_{\pm 1,l+1})$. Treating the rigid masses classically we use a transformation of the original base coordinates and separate the Hamiltonian H into a part H_s which describes solely the dynamics of the sum coordinate $Y_l = Q_{+1,l} + Q_{-1,l}$ and a part H_d which describes the dynamics of the difference coordinate $X_l = Q_{+1,l} - Q_{-1,l}$ (for details see [1]):

$$H_s = \sum_l \left[\frac{1}{2}\dot{Y}_l^2 + \Phi(Y_l - Y_{l-1}) \right] , \tag{1}$$

$$H_d = \sum_l \left[\frac{1}{2}\dot{X}_l^2 + V(X_l) + \Phi(X_l - X_{l-1}) \right] . \tag{2}$$

The two interaction potentials are given through Taylor expansions around the stable groundstate of the system:

$$V(z) = \sum_{n=2,3,...} \frac{1}{n!} v_n z^n , \quad \Phi(z) = \sum_{n=2,3,...} \frac{1}{n!} \phi_n z^n . \tag{3}$$

We have to choose a potential $V(z)$ which should describe the H-bond energy. The physical requirement is that $V(z)$ has a bound state (minimum in z) and unbounded states (both because we can break an H-bond by using a finite amount of energy and because of the fact of the denaturation transition). Consequently $V(z)$ will have an energy barrier which separates the bound states from the unbound ones. If the function $V(z)$ is infinitely often differentiable, then it follows that the frequency of oscillation of an imaginary particle in this potential will equal $\sqrt{v_2}$ for infinitely small energies and decrease to zero (continuously) as the energy is increased up to the barrier height.

If we would consider only small amplitude oscillations of our variables X_l around the groundstate $X_l = 0$ of the system, a linearization of the equations of motion $\ddot{X}_l = -\partial H_d/\partial X_l$ yields phonon lattice waves as exact solutions (within the linearization) with the dispersion relation $\omega_q^2 = v_2 + 4\phi_2 \sin^2(q/2)$, $0 < q < \pi$. Here q is the phonon wave number. Thus we find an optical phonon band with the property $v_2 \le \omega_q^2 \le (v_2 + 4\phi_2)$.

2.2 Nonlinear localized excitations

These ingredients are sufficient for us in order to make the following statements. System (2) supports sets of one-parameter families of time-periodic nonlinear localized excitations (NLEs) of the form

$$X_l(t) = X_l(t + 2\pi/\omega_1) , \quad X_{l \to \pm\infty} \to 0 , \quad 0 < \omega_1 < \sqrt{v_2} . \tag{4}$$

The parameter of each family is the fundamental frequency ω_1. Time-periodic NLEs (4) are *exact* solutions of the equations of motion if the following condition is met: the nonresonance condition $k\omega_1 \ne \omega_q$ has to be fulfilled (here k is an arbitrary integer and ω_q is an arbitrary phonon frequency). This condition can be certainly fulfilled if $\phi_2 \ll 3/4v_2$ and if $\phi_{n\ge 3}$ are zero or small enough. The parameter region of applicability of our statement might be however even larger. The following reasons can be used in support of our statement. The task of finding time-periodic NLEs can be transformed into the task of solving a set of algebraic equations for the Fourier components A_{kl}: $X_l(t) = \sum_k A_{kl} \exp(ik\omega_1 t)$ with $k = 0, \pm1, \pm2, ...$ It was shown that in terms of certain separatrix manifolds the problem is generically solvable if the nonresonance condition is met [3]. Furthermore we have shown that the nonresonance condition can be obtained considering certain effective potentials of the original problem [4]. Finally we mention a recent NLE existence proof by MacKay and Aubry for the case of infinitesimal weak coupling $\Phi(z)$, where an analytical continuation from the uncoupled case $\Phi(z) = 0$ was carried out [5].

The time-periodic NLEs decay exponentially fast in space provided $\phi_2 > 0$ [6]. They are exact solutions and thus never 'decay' in time without perturbations. As for perturbations, we have shown that as long as the perturbations are not too strong, time-periodic NLEs can be locally perturbed into quasi-periodic NLEs (meaning that the time-dependence of a given variable is quasiperiodic). These quasiperiodic NLEs are not exact solutions of the equations of motion [6]. Quasiperiodic NLEs will radiate energy and transform either into time-periodic NLEs or follow more complex scenarios [4]. Still quasiperiodic NLEs are meaningful objects, since their lifetimes can easily exceed 10^6 periods of oscillations [4]. The effects of extended perturbations on time-periodic NLEs will be discussed in detail below.

Time-periodic NLEs can be excited on any lattice site. Because of their robustness to perturbations these excitations can be easily observed in finite temperature simulations of example systems [1,7,8]. One could view NLEs as generalized quasiparticles with finite lifetime (at finite temperatures). On short time scales the quasiparticles behave as being uncoupled. On large time scales the coupling introduces relaxations, which lead to creation and annihilation of NLEs at different lattice sites.

Recently Peyrard and coworkers have proposed a modified model [2,9], which is different from (1),(2) in that the interaction potential Φ contains terms $(X_l + X_{l-1})$ in (2). Analysis of (2) yields then a remarkable change in the thermodynamic properties of H_d - the denaturation transition becomes exactly a first-order phase transition, whereas without these new terms the denaturation transition is only approximately of first order. As it emerges from the analysis [9], the change in the properties of the transition is sharp, even if the additional new terms are switched on perturbatively. Consequently the properties of single excitations will be only smoothly altered with variation of the perturbation. Thus at least for weak enough perturbations the NLE solutions will survive. Indeed numerical simulations [2] have shown that virtually no changes in the dynamical properties of the system appear as long as one does not consider the transition energy (or temperature) itself. Especially NLE structures have been observed both with or without the additional terms.

Estimations for the parameters of our model can be found in the work of Prohofsky [10] and yield $\phi_2/v_2 \sim 0.03...0.05$. These parameter ranges certainly support the existence of very localized NLEs [4]. In terms of the original physical interpretation of our model this means that DNA dynamics supports localized excitations of H-bonds between bases from a base pair. These excitations will have exponential decay in space (along the strands). By itself this statement does not imply any biological interpretation of our results. Of course we could speculate that an existing NLE might act as a precursor of a local DNA-opening. In fact it is even possible that NLEs exist where one H-bond is periodically opened and closed. Still the functioning of DNA is strongly controlled by certain enzymes. Consequently the most important question is whether it is possible to control the properties of the system in the controlled presence of NLEs. In other words, we think it might be much more reasonable to assume that enzymes interact with DNA strands and create NLEs for special purposes. If this hypothesis has some truth, then we

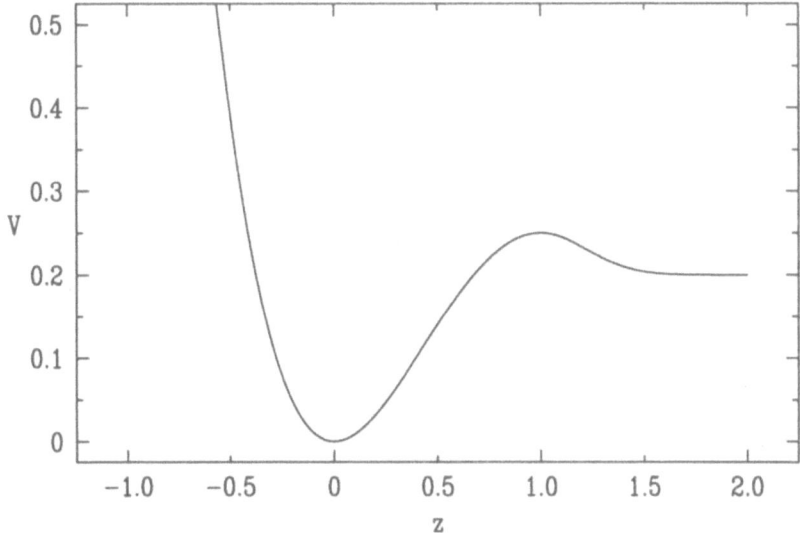

Fig. 1: Dependence of the onsite potential V on z (cf. Eq.(5)).

indeed can assign biological functioning to an object which is described within nonlinear physics of many particles.

3. FUNCTIONING WITH NLEs - PHONON SCATTERING

We can roughly divide the possible functions an NLE fulfills into two classes - biological and physical. The first one is connected with specific chemical and biological processes and is certainly not accessible within our approach. What about the second possibility? Since our toy model is very simple, so are the possible properties of the system. One obvious property is energy or heat transport along the chains (strands). In the following we will show that the presence of an NLE can indeed control the heat flow and thus create a stationary nonequilibrium situation which is not accessible within a usual statistical approach (e.g. of finite temperature simulations).

Let us confine ourselves without any limitation to a simple realization of model (2), such that the details of our results become more concrete. The onsite potential $V(z)$ is choosen as

$$V(z) = \begin{cases} V_{\Phi 4}(z) = \frac{1}{4}((z-1)^2 - 1)^2 & , \quad z \leq 1 \\ V_G(z) = \alpha e^{-\beta(z-1)^2} + \kappa & , \quad z \geq 1 \end{cases} \tag{5}$$

To guarantee smoothness at $z = 1$ up to (including) third derivative we use $0 \leq \kappa \leq 0.25$, $\alpha + \kappa = 0.25$, $2\alpha\beta = 1$. Variations of parameter κ are not essential in the following, since we are not interested in the details of say the denaturation transition. The potential

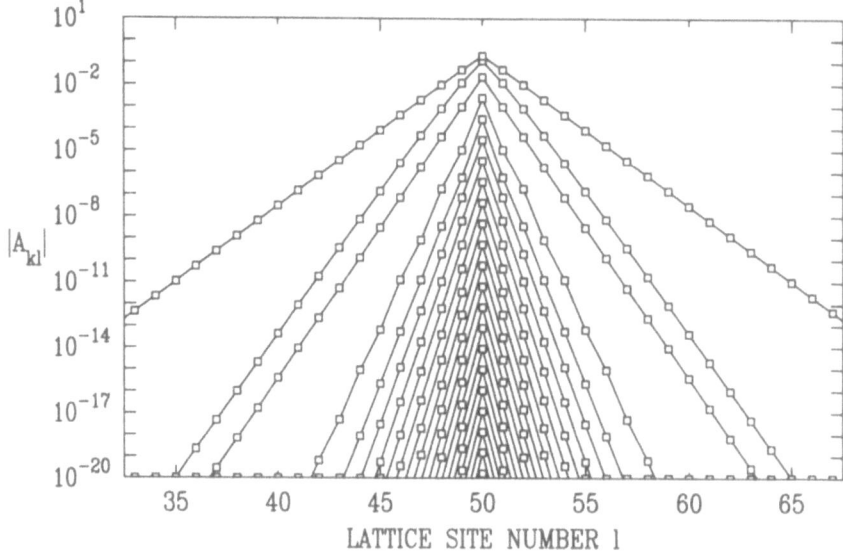

Fig. 2: Numerical result of a time-periodic NLE solution with frequency $\omega_1 = 1.3$ for the model parameters as indicated in the text. The absolute values of the Fourier coefficients A_{kl} of the periodic displacements $X_l(t)$ are plotted on a logarithmic scale versus the lattice site l. The actual data are represented by open squares. Data for equal Fourier numbers k are connected with solid lines. The Fourier numbers from top to bottom are ($k = 1, 0, 2, 3, 4, 5, ...$).

$V(z)$ is shown in Fig.1 for $\kappa = 0.2$. The parameter v_2 (cf. (3)) is equal to $v_2 = 2$. We choose $\phi_2 = 0.1$, $\phi_{n \neq 2} = 0$ in Eq.(3), so that we are in the parameter range as given in [10]. In that case numerical and analytical studies have revealed the existence of NLE solutions [4,11]. An example for a time-periodic NLE solution is shown in Fig.2 by plotting the Fourier coefficients of every displacement variable $X_l(t)$ versus lattice site. The logarithmic decay in space is clearly observed, in perfect agreement with analytical predictions [6]. These data were obtained using a specific discrete mapping which will be described elsewhere. Quasiperiodic NLE solutions have been also reported [4,11]. Their existence can be visualized with the help of Poincare maps [4]. The properties of NLEs can be studied with the help of reduced problems and effective potentials [4].

Let us turn to the question of heat flow control in the presence of a periodic NLE. We create a plane wave of phonons with a given wave number q and small amplitude. We want to measure the transmission coefficient of such a wave by a time-periodic NLE. Since the wave amplitude is assumed to be small, we can account for the scattering problem by adding small perturbations $\delta_l(t)$ to a time-periodic NLE solution $X_l(t)$ and linearizing the equations of motion with respect to the perturbations:

$$\ddot{\delta_l} = -\sum_{l'} \frac{\partial^2 H_d}{\partial X_l \partial X_{l'}} |_{X_l = X_l(t)} \delta_{l'} \quad . \tag{6}$$

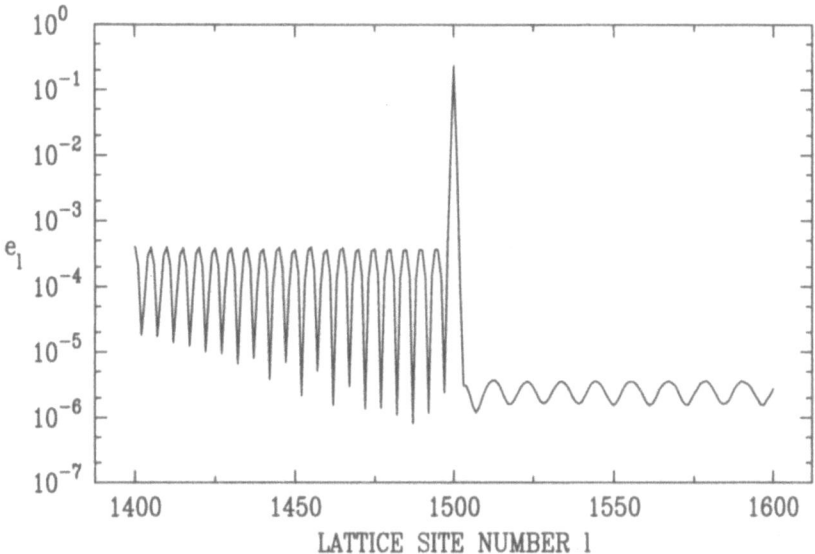

Fig. 3: Discrete energy density distribution versus lattice site l of a scattering experiment after a waiting time of $T = 12000$. The infalling phonon wave has energy density $e_l = 10^{-4}$ and wave number $q = 0.2\pi$. The NLE is positioned at $l = 1500$.

Let us first mention, that even if the nonresonance condition for the time-periodic NLE is met, we can not be sure that the NLE is stable in the presence of phonons. As a stability analysis with respect to phonon perturbations shows [4,11], stability of a time-periodic NLE is given if $k\omega_1 \neq 2\omega_q$. In the case of a stable NLE we expect the process of phonon scattering at an NLE to be elastic. The scattering of a phonon wave (6) is then equivalent to the scattering of a single electron in a tight-binding model [12] with a localized array of time-dependent diagonal defects. We do not know at present how to analytically treat this problem of time-dependent multiple scattering. Consequently we will present numerical data of the scattering. We use a time-periodic NLE with energy $E = 0.256$ and frequency $\omega_1 = 1.177$. The phonon waves are created on one side of the NLE and have initial energy density 10^{-4} per lattice site. After an initial waiting period (in order to make sure that phonon packets with other wave number than the desired one travel away) we measure the transmitted wave over a time period of $T = 12000$. The system consists of 3000 lattice sites which is sufficiently large that we can exclude reflections from the boundary. The ratio of the transmitted to the incoming energy densities (squared transmission coefficient $|t|^2$) is then measured. In Fig.3 we show the discrete energy density distribution e_l as a function of the lattice site number l after waiting time $T = 12000$:

$$e_l = \frac{1}{2}\dot{X}_l^2 + V(X_l) + \frac{1}{2}(\Phi(X_l - X_{l-1}) + \Phi(X_l - X_{l+1})) \ .$$

Here the NLE is positioned at lattice site $l = 1500$ and the incoming phonon wave ($l < 1500$) with $q = 0.2\pi$ travels to the right and is scattered by the NLE. First we find

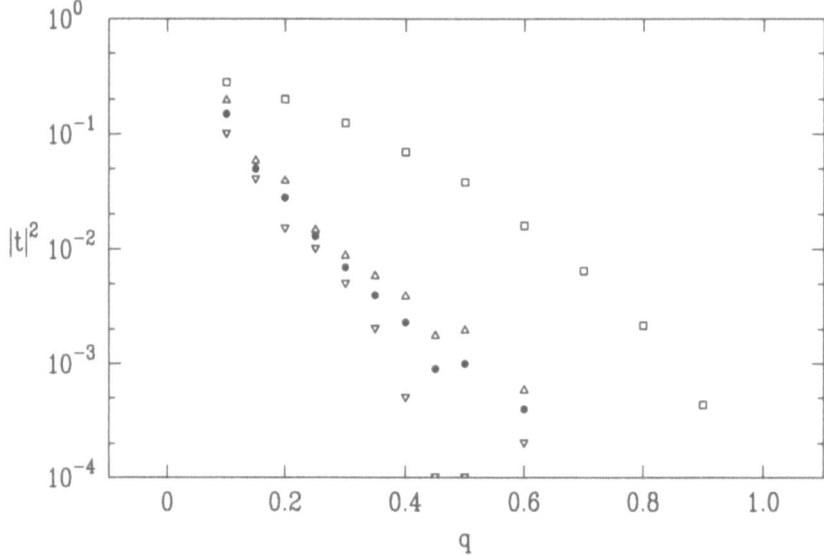

Fig. 4: The squared absolute value of the transmittion coefficient $|t|^2$ as a function of wave number q of the infalling phonon wave. The open triangles show the maxima (up) and minima (down) of $|t|^2$, the filled circles show the time-averaged value of $|t|^2$. The open squares correspond to the static linear Ersatzproblem.

that the transmitted wave energy density is oscillating in time (as it should be because of the oscillating time-dependence of the diagonal defects in the scattering problem) (see Fig.3, $l > 1500$). Secondly we find that the reflected wave does not loose its coherence with the incoming wave - since we observe standing waves due to the interference between incoming and reflected waves (see Fig.3, $l < 1500$). These standing wave structures are indeed stationary, as analogous pictures for different times clearly show. Note that the ordinate in Fig.3 is plotted on a logarithmic scale. Since the corresponding wave length of the phonon is $\lambda = 10$, one should expect the distance between two neighbouring knots of the standing wave (minima in the energy density) to be $\lambda/2 = 5$. That is precisely observed in Fig.3 ($l < 1500$).

But the most important information is the dependence of the squared transmission coefficient on the wave number of the incoming wave as shown in Fig.4. We observe a dramatic decrease of $|t|^2$ over orders of magnitude with increasing wave number q. Only for very small wave numbers (large wavelength) do we find appreciable amounts of transmitted wave intensity. This result might appear to be obvious - since large wavelength phonons do not 'feel' the finite size perturbation by the NLE. This is however wrong, as the simple example of *one* static diagonal defect shows - the transmittion coefficient for $q = 0$ is zero as well as for $q = \pi$ (because of the singularities in the phonon density of states) [12].

If we replace the time-dependent defects in equation (6) by their time-averaged values, we obtain a linear Ersatzproblem. The numerical investigation of the phonon scattering for this Ersatzproblem yields the open squares in Fig.4. Allthough the qualitative q-dependence seems to be reproduced, the quantitative difference between the time-dependent and the time-independent scattering problems is huge. Consequently we are confronted with a rather subtle result of time-dependent multiple scattering events, which leads,as the numerical result clearly shows, to the fact that phonons with almost any wave number are strongly reflected by a time-periodic NLE. Thus we can prevent the heat flux from penetrating certain areas of our chain by creating two NLEs on the borders of desired region or conversely we can trap energy in a region between two NLEs.. This NLE creation does not require an overall rearrangement of the chain as in the case of a kink but only a local excitation of one or a few degrees of freedom is needed.

If we consider larger interaction ranges than nearest neighbour interaction, we do not expect that the NLE properties will change drastically, as long as we have finite range interaction. As analytical considerations indicate [3,6], there is no basic difference between say next nearest neighbour interaction and nearest neighbour interaction with respect to the existence of NLEs.

Let us return to the problem of DNA functioning. One possible role for NLEs in DNA functioning could be an enzyme which could create NLEs in order to prevent certain areas of DNA from being penetrated by heat flux or to store energy in certain areas. This could be an example of functioning of DNA describable on the level of nonlinear physics of many particles. Of course there can be many more possibilities for functioning using NLEs, however to describe these we would have to leave the abstract level of nonlinear physics and become much more specific in our understanding of the system. It makes no sense to speculate about these possibilities without experimental evidence.

Acknowledgements

We thank E.W. Prohofsky (Purdue University) for useful discussions and drawing our attention to Ref.[10] and M. Peyrard (E.N.S. Lyon) for many interesting discussions.

References

[1] Dauxois T., Peyrard M. and Bishop A.R., Phys. Rev. **E47** (1994) 684-695.

[2] Dauxois T., Peyrard M. and Bishop A.R., Phys. Rev. **E47** (1994) R44-R47.

[3] Flach S., preprint (1994).

[4] Flach S., Willis C.R. and Olbrich E., Phys. Rev. **E49** (1994) 836-850.

[5] MacKay R.S. and Aubry S., Warwick Preprints **20** (1994).

[6] Flach S., Phys. Rev. **E** (1994) in press.

[7] Flach S. and Siewert J., Phys. Rev. **B47** (1993) 14910-14922.

[8] Flach S. and Mutschke G., Phys. Rev. **E49** (1994) 5018-5024.

[9] Peyrard M. and Dauxois T., preprint (1994).

[10] Prohofsky E.W., Comments Mol. Cell. Biophys. **2** (1983) 65-85; for recent results see Chen Y.Z. and Prohofsky E.W., Phys. Rev. **E49** (1994) 873-881 and references therein.

[11]Flach S. and Willis C.R., Phys. Lett. **A181** (1993) 232-238.

[12] Economou E.N., Green's Functions in Quantum Physics (Springer Verlag, Berlin, 1990) pp. 71-127.

Chapter II

Proteins, conformation and dynamics.

Besides nucleic acids, proteins are extremely important biological molecules because they do all the work necessary for the functioning of living systems. Some of the problems posed by the understanding of proteins function are similar to the problems encountered for DNA, but new questions appear because of the richness of the configurational space of proteins which can have a very large variety of conformations closely related to their functions. Moreover proteins play also a role in energy transfer and storage in biology. Chapter II deals with their conformation and dynamics. The transfer of energy is treated in chapter III.

As explained by **H. Frauenfelder**, proteins provide beautiful examples of "simple complex" systems. They combine the properties of solids, glasses and liquids, and can be viewed as "physics laboratories" in which many physical phenomena can be studied, ranging from the effects of frustration to "proteinquakes". In his lecture H. Frauenfelder concentrates on the energy landscape and conformational substates, while **A. Garcia** considers the rich dynamics of proteins. The localised nonlinear motions that were discussed in chapter I on very simple models show up here in the full numerical model of a real protein, crambin in aqueous solution. The calculation exhibits also non-localised nonlinear motions with frequencies far away from the values given by normal mode analysis, pointing out once more the need to consider the nonlinearities. The paper introduces a method to extract optimal dynamical variables which could be very useful for building nonlinear model of biomolecules.

As for the case of DNA, the results obtained from theoretical models must be confronted to experimental facts. An interesting approach, pioneered by G. Careri, is to study "model proteins" such as acetanilide, i.e. crystals containing sequences of amide bonds similar to proteins while being more amenable to detailed experiments. Infrared, Raman and neutron diffraction experiments are discussed in the paper of **M. Barthès** which provides a set of results that have to be considered by physicists designing nonlinear models for proteins dynamics. The results on acetanilide are completed by the normal mode analysis and computer modelling calculations of **G. De Nunzio**.

Understanding the way a protein explores its energy landscape requires the development of simple models that do not sacrify the complexity of the configurational space. A step in this direction has been made by **H. Grubmüller, N. Ehrenhofer and P. Tavan** who present a "minimal model" which allows numerical simulations for more than 200 ns, i.e. over times long enough to allow the molecule to explore its conformational space.

This type of approach is an important step toward the development of models which can be studied analytically (at least partly) for a deeper understanding, while retaining the full nonlinearity and the essential features of the dynamics. Chapter III considers such a simple model of proteins which involves the *propagation* of an excitation along the molecule. This is why it was important to check on detailed computer models that long range correlations do exist in a protein, and that a propagation process is possible. This has been done in the simulations presented by **N. Garnier, D. Genest and M. Genest**.

LECTURE 13

Proteins and the physics of complexity

H. Frauenfelder

Los Alamos National Laboratory,
Los Alamos, New Mexico 87545, U.S.A.

The theme of the workshop, "Non linear excitation in biomolecules", gests more general questions: Are all complex systems nonlinear? What are excitations in complex systems? Can investigations of proteins lead to new insights into nonlinear excitations? In these notes, I will try to answer these questions partially by first sketching what proteins are and then describing some aspects of their structure, energy landscape, and dynamics. The notes will be brief, but enough references to books and papers with more details will be given.

The point of view in these notes is that of an experimentalist. The goal is to describe proteins as "simple complex" systems where detailed experiments can be performed, models can be constructed, comparison with molecular dynamics computations are possible, and from where general features of complex systems may ultimately emerge.

The approach is concisely described by a statement made by Stan Ulam, the great Polish-American mathematician, after I described my work to him: " Ah, ask not what physics can do for biology, ask what biology can do for physics."

1. PROTEINS

Nuclei acids and proteins form two essential classes of biomolecule [1]. Vastly simplified one can say that nucleic acids store the information to build proteins, proteins do all the work necessary for the functioning of living systems. Here, we focus on proteins [2]. Proteins are built from twenty different building blocks, the amino acids. All amino acids have the same "backbone", but differ in their side chains. Proteins are built in molecular machines called ribosomes. There, instructed by nucleic acids, of the order of one hundred amino acids are covalently linked into a long linear chain, called the primary structure. This chain then folds into a nearly closed-packed arrangement, called the tertiary structure. This folding process is currently the subject of intense experimental and theoretical work [3,4]. Vastly simplified, the process is sketched in Figure 1. The primary sequence folds, possibly first into a secondary structure and then into the final, space-filling, tertiary structure. This structure is, in some sense, different from solids, liquids, and glasses, because the forces linking the amino acids are covalent and they cannot be broken by thermal fluctuations. The forces stabilizing the tertiary structure and connecting different parts of the primary chain are hydrogen bonds and Van-der-Waals forces. These are "weak", can be broken, and hence give the protein the flexibility necessary for their function. Proteins consequently combine properties of solids, glasses, and liquids.

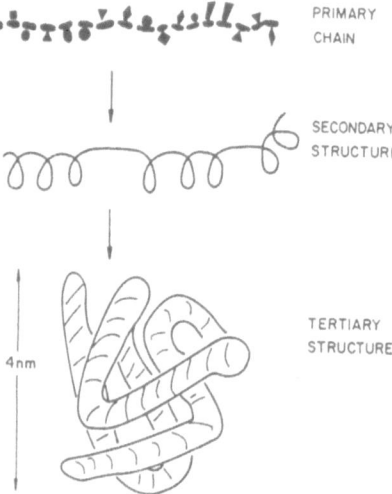

Figure 1. Protein folding. The primary chain folds, possibly first into a secondary structure and finally into the nearly closed-packed tertiary structure. The structure of the working protein is determined by the arrangement in the primary sequence.

The number of different proteins is extremely large. Counting mutations and small changes of the same protein in different systems, the number is probably much larger than 10^{10}. From the point-of-view of the biologist, it is crucial to study diverse proteins to learn how organisms adapt and solve specific problems. For the physicist, however, the situation is different. In order to investigate general phenomena of complex systems, it is more important to study one particular protein in great detail. And, taking a cue from physics, it may be helpful to start with a relatively simple protein. Nearly by accident, we started about 20 years ago with one particular protein, **myoglobin**. In hindsight it is not clear if any other protein would have been just as good a prototype, but in any case myoglobin has turned out to be something like the hydrogen atom of biology.

According to textbooks, myoglobin (Mb) stores dioxygen in muscles and, in fact, it is the protein that gives muscles the red color. Myoglobin is a heme protein; embedded in the folded polypeptide structure shown in Figure 1 is an organic molecule, protoporphyrin IX. For physicists, protoporphyrin (or heme, as it is usually called) is simply a planar molecule with an iron atom in the center. The arrangement is shown schematically in Figure 2. Mb has a molecular weight of about 18 kDalton and is built from about 1200 non-hydrogen atoms. It dimensions are about $3 \times 4 \times 4$ nm^3. Dioxygen enters the protein and binds at the heme iron, and after a certain time returns to the outside again. The remarkable fact is that the binding process is reversible; the iron does not "rust". Myoglobin also reversibly binds carbon monoxide. While the role of CO binding in living systems is unclear (CO has recently been found to be a neurotransmitter), it is technically easier to study the binding process with CO than with O_2. Most experiments are consequently performed with CO as ligand. (Biochemists call the molecule that binds "the ligand".)

Myoglobin is a nearly ideal system for studies of the physics of proteins because it is well known, easy to obtain, and relatively rugged. Not even a physicist can destroy it easily. The heme is a chromophore whose spectrum expresses many protein properties.

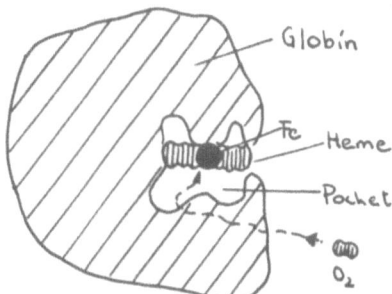

Figure 2. Schematic cross section through myoglobin. The folded polypeptide chain surrounds the heme group. The ligand, dioxygen, moves from the solvent through the protein matrix into the heme pocket and then binds to the heme iron.

Why are proteins ideal systems to study complexity? Could we not learn more, or more easily, from systems such as glasses and spin glasses on the one hand, and more complex systems such as the brain on the other? It may well be that proteins, and biomolecules more generally, offer the best compromise: (i) They contain between 10^3 and 10^4 atoms and hence are large enough to be complex, but small enough so that they can be explored in detail and that experiments and computer simulations can be compared in many instances. (ii) Proteins, moreover, are not random systems; they are the result of 4 Gy of R&D. (iii) Genetic engineering permits changes at the molecular level; the effect of the replacement of a particular amino acid by another one can be investigated. (iv) The number of possible proteins is extremely large ($> 10^{200}$). This fact points to a limitless field, but also implies that a fundamental quantitative understanding of proteins is needed; random tinkering is not enough. (v) Proteins contain many spectroscopic probes placed at well-defined positions. Other probes can be inserted. Studies of the protein properties are aided by these probes.

2. THE APPROACH

How do we study the physics of proteins? The history of physics provides guidance. In nearly all fields, progress in advancing the physics of a particular subject has involved three major steps, the exploration of the spatial structure of the subject, the investigation of the energy levels, and the search for the dynamic laws. In atomic physics, these steps led to the nuclear atom (Rutherford), the hydrogen energy levels (Balmer, Bohr), and the dynamic laws (Heisenberg, Schrödinger, Dirac). Similar steps are found in nuclear, solid-state, and particle physics. The three major areas may also be important in understanding the physics of biomolecules or even, more generally, of complex systems. One level, however, must be added in biomolecules, **function**. We cannot ask why or how an atom functions, but for proteins, this question is crucial and it may help to find the laws that govern structure, energy levels, and dynamics.

3. THE STRUCTURE OF PROTEINS

Three techniques permit the determination of the structure of proteins, X-ray diffraction, neutron diffraction, and nuclear magnetic resonance (NMR). The first two techniques require single crystals, NMR also works with proteins in solutions [5, 6]. The initial experiments required heroic efforts and only the persistence of Max Perutz and John Kendrew led to the structures of hemoglobin and myoglobin. By now, a very large number of protein structures has been determined (> 1000). It is likely that the number of new structures will increase even more rapidly in the future because of three factors: Synchrotron radiation provides intense X-ray beams, counters are improving rapidly, and more sophisticated computers and computer programs speed up the data evaluation.

Computers produce "life-like" space-filling pictures of proteins that appear in *Scientific American* articles and can also be found for instance in a beautiful book by Branden and Tooze [7]. These pictures look very much like aperiodic crystals, as suggested by Schrödinger [8]. **Aperiodicity** is indeed a crucial aspect of proteins, but they are otherwise very different from crystals. The mechanical features of a crystal are rather boring. The main motions executed by a crystal are the vibrations of the atoms; only rarely does something more dramatic happen. Protein, however, have in addition two more properties that are not usually encountered in crystals, frustration and large-scale flexibility.

The concept of **frustration** can most easily be explained with three spins [9]. Assume two spins that favor the antiparallel orientation. A third spin will be frustrated, because it can be antiparallel to only one of the first two. There are thus two possible arrangements, with equal energies; the ground state is degenerate. A similar degeneracy occurs in the neutral kaon system. In proteins, frustration occurs because different side chains may want to occupy the same space and we therefore can expect degeneracy in proteins. We will see later that this expectation is correct; the ground state of a protein is highly degenerate.

Large-scale flexibility is possible because of the spatial arrangement into secondary and tertiary structural elements sketched in Figure 1. As stated earlier, the primary chain is held together by covalent bonds, while the secondary and tertiary structures are stabilized by weaker hydrogen and Van-der-Waals forces. Large scale motions are therefore possible.

The three properties, aperiodicity, frustration, and flexibility, may all be crucial for the existence and the characteristics of nonlinear excitations in proteins. They may all also be important attributes of all complex systems and they may lead to real or perceived disorder.

4. THE ENERGY LANDSCAPE

The textbook representation of a protein gives the impression of a unique and relatively rigid structure. The realization that proteins are frustrated and aperiodic systems forces us to question this picture. Frustration leads to a degenerate ground state; aperiodicity added to it produces disorder. Instead of a simple one or two dimensional description of the energy levels of a protein, we are compelled to introduce the concept of an **energy landscape** [10]. A protein can assume a very large number of somewhat different conformations; each conformation will have a different energy and any two conformations will be separated by a barrier whose height depends on the two conformations. Because a protein can assume a very large number of conformations, the energy landscape will be a hypersurface in a space with many dimensions.

Proteins can usually assume more than one state, because they have to perform functions. A protein can for instance be charged or uncharged. Each of these **states**, in turn, can assume a large number of different conformations. To distinguish these from states, we call them **conformational substates** (CS).

Landscapes appear not only in proteins, but in essentially all fields that attempt to deal with complex systems, such as evolution, economics, artificial intelligence, immunology, and glasses and spin glasses[11]. It may well turn out that biomolecules are the "simplest complex" systems where the properties of energy landscapes can be studied in detail and under controlled conditions. We have justified here the existence of conformational substates on theoretical grounds. In reality, they were first found experimentally. The following subsection describes some of the experimental evidence.

4.1 The Experimental Evidence for Conformational Substates

Evidence for the existence of conformational substates comes from a wide variety of experiments. We select two, ligand binding and spectral hole burning, because they demonstrate the existence of CS convincingly. Additional evidence is described in reference [12].

4.1.1 Ligand Binding to Myoglobin

Myoglobin and many other heme proteins bind dioxygen and carbon monoxide at the heme iron. The covalent bond between the ligand, CO or O_2, and the heme iron can be broken by light. At sufficiently high temperatures, the photodissociated ligand leaves the protein, moves into the solvent, and binds from there as sketched in Figure 2. At low temperatures, however, the ligand remains in the heme pocket and rebinds from there.

The ligand binding process is studied with a conceptually simple approach, flash photolysis, the biophysical analog of nuclear photodissociation. Mb with CO bound to it is embedded in a glycerol-water solvent and placed into a cryostat [13]. The bond between the Fe and the CO is then broken by a laser pulse. The fraction of bound CO is monitored optically. Denote with $N(t)$ the fraction of Mb molecules that have not rebound a CO at the time t after the photodissociation. Already the early experiments indicated that $N(t)$ cannot be described by a single exponential at temperatures below about 200 K. Some rebinding curves from these early experiments are shown in Figure 3. Note that instead of the usual plot of log $N(t)$ versus t, log $N(t)$ is plotted versus log t. In such a plot, a single exponential appears as a rapidly dropping function of universal shape. Results such as shown in Figure 3 have been obtained in a very large number of experiments and with many different heme proteins. It is particularly striking that Azurin (Az) also shows nonexponential rebinding at low temperatures [14]. Mb consists mainly of α-helices, Az of ß-sheets; Mb binds CO, Az NO. The active center in Mb contains an iron atom, in Az a copper atom. Despite these differences, the low-temperature rebinding is very similar.

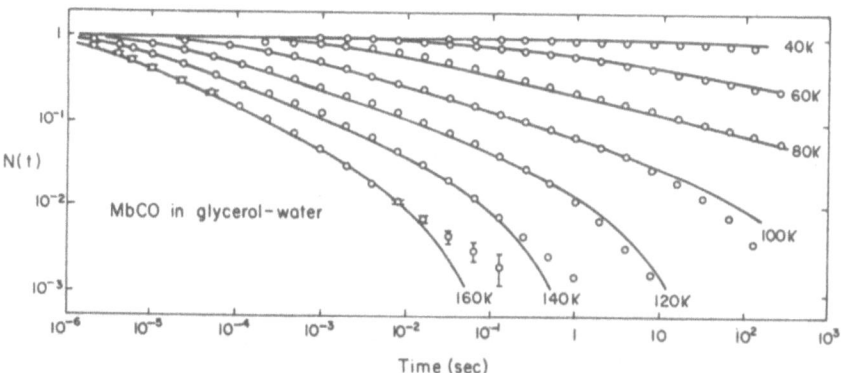

Figure 3. The time course of the rebinding of CO to Mb after photodissociation. $N(t)$ denotes the fraction of Mb molecules that have not rebound a CO at the time t after the photoflash. The data come from early experiments [13]. Present experiments give much better data, but the essential information is unchanged: Rebinding is nonexponential in time.

The data in Figure 3 raise three questions: (i) Is the process leading to the nonexponential rebinding homogeneous or inhomogeneous? (ii) How do we describe the nonexponential rebinding? (iii) Why is the rebinding nonexponential in time?

(i) In a homogeneous system, all protein molecules would show the same behavior and rebinding in each individual protein would be nonexponential in time. In an inhomogeneous ensemble, binding in each protein could be exponential, but different in different proteins. The nonexponentiality would arise only in the ensemble. A multiple-flash experiments permits an unambiguous decision between the two cases [13,15]. In such an experiment, flashing is repeated when a certain fraction of the ligands has rebound. In the homogeneous case, $N(t)$ after the second flash is slower than after a single flash. In the inhomogeneous case, $N(t)$ is the same after a single and after n repeated flashes. Multiple-flash experiments performed on MbCO [13] prove unambiguously that the protein ensemble at low temperatures is inhomogeneous, and that each individual protein molecule rebinds essentially exponentially in time.

(ii) The realization that the MbCO ensemble is inhomogeneous leads to a simple description of the rebinding curves shown in Figure 3 [13]. As model, we introduce a two-well description, sketched in Figure 4. In the bound state, MbCO, the system is in the deeper well A. In the photodissociated state B, the CO has moved into the heme pocket, sketched in Figure 2. The photodissociated state B has recently been observed directly in a beautiful X-ray diffraction experiment [16]. In an individual protein, the barrier between B and A has a unique height H and rebinding is exponential in time, $N(t) = \exp\{-k(H)t\}$. Over a broad range of temperatures, the rate coefficient for passage over the barrier can be approximated by an Arrhenius relation [17,18]

$$k(H,T) = A \ (T/T_0) \ \exp\{-H/RT\}, \tag{1}$$

where A is a preexponential factor, R the gas constant, and T_0 is a reference temperature.

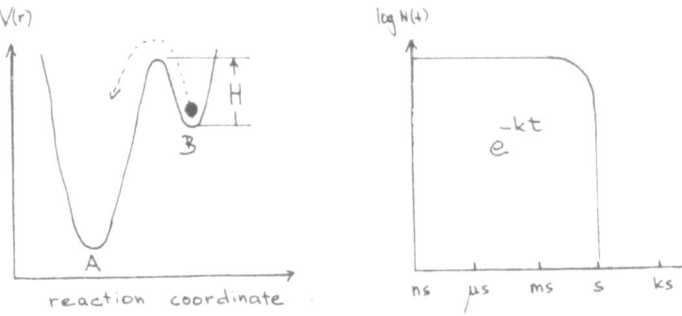

Figure 4. Left: Simplified reaction coordinate for the binding of CO to the heme iron. A denotes the bound state, B is the state with the CO in the heme pocket. H is the barrier height. Right: For a unique barrier height, rebinding is exponential in time.

To describe the nonexponential time course shown in Figure 3, we assume that each protein has a different barrier height and denote with $g(H)dH$ the probability of finding a barrier between H and $H+dH$ in the ensemble. The survival probability then becomes

$$N(t) = \int dH \ g(H) \ \exp\{-k(H,T)t\}. \tag{2}$$

To determine $g(H)$ from the measured $N(t)$, Eq.(2) must be inverted. This inversion, an incomplete inverse Laplace transform, is notoriously unstable. Nevertheless, the data are good enough so that the inversion can be performed, either by some approximation or by the Maximum Entropy method [13,19]. The result yields the probability density $g(H)$ and the preexponential factor A. Often, a Gaussian approximates $g(H)$ well.

(iii) The crucial question for the dynamics and function of proteins, and more generally for the physics of complexity, concerns the cause for the appearance of $g(H)$. A possible explanation is the existence of conformational substates [13,20]. As pointed out above, aperiodicity and frustration together lead to a large number of nearly isoenergetic energy valleys. If proteins in different CS possess different barrier heights, and if the exchange between these CS is much slower that rebinding, the nonexponential rebinding is explained. The connection between structure, substates, barrier height, and rebinding time is sketched in the top four panels of Figure 5.

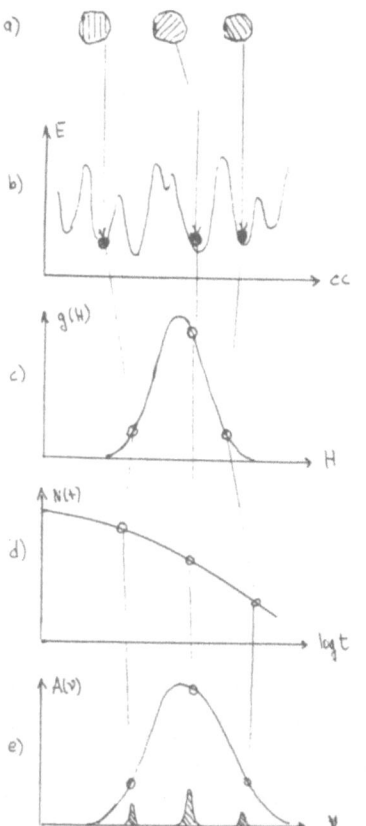

Figure 5. a) Proteins in different conformational substates have somewhat different structures.

b) The different CS correspond to different valleys in the energy landscape. The energy landscape is an energy hypersurface in a high-dimensional space. Shown here is a one-dimensional cross section.

c) The different CS have different activation enthalpies H for the binding step from the heme pocket, B, to the bound state at the heme iron, A (Figure 4). The protein ensemble is characterized by the probability density $g(H)$. The relation between the CS and the activation enthalpy may not be as simple as shown.

d) Proteins with different barrier heights H rebind with different rates. The distribution of activation enthalpies consequently leads to a nonexponential survival probability $N(t)$.

e) Spectral lines in different CS can have different wave numbers as indicated in the figure. The overall shape of the spectral line is therefore not Lorentzian, but Voigtian, a Gaussian superposition of Lorentzians.

4.1.2 Spectroscopic Hole Burning

A chromophore embedded in a protein should see different environments in different CS. Spectral lines in different CS consequently could occur at somewhat different wave numbers, as sketched in Figure 5e. A spectral line in a protein should therefore be inhomogeneously broadened [21,22]. Instead of a Lorentzian, it should have a Voigtian shape, a Gaussian superposition of Lorentzians. Crucial for the experimental observation is the ratio of the natural line width, Γ, to the width caused by the CS. In heme proteins, such as Mb, the excited states have a very short lifetime and Γ is very large. The observation of the inhomogeneous broadening due to the existence of the CS is possible, but the effects seen are often small [23]. The most impressive evidence for the effect of CS comes from experiments using persistent spectral hole burning [24,25,26,27]. A superb example is shown in Figure 6.

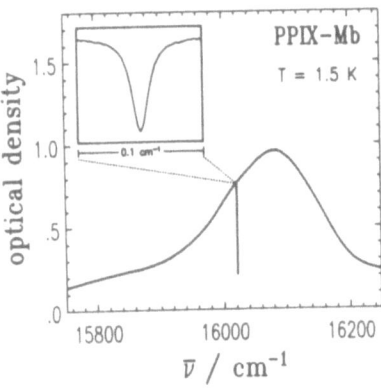

Figure 6. Persistent spectral hole burning in myoglobin. The iron-containing natural heme group with an ultrashort lifetime is replaced by protoporphyrin IX (without iron). The lifetime becomes much longer and the linewidth Γ much smaller. A hole burned into the inhomogeneously broadened band with a laser then is much narrower than the band. The experiment was performed by J. Gafert in the laboratory of J. Friedrich at the University of Bayreuth.

The result shown in Figure 6 has a straightforward explanation in terms of CS, as indicated in Figure 5e. The laser excites proteins in a narrow range of CS. In the excited state or upon deexcitation, small changes in the protein conformation can occur and CS in a narrow range are depleted. At low temperatures, no rearrangement is possible and the hole persists. If the sample is warmed, however, the hole is filled in. The data in Figure 6 show that the ratio of the inhomogeneous width to Γ is about 10^4. The number of substates consequently is very large.

4.2 The Hierarchy of Conformational Substates

Initially we assumed that all CS would be similar (bush-like) [13,20]. A careful look at different experiments shows, however, that the CS must be arranged in a hierarchy [28]. The existence of a hierarchy is already evident from the two experiments discussed above: The nonexponential time dependence of ligand binding shown in Figure 3 extends up to about 200 K. Holes burned at 1.5 K, however, fill in around 15 K. Thus at least two different tiers of CS exist: CS are arranged not like an inverted bush, but like an inverted tree. A sketch of the arrangement of CS in Mb is shown in Figure 7. The figure shows two different types of CS, **taxonomic** and **statistical**. Taxonomic CS are so few in number and so different in properties that each CS can be characterized individually. Statistical CS, in contrast, are so numerous and similar in their properties so that they can only be described by using distributions.

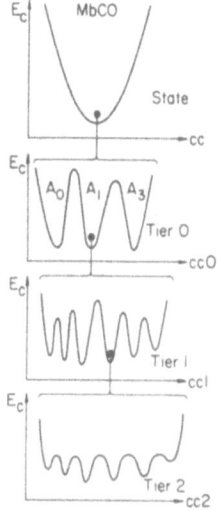

Figure 7. Schematic picture of the energy landscape of MbCO. The figure represents a one-dimensional cross section through the landscape. Shown is the conformational energy, E_c, as a function of a conformation coordinate cc. The landscape is organized in various tiers, characterized by different (average) barrier heights between conformational substates. The state MbCO can assume three different conformations, denoted by A_0, A_1, and A_3. (A substate A_2 may exist, but is difficult to characterize cleanly.) These three substates of tier 0 are called taxonomic, because they can be characterized individually. Each substate of tier 0 can assume a very large number of slightly different conformations. Each of these corresponds to a substate of tier 1. Because the number of these substates is large, their properties must be described by distributions. The furcation continues and may involve substates of tier 2 and 3.

Figure 7 should not be taken too seriously. It illustrates the basic idea, but the details , even in the well studied case of myoglobin, could be different. The detailed study of protein landscapes will take a considerable time and Figure 7 can be compared to a description of atomic energy levels at the time of Balmer. Nevertheless, a few remarks about the various tiers can already be made.

4.2.1 Taxonomic Conformational Substates

Figure 7 shows three taxonomic substates. Many of the properties of these three substates are by now well known. Their clearest attribute is the stretch frequency of the bound CO [29], but they also rebind CO after photodissociation with different rates and their relative energies, entropies, and volumes have been determined. The fact that the three substates are not caused by different primary sequences is established by the observation of transitions among the substates [30].

Even a cursory look through the literature shows that MbCO is not an isolated case; taxonomic substates occur often. Three examples suffice to demonstrate this assertion. In the hemoglobin of Chironimus th. thummi, three "conformation isomers" have been observed [31]. In horseradish peroxidase, two different conformations have been seen [32]. Low-temperature photoreactions in halorhodopsin give evidence for substates [33].

These observations suggest that the occurrence of taxonomic substates in proteins, or more generally in complex systems, may be the rule rather than the exception. Such a possibility may have some theoretical underpinning [34].

Taxonomic substates may be crucial for the control of function. Experiments show that the relative populations of the different taxonomic substates depend on many external factors, such as pH, temperature, pressure, hydration and forces exerted by other proteins [35]. These dependencies provide a compelling mechanism for control [36].

4.2.2 Statistical Substates

Examples for distributions characterizing statistical CS are shown in Figures 5c and 5e. Both the probability density $g(H)$ and the shape of the spectral line are given by smooth distributions. In the case of $g(H)$, the fact that the number of CS is very large emerges from the smoothness of the rebinding curves in Figure 3. The large number of CS giving rise to the Voigtian shape of absorption bands is evident from Figure 6 which shows that the width of an individual component is about 10^4 times smaller than the inhomogeneous width.

The existence of a hierarchy of statistical CS is also demonstrated in the date underlying Figures 3 and 6: The rebinding curves $N(t)$ become exponential only above about 200 K, while the hole in Figure 6 fills in at about 15 K. More evidence for a hierarchy comes from specific heat data that indicate that two-level states occur below about 1 K [37]. The existence of a hierarchy with discrete tiers is thus well established, but the characteristics are still rather uncertain and require far more experimental work.

4.3 Conformational Excitations
Some conceptual problems arise in the discussion of the "ground state" and excitations in the energy landscape. First to the "ground state". Strictly speaking, the CS with the lowest energy should be called the ground state. On cooling infinitely slowly to $T = 0$, only this state should be occupied. But consider transitions from A_0 to A_1 in Figure 7. The rate coefficient for this transition has been measured from 180 to 270 K [30]. Extrapolation of these data suggests that the transition would be slower than the age of the universe at about 100 K. All three taxonomic substates in Figure 7 can be considered as "ground states".

What then are excitations in such a hierarchical energy landscape? The answer depends on the tier. Transitions between the substates of the lowest tier, the "two-level states", can occur at any temperature and they do not concern us here. In the higher tiers, both vibrations and transitions between CS can occur. Transition between CS are always nonlinear; there consequently exist a large variety of nonlinear excitations in a protein.

In the discussion so far, the energy landscape has tacitly been assumed to be fixed. In reality, however, the system is dynamic [38]. The landscape can change if, for instance, a hydrogen bond is broken. Such excitations must also be considered.

4.4 Theoretical and Computational Studies
So far, the discussion has been based nearly entirely on experiments. Computations and theoretical models also support the concept of a rough energy landscape. Molecular dynamics [39] and Monte Carlo simulations [40] give clear evidence for the existence of CS in proteins and provide considerable insight into their properties. The goal is, however, not just to let the computer tell us how proteins behave, but to understand the essential features in terms of reasonably simple models. A beautiful overview of the ideas underlying such models has been given by Peter Wolynes [41]. Dan Stein introduced a spin-glass model [42]. Of particular interest is the random-energy model of Derrida [43]. This model has been used often to describe aspects of the energy landscape and of CS, as for instance in the references [44].

5. SOME REMARKS
The discussion given above has been concentrated on introducing the energy landscape and conformational substates, and to provide some of the underlying experimental facts. Missing in this treatment is the discussion of the dynamics and of the function of proteins. Information on these topics can be found in some reviews [4,10,12,17,18]. As already pointed out above, the entire field is still at a beginning and much work, theoretical, computational, and experimental, remains to be done before even one protein is "fully" understood. A deeper understanding, however, will certainly lead to directed applications

of biological systems to real problems, but it may also give better insight into complex systems in general.

References

1. Stryer L., Biochemistry (Freeman, San Francisco, 1988).

2. Dickerson R.E. and Geis I. The Structure and Action of Proteins (Benjamin, New York, 1969).

3. Chan H.S. and Dill K., *Physics Today*, February 1993, p. 24.

4. Frauenfelder H. and Wolynes P.G. *Physics Today*, February 1994, pp. 58-64.

5. Wüthrich K., NMR of Proteins and Nuclei Acids (Wiley-Interscience, New York, 1986).

6. Clore G.M. and Gronenborn A.M., *Ann. Rev. Biophys. Biophys. Chem.* 20(1991) 29-63.

7. Branden C. and Tooze J., Introduction to Protein Structure (Garland Publishing Inc. New York, 1991).

8. Schrödinger E., What is Life? (Cambridge University Press, Cambridge, 1944).

9. Toulouse G. *Comm. Physics* 2(1977)115-119.

10. Frauenfelder H, Sligar S.G., and Wolynes P.G. *Science* 254(1991)1598-1603.

11. Measures of Complexity. L. Peliti and A. Vulpiani, Eds. (Lecture Notes in Physics 314. Springer, Berlin, 1987). Complexity- Metaphors, Models, and Reality. G.A. Cowan, D. Pines, and D. Meltzer, Eds. (Addison-Wesley, Reading, Mass.,1994). Kauffman S.A, Origins of Order (Oxford University Press, Oxford, 1993).

12. Frauenfelder H., Parak F., and Young R.D. *Ann. Rev. Biophys. Biophys. Chem.* 17(1988) 451-479.

13. Austin R.H., Beeson K.W., Eisenstein L., Frauenfelder H., and Gunsalus I.C., *Biochemistry* 14(1975)5355-5373.

14. Ehrenstein D. and Nienhaus G.U., *Proc. Natl. Acad. Sci. USA 89(1992)9681-9685*.

15. Frauenfelder H., Methods In Enzymology, Vol. LIV, Part E. S. Fleischer and L. Packer, Eds. (Academic Press, New York, 1978)pp. 506-532.

16.Schlichting I., Berendzen J., Phillips G.N., and Sweet R.M., *Nature* 371(1994)808-812.

17. Frauenfelder H. and Wolynes P.G., *Science* 229 (1985) 337-345.

18. Activated Barrier Crossing, G.R. Fleming and P. Hänggi, Eds. (World Scientific, Singapore, 1993).

19. Steinbach P.J., Chu K., Frauenfelder H., Johnson J.B., Lamb D.C., Nienhaus G.U., Sauke T.B., and Young R.D., *Biophys. J.* **61**(1992)235-245.

20. Frauenfelder H., Petsko G.A., and Tsernoglou D., *Nature* **280**(1979)558-563.

21. Cooper A., *Chem. Phys. Lett.* **99**(1983)305-309. Srajer V., Schomacker K.T., and Champion P.M., *Phys. Rev. Lett.* **57**(1986)1267-1270.

22. Agmon N. *Biochemistry* **27**(1988)3507-3511.

23. Ormos P., Ansari A., Braunstein D., Cowen B.R., Frauenfelder H., Hong M.K., Iben I.E.T., Sauke T.B., Steinbach P.J., and Young R.D., *Biophys. J.* **57**(1990)191-199.

24. Friedrich J. and Haarer D., *Angew. Chem. Int. Engl. Ed.* **23**(1984)113-140. Zollfrank J., Friedrich J., Vanderkooi J.M., and Fidy J., *J. Chem. Phys.* **95**(1991)3134-3136. Friedrich J., Gafert J., Zollfrank J., Vanderkooi J., and Fidy J., *Proc. Natl. Acad. Sci. USA* **91**(1994)1029-1033. Friedrich J. in Methods of Enzymology- Biochemical Spectroscopy. K. Sauer Ed.

25. Reddy N.R.S., Lyle P.A., and Small G.J., *Photosynthesis* Research **31**(1992)167-194. Jankoviak R., Hayes J.M., and Small G.J., *Chem. Rev.* **93**(1993)1471-1502.

26. Boxer S.G., Gottfried D.S., Lockart D.J., and Middendorf T.R., *J. Chem. Phys.* **86**(1987)2439-2441.

27. Persistent Spectral Hole Burning: Science and Applications. W.E. Moerner Ed.(Springer, Berlin, 1988).

28. Ansari A., Berendzen J., Bowne S.F., Frauenfelder H., Iben I.E.T., Sauke T.B., Shyamsunder E., and Young R.D., *Proc. Natl. Acad. Sci. USA* **82**(1985)5000-5004.

29. Ansari A., Berendzen J., Braunstein D., Cowen B.R., Frauenfelder H., Hong M.K., Iben I.E.T., Johnson J.B., Ormos P., Sauke T.B., Scholl R., Schulte A., Steinbach J.B., Vittitow J. and Young R.D., *Biophys. Chem.* **26**(1987)337-355.

30. Young R.D., Frauenfelder H., Johnson J.B., Lamb D.C., Nienhaus G.U., Philipp R., and Scholl R., *Chem. Phys.* **158**(1991)315-327.

31. Gersonde K., Sick H., and Wollmer A., *Eur. J. Biochem.* **15**(1970)237-244.

32. Uno T., Nishimura Y., Tsuboi M., Makino R., Iizuka T., and Ishimura Y., *J. Biol. Chemistry* **262**(1987)4549-4556. Sharonov Y.A., Pismensky V.F., and Yarmola E.G., Letters **235**(1988)63-66.

33. Zimanyi L., Ormos P., and Lanyi J.K., *Biochemistry* **28**(1989)1656-1661.

34. Honeycutte J.D. and Thirumalai D., *Proc. Natl. Acad. Sci. USA* **87**(1990)3526-3529.

35. Frauenfelder H., Alberding N.A., Ansari A., Braunstein D., Cowen B.R., Hong M.K., Iben I.E.T., Johnson J.B., Luck S., Marden M.C., Mourant J.R., Ormos P., Reinisch L., Scholl R., Schulte A., Shyamsunder E., Sorensen L.B., Steinbach P.J., Xie A., Young

R.D., and Yue K.T., *J. Phys. Chem.* **94**(1990)1024-1037.

36. Tian W.D., Sage J.T., and Champion P.M., *J. Mol. Biol.***233**(1993)155-166.

37. Goldanskii V.I., Krupyanskii Y.F., and Fleurov V.N., *Dokl. Akad. Nauk SSSR* **272**(1983)978-981. Singh G.P., Schink H.J., Lohneysen H., Parak F., and Hunklinger S., *Z.Phys.B,* **55**(1984)23-26.

38. Berlin Y.A., Drobnitsky D.O., Goldanskii V.I., and Kuz'min V.V., *Chem. Phys. Letters* **189**(1992)316-320.

39. Elber R. and Karplus M., *Science* **235**(1987)318-321. Garcia A., Present volume.

40. Noguti T. and Go N., *Proteins* **5**(1989)97-138. Iori G., Marinari E., and Parisi G., *Europhys. Lett.* **25**(1994)491-496.

41. Wolynes P.G. "Aperiodic Crystals: Biology, Chemistry, and Physics in a Fugue with Stretto." in Proceedings of the International Symposium on Frontiers in Science. Urbana IL 1987, S.C. Chan and P.G. Debrunner, Eds. (American Institute of Physics, New York, 1988)39-65.

42. Stein D., *Proc. Natl. Acad. Sci. USA* **82**(1985)3870-3674.

43. Derrida B. *Phys. Rev. B* **24**(1981)2613-2626.

44. Bryngelson D.J. and Wolynes P.G., *J. Phys. Chem.* **93**(1989)6902-6915. Onuchic J.N. and Wolynes P.G., *J. Chem. Phys.* **98**(1993)2218-2224. Young R.D. and Powell S.W., *J. Chem. Phys.* **101**(1994)1 December.

Multi-basin dynamics of a protein in aqueous solution

A.E. García

Theoretical Biology and Biophysics Group,
T10, MS K710, Los Alamos National Laboratory,
Los Alamos, New Mexico 87545, U.S.A.

ABSTRACT

A molecular dynamics simulation of crambin in aqueous solution shows that motions are characteristic of non-linear systems. We describe typical non-linear excitations, such as intermittency, for various representations of the protein dynamics and structure. The protein backbone dihedral angles show fast correlated transitions from one minimum well to another. Each transition is followed by small overdamped oscillations. Equal-time cross correlations of all (ϕ, ψ) angles show that correlations are extended along the backbone chain. An analysis based on a generalized least squares fitting of the protein fluctuations along vectors show that a small set of molecule optimal dynamic coordinates (MODC) describe most of the protein fluctuations. In addition, the MODC describe a trajectory where the protein conformation jumps from one minimum well to another. An extension of the MODC describing 2- and 3- dimensional cuts of the protein configurational space clearly shows a trajectory around multiple basins of attraction.

1. INTRODUCTION

The dynamics of proteins is closely linked to their function. Proteins and nucleic acids exhibit motions with characteristic time-scales ranging from picoseconds (ps) [1, 2] (normal modes of vibrations) to seconds [3] (as studied by hydrogen exchange experiments). Experimental studies on myoglobin suggest the existence of a hierarchy of motions occurring at various time-scales resulting from an ensemble of nearly degenerate states separated by a distribution of enthalpic energy barriers [4–7]. Theoretical evidence (molecular dynamics and Monte Carlo simulations) for the existence of these substates have been reported [8, 9], although the conclusions of these reports are not free of controversy [10]. The dynamics characteristics of such systems has also been described [11].

Theoretical studies of biomolecular dynamics are mostly concentrated on molecular dynamics simulations of sophisticated [12] models where all the biomolecule's atoms and the aqueous solvent are explicitly included in the calculations. This approach has advantages in the sense that many non-linearities in the system are explicitly included in the modeling. A disadvantage of these models is that by following the trajectory of every atom in the system the length of the simulations is limited by the computational resources. Current atomic-level molecular simulations of biomolecules model the dynamics of proteins up to a few nanoseconds [13]. It seems to be a trend in the field that larger systems and longer simulations are attempted. However, the analysis of the extensive wealth of information provided by these simulations is missing.

In this manuscript we we will present evidence that shows the presence of multi-basin non-linear motion in proteins in the picoseconds time-scale. A method for extracting vectors that best represent the fluctuations in the systems will be discussed. Recent studies by Flach and Willis [14] have motivated the generalization of this method to find the best planes, volumes, etc., describing the fluctuations of the system. We will show that the molecular dynamics trajectory of the protein is clustered around few local minima (basins of attraction), and that transitions among local minima occur within the 310 ps of the trajectory.

2.   DESCRIPTION OF THE SYSTEM

We have studied the dynamics of a small hydrophobic protein, crambin, in aqueous solution, by a molecular dynamics simulation at constant temperature. Crambin is a 46 amino acids amphipatic protein for which high resolution X-ray [15], neutron diffraction [16] and NMR [17–19] data are available. Detailed experimental and theoretical studies of the hydration and dynamics of crambin have been reported in the literature [1, 20–22]. In spite of the detailed experimental and theoretical studies done on crambin, its function is not known.

Crambin is a 46 amino acid protein that contains most structural elements characteristics of larger proteins. Fig.1 shows a ribbon representation of the three dimensional structure of crambin. Starting from the N-terminus and moving along the protein chain we find a β-strand (amino acids 1-4), a loop (amino acids 5-6), a helix (amino acids 7-19), a loop (amino acids 20-22), another helix (amino acids 23-30), another β-strand that makes hydrogen bondings with the first β strand to form a β sheet (amino acids 32-35), and a turn (amino acids 41-44). Three disulfide bonds are formed by Cys(3)-Cys(40), Cys(4)-Cys(32), and Cys(16)-Cys(26). Bacause of these disulfide bonds the connectivity of the amino acid chain is not well described by a quasi-one-dimensional chain.

In this simulation study, crambin was contained in a box of dimension 42.11 × 36.85 × 29.34 Å containing 1315 water molecules. The initial conformation of the protein was obtained from the crystallographic coordinates reported by Hendrickson and Teeter [15]. The system contains 4353 atoms; 408 in the protein and 3945 in the solvent. The system was equilibrated during a period of 24 ps. A production period of 310 ps has been obtained. A previous description of the dynamics was reported for the first 216 ps after equilibration [11]. Details about the system and simulation have been described in a previous paper [23]. In the following analyses the dynamics of the protein will be interpreted as that of an open system coupled to an external bath (the solvent) where the protein exchanges energy and momentum with the solvent.

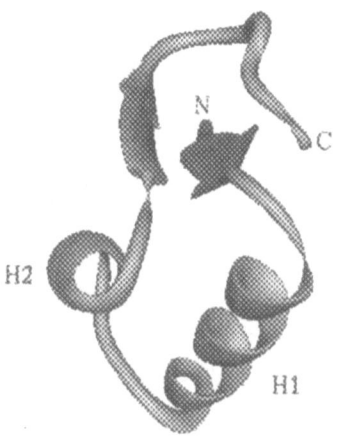

Figure 1: Ribbon representation of the secondary and tertiary structure of crambin. The labels N and C label the amino and carboxy termini, respectively. The labels H1 and H2 show two helical regions.

3.   RESULTS AND DISCUSSION

3.1   Localized Non-Linear Motions

Analysis of the distributions of occurrence and the time dependence behavior of the protein backbone dihedral angles [24], (ϕ, ψ), suggest that the dynamics of crambin is typical of a system with multiple potential energy minima.

Figs. 2a-b show histograms of the occupancy of the backbone ϕ and ψ dihedral angles during the first 216 ps of simulation. The backbone dihedral angles ω remain in a *trans* conformation throughout the simulation. The ϕ and ψ dihedral angles for residues 2-3,33-35 (forming part of β-strands) and 19-21,40-43 (forming parts of *turns*) show bi-modal distributions. The helical regions of the protein, residues 7-18 and 23-30, show sharp unimodal distributions. Time-series of some angles found to have multimodal distributions are shown in Figs. 3 and 4. These time- series are characteristic of systems showing intermittency [25]. That is, there are many fast flips from one conformation to another, with rapid underdamped oscillations between. Figs. 3a-3d show the time dependence of the (ϕ, ψ) angles near the residues Cys(2) and Cys(3) that sample bimodal distributions. The average value for ψ_3 is around 90°, and the standard deviation is around 50°. However, $\psi = 90°$ is hardly ever adopted, and the fluctuations around each minimum are only 15°. Equal times transitions in ψ_3, ϕ_4, shown in Figs. 3b and 3c, are

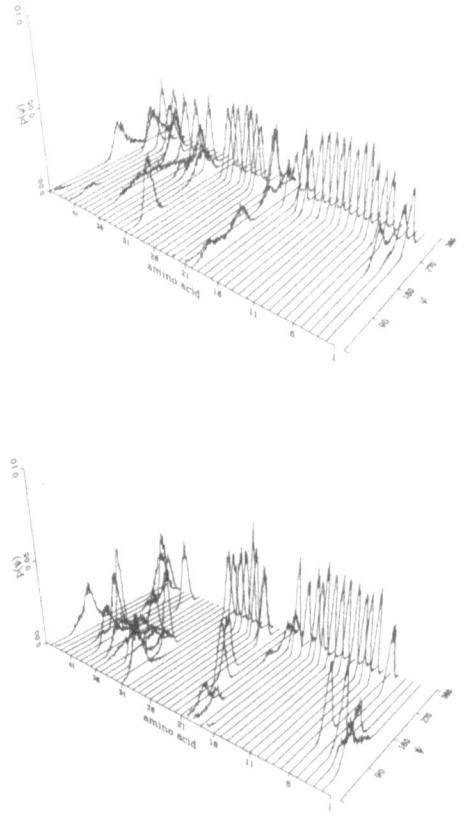

Figure 2: Histograms of the fraction of occurances of the backbone dihedral angles (a) ϕ (on top) and (b)ψ (in degrees) of crambin during 216 ps of simulation. The amino acids sequence is labeled by numbers.

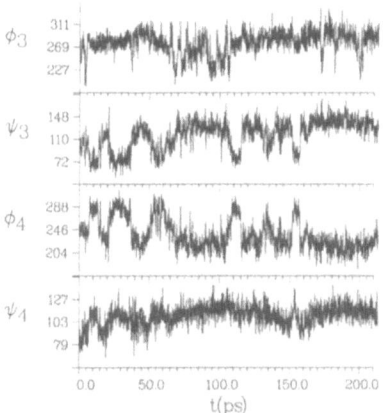

Figure 3: Dihedral angles ψ and ϕ (in degrees) as a function of time (in ps) for the amino acids Cys(4) and Thr(4). Figures are refered as a-e, from top to bottom

anti-correlated. Notice that when ψ_3 changes abruptly from one conformation to another, ϕ_4 also changes abruptly in the opposite sense. Some of these transitions also seem to be coupled to the other two neighboring dihedral angles, ϕ_4 and ψ_3, shown in Figs. 4a and 4d, but not as strongly as ψ_3 and ϕ_4. The strongly coupled transition in ψ_3, ϕ_4 will localize the stress induced by one transition to two residues [26]. Similar behavior is exhibited at other backbone positions. For example, Fig.4a-e show the ψ, ϕ angles for residues Pro(19), Gly(20), and Thr(21) that form a *turn*. The same anti- correlations are shown between different angles. However, the transitions are different. Here the coupling between angles is extended to five consecutive ψ and ϕ angles. The ring structure of the proline residue Pro(19) does not allow ϕ_{19}, Fig.4a, to change from g^- to g^+ or t, but two separate values of ϕ_{19} are sampled around g^-. Notice that the pairs of angles (ϕ_{19}, ψ_{19}), (ψ_{19}, ϕ_{20}), and (ψ_{20}, ϕ_{21}) are anti-correlated, but (ϕ_{20}, ψ_{20}) are not anti-correlated over the whole trajectory. These motions have the collective effect of changing the *turn* conformation.

The equal-time cross-correlation matrix shows that correlations extend through the chain to non-nearest-neighbor residues, although with diminished cross-correlation values. Fig. 5 shows the equal-time cross-correlation matrix for all (ϕ, ψ) dihedral angles. The matrix elements range from -1.0 (perfect anti-correlation) to 1.0 (perfect correlation). Regions corresponding to the matrix elements are shaded from white (-1.0) to black (+1.0). The cross-correlations matrix elements among nearest-neighbors (elements near the diagonal) are large and negative, showing strong anti-correlation. Large cross-correlation matrix elements among non-nearest-neighbors dihedral angles are obtained for amino acids involved in non-alpha helical regions. Regions of different secondary structure are easily distinguished in the cross-correlation matrix. Helical regions show a strong anti-correlation between neighboring angles, a moderate correlation between next-nearest neighbors, and a weak anti-correlation to amino acids forming a backbone hydrogen bond. These patterns can be easily seen in Fig. 5, near the diagonal, for amino acids 7-15, and

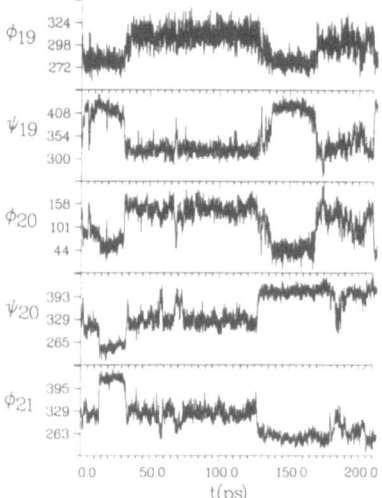

Figure 4: Dihedral angles ψ and ϕ (in degrees) as a function of time (in ps) for the amino acids Pro(19), Gly(20), and Thr(21). Figures are refered as a-e, from top to bottom.

20-28. Non-helical regions show a cross-correlation pattern among non-nearest-neighbor amino-acids that extends over the whole protein. Amino acids that are not directly connected (neither by covalent bonds nor in van der Waals contact) are correlated. These patterns can be seen in Fig. 5, amino acids 16-20 and 31-45.

3.2 Delocalized Non-Linear Motions

The inter-dependence of local structural-variables describing collective, delocalized excitations is not trivial. However, the description of the dynamics of a protein in terms of non-structural variables is desired. To do so we need to find a measure that will represent the fluctuations of the system. We have employed the N-particle root-mean-square (rms) distance [27], $d(t, t')$, between evolving protein configurations. A large rms distance between configurations at short $t - t'$ are indicative of fast configurational changes.

The distance matrix $d(t, t')$ between pairs of conformations at t, t', sampled every 2 ps, during 310 ps of simulation, is shown in Fig. 6. A darker gray shading implies a larger rms distance between pairs of configurations. A lighter gray shading implies a small rms distance between pairs of configurations. The configurations of the protein during the first 50.0 ps of the trajectory are far away from other configurations in the trajectory. The rms distance smoothly increases from 0 to values near 1.0 Åin a time near 50.0 ps. Oscillations between larger (1.5 - 2.0 Å) and smaller (1.0 Å) rms distances occur also at intervals of about 50 ps. Normal mode analysis of proteins show the lowest frequency modes to have periods in the order of magnitude of a few ps [1, 2]. That is, the motions responsible for these oscillations are not normal modes.

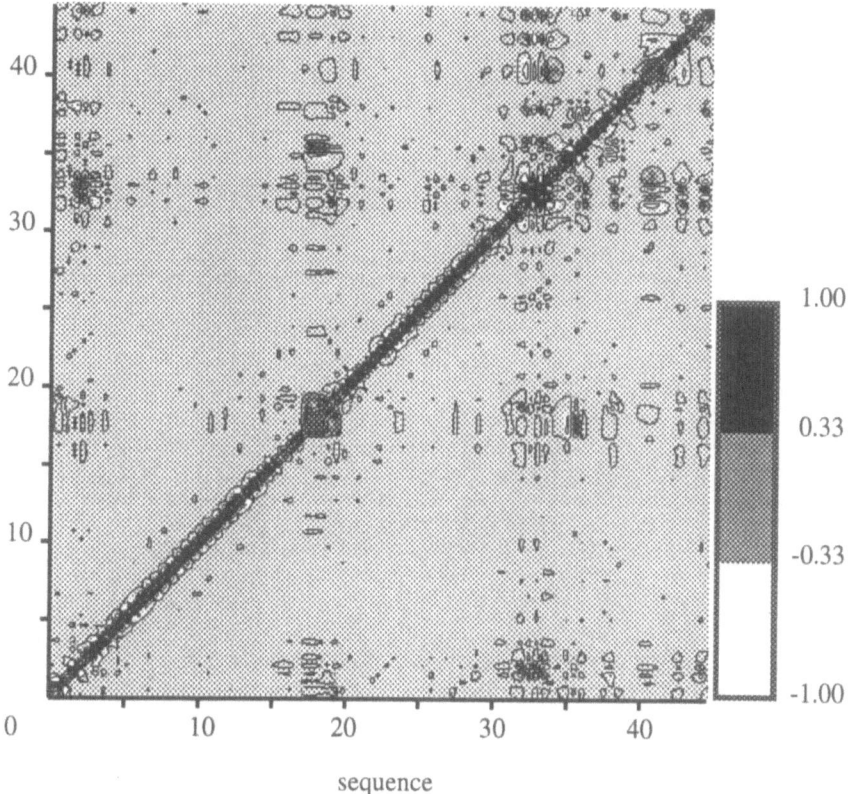

Figure 5: Equal-time cross-correlation matrix among all backbone dihedral angles. Regions representing elements of the matrix are shaded from white (-1.0) to black (+1.0). Contours are drawn to guide the eye toward regions showing correlations (or anti-correlations).

3.2.1 Tree Analysis

The results shown above suggest that the *rms* distance can be used as an effective measure to detect conformational transitions among local minima. The information contained in $d(t, t')$ is enough to built a hierarchical representation of all configurations adopted by the system. The branching of such a tree will be indicative of the proximity of one configuration to another. To built the hierarchy we use the following clustering algorithm [28]:

1. We start with N configurations and a distance matrix, $d(t, t')$, containing the distance among all pairs of configurations. At this stage, each configuration belongs to a separate cluster.

2. We join two distinct ($d \neq 0.$) configurations for which $d(t, t')$ is the smallest into one cluster. Now we have $N - 1$ clusters. To built the new distance matrix we take $d^{(N-1) \times (N-1)}(new, k) = inf[d(i, k), d(j, k)]$. This step is repeated $N - 1$ times, until only

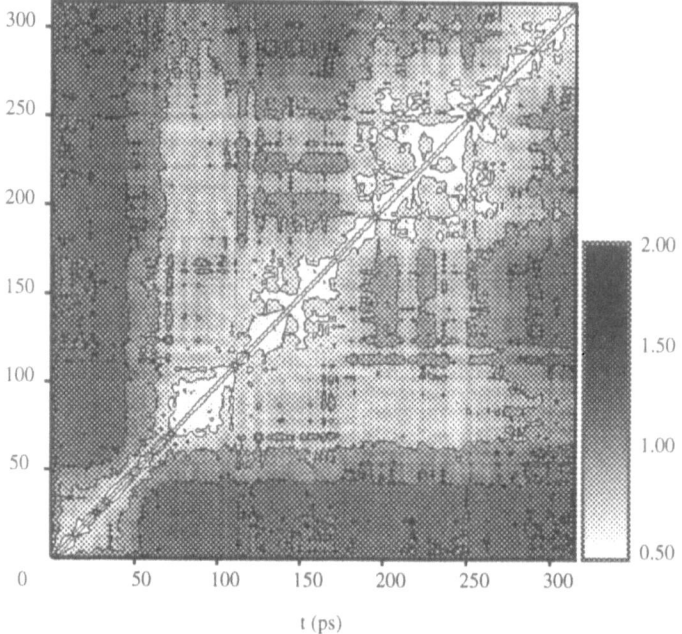

Figure 6: Contour plot of the *root-mean-square* distance between pairs of conformations adopted by the protein every 2 ps along the last 310 ps of the molecular dynamics trajectory. Regions surrounded by the contours are shaded from white ($d \approx 0.50$ Å) to black ($d \geq 2.0$ Å). The largest *rms* distance is 2.45 Å.

one cluster remains.

This hierarchy can be graphically represented by joining each pair of newly clustered configurations by a line of length proportional to the distance between the two clustered structures. The resulting hierarchy can be indexed by the distance between clusters. Fig. 7 shows a radial representations of the hierarchy obtained by this algorithm. We have added labels indicating the time (in ps) at which the configuration represented in the tree occured in the molecular dynamics trajectory. All configurations belong to a cluster with a branch point (labeled **O**) near the center of the diagram. This point represents the stem of the tree in a hierarchical representation. Each branch emerging from this point represents a family of structures that are closely related, i.e., they represent configurations in nearby local minima, while members of different families are configurations in far away minima. This tree conforms to the ideas presented by H. Frauenfelder [4], where a hierarchy of structures exist and transitions between structures in nearby minima are fast, while transitions to far away states are reached through of multiple jumps to nearby minima. We believe that the tree presented here is just the bottom of this hierarchy; i.e., it goes from structures differing in the position of a few atoms, to structures differing in the relative orientation of helices and turns. The complete hierarchy may extend from folded structures to structures that exhibit completely different folding or unfolded structures. A complete tree may show that the stem represented by **O** is only one branch of a larger tree. Such a hierarchy cannot be completely sampled by molecular dynamics

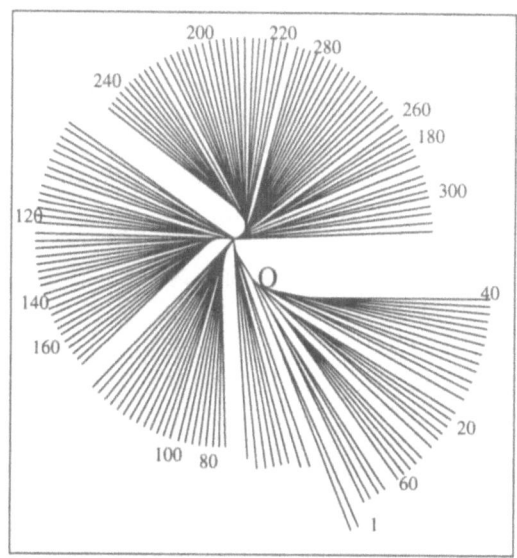

Figure 7: Radial tree representation of structures in different clusters. The numbers around the tree show the time (in ps) at which the structure occured in the trajectory.

simulations of systems that explicitly include all atoms and pretend to follow the history of every molecule in the solvent. The hierarchy presented here can be proven to satisfy ultrametricity. That is, the distance between clusters satisfy $d(i,j) \leq min[d(i,k), d(j,k)]$.

It is our hypothesis that the oscillations shown in Fig. 6 and the branching of the tree in Fig. 7 represent collective non-linear motions. To show this we will define a set of directions, $\vec{m}$, in the $3N$ dimensional space of the protein, that best represents the mean square fluctuations of the protein atoms around their equilibrium positions. If the oscillations shown in $d(t,t')$ are non-linear excitations, the projections of the protein trajectory along these directions should exhibit a dynamical behavior similar to that shown by the *turn* and *β-strands* dihedral angles. We expect to observe multicentered oscillations, rapid transitions from one center to another, and damped quasi-harmonic oscillations around each center.

3.3   Molecule Optimal Dynamical Coordinates (MODC)

At this point it is desirable to establish the nature of the conformational space sampling (one-basin-of-attraction-quasi-harmonic motion versus multiple-basins-non-linear motions) performed during the molecular dynamics simulation of a protein in solution. The method involves the construction of a set of directions $\vec{m}^{3N}$ in the 3N dimensional conformational space that can be most efficiently used to describe the structural fluctuations of the molecule under study. This method has been previously described [11, 33]. A generalization of the method to represent 2-dimensional (planes) and 3-dimensional (volumes) cuts of the (coordinates) configurational space that better represent the dynamics of the system will be presented. These coordinates are specific to the molecule and trajectory sampled during a molecular dynamics simulation.

The directions $\vec{m}\ ^{3N}$ are determined by minimizing the mean square distances of the $\{\vec{r_i}^{3N}\}$ configurations *normal* to $\vec{m}\ ^{3N}$, such that most of the fluctuations will be along $\vec{m}\ ^{3N}$. The distance between a point $\vec{r_i}$, that here represents a biomolecule conformation, and a line with direction $\vec{m}$, passing through the point $\vec{y_0}$, is given by

$$d_i^2 = (\vec{r_i} - \vec{y_0})^2 - [(\vec{r_i} - \vec{y_0}) \cdot \vec{m}]^2.$$

The average square distance between a set of S points representing all the trajectory points of the biomolecule is then given by:

$$d^2 = \frac{1}{S} \sum_{i=1}^{S} d_i^2 = \frac{1}{S} \sum_{i=1}^{S} (\vec{r_i} - \vec{y_0})^2 - [(\vec{r_i} - \vec{y_0}) \cdot \vec{m}]^2. \tag{1}$$

The least square distance is obtained by finding the 6N parameters $\vec{y_0} = \{y_{0\alpha}\}$, and $\vec{m} = \{\vec{m}_\alpha\}$, with $\vec{m} \cdot \vec{m} = 1$, that minimize d^2. That is, we have to minimize a functional of the trajectories, $r_i(t)$, and a function of $\vec{m}$, $\vec{y_0}$ and λ,

$$f(\vec{m}, \vec{y_0}, \lambda) = \frac{1}{S} \sum_{i=1}^{S} \{(\vec{r_i} - \vec{y_0})^2 - [(\vec{r_i} - \vec{y_0}) \cdot \vec{m}]^2\} + \lambda[\vec{m} \cdot \vec{m} - 1] \tag{2}$$

where λ is a Lagrange multiplier. An extreme value of d^2 is given by a set $\vec{z} = (\vec{m}_\alpha, y_{0,\alpha}, \alpha = 1, ..., 3N, \lambda)$ that gives $\nabla_{\vec{z}} f(\vec{z}) = 0$. The gradient of $f(\vec{m}, \vec{y_0}, \lambda)$ gives:

i) with respect to $\vec{y_0}$:

$$\nabla_{\vec{y_0}} f = \frac{2}{S} \sum_{i=1}^{S} \{-(\vec{r_i} - \vec{y_0}) + [(\vec{r_i} - \vec{y_0}) \cdot \vec{m}]\vec{m}\} = 0 \tag{3}$$

that implies

$$\vec{y_0} = \frac{1}{S} \sum_{i=1}^{S} \vec{r_i} \ , \tag{4}$$

That is, $\vec{y_0}$ is the average over all configurations;

ii) with respect to λ:

$$\nabla_\lambda f = \vec{m} \cdot \vec{m} - 1 = 0 \tag{5}$$

that normalizes the vector $\vec{m}$;

iii) with respect to m_α:

$$\nabla_{m_\alpha} f = -\frac{1}{S} \sum_{i=1}^{S} \{(r_i - y_0)_\alpha (r_i - y_0) \cdot \vec{m}\} + \lambda m_\alpha = 0. \tag{6}$$

We can re-write the right hand side of Eq. 6 as

$$\frac{1}{S} \sum_{\beta=1}^{3N} \sum_{i=1}^{S} (r_i - y_0)_\alpha (r_i - y_0)_\beta m_\beta = \lambda m_\alpha. \tag{7}$$

Defining

$$\sigma_{\alpha,\beta} = \frac{1}{S} \sum_{i=1}^{S} (r_i - y_0)_\alpha (r_i - y_0)_\beta \tag{8}$$

where $\sigma_{\alpha,\beta}$ is positive semi-definite, we obtain

$$\sigma \cdot \vec{m} = \lambda \vec{m},$$ (9)

that is the eigenvalue equation for σ. σ has 3N eigenvalues, λ_i, and 3N eigenvectors, $\vec{m}_i$.

To find out the eigenvectors $\vec{m}_i$ that minimizes d^2, we evaluate d^2 for each line defined by the direction $\vec{m}_i$ and $\vec{y}_0$. That is,

$$d^2(\vec{m}_k) = \frac{1}{S}\sum_{i=1}^{S} d_i^2 = \frac{1}{l}\sum_{i=1}^{S}(r_i - y_0)^2 - [(r_i - y_0) \cdot \vec{m}_k]^2.$$

$$= \sum_{\alpha=1}^{3N}[(\frac{1}{S}\sum_{i=1}^{S}(r_i - y_0)_\alpha^2 - \sum_{\alpha=1,\beta=1}^{3N}(r_i - y_0)_\alpha(r_i - y_0)_\beta \vec{m}_{k,\alpha}\vec{m}_{k,\beta}]$$

$$= Tr(\sigma) - \vec{m}_k \cdot \sigma \cdot \vec{m}_k = Tr(\sigma) - \lambda_k .$$ (10)]

The eigenvector corresponding to the largest eigenvalue, corresponds to the direction of the line passing through the average conformation, $\vec{y}_0$, that best represents the predominant motions in the protein.

Notice that Eqs. 8 and 9 are closely related to the definitions used in the quasi-harmonic approximation [29, 31, 32]. In the quasi-harmonic approximation, the eigenvalue system solved involves the matrix,

$$K_{\alpha,\beta} = kT\sqrt{a_\alpha a_\beta}\sigma_{\alpha,\beta}^{-1} ,$$ (11)

where σ is defined by Eq. 8 , $\sigma_{\alpha,\beta}^{-1}$ refers to an element of the inverse of the matrix σ, and a_α is the mass of atom α. The difference between quasi-harmonic analysis and the analysis presented here is that we do not assume unimodal distributions of the atomic fluctuations.(i.e., motions in a single basin of attraction or in other words, around a single minimal energy structure). The quasi-harmonic approximation uses the relation between the mean square displacement and the eigenfrequencies of a harmonic system to identify a set of temperature dependent frequencies. These eigenfrequencies will, under the assumption of harmonicity, determine the thermodynamics of the system in a closed form. The accuracy of the results strongly depend on the assumption of quasi-harmonicity. Any approach that relies on a quadratic form of the Cartesian displacements (i.e, correspondence analysis, [28], quasi-harmonic analysis, etc.)of the molecule will end with the matrix σ, or its inverse. The significance of the eigenvectors and eigenvalues of σ will strongly depend on the model used to interpret them.

The above formalism can be easily extended to define the best (in the least square sense) D-dimensional subspaces that describe the motions of the protein. We will present results for D=2 and 3. This generalization can be done by defining the distance of a configurational point, $r_i(t)$, from a D-dimensional subspace as

$$d_i^2 = (r_i - y_0)^2 - \sum_{k=1}^{D}[(r_i - y_0) \cdot \vec{m}_k]^2,$$ (12)

where $\vec{m}_k$ are D vectors spanning the D-dimensional subspace. Then, Eq. 1 can be replaced by

$$d^2 = \frac{1}{S}\sum_{i=1}^{S} d_i^2 = \frac{1}{S}\sum_{i=1}^{S}(\vec{r}_i - \vec{y}_0)^2 - \sum_{k=1}^{D}[(\vec{r}_i - \vec{y}_0) \cdot \vec{m}_k]^2.$$ (13)

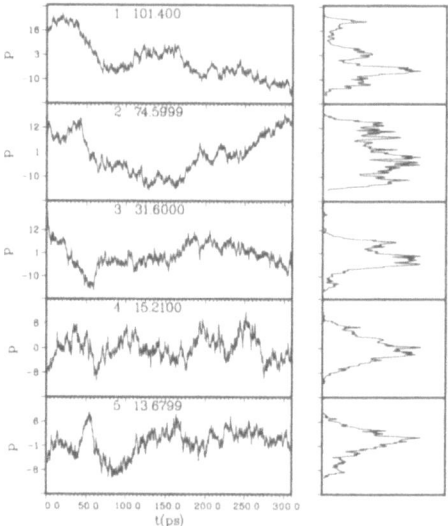

Figure 8: Projection $p_i(t)$ of the 310 ps molecular dynamics trajectory along the five largest eigenvalues of Eq. 9 are shown on the the left-hand-side plots. We refer to the figures as a-e, from top to bottom. The-right-hand-side plots show histograms of the frequency of occurence of all values of $p_i(t)$ for the corresponding vectors. $p_i(t)$ are given in Å, and t in ps. The labels on top of each curve show the eigenvalue ordering (from large to small) and the corresponding eigenvalues, λ (in Å^2).

and Eq. 2 is replaced by

$$f(\{\vec{m}_k\}, \vec{y}_0, \{\lambda_k\}) = \frac{1}{S} \sum_{i=1}^{S} \{(\vec{r}_i - \vec{y}_0)^2$$

$$- \sum_{k=1}^{D} [(\vec{r}_i - \vec{y}_0) \cdot \vec{m}_k]^2\} + \lambda_k [\vec{m}_k \cdot \vec{m}_k - 1] + \sum_{k,l \neq k}^{D} \lambda_{k,l} (\vec{m}_k \cdot \vec{m}_l) , \qquad (14)$$

where λ_k and $\lambda_{k,l}$ are Lagrange multipliers constraining $\vec{m}_k$ to be orthonormal. Following the procedure leading to Eqs. 3-10, we find that Eq. 10 can be generalized to

$$d^2(\{\vec{m}_k\}) = Tr(\sigma) - \sum_{k=1}^{D} \lambda_k . \qquad [15]$$

Here $\{\vec{m}_k\}$ represent any subset of D eigenvectors of σ. This equation shows that the best planes and volumes are spanned by the eigenvectors of σ with the lowest two and three eigenvalues, respectively. The fitness of each subspace will depend explicitly on the specific eigenvalues of σ. To use Eqs. 10 and 15 we need to find the highest eigenvalues and corresponding eigenvectors of σ. Once the eigenvalues and eigenvectors are calculated, the molecular dynamics trajectory is projected along the eigenvectors,

$$p_i(t) = r(t) \cdot \vec{m}_i. \qquad (16)$$

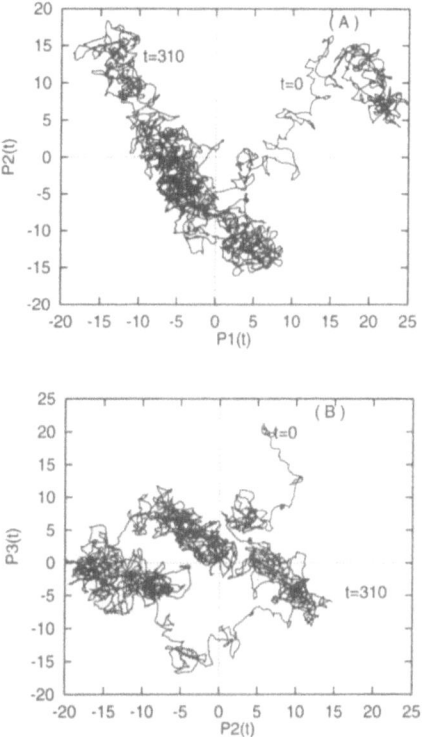

Figure 9: Projection of the 310 ps molecular dynamics trajectory on the plane spanned by directions (a) $\vec{m}_1$ and $\vec{m}_2$, and (b) $\vec{m}_2$ and $\vec{m}_3$.

Plots of $p_i(t)$ vs. t show the history (time series) of the trajectory along each direction. Two and three dimensional plots of (p_i, p_j) and (p_i, p_j, p_k) show 2D and 3D cuts of the configurational space sampled by the protein. Eigenvectors and eigenvalues are computed from the simulation data by calculating σ in Eq. 8.

Figs. 8a-e show the projection of the trajectory along the 5 first five directions (left) and the histograms of the occurence of all values $p_i(t)$ for the same vectors. The histograms of the population distributions can be fitted to multi-centered distributions with three (Figs. 8a-c) clearly distinguishable centers. Each center is indicative of different basins of attraction. A quasi-harmonic approximation is clearly violated for the first three directions. The projection corresponding to the largest eigenvalue, $\lambda = 101.4$ Å^2, shows a trimodal distribution. The distribution centered at $p_1(t) \sim 20$ Åis only sampled during the first 40 ps of the molecular dynamics trajectory. A transition from that basin to one centered at $p_1(t) \sim -7$ Åoccurs during the next 10-20 ps. Configurations near this basin of attraction are sampled for the next 40 ps. Another fast transition occurs toward the distribution centered at $p_1(t) \sim 3$Å. Another transition to the distribution with center near $p_1(t) \sim -7$ Å occur within the next 65 ps. Notice that the transitions from one basin of attraction to another occur fast (less than 5 ps) while the mean time between

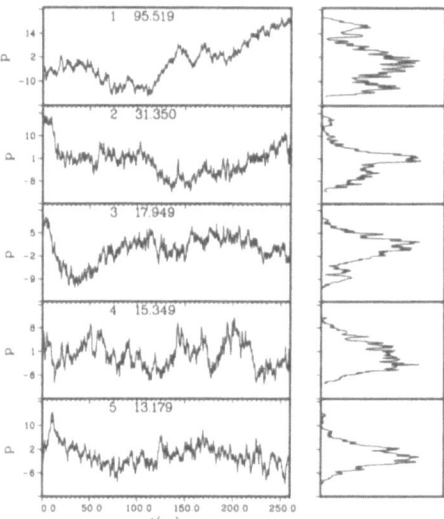

Figure 10: Projection $p_i(t)$ of the last 260 ps molecular dynamics trajectory along the five largest eigenvalues of Eq. 9 are shown on the the left-hand-side plots. We refer to the figures as a-e, from top to bottom. The right-hand-side plots show histograms of the frequency of occurence of all values of $p_i(t)$ for the corresponding vectors. $p_i(t)$ are given in Å, and t in ps. The labels on top of each curve show the eigenvalue ordering (from large to small) and the corresponding eigenvalues, λ (in Å^2.)

transitions is long (40-70 ps). The nature of the trajectories described here were apparent from the calculation of $d(t,t')$ matrix shown in Fig. 5 and the tree analysis shown in Fig. 6.

Similar patterns of fast inter-basin transitions followed by overdamped oscillations (and possibly transitions to other local minima within each basin of attraction), are observed for the other four directions in Figs. 8b-e. The *rms* fluctuations of the coordinates during the simulation are 0.98 Å, with $tr(\sigma) = 394$ Å^2. The first five directions describe 60. % of the fluctuations, with the first direction alone describing 25%. Projections of the trajectories on 2-dimensional subspaces of the configurational space will better characterize the nature of the motions described in Fig. 8. Figs. 9a-b show projections of the trajectory on planes spanned by the directions $\vec{m}_1$ and $\vec{m}_2$ (with the largest eigenvalues) and $\vec{m}_2$ and $\vec{m}_3$ (the best planes that exclude the direction $\vec{m}_1$). The initial (t = 0 ps) and final (t = 310 ps) positions of the trajectory on the planes are labeled in the figures. The distribution of conformations in Fig. 9a show four basins of attraction with centers near $(p_1, p_2) = $ (I) (20,10), (II) (5, -12), (III) (-7,-5) and (IV) (-12,10). These points are chosen to label the four basins and do not carry any other significance. Basin **I** contains the initial configuration and is well separated from the other three basins. The other three basins are densely sampled during the trajectory. Together with Figs. 8a-b we can

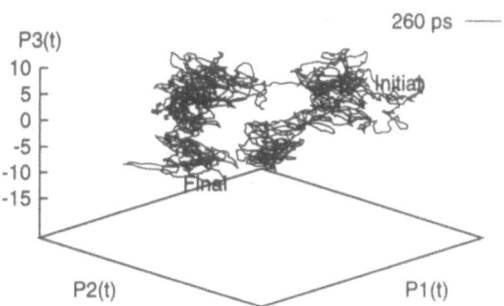

Figure 11: Projection of the last 260 ps molecular dynamics trajectory on the 3-dimensional subspace spanned by directions $\vec{m}_1$, $\vec{m}_2$ and $\vec{m}_3$.

see that the protein samples basin **I** and undergoes a transition to the other basins. The protein will remain there for the remaining 260 ps of the simulation. Transitions to each of the other basins (**II, III,** and **IV**) are observed.

During the last 260 ps of the simulation $p_1(t)$ samples values between -20 and 10 Å, while $p_2(t)$ samples the same range of values as in the whole trajectory. Fig. 9b shows a projection of the trajectory on the plane spanned by the $\vec{m}_2$ and $\vec{m}_3$ directions. The densely sampled distribution of p_2 values is well separated into the third dimension, $\vec{m}_3$. Four densely occupied basins of attraction are observed. A closed pathway interconnecting these basins can also be seen.

The large change in $p_1(t)$ from $p_1 \sim 20$ Å to $p_1 \sim 3$ to -7 Å, and since $p(t)_1 \sim 20$ Å is not sampled again during the simulation lead us to consider this transition as a transient in the dynamics from a metastable initial conformation to a lower-energy conformation. This was first observed in a previous article [11] when a 216 ps trajectory was analyzed. An analysis of a 310 ps trajectory shows the same features. Therefore, we identify this transition as a transient and the first 50 ps of the trajectory are going to be considered as part of the system equilibration. The remaining 260 ps trajectory (after 74 ps equilibration) is analyzed next. Fig. 10 shows the projection of this trajectory along the five largest eigenvalues of Eq. 9. Multicentered distributions of the occurence of $p_i(t)$ are shown by the first three direction (Figs.10a-c). Patterns of fast inter-basin transitions followed by overdamped oscillations observed in Figs. 8a-e are also present. [It must be clarified that the MODC analysis yields different eigenvalues and eigenvectors when different segments of the trajectory are analyzed.]

The projection of the equilibrated trajectory along the first three directions (i.e., the best 3-dimensional volume) is shown in Fig. 11. Four basins of attraction can be distinguished as they are densely sampled during the trajectory. This volume describes 47 % of

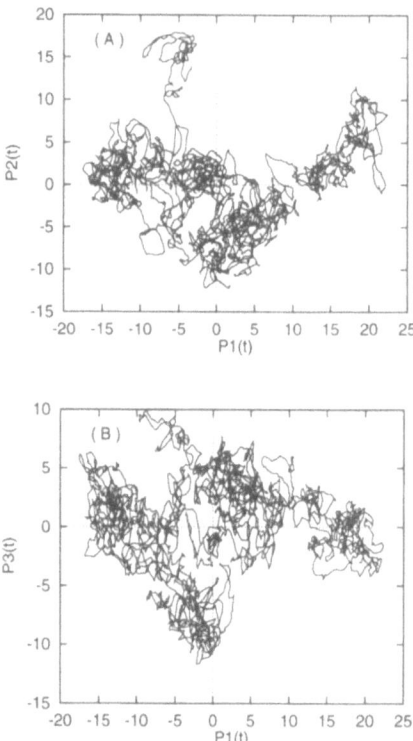

Figure 12: Projection of the last 260 molecular dynamics trajectory on the plane spanned by directions (a) $\vec{m}_1$ and $\vec{m}_2$, and (b) $\vec{m}_1$ and $\vec{m}_3$.

the total fluctuations during the last 260 ps trajectory. Projection of this volumes on the (p_1, p_2), and (p_1, p_3) planes are shown in Fig. 12. Figs. 12a-b show a clear partitioning of configurational space similar to the one shown in Fig. 9b. The planes (p_1, p_2) and (p_1, p_3) are the two planes that best describe the fluctuations of the system, representing 41% and 37% of the total fluctuations. The projection of the trajectory on the (p_2, p_3) plane (not shown) does not show a clear partitioning of the configurational space. The MODC analysis provides a quantitative tool to dissect the multi-dimensional configurational space into smaller subspaces that best describe the fluctuations of the system and the topology of the energy surface sampled during a trajectory.

4. CONCLUSIONS

We have used the molecular dynamics technique mainly as a convenient (although not efficient) method of exploring the configurational dynamics of a small protein in solution. From the results presented here we can conclude that non-linear motions describing oscillations around multicentered distributions are responsible for most of the atomic fluctuations sampled by a protein on a 100 ps time-scale. These atomic fluctuations are not well described by large fluctuations of individual atoms or small groups of atoms, but by concerted motions of many atoms. These modes are non-linear in the sense that they describe transitions among different basins of attraction. The mean time between transitions is around 50 ps. The signature of these non-linear modes can be seen in various local structural variables (dihedral angles) and global variables (rms distance between all pairs of configurations and clustering analysis). A method for extracting optimal dynamical coordinates that best describe the protein fluctuations has been described. A generalization of this method to identify small (1-3) dimensional subspaces of the configurational space has been used to show, for the first time, a conclusive description of the protein dynamics within the context of multi-basin dynamics.

Acknowledgements

I wish to offer special thanks to Prof. J. A. Krumhansl for his interest on this work, comments and suggestions. I also want to thank Drs. G. Hummer, Javier Sobehart, and Raphael Blumfield, members of the Los Alamos National Laboratory (LANL) Center for Non-Linear Studies (CNLS) Protein Dynamics Working Group (PDWG) for comments and suggestions. This work has been supported by the US-DOE under LANL LDRD-PD research funds.

References

[1] Levitt, M., C. Sander, and P.S. Stern, *J. Mol. Biol.*, **181**, 423 (1985).

[2] Noguti, T. and Go, N., *Nature*, **296**, 433 (1982);

[3] Englander, S.W., and Kallenbach, N.R., *Quarterly Rev. Biophys.*, **16**, 521 (1984).

[4] Frauenfelder, H., Siglar,H.A., and Wolynes, P., *Science,* **254**, 1598 (1991)

[5] Frauenfelder, H., Steinbach, P.J., and Young, R.D., *Chemica Scripta*. **29A**, 145 (1989).

[6] Ansari, A. et al., *Proc. Natl. Acad. Sci. (USA)*, **82**, 5000 (1985)

[7] Iben, I.E.T. et al., *Phys. Rev. Lett*, **62**, 1916 (1989).

[8] Elber, R. and Karplus, M., *Science*, **235**, 318 (1987)

[9] Go, N. and Noguti, T., *Chemica Scripta*, **29A**, 151 (1989).

[10] Austin, R.H., in *1992 Lectures in Complex Systems*, L. Nadel and D.L. Stein, Editors. (Addison-Wesley, New York. 1993.) pp. 353-400.

[11] García, A.E. *Phys. Rev. Lett.* **68**, 2696 (1992).

[12] Webster Dictionary defines so.phis.ti.cat.ed : aj [ML sophisticatus] 1: not in a natural, pure, or original state : Adulterated {a   oil} 2: deprived of native or original simplicity : as 2a: highly complicated : Complex { instruments} 2b: Worldly-wise,

Knowing {a   adolescent} 3: devoid of grossness : Subtle : as 3a: finely experienced and aware {a   columnist} 3b: intellectually appealing {a   novel} - so.phis.ti.cat.ed.ly av.

[13] Steinbach, P.J. and B.A. *Proc. Natl. Acad. Sci. (USA)*, **90**, 9135 (1994).

[14] Flach, S. and Willis, C.R., *Phys. Rev. Lett.*, **72**, 1777 (1994).

[15] Hendrickson, W.A. and M.M. Teeter, *Nature*, **290,** 107 (1981).

[16] Teeter, M.M., *Proc. Natl. Acad. Sci. (USA)*, **81**, 6014 (1984).

[17] Llinás,M., A. de Marco, and J.T.J. Lecomte, *Biochem.*, **19**, 1140 (1980).

[18] Usha, M.G. and R.J. Wittebort, *J. Mol. Biol.*, **208**, 669 (1989).

[19] Vermeulen,J.A.W.H., R.M.J.N. Lamerichs, L.J. Berliner, A. de Marco, M. Llinás, R. Boelens, J. Alleman, and R. Kaptein, *Febs*, **219**, 426 (1987). R.M.J.N. Lamericks, L.J. Berliner, R. Boelens, A. de Marco, M. Llinás, and R. Kaptein, *Eur. J. Biochem.*, **171**, 307 (1988).

[20] Whitlow, M. and M.M. Teeter, *J. Amer. Chem. Soc.*, **108**, 7163 (1986).

[21] Teeter, M.M. *Ann. Rev. Biophys. Biophys. Chem.*, **20**, 577 (1991).

[22] Teeter, M.M. and D.A. Case, *J. Phys. Chem.*, **94**, 8091 (1990).

[23] García, A.E. & Stiller, L. *J. Comp. Chem.*, **12**, 1 (1993).

[24] Schulz, G.E. and R.H. Schirmer, *Principles of Protein Structure*, (Springer-Verlag, NY, 1978.)

[25] Krumhansl, J.A., in *Computer Analysis for Life Science*, C. Kawabata and A.R. Bishop, Eds., (Ohmsha, LTD, Tokyo, Japan, 1985.) pp. 78-88.

[26] Krumhansl, J.A., in *Proceedings in Life Sciences: Protein Structure, Molecular and Electronic Reactivity*, R.H. Austin, et al., Eds., (Springer Verlag, New York, 1987.)

[27] McLachalan, J., *J. Mol. Biol.*, **128**, 49 (1979).

[28] Lebart, L., Morineau, A., and Warwick, K. M., *Multivariate Descriptive Statistical Analysis*, (John Wiley & Sons, New York, 1984).

[29] Karplus, M. and Kushick, J.N., *Macromolecules*, **14**, 325 (1981);

[30] Ichiye, T. and Karplus, M., *Biochemistry*, **27**, 3487 (1988).

[31] *Proteins: A Theoretical Perspective of Dynamics, Structure, and Thermodynamics.* Brooks, C.L., Karplus, M. & Montgomery-Pettitt, B., *Advances in Chemical Physics*, **LXXI**, (John Wiley & Sons, New York, 1988.)

[32] Levy, R.M., Karplus, M., Kushick, J.N. and Perahia, D., *Macromolecules*, **17**, 1370 (1984).

[33] García, A.E., Soumpasis, D.M. and Jovin, T.M., *Biophys. J.*, **66**, 1742 (1994).

Nonlinear excitations in molecular crystals with chains of peptide bonds

M. Barthes

GDPC, Université Montpellier II, cc26,
34095 Montpellier cedex 05, France

1 - INTRODUCTION

In some crystals containing chains or networks of >N-H·····O=C< hydrogen bonds , several vibrations of the amide group are very anharmonic or even anomalous . The best known and the most studied among these crystals is acetanilide ($C_6H_5NHCOCH_3$ or ACN).

Recently , some other crystals with chains of hydrogen bonds have been shown to possess also very non harmonic or supplementary modes related to motions of the N, H, C, O atoms. Some of these systems have,like ACN,parallel chains of peptide bonds, and others like the alanines, a three -dimensional network .

Most of the actual theories which attempt to explain these anomalous behaviours imply nonlinear excitations, called Davydov-like solitons or localized "polaron -like" modes. These interpretations are still submitted to discussion . Conformational substates of the amide proton are alternatively invoked , or Fermi resonance , or transition dipole coupling in some cases . None theoretical model takes into account the whole experimental situation . On the other hand , a direct proof of the existence of non linear excitations in these systems has not yet been obtained

The motivation for this research field comes from the fact that >N-H·····O=C< networks are present in all polypeptides and alpha-helix proteins.Such hydrogen bonds exist in DNA , contribute to link the two strands and are involved in the opening and replication processes .

So, it is important to understand the dynamics of atoms belonging to this bond, particularly that of the amide proton, and also to check the part of localized polaronic or solitonic excitations , often invoked to model energy transport in proteins , DNA conformational changes and denaturation .

In this paper we present mostly spectroscopic data on the discussed , or "unconventional" features of the amide vibrations ,and recent structural data .The above mentioned theories are not exposed into the details but only their adequation to the experimental observations is discussed .

2 - ACETANILIDE AND DERIVATIVES

Acetanilide ($C_6H_5NHCOCH_3$ or ACN) is a crystal with some one-dimensional character . It contains parallel chains of hydrogen-bonded ...H -N-C-O···· amide groups (Fig.1) like

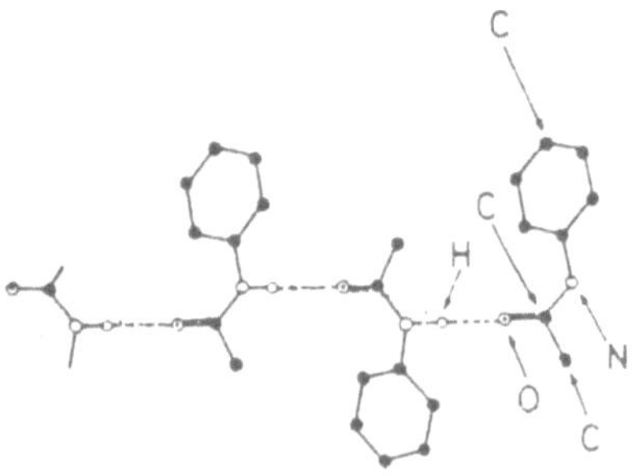

M. Barthes

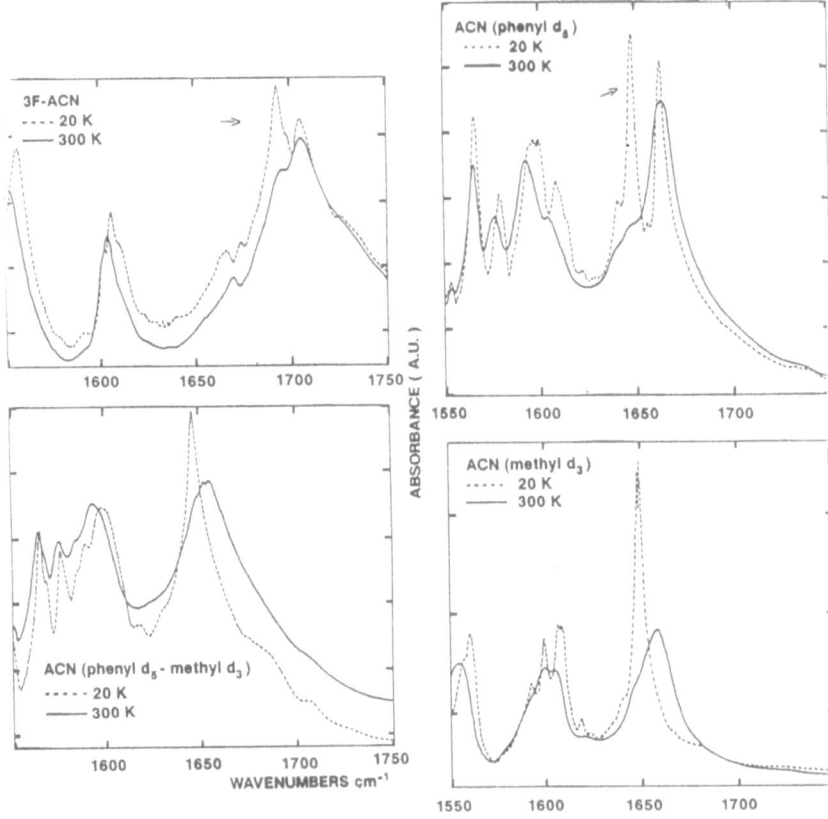

Fig. 1 - Projection of one molecular chain of ACN on the (a,b) plane

Fig. 2 - Infrared spectra of the Amide-1 band in four derivatives of ACN . The anomalous peak is indicated by
an arrow .

polypeptides and alpha-helix proteins. ACN displays some anomalous infrared and Raman modes(1),whose origin is subject of controversy .

The extra-intensities (at 1650 cm-1) observed at low temperature in the region of the amide-1 mode (or C=O stretching mode , at about 1665 cm-1), have been tentatively explained in terms of Davydov- soliton theory (2) , vibron soliton (3) , "polaronic" localized modes(4) (5),nonlinearly coupled oscillators(6). All these theories involve the self-trapping of the vibrational energy of the amide-1 by coupling with low frequency phonons.The self-trapped mode has the properties of a nonlinear coherent excitation or a quasi-soliton,assumed to have a very low decay.This excitation would be in theory,able to propagate along the chain to transfer energy without loss and, in a further step, to provide a plausible mechanism for the high efficiency of energy transport in biological molecules,an unsolved problem in bioenergetics (7). This is the reason why solving the acetanilide problem, i.e. prooving the existence or the lack of self-trapped vibrational excitations, is an attractive challenge .
Alternative interpretations of the anomalous modes in ACN have been proposed , involving Fermi resonance (8) , and more recently the hypothesis of two slightly non-degenerate configurations of the amide proton (9) in a double well potential as a conclusion of a transient IR spectroscopy , carried out by using the pump-probe technique , and indicating that the 1650 cm-1 band does not show a long -lived feature expected for a self-trapped state .

Our recent infrared investigation has been undertaken with the aim of testing the different theoretical models of the acetanilide specific optical anomalies .Polarised Raman scattering measurements have been performed at many temperatures and frequencies . Neutron diffraction crystallographic structure of ACN at room and liquid helium temperature is presented also, giving a special attention to the location of the amide proton,with intent to check the conjecture(9) of two positions for the amide hydrogen.

2.1 Infrared Results

Absorption measurements of several selectively deuterated ACN and of the methyl-three-fluorinated ACN ($C_6H_5NHCOCF_3$, or 3F-ACN) have been carried on from 300K to 20K (10) , using powdered samples .
A detailed study of the amide-1 region indicates different behaviours among the derivatives : besides the "unconventional" mode first observed by Careri et al in pure ACN[1] at about 1650 cm-1,near the C=O stretching at 1665 cm-1, our new data are summarized in fig 2 . The amide-1 mode exhibits a classical thermal behaviour in the methyl-deuterated derivative (or d_3) and also in the methyl-and-phenyl-deuterated derivative (or d_8): its frequency decreases with decreasing temperature, no extra-peak is observed, the integrated intensity is approximately constant (within about 15%). On the contrary, if only the phenyl group is deuterated in the molecule (or d_5) ,the unconventional mode appears like in pure ACN ,with an increasing intensity with decreasing temperature .In this case also, the energy of both the C=O stretching and the anomalous mode remain roughly constant between 300 K and 20 K . The same anomalous band is observed in the methyl fluorinated derivative (or 3F-ACN) (Fig.2). Following (2)(5) the energy shift between the fully delocalized exciton (or standard amide-1 frequency) and the assumed self-trapped state or "soliton" (frequency of the anomalous band) should reflect the strength of the exciton-phonon coupling at the origin of the nonlinear term in the hamiltonian . the experimental values are $\Delta E = 16.5$ cm-1 in ACN , $\Delta E = 15$ cm-1 in d_5 , and $\Delta E = 11$ cm-1 in 3F-ACN .
Other non harmonic behaviours are observed : the intensity of the amide-V mode (out-of-plane N-H bending , or $\gamma(NH)$ at about 770 cm-1) exhibits a strongly non-harmonic temperature dependence , as well as the "H-bond strain" in the range 100-110 cm-1 . In all derivatives , a very strong intensity is observed when lowering the temperature, at about

M. Barthes

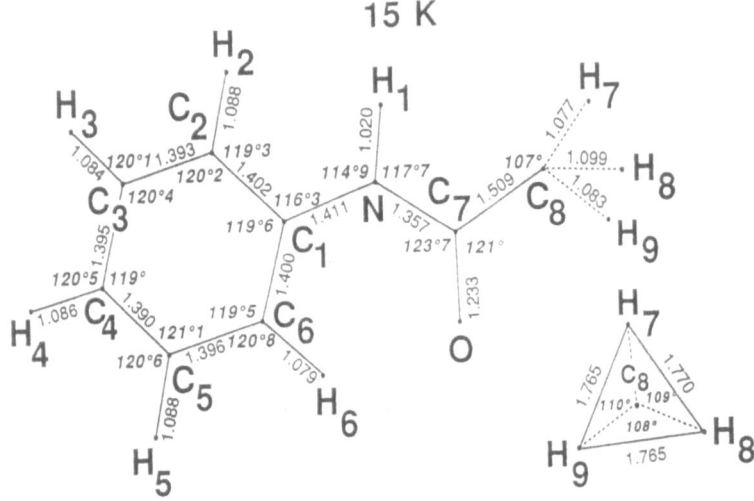

Fig. 3 - Distances and angles in the ACN molecule at 15K .

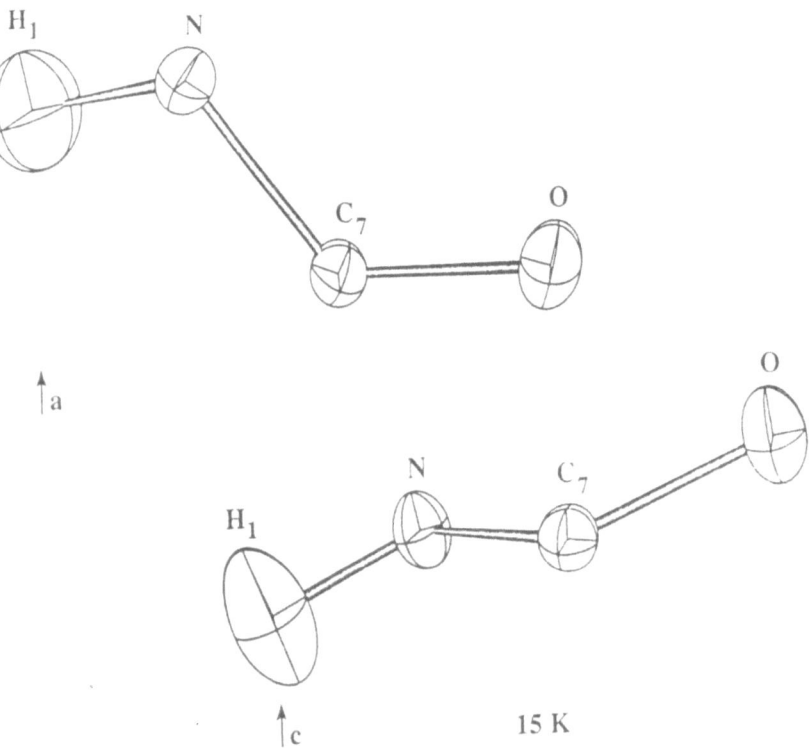

Fig. 4 - Thermal ellipsoïds for the amide group (O-C-N-H) at 15K in view down a and c axes .

twice the amide-V frequency or less (around 1500 cm-1)(10) and is assigned to a very intense overtone.

2.2 Raman scattering

After the pionering experiments by G.Careri et al.(1) , the measurements(8) on the N-D and ^{13}C-O substituted species suggested a temperature tuning of a Fermi coupling as an explanation of the anomalous intensity changes observed in the IR and Raman spectra of the amide-I region.

More recently low-frequency Raman spectra on fully deuterated ACN have shown a new band in the 70-120 cm-1 region(11) with a strong temperature dependence of its frequency . Another study shows the splitting in the B_{1g} polarization of a band at 142 cm-1 at 20K(12) which is assigned , in agreement with data from neutron scattering(13)(14) to a methyl torsion . Measurements on a d_8 single crystal allow to observe a band at about 1495 cm-1 which increases in intensity on cooling(15) , and whose origin is still under discussion :

new "polaronic" state , or overtone of the γ(NH) transition ?

New Raman measurements are now underway , related to molecular dynamics simulations (16).High-pressure Raman study of the amide-I vibrational mode demonstrates the existence of a new band at 30 cm-1 below the amide-I phonon(17) .

2.3 Low temperature neutron diffraction

Previous X-ray measurements (18)(19)(20) have determined the structure (space group D^{15}_{2h}), the absence of phase transition above 113K , and the location of the amide proton near the amide nitrogen in a simple potential well along the chain direction (b) ; however the existence of a multiple well potential in the other space directions could not be a priori ruled out . The neutron diffraction measurements presented here take advantage of the very high neutron scattering cross-section of hydrogen which allows to expect a better accuracy in the proton position determination .

The space group at 15K and 295K are found to be the same , in agreement with previous X-ray determinations .So, the possibility of a structural phase transition taking place above 15K is now ruled out . The X-ray and neutron values for the C=O distance at low temperature are the same, namely 1.233 Å , so the maxima of the nuclear and electronic densities are in good coincidence . The intermolecular hydrogen-bond distance N-H...O' is 2.903Å at 15K and 2.943Å at room temperature . The N-H distance is 1.020(2) Å at 15K. This means that the proton is firmly localized near the nitrogen atom .

The intramolecular O...H6 distance (2.232 Å at 15K, 2.257 Å at 295K) is unusually short , about - 0.3Å shorter than the expected Van der Waals contact , and may affect the atomic vibrational parameters for the two atoms . (Fig .3)

This fact should be taken into account in the interpretation of the dynamics of ACN , especially of the unexplained mode near the amide-1 : the oxygen may be weakly connected to the phenyl H6 proton of its own molecule .

The NH...O' angle is 170°7(.1) at 15K and 172.3°(.4) at RT . The molecule undergoes significant conformational change between 295K and 15K . Comparing with available X-ray data at 113K , this conformational change seems to be a continuous function of the temperature .

Thermal ellipsoids for the amide group (H-N-C=O) and Fourier maps have been plotted at 15 and 295K . The vibrational amplitude of the H1 (amide) proton is maximum in a direction (roughly the c axis) normal to the bond .(Fig.4)

The Fourier maps plotted in fig.5 show that *the cut through the proton position, along the direction in which the thermal ellipsoid is elongated , indicates that there is only one minimum for the local potential . However substantial delocalization is obvious , and two minima separated by less than 0.1 Å would not show up in these maps .*

Only one static conformation of the amide proton, the same for both temperatures , is obvious in the plane of the amide group .(Fig.6)

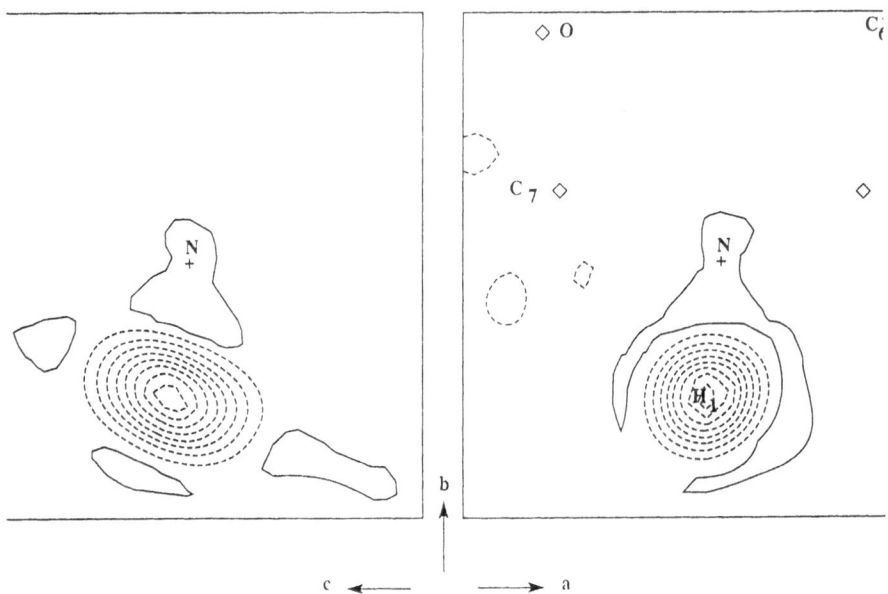

Fig. 5 - Difference Fourier maps showing the nuclear density corresponding to the amide proton (H1)

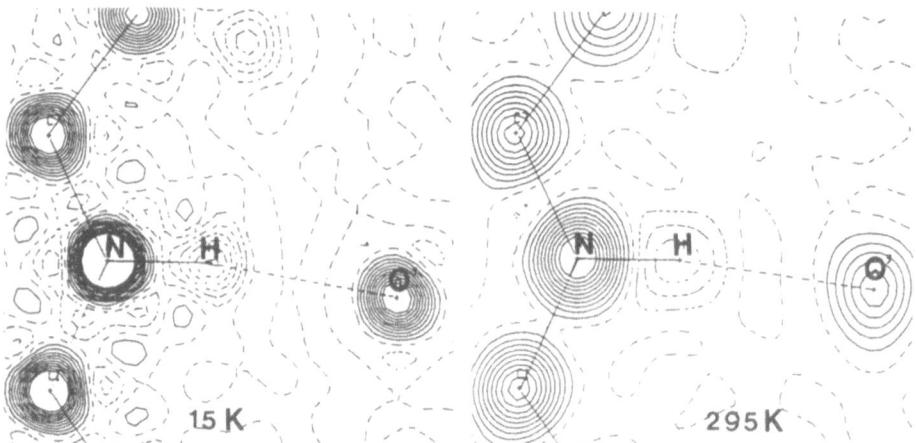

Fig.6 - Nuclear density maps in the best plane of the amide group at 15K and 295K.The hydrogen bond
N -> O' is approximately horizontal

2.4 Discussion

Several arguments have been already invoked in [1] to rule out the interpretation of the extra-amide-1 peak by a Fermi resonance (FR). However, the disappearance of the suplementary peak at 1650 cm-1 in both the methyl-deuterated derivative (d_3) and the amide-deuterated ACN (d_1) (1)(8) suggests a FR involving a mode combining methyl and amide vibrations .

Nevertheless, it was not possible to fit the data with the predicted relation between energies and intensities of the interacting modes(10) in an assumed FR. Secondly it is unlikely to observe at one and the same frequency a FR both in IR and Raman, as it is the case in ACN . Thirdly , our measurements show that the supplementary peak is observed as well in the 3F-ACN derivative,in which the H-bond strength is weaker, the amide frequencies are different, and without any CH_3 . Moreover , the supplementary peak possesses overtones (21).This last feature is unexpected when the fundamental is the result of a FR ,but very well known in the case of localized modes in alkali halides(22) , formally analog to "polarons"(23) .

So the explanation of the optical anomalies in term of a Fermi coupling seems very unlikely. It needs now to examine other theories , that explain the extra-peak by the existence of non degenerated conformational substates for the amide proton or a nonlinear excitation like a soliton or a "polaron" .

The above presented neutron diffraction measurements give informations on the low temperature localization of the hydrogen atoms in the ACN molecule . No evidence is found for a double-well potential for the amide hydrogen at any temperature , within a resolution of about 0.1Å , contrary to the suggestion in (9) .

From these new structural data it appears that the interpretation of the "unconventional" amide-1 infrared and Raman mode could not rest on the existence of conformational substates, as did the C=O stretching mode in myoglobin (24) .

Dynamical nonlinear effects resulting from the coupling of internal modes with phonons could be more relevant as a possible origin of these anomalies, because they remain the only explanation which has not been ruled out by experiments as yet , but no direct proof of their existence has been established for the moment .

The energy shift of the 1650 cm-1 band , its intensity dependence on temperature , the existence of its overtones , are all in agreement with the predictions of the "polaron" model . However , the non-harmonic behaviour of the other amide modes are not taken into account by the theory .

Normal modes analyses and phonon dispersion calculations are now underway by computer simulation techniques , using the program CHARMM and will be compared with all available optical and neutron scattering data (16)(25).

3 - N-METHYLACETAMIDE

N-methylacetamide (NMA or CH_3-NHCO-CH_3) is the simplest molecule to posses a peptide group . In its crystal structure , NMA is arranged in chains , each molecule being hydrogen bonded head -to -tail to its neighbour. The chain structure of NMA is similar to that in acetanilide but the hydrogen bond is slightly stronger ; the NH...O' distance is about 2.8 Å. This simple molecule may be taken as a starting point for peptide studies ; some force fields calculated for NMA are expected to be transferable to polypeptides and proteins

NMA has been the subject of detailed infrared and Raman investigations(26)(27), of inelastic neutron scattering studies (INS)(28) and of ab initio calculations (29)(30) .

A study of the temperature dependence of the infrared spectra of NMA has been performed(27) . On the contrary of ACN, no supplementary band has been observed near the amide-1 band .However an additional "A" band, on the small wave number side of the amide-II band(Fig.7) is observed around 1525 cm-1 with an increasing intensity as the temperature is lowered .

There is a resemblance between this band and the unconventional vibrational band accompanying the amide-1 band in ACN. The same mechanism , that is the "polaron" model , is suggested to be responsible for the additional amide-II mode in NMA. The predictions of this model(4) are compatible with the data of the temperature dependence of the integrated intensity , I , of this mode whiwh can be fitted with the law

$$Ln\ I = const\ x\ (- AT^2)\ .$$

The hypothesis of a Fermi resonance between amide-II and an overtone of amide-V has been examined and rejected .

The existence of a double conformation of the NMA molecule as a possible origin for the additional "A" band has not been investigated by crystallographic techniques: unfortunately the crystal structure of NMA is unknown below the structural transition at 10°C , and is poorly refined at room temperature (31) .

Extensive INS studies have been performed on NMA and its deuterated derivatives (28). The dependence of the scattering function on the momentum transfer has been used to obtain the shapes of the potential functions of the various vibrations, and to make distinction between single and double minimum potentials (32). So a double minimum potential has been discovered in the bond direction and proton transfer exists between both wells .

Another surprising feature is also found in the frequency range of 1300-1800 cm-1 in incoherent INS measurements .INS spectra of four isotopic derivatives of NMA at 20K are studied and the frequencies compared with those from IR and Raman measurements ."The quantitative simulation of the INS intensities in the harmonic force-field approximation shows that the proton dynamics for the NH proton are different from those proposed previously...The INS intensity in the 1300-1800 cm-1 region is too great , compared to the intensity of the out-of-plane bending mode,to be represented with only the in-plane bending mode (mixed with skeleton modes...) and the overtone of the γ(NH) mode. The data clearly say that there is an additional mode in this region,i.e. the ν(NH) mode .This conclusion...reveals a great weakening of the NH bond , a strong ionic character for the hydrogen bond and proton transfer."(28)

This new assignment of the NH stretch at 1575 cm-1 instead of 3250-3100 cm-1 which is assigned to overtones , seems consistent with a dynamical exchange of the amide proton between the amide-like(>NH..O=C<) and the imidol-like(>N..HO-C) forms .

Following these new surprising data , there is now a stronger motivation to achieve the crystallographic structure determination of a NMA single crystal by neutron diffraction at low temperature .Such a weakening of the N-H bond should have for consequence an increase of the N-H distance, which should be measured by neutron diffraction , at low temperature on a single crystal .

Secondly it appears that the comparaison with ACN could be very fruitful.Both compounds have very similar structures but different properties : the hydrogen bond distance differs by less than about 0.1Å in ACN and NMA,however no evidence for a proton transfer has been found in ACN. In NMA the amide proton dynamics are isolated from the dynamics of the molecular backbone.On the contrary , it is not the case in ACN where the dynamics of the amide and methyl group influence each other(14). About the surprising assignment of the NH stretching mode in NMA, it is worth to remark however that one model of the molecular dynamics of ACN(6) attributes the anomalous band at 1650cm-1 to a coupling of the motion of the amide proton - and not to the CO stretch - with low-frequency phonons . Perhaps an analogous treatment applied to NMA could help to better understand the assignment of the NH mode in the 1575 cm-1 region.

Moreover it will be interesting also to fit the INS spectra of NMA with molecular dynamics simulations including not only the internal molecular vibrations but also the intermolecular potentials (like CHARMM) so as to take into account the coupling of amide modes with low frequency phonons.

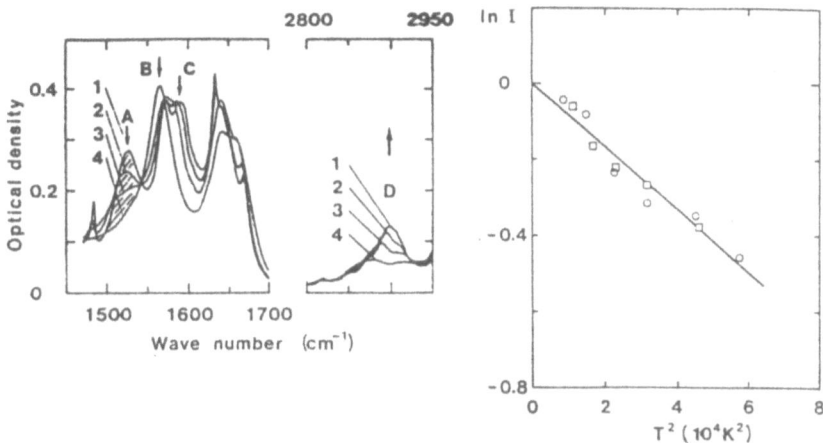

Fig.7. Infrared spectra of the amide-II region in N-methylacetamide.From (27)

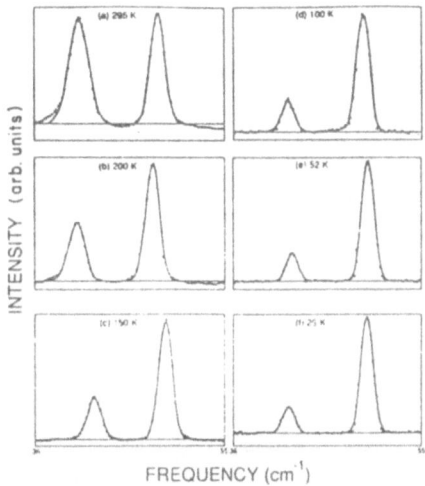

Fig.8 - Anomalous temperature dependence of the intensity of two Raman modes in L-Alanine.From (33)

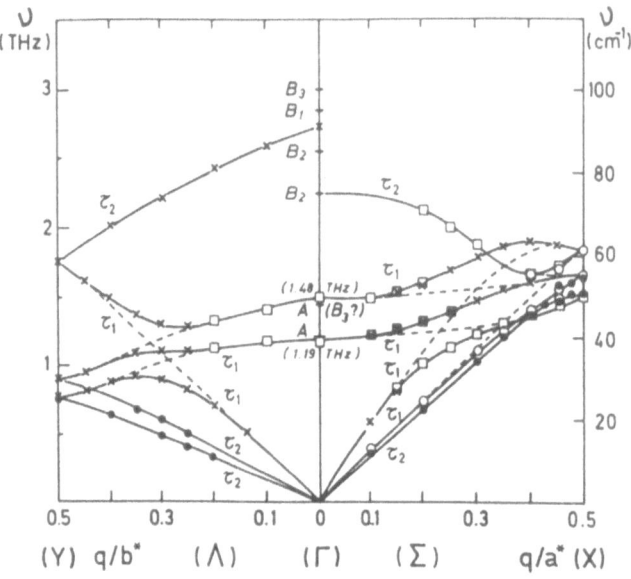

Fig.9 - Dispersion branches of phonons in L-Alanine . From (34)

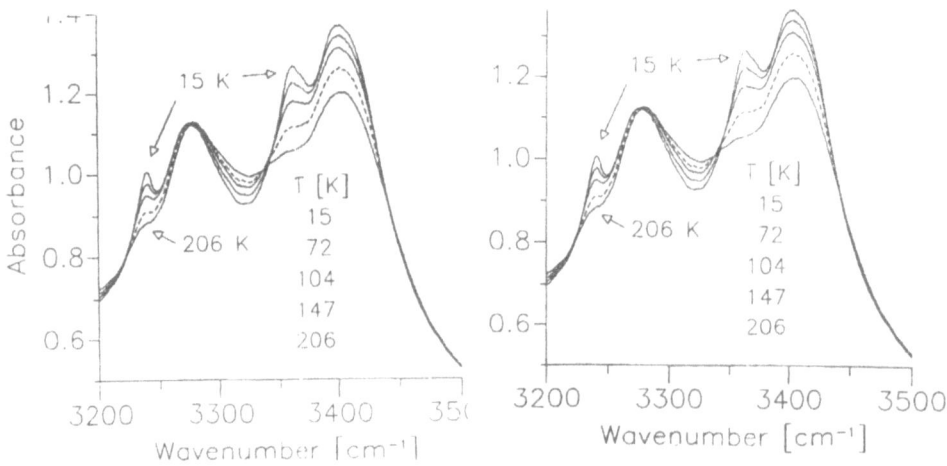

Fig.10 - Temperature dependent sidebands near the N-H stretching in Tryptophane-Alanine and Tyrosine-Alanine .From (35)

4- L-ALANINE   AND   POLYALANINES

An anomalous temperature dependence of the intensity of the two lowest Raman modes at 42 and 49 cm-1 was observed in single crystals of L-Alanine (33) (Fig.8).The sum of the intensities obeys Maxwell-Boltzmann statistics from 20K to 340K, but the intensities of the individual lines are anomalous.This behaviour is explained by assuming that both lines share the same degrees of freedom but that a mode instability is triggered abruptly .This instability possibly indicates the existence of dynamic localization of vibrational energy.

More recently the dispersion of the low frequency lattice vibrations have been studied by coherent INS (Fig.9). There is evidence (anticrossing) for interaction between the two lowest frequency optical phonons and the longitudinal acoustic mode (34) .The anticrossing effect may be related to the anomalous temperature dependence of the two lowest frequency optical phonons .Phonon dispersion relations are derived from normal mode analyse using the program CHARMM .The so calculated dispersion curves are in reasonable agreement with experiment .Further neutron experiments are planned to determine the dispersion curves at low temperature .

The infrared spectra of two poly-alanines , tryptophane-(alanine)$_{15}$ and tyrosine-(alanine)$_{15}$ have been recently investigated(35). The anomalous red-shifted sideband of amide-I is not observed. In this frequency range the temperature behaviour of both polypeptides differ completely from that of ACN. In the NH stretch region the situation is quite different . Both polypeptides have two broad absorption bands at about 3275 cm-1 and 3410 cm-1 (Fig.10), assigned to the mixing of the basic NH stretch to the overtone of amide-II . Both bands have weaker red-shifted and strongly temperature dependent sidebands. The temperature dependence of their integrated intensities I is fitted with a law

Ln I α - AT^2 , in consistence with the predictions of the "polaron" model, and suggest a coupling of the NH stretch to low frequency phonons .

The occurence of two sidebands in each spectrum is explained by assuming that the NH "polaron" mode , as the basic NH mode , is also mixed through Fermi coupling to the amide overtone.

It is moreover proposed that the lattice modes coupled to excitons involve the side groups attached to the main chains of hydrogen-bonded peptide units . More experiments and analysis are needed to check these interpretations.

5 - NUCLEOTIDES , POLYPEPTIDES AND PROTEINS

The vibrational spectroscopy of some polypèptides and proteins has been extensively reviewed (36) .It appears that in a number of these substances , the absorption band associated with the amide-I vibration is split , the shift between the components could attain 60 cm-1 .

This splitting is accounted for by the effect of the transition dipole coupling term in the potential . The integrated intensities are also related to the transition moments , but in some of these compounds the temperature evolution of these amide spectra has not been systematicaly studied .As an example , the observed splitting of amide-I in polyglycine-I is about 50 cm-1 , and is well accounted for by transition dipole coupling interactions. Moreover, INS investigations on this compound have recently demonstrated that the amide proton is delocalized between two equivalent sites in a symetric double minimum potential (32).

In some pyrimidic bases like polycrystalline uracil and thymine, ab initio calculations were used to determine semiempirical force fields by fitting to the Raman and IR spectra .The Fermi resonance of the overtones of the out-of-plane deformation vibrations of oxygens with their stretching modes have been proposed as an explanation for the band splitting observed in the 1600-1800 cm-1 region of uracil(37).In thymine (that is an uracil molecule in which a methyl group is substituted) no more splitting of the C=O stretching mode is observed .

6 - CONCLUSION

When a pendulum has more than two degrees of freedom , and a coupling between some of these variables , it adopts a chaotic motion .

In an organic molecule , or a hydrogen bonded chain of molecules, the N,H,C,O atoms of the peptide unit can be viewed as well as nonlinear or chaotic oscillators due to the number of interactions that they experiment . It is not prooved as yet that nonlinearities give rise to coherent excitations like solitons at the molecular level . The experimentally observed specific vibrational properties of the amide group in a molecular chain are not in contradiction with a model of a vibrational "polaron" , that is an exciton "dressed" with phonons , but a direct proof of the existence of this "polaron" has not been found.

Such a localization of the vibrational energy induces dynamic local deformations of the neighbouring lattice, which result in a local multiple well potential for the vibrating atom . In turn a double-well potential acting on the amide proton results from nonlinear interactions (anharmonic and higher order dipole moment effects) . Either description of the spectroscopic anomalies is nonlinear .

It is clear now that the splitting of the CO mode is a feature shown by a number of peptide chains . Actually , the comparison of slightly different peptide chains,like NMA and ACN seems very fruitful, especially.the spectroscopy of deuterated derivatives , and the low temperature structure determination .

Why does proton transfer occurs in NMA and not in ACN whose hydrogen bond possesses a comparable length and strength ? Is the bond between N and H atom considerably weaker in NMA than in ACN ?

The direct search for a dynamic disorder associated to a polaronic state is also a very needed experiment , as well as time resolved infrared spectroscopy , with the aim to observe the possible ways to excite the amide modes , the transfer of amide excitation along a chain , the lifetime of this excitation. The H^1 NMR measurement of the chemical shift and its temperature evolution will give informations on the local electronic density .

The needed theories, from the point of view of an experimentalist , could be calculations involving the whole amide group as a nonlinear oscillator and giving predictions not only for the amide-I mode but also for the other amide modes, in the case of ACN or NMA for example .Calculations could focus more precisely on the electric dipole interaction between neighbouring molecules and its evolution with temperature , taking advantage of the new data from structure determination .

The increasing amount of investigations devoted to this domain contributes to a better knowledge of the force fields involved in the interatomic and molecular interactions, of the peptide bond and of proton transfer, which are of considerable importance to much of biology.

Summary

The non-harmonic or anomalous behaviour of some amide vibrations in acetanilide , N-methylacetamide , poly-alanines and some peptides is briefly reviewed . The interpretation of these special properties is a subject of controversies . The relevance or insufficiency of the main theoretical models of the amide anomalies are discussed in connection with the experimental data .

As a conclusion , the need for more complete theories will be evoked , and correlatively the elaboration of new experiments with the aim of a direct search of nonlinear excitations .

Acknowledgments

This work is supported by NATO under grant n° 910281 and by the cooperation CNRS/NSF.

References

1- G.Careri,U.Buontempo, F.Galluzi, A.C.Scott, E.Gratton and E.Shyamsunder Phys.Rev.B 30, 4689, 1984.

2- J.C.Eilbeck , P.S.Lomdahl and A.C.Scott - Phys.Rev. B 3Q, 47O3, 1984.

3- S. Takeno - Prog. Theor. Phys. 75,1,1986.

4- D.M.Alexander and J.A.Krumhansl - Phys. Rev.B 33, 7172, 1986.

5- A.C. Scott , I. J.Bigio and C.T. Johnston - Phys. Rev. B 39,12883 , 1989

6- A. Campa , A. Giansanti and A. Tenembaum - Phys. Rev. B 36 , 4394 , 1987 ; A.Tenenbaum, A. Campa and A. Giansanti - Phys.Lett.A 121, 126, 1987.

7- "Davydov's Soliton Revisited - Self-Trapping of Vibrational Energy in Protein" - Ed. P.L. Christiansen and A.C. Scott - Plenum Press 1990

8- C.T.Johnston , B.J.Swanson - Chem. Phys. Lett. 114, 547, 1985.

9- W. Fann , L. Rothberg , M. Roberson , S. Benson , J. Madey, S. Etemad and R.Austin - Phys. Rev. Lett. 64 , 607 , 1990

10- M.Barthes , H. Kellouai , G. Page , J. Moret , S. W. Johnson and J. Eckert - Physica D 1993 68 , 45 , 1993. See also H. Kellouai's Thesis - Université de Montpellier , juin 1993 .

11- J.L.Sauvajol et al. - J. Ram. Spectr. 20 , 517 , 1989 .

12- C.T.Johnston et al. - J.Phys.Chem. 95 , 5281 , 1991.

13- M. Barthes ,R. Almairac, J.L. Sauvajol ,J. Moret,R. Currat ,J. Dianoux - Phys.Rev.B 43 , 5223, 1991 .

14- M.Barthes,J.Eckert,S.Johnson,J.Moret,B.Swanson,and C.Unkefer - J.Phys.I France 2 , 1929 , 1992.

15- J.L.Sauvajol et al.- Sol.St.Comm. 77, 199, 1991

16- G.De Nunzio - to be publ.

17- M.Sakai,N.Kuroda and Y.Nishina - Phys.Rev.B 47,150, 1993

18- C.J. Brown and D.E. Corbridge - Acta Cryst. 7 , 711 , 1954 .

19- C.J.Brown - Acta Cryst. 21, 442 , 1966

20- H.Wasserman ,R..Ryan and S.Layne - Acta Cryst. ,C41 ,783 , 1985

21- A.C.Scott et al. - Phys.Rev.B 32, 5551, 1985.

22- R.J.Elliott et al. - Proc.Roy.Soc.A289, 1, 1965

23- D.B.Fitchen - in " Physics of Color Centers" ed.W.B.Fowler -Acad.Press 1968.

24- I.Iben , D. Braunstein , W. Doster , H. Frauenfelder - Phys.Rev.Lett. 62 , 1916 , 1989

25- L. Hayward - to be publ.

26- F.Fillaux and M.H. Baron - Chem.Phys. 62, 275, 1981.

27- G. Araki et al. - Phys.Rev.B - 43,12662, 1991.

28- F. Fillaux, J.P. Fontaine, M.H. Baron, G.J. Kearley and J. Tomkinson - Chem.Phys. 176, 249, 1993.

29- H. Guo and M. Karplus - J. Phys.Chem.96, 7273, 1992 .

30- N. Ostergard - Thesis -The Technical University of Denmark - Sept. 1991.

31- L. Katz and B. Post - Acta Cryst. 13, 624, 1960 .

32- G.J.Kearley et al. - Science 264, 1285, 1994 .

33- A. Migliori et al. - Phys.Rev.B 38, 13464, 1988 .

34- D.Durand et al. - Biopolymers 33,725, 1993 .

35- V.Helenius et al. - to be publ.
   - R. Lohikoski et al. in "Time Resolved Vibrational Spectroscopy VI" A.Lau and F.Siebert Eds, Springer-Verlag , Berlin 1994.

36- S. Krimm and J. Bandekar - Adv. Prot. Chem. 38, 181, 1986.

37- J. Florian and V. Hrouda - Spectrochimica Acta 49A, 921, 1993

- - - - - - - -

Low temperature Raman spectra of acetanilide and its deuterated derivatives: comparison with normal mode analysis

G. De Nunzio

Groupe de Dynamique des Phases Condensées,
Université Montpellier II, Place E. Bataillon,
34095 Montpellier cedex 5, France

1   INTRODUCTION

1.1   Bioenergetics

Bioenergetics is the study of energy production and transferring in living cells.

The hydrolysis of ATP (adenosine triphosphate) provides the energy for the majority of the biological processes; the molecule of ATP, after reacting with a molecule of water, releases 0.42 eV of energy:

$$\mathbf{ATP^{4-} + H_2O \rightarrow ADP^{3-} + HPO_4^{2-} + H^+} + energy.$$

This energy is often released far from the site where it will be used, so the problem is: what is its mechanism of propagation?

Proteins seem to play a principal role in the transfer process. One hypothesis is that this energy is transferred by exciting a vibrational state in the protein, and the *Amide I* excitation (stretching of the $C = O$ bond of the peptide groups) has been proposed. The vibrational energy of the *Amide I* mode is about 0.21 eV, i.e. half the energy released by the ATP hydrolysis, but the problem is the lifetime of this and of other typical vibrations in proteins, about 10^{-12} sec: too short to be important in biological processes.

To overcome this and other problems, the physicist A. S. Davydov (see e.g. [1], where references for erlier works are given) has proposed a model based on the following points:

- the transfer of the energy released by the hydrolysis of ATP is accounted for by the propagation of a soliton through the chains of peptide groups created by hydrogen bonds (see, for example, the case of the α-helix protein structure with three hydrogen-bonded channels);

- the soliton is the result of a non-linear interaction of the *Amide I* vibrational excitation and the deformation in the protein structure caused by the presence of the excitation.

We shall show only the starting point of the theory, i.e. the Hamiltonian of a hydrogen-bonded chain of molecules, and some conclusions. For an extended review of the theory of Davydov's soliton, see [2], where a fairly complete bibliography is given.

1.1.1 Davydov's Soliton

We present here the Hamiltonian of the basic theory of Davydov's soliton for one linear chain of masses.

Its application to proteins (α-helix) is straightforward in principle (it corresponds to considering more than one chain and thus the interactions between chains) but difficult in practice, and only numerical solutions are possible.

The Hamiltonian is:

$$H = H_{ex} + H_{ph} + H_{int} \tag{1}$$

where:

$$H_{ex} = \sum_n [E_0 \, B_n^+ B_n - J(B_{n+1}^+ B_n + B_{n-1}^+ B_n)] \tag{2}$$

$$H_{ph} = \sum_n [\frac{p_n^2}{2m} + \frac{w}{2}(u_{n+1} - u_n)^2] \tag{3}$$

$$H_{int} = \chi \sum_n (u_{n+1} - u_n) B_n^+ B_n \tag{4}$$

H_{ex} represents the *Amide I* excitation energy, H_{ph} is the phonon Hamiltonian and H_{int} is the interaction Hamiltonian.

B_n^+ and B_n are creation and annihilation operators for the *Amide I* excitation.

u_n e p_n are position and momentum operators for the n-th molecule; w is the elastic constant of the lattice; J represents dipole-dipole interaction between nearest neighbors; χ is the exciton-phonon non-linear coupling constant; m is the mass of each point in the chain.

The solution of the equations derived from this Hamiltonian gives, after considering stationary solutions, the **discrete non-linear Schrödinger equation:**

$$i(\hbar/J)\dot{\phi}_n + (\phi_{n+1} - 2\phi_n + \phi_{n-1}) + \sigma_0^{-1}|\phi_n|^2\phi_n = 0 \tag{5}$$

where $\sigma_0 = wJ/\chi^2$.

This equation has two different kinds of solutions: an excitonic, delocalized one, with energy $E_{ex} = E_0 - 2J$, and a solitonic one with energy $E_{sol} = E_{ex} - E_B$, where E_B is the binding energy of the soliton.

So the soliton has lower energy and is therefore stable.

1.2   Where ACN comes in

Crystalline Acetanilide (fig. 1) is an orthorhombic crystal (space group $Pbca - D_{2h}^{15}$) characterized by soft hydrogen-bonded chains of molecules parallel to the b axis of the crystal.

Acetanilide (briefly **ACN**) is an anharmonic solid with a one-dimensional character, thus it is a likely candidate to study non-linear excitations in organic compounds, with the advantage of working on a crystallized material.

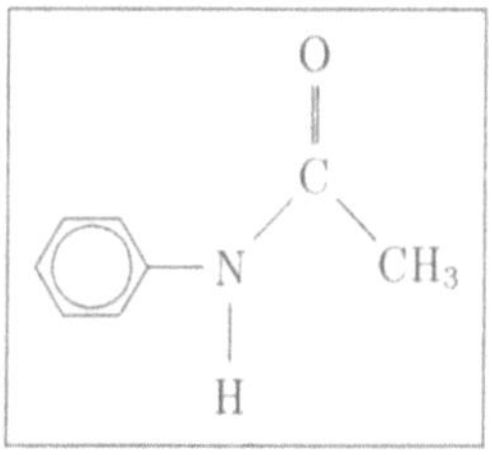

Figure 1: **A molecule of acetanilide.** One can recognize the three atomic groups that form the molecule: the phenyl, the peptide and the methyl group.

Structural similarities between ACN and polypeptides (see, e.g., [2, page 28]) lead to consider ACN as a model system for the investigation of physical phenomena in proteins.

1.3   Optical anomalies in ACN

About twenty years ago (1973, G. Careri [3]) an 'anomalous' excitation was discovered in the infra-red and Raman spectra of crystalline Acetanilide, with a wave number of about 1650 cm^{-1}, i.e. in the same frequency range as the C = O stretching mode (*Amide I*, 1665 cm^{-1}).

This band is considered 'anomalous' because:

- it is not possible to assign it to a particular mode of vibration by group theory techniques;

- it has the unusual temperature behaviour shown in figure 2, i.e. its intensity strongly increases with decreasing temperature.

Some other band in ACN has similar characteristics: see, e.g., [5, 6].

1.4   Theories and experiments

The existence of these 'anomalies', and the interest in the study of ACN as a model for proteins, has originated a great many theoretical works as well as experimental researches.

As far as **theory** is concerned, several models have been proposed, such as Davydov's soliton model, pseudo-polaronic models and many others (for a survey of the theories, and a complete bibliography, see [7]).

As to the **experimental** aspects of the problem, the work in progress at the GDPC (*Groupe de Dynamique des Phases Condensées*, Montpellier) consists in the study of different products derived from ACN by selective deuteration or fluoration of the molecule:

a) $C_6H_5 - CONH - CH_3$ (ACN)   e) $C_6D_5 - COND - CD_3$ (ACN-D9)
b) $C_6H_5 - CONH - CD_3$ (ACN-D3)   f) $C_6H_5 - CONH - CF_3$ (ACN-F3)
c) $C_6D_5 - CONH - CH_3$ (ACN-D5)   g) $C_6F_5 - CONH - CF_3$ (ACN-F8)
d) $C_6D_5 - CONH - CD_3$ (ACN-D8)

It is important to study partially and totally deuterated or fluorated samples for two reasons:

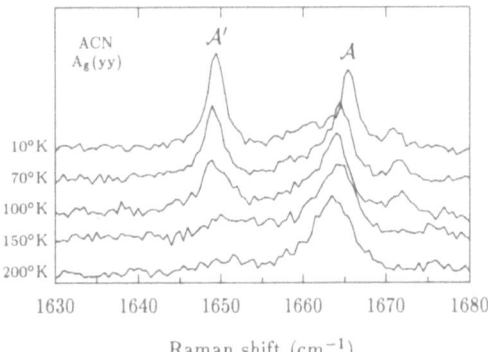

Figure 2: **The 'anomalous' band** (at about 1650 cm^{-1}, labelled $\mathcal{A}'$) at different temperatures (Raman spectra, from [4]). The peak $\mathcal{A}$ is the *Amide I* band, mainly C = O stretching.

- the isotopic effect on the peak frequencies may lead to the identification of the vibrations;

- the modification of the force constant of the hydrogen bond has an important influence on the 'anomalous' peaks: e.g. all of the anomalies disappear when the hydrogen bond is deuterated.

Several techniques of analysis are used: infra-red, neutron, Raman, Brillouin and NMR spectroscopy.

2 NORMAL MODE ANALYSIS

In order to be able to distinguish between normal and 'anomalous' features in the spectra of Acetanilide, we have begun a study of the compound from the point of view of Molecular Dynamics, in collaboration with the 'Département de Biologie Cellulaire et Moléculaire' of Saclay (France).

The first step has been the analysis of the normal modes of ACN and of its deuterated derivatives, and a comparison of the results with Raman spectra.

We have used the program CHARMm in its Normal Modes option.

2.1 The program CHARMm

CHARMm (©*Polygen Corporation - 200 Fifth Avenue - Waltham, MA 02254*) is a general software application for modelling the structure and the dynamics of a molecular system. CHARMm uses an empirical energy function for energy minimization, vibrational analysis and molecular dynamics simulation.

We briefly outline here the three stages involved in the simulation of a molecular crystal like ACN.

1. At the very start, some **preliminary steps** are performed, namely:

- the building of a **single molecule**, starting from the structural data found in the literature;

- the **minimization** of its total energy in the crystalline environment;

- the building of a **unit cell**;

- the building of a piece of **crystal** of sufficient dimensions by translating the unit cell in the *a*, *b* and *c* directions.

2. The next phase is the **vibrational analysis**. A harmonic calculation of the normal modes of vibration of the crystal is performed: for each mode found, we get the corresponding eigenvalue and eigenvector, and the potential energy distribution (*ped*). At this point, a process of **refinement** of the force constants, mainly by trial and error, begins. The normal modes are compared with the peak positions from low temperature i.r. and Raman spectra, and a better agreement with experiments is looked for, by slightly modifying the values of the force constants.

3. The third stage in the simulation of the ACN crystal is the **molecular dynamics** phase. Now the interactions between the atoms cease to be harmonic, and the full anharmonic CHARMm energy field is used. The molecular system evolves awhile (something like 10^1–10^2 picoseconds) and time series of the coordinates and the velocities of the atoms are produced (*trajectories*). The trajectories may be later analysed (by correlation functions and Fourier transform) to obtain power spectra.

2.2 Results

At present the second stage, particularly parameters refinement, is in progress. We are taking into consideration some of the selectively deuterated samples, particularly ACN, ACN-D5, ACN-D8 and ACN-D9.

All of these compounds have been extensively studied at the GDPC [4] by means of Raman spectroscopy. The spectra have been only partially published: papers will soon follow, describing in detail the effects of selective deuteration on Raman spectra.

At the time being, as far as CHARMm is concerned, the simulation of the deuteration process is obtained by doubling the mass of some ^{1}H hydrogens, transforming them into ^{2}H's; no attempt to redistribute the partial charges on the atoms (simulating different distributions of the electronic clouds) is done. This is obviously a limit of this kind of classical calculation of molecular dynamics: only *ab initio* calculations could overcome it.

We present here, as an example, some comparisons between the calculated 'normal mode spectra' and Raman spectra at low temperature (about 10°K).

2.2.1 The range 1400–1700 cm^{-1}

A good agreement with experimental data has been obtained for some high frequency regions; we include as an example (fig. 3) the range 1400–1700 cm^{-1}, i.e. the part of the spectrum where we find (in ACN and ACN-D5) the anomalous band (at about 1650 cm^{-1}). The results for the four compounds are shown.

We have labelled the main peaks in the figure with 'calligraphic' letters, such as $\mathcal{A}$, $\mathcal{B}$, etc. Of course, we cannot expect a perfect agreement between normal modes and Raman spectra, in that the former are calculated in a harmonic approximation, while

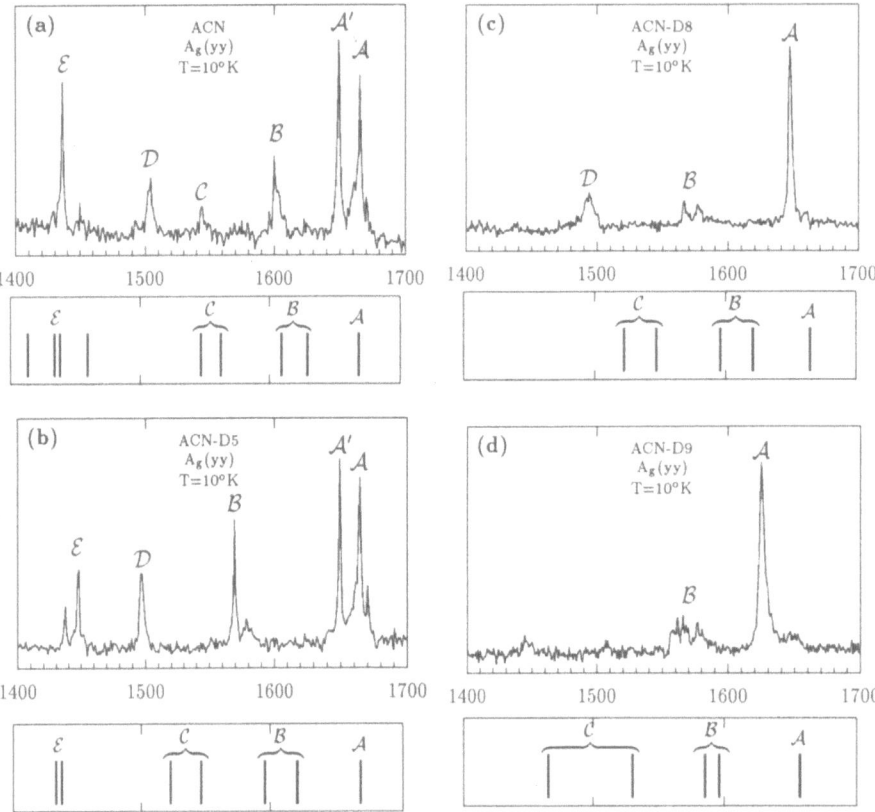

Figure 3: **Comparison between polarized Raman spectra and normal modes** in the range 1400–1700 cm^{-1}. The Raman spectra of the four compounds are shown: ACN, ACN-D5, ACN-D8 and ACN-D9. The x axys shows the Raman shift measured in cm^{-1}; the y axys reports the Raman intensity for monocrystalline samples at low temperature (about 10°K). We only show (among the four D_{2h}^{15} Raman active polarizations A_g, B_{1g}, B_{2g}, B_{3g}) the A_g spectra, in that this is the polarization of the 'anomalous' peak at about 1650 cm^{-1}. Beneath each experimental spectrum we show the position of the normal modes in the same frequency region. See the text for more details.

the latter are anharmonic in nature and contain contributions of second order Raman scattering.

Moreover, as presently deuteration means only mass change, the agreement is even worse when we examine a selectively deuterated compound, showing the limits of this assumption.

Nevertheless, as we have already stated in (2.1), this kind of analysis is important as a preliminary stage to molecular dynamics; moreover, it lets us discriminate between (almost) harmonic peaks (that we can find by the normal mode calculation) and more or less anharmonic (and eventually 'anomalous') ones.

Now, back to figure 3. We shall examine each peak in the experimental spectra and then (eventually) find the corresponding normal mode.

$\mathcal{A}$ This is the mode *Amide I*; it is normally attributed to the stretching of the $C = O$ bond. The analysis of the *ped* of the molecule in this particular mode confirms the attribution, while stating, at the same time, that there is a contribution by a kind of 'in-plane bending' of the whole methyl; some other movements of the peptide group are also involved. For ACN and ACN-D5, we see that the calculated wave number of the *Amide I* is fairly correct; on the contrary, the other two cases (ACN-D8 and ACN-D9) show a very poor agreement with the experimental value; probably a redistribution of charges, as already stated, should be taken into account.

$\mathcal{A}'$ The 'anomalous' band is labelled $\mathcal{A}'$. As we can see, it only exists when the methyl group is not deuterated (figs. 3.a and 3.b), while the phenyl deuteration is ininfluent. Very important, on the contrary, is the peptide proton: its deuteration immediately kills $\mathcal{A}'$ peak (see infrared spectra of ACN-D1 [8]). The normal mode calculation, as it was expected, cannot see this 'anomalous' peak.

$\mathcal{B}$ This band seems to be a badly resolved doublet; it is a phenyl band, as it is evident from the comparison of the four compounds: phenyl deuteration moves $\mathcal{B}$ towards lower frequencies. The corresponding normal mode show a *ped* with phenyl contribution, but also with some peptide stretching and $N - H$ in-plane bending.

$\mathcal{C}$ This band in Raman spectra is generally considered to be the *Amide II* mode, i.e. the in-plane bending of the $N-H$ bond. In the normal mode 'spectrum' we find a doublet with a *ped* compatible with this attribution. Also some phenyl excitations contribute to this mode, therefore in ACN-D5 and ACN-D8, i.e. in phenyl deuterated samples, this doublet red-shifts (its Raman intensity becomes too low to be seen in the experimental spectra).

$\mathcal{D}$ Another interesting band. This is a strongly anharmonic peak, present whenever the peptide hydrogen is not deuterated. Its anharmonicity is reported, for example, in [9] for the case of ACN-D8. Like the $\mathcal{A}'$ peak, its intensity strongly increases when the temperature decreases. In [8], $\mathcal{D}$ is considered the overtone of the so called γ_{N-H} mode (the out-of-plane bending of the $N - H$ bond) at 772 cm^{-1} (in ACN). The normal mode calculation is not able to see the peak $\mathcal{D}$, confirming that anharmonicity plays an important role in its existence.

$\mathcal{E}$ This is a doublet, evidently of methyl origin. It exists in ACN and ACN-D5, but not in methyl-deuterated compounds. The analysis fully confirms this attribution. The normal modes placed on the two sides of the mode $\mathcal{E}$ (in ACN), have phenyl origin; they are probably too week to be seen in Raman spectra.

2.2.2 Some other frequency range

For the **low frequency** spectra, only a very rough agreement is by now possible. The difficulty in trying to put things at the right place comes from the fact that low frequency modes depend on the movement of whole atom groups; therefore it is extremely difficult to make parameter refinement, because the right combination of force constant changes should be found, paying attention not to destroy the correspondence established for high frequency peaks

On the contrary, **very high frequency** spectra, above 2700 cm^{-1}, are fairly well described by the normal mode analysis. In this region we find the N − H stretching mode and some phenyl or methyl C − H stretching modes.

For the sake of brevity, we do not show these or other frequency ranges, which will be analysed in detail in subsequent papers.

Acknowledgements

The author is grateful to Prof. Jeremy Smith (Section de Biophysique des Protéines et des Membranes, Département de Biologie Cellulaire et Moléculaire, Centre d'Etudes de Saclay), and to Larry Hayward (Biochemistry Dept., University of Edinburgh) for the helpful discussions and for the building of the ACN molecule.

References

[1] A. S. Davydov, Solitons in Molecular Systems, 2nd edition (Reidel, Dordrecht, 1991)

[2] A. C. Scott *Phys. Rep.* **217** N°1 (1992) 1–67

[3] G.Careri, Search for Cooperative Phenomena in Hydrogen-Bonded Amide Structures, in 'Cooperative phenomena' (Hagen & Wagner, Springer, Berlin, 1973) pp. 391–394

[4] G. De Nunzio, J. L. Sauvajol, unpublished results.

[5] G. B. Blanchet, C. R. Fincher Jr *Phys. Rev. Lett.* **54** (1985) 1310

[6] A. C. Scott, E. Gratton, E. Shyamsunder, G. Careri *Phys. Rev. B* **32** (1985) 5551–5553

[7] P. S. Lomdahl, Soliton models of protein dynamics, in 'Soliton theory: a survey of results' (A. P. Fordy, Manchester University Press, Manchester, 1990) pp. 209–232

[8] H. Kellouai, 'Modes non-armoniques de l'Acetanilide: absorption infrarouge. Approche nonlinéaire', Ph.D. Th. (Université Montpellier II, Montpellier, 1993)

[9] J. L. Sauvajol, G. De Nunzio, R. Almairac, J. Moret, M. Barthes *Sol. St. Comm.* **77** **(3)** (1991) 199–205

Conformational dynamics of proteins: beyond the nanosecond time scale

H. Grubmüller, N. Ehrenhofer and P. Tavan

Institut für Medizinische Optik, Universität München, Theresienstr. 37, 80333 München, Germany

1. INTRODUCTION

Protein motions and functional processes in proteins occur on a wide range of time scales. The fastest atomic motions take place on a femtosecond time scale. Fast biochemical reactions like the primary steps in photosynthesis last few picoseconds. Most biochemical reactions like enzymatic processes take much longer — microseconds or even few milliseconds. They are often accompanied by larger structural rearrangements in the protein, called conformational transitions[1], which are characterized by transition times of nanoseconds or much longer. A prominent example for an extremely slow conformational transition, with a transition time of many years, is the one which is believed to be responsible for the pathogenic effect of prions[2]. Often, conformational transitions constitute functional important motions, as for the gating of channel proteins or in protein folding.

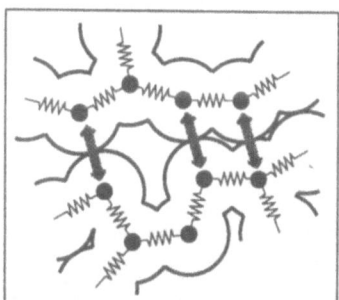

Figure 1: Molecular dynamics (MD) simulations serve to describe atomic motions in proteins[3] and other biological macromolecules. In this semiclassical approach, a protein is modeled as a set of point masses (atoms). The interatomic forces are described by a semiempirical energy function, and the molecular dynamics is determined by numerical integration of the Newtonian equations of motion. Typical forces included in the energy function are chemical binding forces ('springs') as well as noncovalent interactions like Pauli repulsion (grey arrows), Van der Waals- and Coulomb-forces (see, e.g., Ref.[4]). A good review on the MD-method is Ref.[5].

Most of these processes are not understood on an atomic level. Therefore, adequate descriptions of their molecular dynamics (MD) are required in the attempt to derive function from structure. Due to the structural complexity of proteins and a corresponding lack of well-founded coarse-grained effective models for the dynamics, the method of MD-simulation[6, 3] currently is the only approach, to which some reliability can be assigned. That method (see Figure 1) conceives a macromolecule as a classical many-body system of 'atoms' and describes the quantum-mechanical forces like the chemical binding forces, which are caused by the electronic degrees of freedom, by a semi-empirical force field. Accordingly, the molecular dynamics is simulated by integration of the Newtonian equations of motion.

However, the enormous computational task associated with MD-simulation of biological macromolecules entails an upper limit to the time scale of dynamical processes accessible by this method: At present, only few nanoseconds of protein dynamics can be simulated. Extrapolating the past increase of available computer power, one is forced to the conclusion, that we will not be able to enter the biochemically most relevant microsecond time scale within the next ten years, if we cannot manage to *drastically reduce* the necessary amount of computations. Also, protein folding simulations, which might contribute to solve the challenging 'folding problem', will be out of range for quite a while.

Given this situation, it is worthwhile to look for new concepts in order to simplify the description of protein dynamics. Clearly, the identification and elimination of *irrelevant* degrees of freedom would be particularly helpful. In this report, we will describe the stage, on which, as far as we believe, the necessary simplifications will take place. Transitions between conformational substates, which are ubiquitous in proteins[7] play a central role here, as they mediate between detailed atomic motions, as described by MD-simulations on the one hand and, on the other, the biochemical reactions we want to study. For this reason, those degrees of freedom, which describe the conformational dynamics of proteins — conformational coordinates — are of special interest. Unfortunately, conformational transitions have rarely been observed in MD-studies, since the latter are usually too short. Hence, it is not clear how to identify suitable conformational coordinates.

Having introduced the concepts of 'configurational space' and 'projected configurational space densities', we will suggest two methods to construct conformational coordinates. The first one describes conformational transitions on time scales above nanoseconds and is based on a neural clustering algorithm. The second method aims at a description of conformational motion *within* configurational substates on a faster time scale.

2. A SIMPLIFIED PROTEIN MODEL

In order to illustrate these methods, we use a simplified protein model[8] (Fig. 2) with a reduced number of degrees of freedom. It has been designed to fulfill two contradicting requirements: (i) the model should be structurally and dynamically similar to real proteins; in particular, it should exhibit conformational transitions; (ii) the model should have as few degrees of freedom as possible in order to allow extended MD-simulations. In that sense, it is a *minimal model*.

Figure 2: Simplified protein model used in the analysis described in the text. The model consists of 100 point masses (not drawn), which are covalently linked by chemical bonds. Like the residues in a real protein, the 100 model residues differ in their interaction properties. A stable tertiary structure of the model system (shown as a 'ribbon-plot') has been obtained from a simulated folding process of 5 ns duration.

Accordingly, a primary structure consisting of 100 amino acids was defined, where each of the residues is modeled as a single van der Waals sphere. Covalent and non-covalent interactions were added to account for the dominant forces known to stabilize the tertiary structure of proteins, such as chemical binding forces or hydrophilic/hydrophobic inter-actions. Starting from a 'denatured' configuration, a folding process of 5 ns length has been simulated, which yielded the stable and equilibrated tertiary structure sketched in Fig. 2. As shown in Ref.[9], the model meets the above two criteria.

MD-simulations of a total of 232 ns have been carried out on that protein model. We will use results of these simulations to illustrate concepts and methods to study the conformational dynamics of proteins.

3. CONFIGURATIONAL SPACE AND CONFORMATIONAL SUBSTATES

We start with the concept of *configurational space*, which is closely related to the concept of phase space. Fig. 3 illustrates the basic notions.

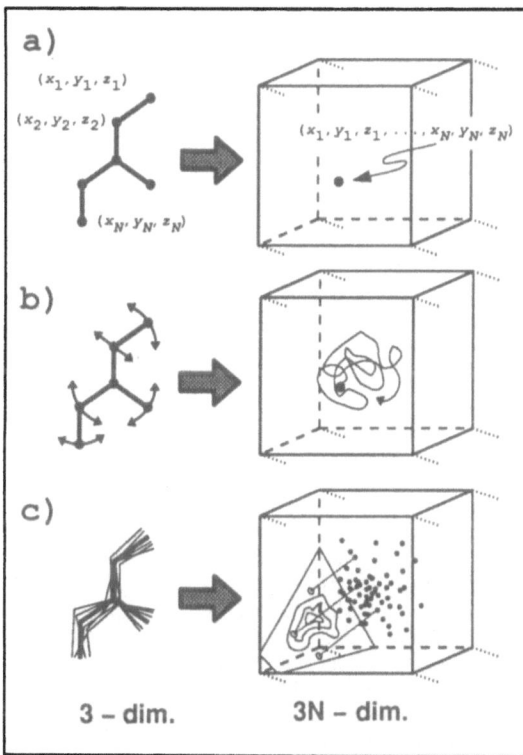

Figure 3: Protein structure and dynamics can be analyzed in configurational space. (a): A given protein structure described by N atomic positions (x_1, y_1, z_1), (x_2, y_2, z_2), ..., (x_N, y_N, z_N) (left side) is mapped into configurational space by merging the N vectors into one single vector $(x_1, y_1, z_1, x_2, y_2, z_2, \ldots, x_N, y_N, z_N)$ with dimension $3N$ (right side). (b): Accordingly, the concerted motion of all atoms in a protein can be described by one single high-dimensional trajectory in configurational space. (c): Snapshots of protein dynamics, e.g. derived from an MD-simulation, are mapped into a 'cloud' of configurations, where each point in configurational space resembles the particular geometry of the protein at one instance in time. If a sufficiently large sample of configurations is available, one can estimate the configuration space density, which is generated by the molecular dynamics. By projecting this high-dimensional density on a suitable low-dimensional subspace one can gain insights into the overall structure of this density (contour lines).

At an instant of time, the three-dimensional structure of a protein consisting of N atoms is defined by N atomic positions, i.e., N three-dimensional vectors (Fig. 3 (a), left side). Alternatively, the structure can be represented by one single $3N$-dimensional vector, which is obtained by combining the N atomic positions. In this picture, *one single point* in high-dimensional configurational space defines *all* atomic positions of the protein (Fig. 3 (a), right side). For typical proteins, configurations space has some 10 000 dimensions; for the simplified protein model described in the preceding section, configurational space is of dimension 300.

The conceptual advantage of looking at a protein in configurational space becomes clear if we switch from protein structure to dynamics. Then, the concerted motion of all atoms of the protein (Fig. 3 (b), left side) is represented by one single trajectory in configurational space (Fig. 3 (b), right side). An MD-simulation typically provides a set of 'snapshots' of that atomic motion (Fig. 3 (c), left side), which appears in configurational space as a 'cloud' of configuration vectors (Fig. 3 (c), right side). In the limit of large numbers of configurations a 'configuration space density' ρ, as determined by the protein dynamics, can be derived.

Now consider a few (m) degrees of freedom, $c_i(x_1, x_2, \ldots, x_N)$ $(i = 1 \ldots m)$, called 'conformational coordinates'. For a given protein, these degrees of freedom shall be chosen such, that they reflect the conformational dynamics of that protein. Then, a coarse-grained effective description of that (microcanonical) system with Hamiltonian $\mathcal{H}(x_1, x_2, \ldots, x_N)$ can be achieved by explicitly considering the dynamics of only the c_i. If the remaining $3N - m$ degrees of freedom are regarded as a heat bath, the resulting system of reduced dimension belongs to a *canonical ensemble*.

The free energy landscape of that sub-system, $F(c_1, \ldots, c_m)$, determines, together with the heat bath, the dynamics of the conformational coordinates $c := \{c_i\}$. $F(c_1, \ldots, c_m)$ can be derived from the configurational space density ρ by

$$F(c) = -k_b T \ln \rho_c(c), \tag{1}$$

where ρ_c is a canonical conformational space density, obtained by projecting ρ on the m-dimensional subspace spanned by the conformational coordinates c_i (Fig. 3 (c), right side),

$$\rho_c(c_1, \ldots, c_m) = \int dx_1 dx_2 \ldots dx_N \rho(x_1, x_2, \ldots, x_N) \delta(c - c'). \tag{2}$$

As an illustration, Figure 4 shows a contour-plot of the free energy landscape of our protein model, which has been determined from $\rho_c(c_1, c_2)$ according to (1). The inset of Fig. 4 illustrates the shape of the free energy landscape F by plotting its value in units of $k_b T$ (at $T = 300K$) along an arbitrarily chosen 'reaction coordinate' (bold line) passing through the three minima. We did not yet describe how the two conformational coordinates c_1 and c_2 have been defined; we will do that in the following section.

The energy landscape in Fig. 4 exhibits three distinct minima, denoted as 'B', 'C', and 'D', respectively. As suggested on the basis of experimental data by Frauenfelder[10] such regions of low *free* energy in conformational space, separated by free energy barriers, generally define distinct *conformational substates* of a protein. Accordingly, we define our model to be in substate 'B', 'C', or 'D', respectively, if its conformation lies in the corresponding region of the free energy landscape. Like the Brownian motion of a particle coupled to a heat bath, the dynamics of the system within the energy landscape shown in Fig. 4 is diffusive. Occasionally, the fluctuating forces generated by the heat bath drive the system across one of the energy barriers and induce a conformational transition, which reveals itself as a rapid change in the sterical structure of the model.

Note, that the above definition of conformational substates differs from the approach commonly employed for theoretical explorations of substate hierarchies[11, 12]. In these studies, the distribution of thermally accessible local minima of *potential energy* within configurational space is studied, and it is assumed, that these local minima or clusters thereof can provide information on the distribution of conformational substates. At low temperatures, where entropic contributions are small and safely can be neglected, the *potential* energy landscape definitely can serve as a tool for the analysis of conformational substates. However, at room temperature the suggested relation between accessible minima of potential energy and conformational substates, defined as minima of *free* energy, is questionable. In contrast, our approach allows the study of conformational substates *at physiological temperatures*, as it refers to the *free* energy landscape within conformational space. Therefore, if applied to realistic protein models, our approach enables comparisons of theory and experiment. Admittedly, much larger computational effort is involved in such analysis, because a sufficiently dense sampling of phase space by extended simulations is required for the determination of ρ_c. At present, that computational effort restricts our method to studies of simplified protein models.

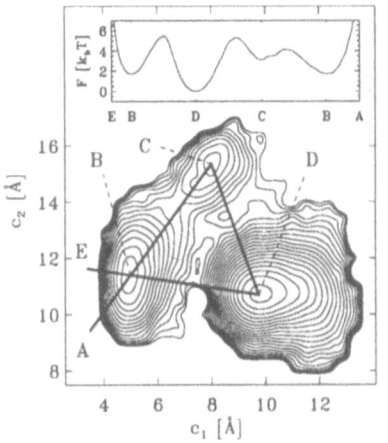

Figure 4: Free energy landscape derived from a 100 ns simulation of a simplified protein model, shown as a contour plot. It has been determined by projecting the configuration space density ρ onto onto the two-dimensional subspace defined by the conformational coordinates c_1 and c_2 (see text). The energy surface exhibits three minima, denoted by 'B', 'C', and 'D', which define three configurational substates. The inset shows a cross section through the energy landscape along a hypothetical reaction coordinate E-B-D-C-B-A in units of k_bT at 300 K.

We now return to the question, how suitable conformational coordinated can be determined. Clearly, the quality of a dimension-reduced description of protein dynamics depends on the choice of conformational coordinates. We studied two methods[13], which work on different levels of coarse graining. The first method, sketched in the following section, serves to identify conformational substates on the basis of MD-data and aims at a description of conformational transitions. The second method, described in Section 5, can be used to construct an effective description of the molecular dynamics *within* a conformational substate.

4. CONFORMATIONAL COORDINATES FROM NEURAL CLUSTERING

We defined conformational substates as regions in configurational space of low free energy, or, according to Eq.1, of high configuration space density. Hence, statistical data analysis should provide tools that can be used to find clusters in a 'cloud' of high dimensional configuration vectors. As conformational substates are known to be hierarchically structured[14], such a tool should be able to reveal cluster hierarchies.

For that purpose we used a neural cluster algorithm ('minimum free energy clustering'), which has recently been proposed[15]. We will not go into details of this algorithm here, but show merely the results of its application to a 100 ns MD-simulations of our simplified protein model.

The algorithms takes as input a (typically large) set of configurational vectors. In our example, a total of 390 625 configurational vectors were used, sampled at intervals of 256 fs. Each of the six pictures in Fig. 5 shows as a contour plot a projection of the configurational space density defined by these configurational vectors similar to the one shown in Fig. 4. Here, too the three substates A, B, and C, respectively, can be identified (Fig. 5 (a)). The algorithm then finds minima in a 'blurred' energy landscape, which is, approximately, a convolution of the free energy $F = -k_b T \ln \rho$ with a multidimensional Gaussian of width σ. By variation of the length scale σ, one can conveniently analyze the hierarchical structure of substates.

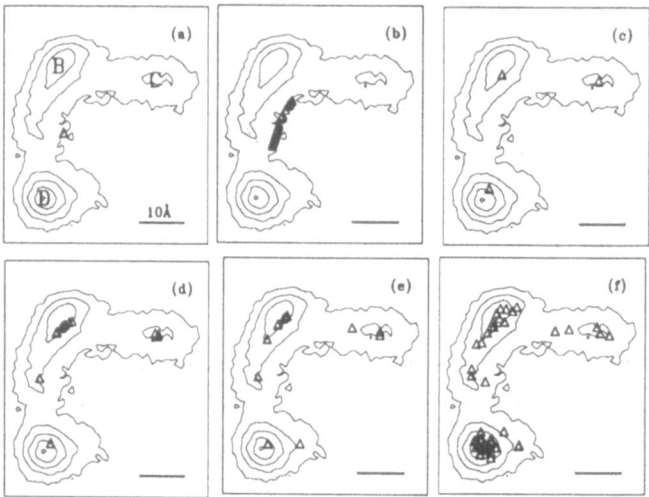

Figure 5: The structure of high dimensional configuration space densities can be analyzed using hierarchical neural clustering algorithms. The sequence of six pictures (a-f) shows how such an algorithm can resolve structures of the density on decreasing length scales [13.0 Å (a), 11.5 Å (b), 7.0 Å (c), 5.7 Å (d), 5.0 Å (e), 3.0 Å (f)]: Regions of high configuration space density, as detected by the clustering algorithm, are marked as triangles; superimposed is a contour plot of a suitable configuration space density projection similar to that shown in Fig.4.

The sequence of six pictures in Fig. 5 shows the output of the clustering algorithm (triangles) for different length scales σ. For σ as large as 13 Å (Fig. 5 (a)) the three clusters cannot be resolved and are identified as one single cluster positioned at the center of mass of the data set. In picture (b), as σ is decreased, a bifurcation occurs, and the single minimum splits into three distinct minima representing the three conformational states (c). Below $\sigma = 5.7$ Å the conformational states exhibit further substructures, which we will not discuss here.

Now we are prepared to define the conformational coordinates c_1 and c_2, which we used in Figs. 4 and 5 in order to visualize the high dimensional configurational space density ρ: The three free energy minima identified with the neural clustering algorithm define a plane, onto which ρ has been projected. Accordingly, the conformational coordinates c_1 and c_2 have been chosen to lie within this plane. Hence projections of ρ on that subspace optimally resolve the three substates. Note that it would have not been possible to identify number and location of the three substates by just looking at a picture like Figs. 4, since such plot can only be obtained, if the location of the substates is known beforehand.

The substate picture permits a drastically simplified description of protein dynamics, if the dynamics *within* each substate can be considered irrelevant. For our simplified protein model, the low frequency conformational transitions can be described as a three state Markov process (Fig.6), which has been shown employing a statistical analysis of the nine transition rates $k_{\alpha\beta}$, $\alpha, \beta \in \{B, C, D\}$. These rates were obtained from a 232 ns MD-simulation using the conformational coordinates defined above[8].

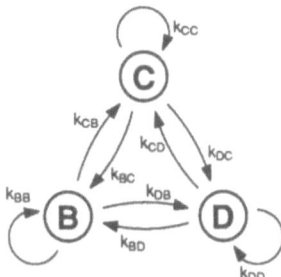

Figure 6: The low frequency conformational dynamics of the protein model can be modeled as a Markov process with three states, B,C and D, and nine transition rates, $k_{\alpha\beta}$, $\alpha, \beta \in \{B, C, D\}$. A statistical analysis of 236 conformational transitions observed during a 232 ns simulation showed that the dynamics of the model, if viewed on this coarse grained level, is indeed memory-free.

5. CONFORMATIONAL COORDINATES FROM PRINCIPAL COMPONENT ANALYSIS

We now turn our attention towards the conformational dynamics *within* one conformational substate, which has been neglected in the above Markov model. For our protein model, we chose substate D (cf. Fig. 4). Accordingly, we select only those configurational vectors out of the 100 ns simulation for further analysis, which are closest to that substate.

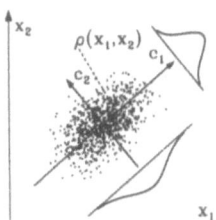

Figure 7: A two-dimensional example illustrates the method of principal component analysis (PCA): For a given set of vectors (points), which approximate a density $\rho(x_1, x_2)$, PCA yields as eigenvectors the direction of largest variance (c_1), the direction of second largest variance (c_2), and so on. The corresponding eigenvalues ($\sigma_1, \sigma_2, \ldots$) measure the variances of the data set along these directions. In the figure, these variances are indicated by the two Gaussians.

The structure of the free energy minima exhibited in (cf. Fig. 4) suggests that the corresponding density clusters can be described well by a — possibly multivariate — Gaussian distribution in configurational space. Under that assumption a principle component analysis (PCA)[16] can provide further information on the structure of that 'cloud'. As a reminder, Fig. 7 illustrates the PCA-method using a two-dimensional data set.

The results of a PCA of substate D are shown below. In the following, the eigenvalues, σ_i, and the corresponding eigenvectors, c_i, shall be ordered in decreasing size. Fig. 8 shows a sample of six projections of ρ along the eigenvectors c_1 (largest variance), c_4, c_{50}, c_{120},

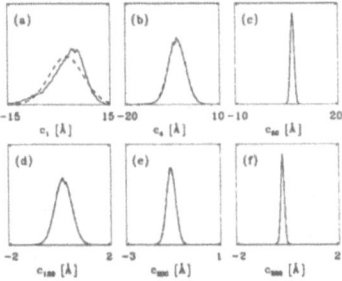

Figure 8: A PCA-analysis was carried out on substate 'D' of the simplified protein model. The figure shows a selection of six out of a total of 300 projections of the configuration space density along the eigenvectors, ordered in decreasing variance; c_1 (a) (largest variance), c_4 (b), c_{50} (c), c_{120} (d), c_{200} (e), and c_{280} (f), respectively. A Gaussian fit is shown with dashed lines for each of the projections. Note that the scale of the abscissa in pictures (d-f) differs from that in pictures (a-c).

c_{200}, and c_{280} (smallest variance), respectively. As can be seen from the fits (dashed), all projections are nearly Gaussian. which assures, that PCA can be reasonably applied.

Fig. 9 shows a histogram of the eigenvalues σ_i, $i = 1, 2, \ldots, 300$ obtained from the PCA-analysis. These eigenvalues measure the variance of substate-'cloud' along those directions in configurational space, which are given by the corresponding eigenvectors.

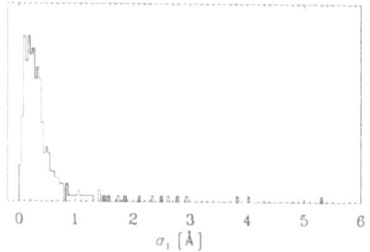

Figure 9: A histogram of standard deviations σ_i, derived from the 300 eigenvalues given by the PCA analysis described in the text reveals, that the vast majority of the σ_i is smaller than 0.5 Å, i.e., the configuration space density is 'flat' in most directions of configurational space. Only few extensions are larger than 2 Å and contribute to the large-amplitude motions observed in the protein model.

The surprising observation is that the cluster exhibits large extensions (> 1 Å) only along very few directions, whereas it is 'flat' (< 0.5Å) in 90% of the directions. Obviously, only few degrees of freedom (namely those given by the 'large' eigenvectors) contribute significantly to the large amplitude conformational dynamics of the protein model. Can these few degrees of freedom be considered as *relevant* degrees of freedom for the conformational dynamics? If so. can the remaining 90% degrees of freedom be considered as 'noise', which need not be included explicitly in a protein dynamics description?

We cannot answer that questions yet; work is in progress. It is clear, however, that the PCA serves to separate time scales, a prerequisite for a dimension-reduced, effective description. As an example, Fig. 10 compares part of our MD-simulation projected on a 'large' eigenvalue, c_3, (upper left picture) with a projection on a 'flat' eigenvalue, c_{290}, (upper right). The bottom row of Fig. 10 shows blow ups of the small rectangles in the upper two pictures.

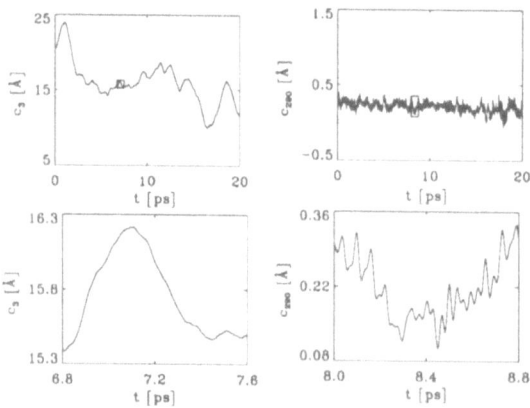

Figure 10: *Upper row:* Plots of the protein dynamics, projected onto eigenvectors c_3 (left side) and c_{290} (right), respectively, reveal that slow motions occur predominantly along directions in configurational space characterized by large variances, whereas fast motions are described by directions of small variance. *Lower row:* Blow-ups of part of the dynamics (marked by small rectangles in the upper two pictures) demonstrate that virtually no high-frequency contributions of the molecular dynamics are present in the large variance collective motions.

As can be seen, the high frequency parts of the dynamics are only present in the 'flat' degrees of freedom, whereas the low frequency parts are captured by the 'large' eigenvalues. Vice versa, no high frequency fluctuations are visible in the 'large' eigenvalue dynamics, and no low frequency fluctuation occurs in the 'flat' directions.

In order to quantify this observation, Fig. 11 shows Fourier spectra of five projections on the eigenvectors c_{293}, c_{250}, c_{200}, c_{100}, and c_1, respectively. Again, the predominant contributions to the atomic fluctuations (hatched areas) occur in the high frequency part in the case of the 'flat' degrees of freedom, whereas the dynamics along 'large' eigenvalues is exhibits merely low frequency contributions.

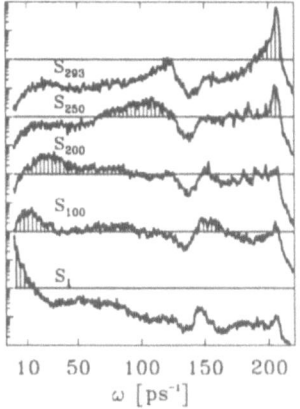

$$\omega \ [\text{ps}^{-1}]$$

Figure 11: Predominant contributions to atomic fluctuations (hatched areas) are shifted from high frequencies to low frequencies as one proceeds from small variances in configurational space to large ones. The figure shows five Fourier spectra of the protein dynamics, which has been projected onto the eigenvectors *(from top to bottom)* c_{293}, c_{250}, c_{200}, c_{100}, and c_1, respectively.

6. SUMMARY AND CONCLUSION

We illustrated concepts which aim at a reduction of the computational effort, that has to be spent in the attempt to understand protein function on the basis of protein dynamics. We illustrated these concepts using extended molecular dynamics simulations of a simplified protein model. Based on a thermodynamic formulation of the concept of conformational substates in terms of minima in *free* energy, two methods to define conformational coordinates, which serve to separate relevant from irrelevant degrees of freedom, were explained and applied to a simplified protein model.

The first method, the application of a neural clustering algorithm, allowed to identify substate hierarchies and led, in the case of the simplified protein model, to a Markov model for conformational transitions.

The second method served to analyze the dynamics within a conformational state. It was shown that only few degrees of freedom contribute to the large fluctuations in the protein model. At the same time, these large fluctuations embody the low frequency contributions to the protein dynamics, whereas all other degrees of freedom carry the high frequency aspects. These findings should enable the construction of *effective* descriptions for protein dynamics, which permit longer simulation times and thereby extend the range of biochemical processes that can be studied theoretically.

Acknowledgements

This work was supported by the Deutsche Forschungsgemeinschaft (SFB 143/C1).

References

[1] A. Ansari, J. Berendzen, D. Braunstein, B. R. Cowen, H. Frauenfelder, M. K. Hong,

I. E. T. Iben, J. B. Johnson, P. Ormos, T. B. Sauke, R. Scholl, A. Schulte, P. J. Steinbach, J. Vittitow, R. D. Young, *Biophys. Chem.* **26** (1987) 337.

[2] D. A. Kocisko, *et al.*, *Nature* **370** (11 Aug 1994) 471.

[3] J. A. McCammon, B. R. Gelin, M. Karplus, *Nature* **267** (1977) 585.

[4] B. R. Brooks, R. E. Bruccoleri, B. D. Olafson, D. J. States, S. Swaminathan, M. Karplus, *J. Comp. Chem.* **4** (1983) 187.

[5] W. F. van Gunsteren, H. J. C. Berendsen, *Angew. Chem. Int. Ed. Engl.* **29** (1990) 992.

[6] M. Levitt, S. Lifson, *J. Molec. Biol.* **46** (1969) 269.

[7] H. Frauenfelder, S. G. Sligar, P. G. Wolynes, *Science* **254** (1991) 1598.

[8] H. Grubmüller, P. Tavan, *J. Chem. Phys.* (Sept. 1994). In press.

[9] H. Grubmüller, thesis, Technische Universität München (Jan. 1994).

[10] H. Frauenfelder, *Biophys. J.* **47** (1985) 35.

[11] R. Elber, M. Karplus, *Science* **235** (1987) 318.

[12] N. Go, T. Noguti, *Chem. Scr.* **29A** (1989) 151.

[13] N. Ehrenhofer, thesis, Ludwig-Maximilians-Universität München (1994).

[14] A. Ansari, J. Berendzen, S. F. Browne, H. Frauenfelder, I. E. T. Iben, T. B. Sauke, E. Shyamsunder, R. D. Young, *Proc. Natl. Acad. Sci. USA* **82** (1985) 5000.

[15] D. R. Dersch, P. Tavan, "Control of annealing in minimal free energy vector quantization", Proceedings of the IEEE International Conference on Neural Networks ICNN'94, S. K. Orlando, Florida, June 28–July 2, 1994, Rogers and D. W. Ruck, Eds. (IEEE, Piscataway, U.S.A, 1994) pp. 698–703.

[16] K. V. Mardia, J. T. Kent, J. M. Bibby, "Multivariate Analysis" (Academic Press, London, 1979).

Motions and correlations of the transmembrane domain of a protein receptor studied by molecular dynamics simulation

N. Garnier, D. Genest and M. Genest

Centre de Biophysique Moléculaire et Université d'Orléans,
1A avenue de la Recherche Scientifique,
45071 Orléans cedex 2, France

1. INTRODUCTION

Internal motions and dynamics of macromolecules are essential to understand their structure and their functions [1]. Such internal motions might mediate conformational change transmission or non covalent interactions for oligomerization processes, two mechanisms dominantly hypothesized for signal transduction of transmembrane receptors. The receptors of tyrosine kinase family, implicated in cellular transformation processes are identified as potent oncogenes and structural changes may result from ligand binding or single amino acid replacement. Such changes are associated with increased kinase activity, leading to oncogenic potential. In the case of the c-erbB2 encoded protein, the activating mutation is located in the transmembrane domain at position 659 [2] and the single Val to Glu amino acid replacement results in constitutive activation of the intrinsic protein kinase activity of the receptor and transformation [3]. Because c-erbB2 is a single membrane spanning protein, one expects the transmembrane segment as a module for signal transmission.

To give insight of the role of the transmembrane segment in such mechanisms, we have undertaken a study of internal motions of these protein portions [4] using molecular dynamics simulation methods [5]. We have shown the dynamical character of the wild transmembrane portion of c-erbB2 which differs from the rigidity of the corresponding mutated portion. In addition, the analysis of angular time correlation functions and the analysis of the response of the wild type transmembrane α helix to a conformational change evidence first, concerted motions in the wild transmembrane peptide and second, the existence of the propagation phenomenon. Essential results are here reported.

2. MATERIALS AND METHODS

2.1 Molecular dynamics simulation

Molecular dynamics simulations have been performed on the hydrophobic intramembrane polypeptides ended with four polar intracellular residues. The sequence includes residues 651 to 679 of the protein with Val^{659} for the wild type peptide and Glu^{659} for the mutated one. The N and C terminal extremities were capped by end groups, leading to the following sequences of 29 residues: CH_3CO-Leu-Thr-Ser-Ile[4]-Ile-Ser-Ala-Val-Val[9][659]/Glu[9]-Gly-Ile-Leu-Leu-Val-Val-Val-Leu-Gly-Val-Val-Phe-Gly-Ile-Leu-Ile[25]-Lys-Arg-Arg-Gln-$NHCH_3$.

A classical protocol of molecular dynamics simulation in vaccum, using the GROMOS program [6] has been applied. The two peptides, initially built with the standard α helix structure, were first energy minimized before to be progressively heated to 300°K and then equilibrated during 50 ps. The two extremities of the polypeptide chains were maintained in helical structure by using a constraint potential applied to standard α helix interatomic distances to simulate the short range interactions with the extramembrane domains not taken into account. A 160 ps production phase was performed for each peptide and atomic coordinates were stored every 0.04 ps to build up a trajectory file containing 4000 configurations. Details on molecular dynamics simulation protocol are given elsewhere [4]. All calculations have been performed on Silicon Graphics Workstations.

2.2 Correlations

To analyse in details the internal dynamics of the two transmembrane peptides, we have first evaluated the correlation matrix for the backbone dihedral angle fluctuations and second, we have evaluated the time dependence of these correlations. The time correlation function $C_{(i,j)}(m\tau)$ between the fluctuations $\Delta\theta_i$ and $\Delta\theta_j$ of the θ_i and θ_j angles is defined by the following expression:

$$C_{(i,j)}(m\tau)= \frac{\frac{1}{N'}\sum_{n=1}^{N'} \Delta\theta_i(t_n)\Delta\theta_j(t_n+m\tau)}{(\frac{1}{N'}\sum_{n=1}^{N'} \Delta\theta_i^2(t_n))^{\frac{1}{2}} (\frac{1}{N'}\sum_{n=1}^{N'} \Delta\theta_j^2(t_n+m\tau))^{\frac{1}{2}}} \quad (1)$$

where τ is the time interval used for coordinate storage (0.04 ps), m=1,....500, $t_n=\tau.n$ (n=1,...N') where N'=4000-m. θ_i represents the Φ_i or Ψ_i backbone dihedral angle of residue number i. For m=0 the expression reduces to equal time correlation coefficient, that means elements of the correlation matrix.

2.3 Propagation of a local perturbation

This study was undertaken on the wild type transmembrane peptide.

To evidence the propagation of a conformational change along the peptidic chain, we have determined the difference between the peptide dynamical behavior observed along two short 2 ps MD simulations: one being performed under local perturbation, the second being performed without perturbation. The initial structure was the last configuration obtained from the previous 160 ps simulation. The perturbation has been applied at residue Ile[4] by constraining the Ψ dihedral angle to 90°, producing a kink at the beginning of the helix. Time step for the integration was 0.002 ps using the SHAKE algorithm [7] and coordinates were stored every 0.004 ps producing a trajectory of 500 configurations.

Starting from different initial conditions for velocities, the procedure was repeated 20 times. The mean difference between the values of each backbone dihedral angle measured at equilibrium (Not Perturbated (NP)) and out of equilibrium (Perturbated (P)) is calculated as a function of time using the expression:

$$<\theta_i^{NP}(t) - \theta_i^P(t)> = \frac{1}{20} \sum_{n=1}^{20} (\theta_{i,n}^{NP}(t) - \theta_{i,n}^P(t)) \qquad (2)$$

where θ_i represents the Φ_i or Ψ_i backbone dihedral angle of residue i, $t=k \times 0.004ps$ (k=1,...500).

3. RESULTS

Analysis of the root mean square (RMS) deviation of the Φ and Ψ backbone dihedral angles over the 160 ps MD simulations of the two transmembrane peptides (Figure 1) evidence their different dynamical behavior.

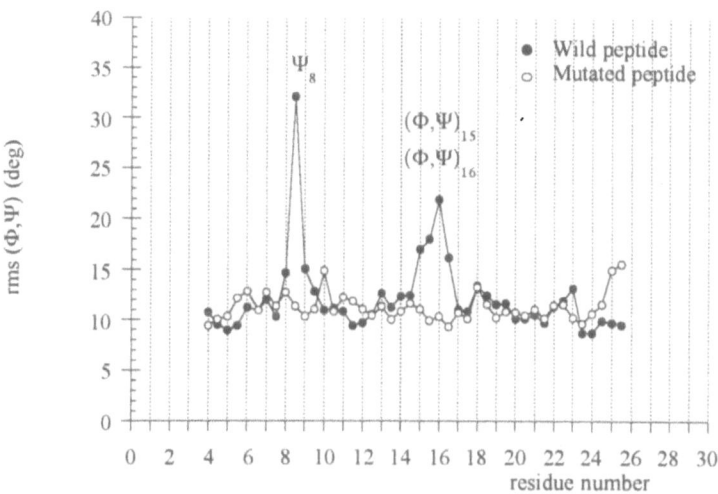

Figure 1: Root mean square (RMS) deviations of Φ and Ψ dihedral angles calculated for the whole transmembrane residues (Ile[4]-Ile[25]) along the 160 ps MD simulations for the wild and mutated peptides.

Two labile zones located at Val[8], at the junction of the mutation point, and at Val[15]-Val[16] are detected for the wild peptide. At the opposite, the low RMS values obtained for the mutated sequence suggest a rigid behavior. To illustrate this dynamical difference between the two peptides, the temporal evolution of two dihedral angles is given in Figure 2. The large fluctuations observed for dihedral angles of the wild peptide correspond to conformational transitions associated with important local structural changes. Hydrogen bond network of the α helix is, in that case, strongly perturbed leading to a bulge of one helix turn.

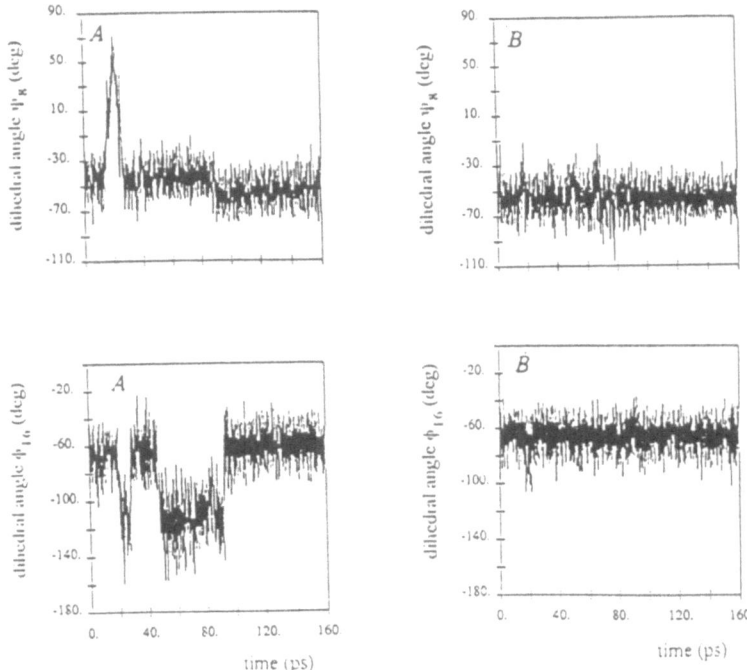

Figure 2: Time evolution of the dihedral angles Ψ_8 and Φ_{16} along a 160 ps trajectory for the wild peptide (**A**) and the mutated peptide (**B**),(taken from ref [4]).

Examination of the correlation matrix between dihedral angles fluctuations evidence long range correlated motions between the Val[8] and Val[15]-Val[16] regions for which conformational changes were previously observed (to be published elsewhere). An example of time dependent correlation is given in Figure 3 showing the correlation function between the Φ_9 dihedral angle fluctuations, defining the Val[8]-Val[9] junction, and Ψ_{16}. The correlation between these angular motions is maximum after about 5 ps and decrease a few picoseconds later. This result tends to demonstrate the existence of a propagation process throughout the transmembrane peptide.

For the mutated peptide, concerted motions of group of residues do not exist, except anti-correlations between neighbor residues.

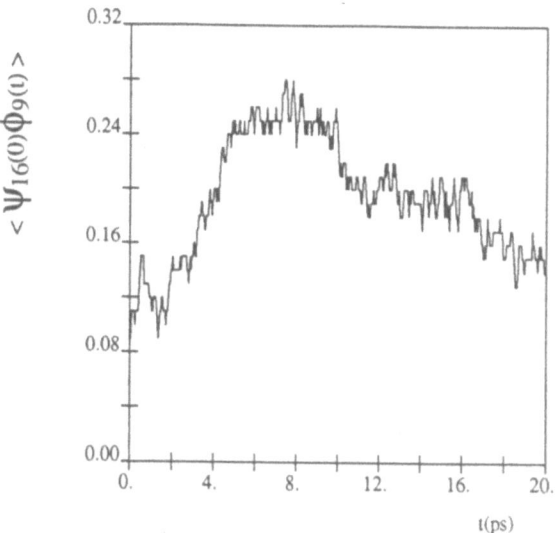

Figure 3: Time correlation function $<\Psi_{16}(0)\Phi_9(t)>$ for the wild peptide.

A third analysis relies on the qualitative study of the propagation of a conformational change along the peptidic chain. The propagation phenomenon is illustrated in Figure 4 for the Φ dihedral angles of residues Val[8] and Leu[24], which gives the time evolution of the mean difference between the values obtained in MD simulations performed with and without perturbation. These curves demonstrate that residue Val[8], located at the beginning of the chain and adjacent to the mutation site, is affected by the perturbation after 0.1 ps while residue Leu[24], close to the C terminal is affected about 0.4 ps later.

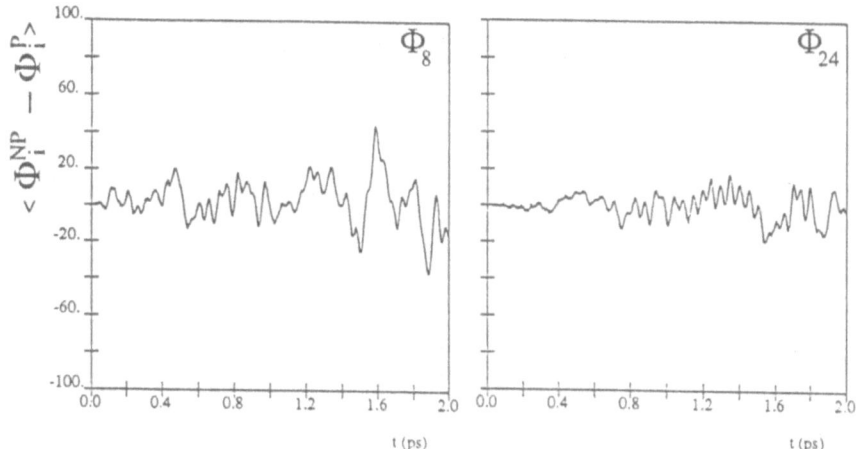

Figure 4: Time evolution of the quantities $<\Phi_8^{NP}(t) - \Phi_8^{P}(t)>$ and $<\Phi_{24}^{NP}(t) - \Phi_{24}^{P}(t)>$ in degrees.

4. DISCUSSION AND CONCLUSION

Analysis of 160 ps molecular dynamic simulations carried out on peptides including the transmembrane sequence of the c-erbB2 proto oncogene protein and its V659E mutant have shown a noticeable difference in their flexibility. Two flexible zones located at the mutation site and two helix turns farther have been revealed for the wild peptide whereas the mutated peptide is characterized by an intrinsic rigidity.

The long range correlations between these two regions far apart in the sequence have been clearly demonstrated from angular time correlation functions analysis. Such correlated motions certainly result from local structural changes induced by dihedral transitions producing displacement of Val^8 and Val^{16} in opposite direction (anti-correlation). However, the global helical structure is maintained by weak long range anti-correlation between central residues of the helix and one residue Ile^{23} located close to the chain extremity (result not shown). These long range correlations might be mediated by non covalent interactions between these residues placed along the same face of the helix.

The propagation phenomenon of a structural change along the peptidic backbone that we have qualitatively shown, may also contribute to the occurence of correlated displacement of group of residues.

None of these concerted motions have been observed for the mutated transmembrane portion, and such a difference in their behavior can be attributed essentially to the single amino acid replacement in position 9, which contributes to strongly perturb the flexibility of the chain. This allows to suppose the important role of the nature of the residues constituting the polypeptide chain and their sequence in the internal dynamic.

From these results, it is quite conceivable that transmission of information required after ligand binding at the extracellular part of the protein involves both the propagation of conformational change induced upon ligand fixation and correlated motions between residues of the transmembrane segment.

References

[1] Karplus M. and Petsko G.A., *Nature* **347** (1990) 631-639.

[2] Bargmann C.I., Hung M.C. and Weinberg R.A., *Cell* **45** (1986) 649-657.

[3] Bargmann C.I. and Weinberg R.A., *Proc. Natl. Acid. Sci.* USA **85** (1988) 5394-5398.

[4] Garnier N., Genest D., Hebert E. and Genest M., *J.Biomol.Struct. & Dynamics* **11** number 5 (1994) 983-1002.

[5] van Gunsteren W.F. and Berendsen H.J.C., *Angew.Chem.Int.Ed.Engl* **29** (1990) 992-1023.

[6] van Gunsteren W.F., Groningen Molecular Simulation System, BIOMOS Biomolecular Software b.v. Groningen 1987.

[7] Ryckaert J.P., Ciccoti G., Berendsen H.J.C., *J. Comput. Phys* **23** (1977) 327-341.

Chapter III

Energy and charge transport.

A complex system cannot function without a continuous flow of information to control that all parts work together in a coherent manner. This is true for our modern societies which are overflooded by information, but this is also true for a living organism which is even more complex. At the microscopic scale, the information is stored and transported under the form of energy, for instance vibrational energy, or electric charge. This chapter addresses this question from various viewpoints.

In his lecture **A. C. Scott** shows how the concept of solitary wave occurs naturally when one considers the propagation of a signals which is localised in space. Then he discusses two generic types of solitary waves, those that balance the release and dissipation of energy like a nerve impulse, and those that conserve energy. The second category includes the "polaron", or "conformon" in proteins which is closely related to the soliton concept. Its application to energy transport in proteins, first suggested by Davydov, has been the object of a long debate which is reviewed by A. Scott, including quantum and thermal effects. He points out that the same concept could also be applied to charge transport and other biological processes like ionic pumps or muscular contraction. The paper by **L. Cruzeiro-Hansson and V.M. Kenkre** consider more precisely the two-quantum states of the Davydov model which are particularly important because their energy matches almost exactly the energy released by ATP hydrolysis, one of the major energetic reactions of biophysics. The authors suggest also experiments that could provide evidence for some aspects of the Davydov mechanism. On the contrary **D. Brown** adopts a completely different point of view in his attempt to go beyond the semi-classical character of the soliton models and he builds polaron-Wannier states which could be the quantum counterpart of the soliton. But if one expects to detect the solitons, it is necessary to know their signature, and in particular their form factors in a scattering experiment. The calculation is very difficult for Davydov solitons, but, as shown by **A. Neuper and F. Mertens**, it can be performed for the Yomosa model when the energy transport in proteins is described by Toda solitons.

The Davydov model was designed to describe some aspects of muscle contraction where there is an obvious problem of energy transfer. But efficient energy transfer exists in other areas of biophysics, and in particular photosynthesis which is the most important energy conversion process on earth. The common principles which all organisms use to perform solar energy conversion are explained by **W. Mäntele.** Most of the processes are located at lipid bilayer membranes. A model system for photosynthesis is presented by **O.**

Bang, P. Christiansen, K. Rasmussen and Y. Gaididei who consider the example of Scheibe aggregates which function as energy funnels for sunlight to be used in photochemical processes. Here too ideas coming from nonlinear science appear to be very fruitful. It is known that, in many multidimensional systems, a nonlinear self focusing of energy occurs. Following closely the derivation of the Davydov model of energy transport in proteins, the paper shows that it is possible to derive a two-dimensional model for the properties of the excitons in a light-collecting molecular system.

Charge transport, and particularly proton transport, is another fundamental process of molecular biology. **G. Careri, F. Bruni and G. Consolini** provide a beautiful illustration of the power of the reductionist approach, which often characterises physical studies, by showing how the concept of the percolation transition of statistical physics is tightly connected to the emergence of biological function in nearly dry biosystems. Nonlinear excitations are not directly involved at this level but this first part of the paper shows how simple models can be useful in bringing unifying concepts. In the second part, the paper discusses experimental results which are perhaps the first direct evidence of collective proton transport in hydrogen-bonded chains. They strongly suggest that, below 278 K, a new species contribute to proton relaxation of hydrated proteins. Although this has still to be demonstrated, it could be the "soliton" or "polaron-soliton" which is discussed in the following papers by **M. Peyrard**, and by **G. Tsironis**. The first of the two includes a review of the models proposed for collective proton transport, followed by the study of a particular model which uses parameters derived from ab-initio calculations of water chains and includes thermal fluctuations. The preliminary results agree qualitatively with the experiments of G. Careri. A similar point of view is presented by G. Tsironis who directs his attention to quantum tunnelling effects. As the experimental results obtained by G. Careri on ice tend to show the great importance of tunnelling, even for "heavy particles" like protons (much heavier than electrons), the study of quantum effects is important.

Modelling hydrogen bonded systems in molecular biology requires however a basic knowledge of the structure and dynamics of water in these systems. Considering two particular systems, **H. Middendorf** illustrates the possibilities of neutron scattering techniques to provide spatio-temporal information. The results show in particular that water in biological system cannot be considered as either bound or free. Although it may complicate significantly modelling, going beyond this two-state description will have to be considered in theoretical studies.

Solitary waves in biology

A.C. Scott

Department of Mathematics, University of Arizona,
Tucson, AZ 85721, U.S.A.
and
Institute of Mathematical Modelling,
Technical University of Denmark,
2800 Lyngby, Denmark

1   INTRODUCTION

"Have soliton, will travel (across professional boundaries)" is the caricature that springs to many a biologist's mind when he or she thinks of the physicists and applied mathematicians who attempt to apply their simple theories from nonlinear dynamics to the teeming world of life. "Don't they realize," the biologist wonders, "how complicated living things really are?"

Leaving aside the technical contributions that physics has made to biology (conservation of energy, x-ray and neutron structure determinations, nuclear magnetic resonance, computed tomography, etc.), the theoretical insights of modern nonlinear dynamics are of some interest. Almost every cause and effect relationship in biology is nonlinear, and many biological structures—nerve and muscle fibers, the alpha-helical regions of protein, DNA, and the cytoskeleton, to name a few—are extended in one spatial dimension. Let us measure this dimension with the variable x. To the applied mathematician this means that one can seek *traveling wave* solutions to the underlying nonlinear partial differential equations of the form

$$f(x - ut),$$

where t is time and u is the speed of the wave. When f approaches zero as x goes to large positive and negative values, the solution is called a *solitary wave*.

Solitary waves come in two generic types:

- Those that balance the release and dissipation of energy, like a nerve impulse or the flame of a candle, and

- Those that conserve energy.

Both types of solitary waves are considered in this chapter. The nerve impulse is described in the next section, followed by a discussion of energy conserving solitary waves (or *solitons*) in protein. In each example, emphasis is placed on independent determinations of the parameters that enter into the models

2 THE NERVE IMPULSE

2.1 Linear and nonlinear diffusion

In a "space-clamped" observation of the switching of a nerve membrane [8], the membrane voltage, $v(t)$, is independent of the space coordinate x. The dynamics of this voltage is governed by the ordinary differential equation

$$c\frac{dv}{dt} + j_i = 0 \,, \tag{1}$$

where c is the membrane capacitance and j_i is the ionic current per unit length of the fiber. Since cv is the electrical charge stored per unit length of the fiber, Equation (1) says that ionic charge flowing across the membrane is balanced by the charge in the membrane capacitance.

For a normally functioning nerve fiber the membrane voltage is a function of both time (t) and distance along the fiber (x). In this case electric current flows through the tube of the fiber in the x-direction, and $v(x,t)$ is governed by the *nonlinear diffusion equation*

$$\frac{\partial^2 v}{\partial x^2} - rc\frac{\partial v}{\partial t} = rj_i \,, \tag{2}$$

where r is the electrical resistance per unit length of the fiber tube [44]. Evidently Equation (2) reduces to Equation (1) when v is independent of x.

Since the ionic current is a nonlinear function of voltage, solutions of Equation (2) can be quite complicated. It is useful, therefore, to look at special solutions that have the form of a traveling wave. Thus

$$v(x,t) = \tilde{v}(x - ut) \,, \tag{3}$$

where the wave speed—u—is a free parameter in the calculation. Under this assumption, Equation (2) reduces to the second order ordinary differential equation [44]

$$\frac{d^2\tilde{v}}{dx^2} + rcu\frac{d\tilde{v}}{dx} = rj_i \,, \tag{4}$$

which can be solved by standard numerical methods for particular values of the pulse velocity u.

A traveling wave solution of the form given in Equation (3)—shown in Figure(1)—corresponds to a nerve impulse, where u is its conduction velocity. It is interesting to note that nonlinear diffusion was also suggested by Turing as a basis for morphogenesis in biological systems [48].

2.2 The Hodgkin-Huxley equations

In the early 1950's A.L. Hodgkin and A.F. Huxley made careful measurements of the sodium and potassium components of membrane currents of the giant axon of the squid under space clamped conditions and constructed the following phenomenological expressions for the total ion current [31]:

$$j_i = G_K n^4(v - V_K) + G_{Na}m^3h(v - V_{Na}) + G_L(v - V_L) \,, \tag{5}$$

where

$$\frac{dn}{dt} = \alpha_n(1 - n) - \beta_n n \,,$$

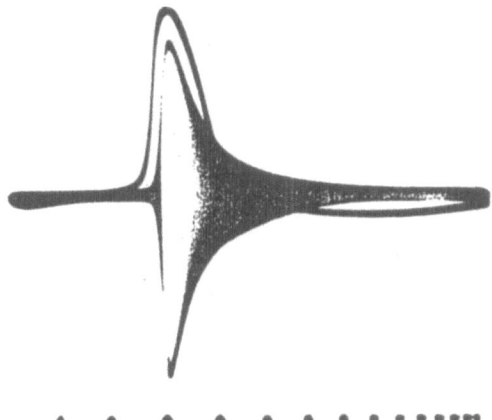

Figure 1: The time course of a nerve impulse on the giant axon of squid, from Cole and Curtis (1938). The time marks are at one millisecond intervals.

$$\frac{dm}{dt} = \alpha_m(1 - m) - \beta_m m \;, \tag{6}$$

and

$$\frac{dh}{dt} = \alpha_h(1 - h) - \beta_h h \;.$$

In these equations m is a "sodium turn-on" variable, h is a "sodium turn-off" variable and n is a "potassium turn-on" variable, and

$$\alpha_n = \frac{0.01(10 - v)}{[\exp(10 - v)/10 - 1]} \;,$$

$$\beta_n = 0.125 \exp(-v/80) \;,$$

$$\alpha_m = \frac{0.1(25 - v)}{[\exp(25 - v)/10 - 1]} \;,$$

$$\beta_m = 4 \exp(-v/8) \;, \tag{7}$$

$$\alpha_h = 0.07 \exp(-v/20) \;,$$

$$\beta_h = \frac{1}{[\exp(30 - v)/10 + 1]} \;,$$

in units of milliseconds^{-1}, where the membrane voltage v is measured in millivolts with respect to the resting potential. If v_{12} is the voltage difference between the inside and outside of the cell, then

$$v \equiv v_{12} + 65 \;.$$

Values for the parameters in Equation (5) are given in Table 1. The physical scientist should be aware that the values of these parameters measured on particular axon

Parameter	Value	Units
G_{Na}	36	millimhos/cm^2
G_{K}	120	millimhos/cm^2
G_{L}	0.3	millimhos/cm^2
V_{K}	-77	millivolts
V_{Na}	$+50$	millivolts
V_{L}	-54.4	millivolts

Table 1: Membrane permeability parameters for the giant axon of the squid.

membranes will exhibit "normal physiological variation". Thus Table 1 describes a "typical" axon from which a particular axon may differ by 10% to 20%.

Equations (2), (5), (6), and (7) are the celebrated "Hodgkin-Huxley" (or H-H) equations. Using these equations and assuming the solution to be a traveling wave, they were able to compute (using a mechanical hand calculator!) a numerical solution that agrees with the experimental observations shown in Figure 1. In the wake of repeated testing over the past 40 years, the H-H equations have been established beyond reasonable doubt as the fundamental equations of neurodynamics.

In Equation (5) the first (G_{K}) term accounts for sodium ion current and the second (G_{Na}), while the last term (G_{L}) accounts for all other ions under the heading of "leakage". To understand how this system works note that Equations (6) can also be written in the form:

$$\frac{dn}{dt} = -\frac{n - n_0(v_{12})}{\tau_n(v_{12})} \, ,$$

$$\frac{dm}{dt} = -\frac{m - m_0(v_{12})}{\tau_m(v_{12})} \, , \qquad (8)$$

and

$$\frac{dh}{dt} = -\frac{h - h_0(v_{12})}{\tau_h(v_{12})} \, ,$$

where, from Equations (7), $n_0(v_{12})$, $\tau_n(v_{12})$, $m_0(v_{12})$, $\tau_m(v_{12})$, $h_0(v_{12})$, and $\tau_h(v_{12})$ are the functions of membrane voltage indicated in Figure 2a. ¿From this figure it is evident that sodium turn-on (mediated by m) is about an order of magnitude faster than potassium turn-on and sodium-turn off (mediated by n and h). The switching of the membrane can now be described as follows:

- At rest the sodium ion permeability

$$G_{\mathrm{Na}} m_0^3(-65) h_0(-65)$$

  is small.

- As the membrane voltage, v_{12}, is increased from -65 millivolts, sodium channels open (m increases) on a time scale of $\tau_m \sim 0.4$ milliseconds.

- This influx of sodium ions brings the membrane voltage to a level of $+55$ millivolts, and potassium ions turn on ($n \to 1$) as sodium ions turn off ($h \to 0$).

- The efflux of potassium ions carries the membrane back to its resting value on a time scale of a few milliseconds.

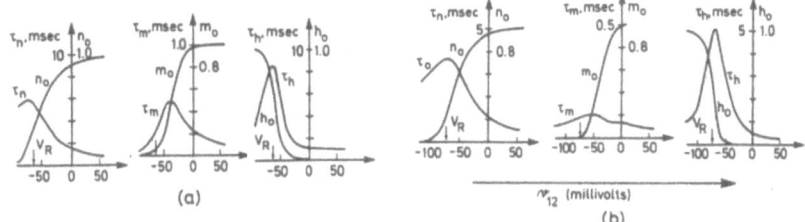

(a)   (b)

Figure 2: The functional dependence of Hodgkin-Huxley parameters on membrane voltage. (a) For a typical squid giant axon, and (b) For the active node of a frog's sciatic nerve. Redrawn from Cole, 1968.

Parameter	Value	Units
G_{Na}	0.104	micromhos
G_K	0.57	micromhos
G_L	0.025	micromhos
V_K	-75	millivolts
V_{Na}	$+47$	millivolts
V_L	-75	millivolts

Table 2: Membrane permeability parameters for an active node of a frog's sciatic nerve with an area af about 20 micron^2.

This behavior is not a special property of the squid axon. Figure 2b presents a corresponding picture for the sciatic nerve of a frog. In this case the measurement was made on a small active node because each nerve fiber is largely covered with an insulating sheath (called myelin) to increase the conduction velocity of a pulse without increasing the nerve diameter. Corresponding values for the membrane permeability parameters of a typical active node are given in Table 2. Comparison of the data in Figure 2 and Tables 1 and 2 shows that the properties of nerve from different phyla are similar.

In agreement with experiments, the Hodgkin-Huxley theory shows that a certain quantity of electric charge must be delivered to an axon in order to initiate a nerve pulse. The candle provides a metaphor for this threshold effect; thus in order to light a candle (fire an axon) a certain temperature difference (electric current) must be applied to the wick (end of the axon) for a certain length of time, and the product of temperature difference (current) times time is heat (electric charge). The ratio of the size of a fully developed nerve pulse to the size of a threshold pulse is called the *safety factor* of the nerve fiber.

In summary, the salient features of a nerve axon are these [44]:

- The dynamics of a voltage pulse is governed by the *nonlinear diffusion equation*, where the ionic current is given by the Hodgkin-Huxley equations.

- The nonlinear diffusion equation has a traveling wave (or pulse) solution of the form indicated in Figure 1. This solution is a nerve impulse.

- Certain threshold conditions must be satisfied to launch a nerve impulse on an axon.

- The nerve impulse has "all or nothing" character.

- Like the flame of a candle, the nerve impulse does not "conserve energy"; instead it "balances power" and propagates at a speed that is fixed by the parameters of the system.

3   SOLITONS IN PROTEIN

The concept of a *polaron* as a means for charge transport in a crystal was first suggested by Landau in 1933 [39] and studied extensively in the early 1950's by Pekar [41], who seems to have coined the term. The basic idea is as follows: A local charge distorts the crystal structure in its vicinity, and this distortion, in turn, creates a "potential well" for the charge, which prevents its dispersion. The picture is rather like that of a child standing on a trampoline, where the depression of the surface requires that a (gravitational) *binding energy* be overcome before the child can step off.

3.1   Polarons and conformons

The polaron idea is closely related to that of a *conformon*, a term that was coined twice by biologists in 1972—once by Green and Ji [28] and again by Vol'kenstein [49]—and the same term was used in 1973 by Kemeny and Goklany [35, 36, 37] to express a concept that they had been independently developing. Since these authors have defined the conformon in different ways, the following comments may be helpful:

- Green and Ji defined the conformon (in connection with a model for the mitochondrion) as "the free energy associated with a localized conformational strain". It was assumed to be a mobile "packet of energy" that moves along the alpha-helical structure of a protein.

- Vol'kenstein defined the conformon as "the displacement of an electron or of the electronic density in a macromolecule [that] produces the deformation of the lattice, i.e., the conformational change. It can be treated as an excitation of the long wave phonons, and the system electron plus local deformation becomes like [a] polaron."

- Kemeny and Goklany were motivated to explain the unexpectedly high electrical conductivity of biological substances. They state: "Vol'kenstein's mechanism and ours are very similar.... Although the applications may be very different, it still seems to be worthwhile to retain Vol'kenstein's term, and we will call the activated carrier plus the accompanying changes carrying energy and entropy the conformon.... The concept of the conformon is a generalization of the small polaron concept."

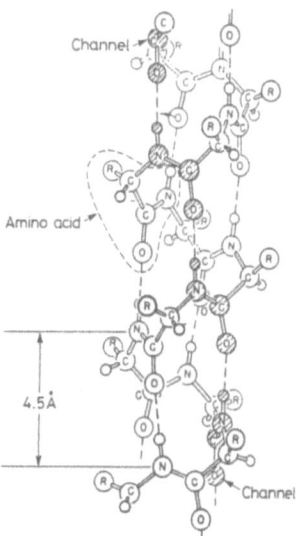

Figure 3: An atomic model of the alpha-helical structure in protein. One "channel" is cross-hatched.

3.2 Vibrational energy transport

At the same time that Green and Ji, Vol'kenstein, and Kemeny and Goklany were proposing their conformons as means for transport of free energy and charge in biology, Davydov applied the analytical picture of Landau's polaron to the alpha-helical structure of protein [12, 16] to describe what is essentially the conformon of Green and Ji. With reference to Figure 3, turns of the helix are numbered with the index j and $\{A_j\}$ is a set of time dependent variables that obey the differential equations

$$i\hbar\frac{dA_j}{dt} = \underbrace{[E_0 + \chi\overbrace{(\beta_{j+1} - \beta_j)}^{R-R_0}]}_{E} A_j - J(A_{j+1} + A_{j-1}),\tag{9}$$

in which every parameter can be experimentally determined. The meaning of A_j depends upon whether one is thinking about the problem classically or quantum mechanically; thus:

- **Classical perspective.** A_j is the complex mode amplitude of a CO stretching (Amide-I) oscillation at the jth turn of the helix. In other words, $A_j = x_j + ip_j$, where x_j is the extension of a CO oscillator and p_j is its momentum.

- **Quantum perspective.** A_j is the probability amplitude for finding a quantum (boson) of CO oscillator energy in the jth turn of the helix. In other words $|A_j|^2$ is the probability of finding an Amide-I boson.

The parameters in Equation (9) are defined as follows

- $E_0/\hbar$ is the site frequency (radians per second) of an isolated CO stretching oscillation (where $\hbar$ is Planck's constant),

- J is the energy of interaction between neighboring CO oscillators,

- β_j is the longitudinal displacement of an amino acid, and

- χ is an anharmonic parameter that relates changes in the CO stretching frequency to the extension of the adjacent hydrogen bond.

With reference to Figure 3, the resting value of the displacement between two amino acids is $R_0 = 4.5\text{Å}$. If the helix is stretched, this displacement becomes

$$R = R_0 + \beta_{j+1} - \beta_j$$

and the anharmonic parameter χ is defined as

$$\chi \equiv \frac{dE}{dR} \tag{10}$$

as is seen from the structure of Equation (9).

At this point a few "reader friendly" remarks may be appropriate. First of all think classically and suppose that χ is zero (no anharmonicity) and also that J is zero (no dispersion). Then the solution of Equation (9) is simply

$$A_j \propto \cos \omega_0 t + i \sin \omega_0 t \,,$$

as one expects for a simple harmonic oscillator of frequency $\omega_0 = E_0/\hbar$. Now let J (the dispersive parameter) enter the picture. In this case one would have the spectral broadening of alpha-helix, which has been measured by Chirgadze and Nevskaya and used to determine the value of J [7].

If we allow the anharmonic parameter (χ) to be nonzero, it is necessary to introduce another equation in order to follow the dynamics of the helix (the β_j's). This equation is

$$3M\frac{d^2\beta_j}{dt^2} - 3w(\beta_{j+1} - 2\beta_j + \beta_{j-1}) = \chi(|A_j|^2 - |A_{j-1}|^2) \,. \tag{11}$$

With no anharmonicity $(\chi = 0)$, Equation (11) is merely the equation for longitudinal sound waves on the helix. Also:

- M is the mass of a single amino acid. There are three of them in each turn of the helix.

- w is the spring constant of a single hydrogen bond. From Figure 3 there are three of these between adjacent turns of the helix.

When $\chi \neq 0$ (anharmonicity is present) the form of the term on the right side of Equation (11) is fixed by the requirement of energy conservation.

Taken together, Equations (9) and (11) have solitary wave solutions in the sense that was defined in the introduction. Unlike the solitary waves on nerve, however, these can travel at any speed from zero up to that of longitudinal sound on the helix, which is

$$u_{\max} = R_0\sqrt{\frac{w}{M}}$$

where R_0 is the spacing between turns of the helix. The reason for this difference is that the nonlinear wave motion on the helix conserves energy, while that on nerve balances power.

Parameter	Value	Units
χ	35–62	piconewtons
$\tilde{w} \equiv 3w$	39–58.5	newtons/meter
$\tilde{M} \equiv 3M$	5.7×10^{-25}	kg
E_0	3.34×10^{-20}	joules
J	1.55×10^{-22}	joules

Table 3: Parameter values for the alpha-helix.

If we are interested in a solitary wave that travels slowly compared with $u_{\max}$, then the kinetic energy of the longitudinal sound system can be neglected, and the $3M d^2\beta_j/dt^2$ term can be dropped from Equation (11). In this approximation, Equation (11) reduces to

$$\beta_{j+1} - \beta_j = -\frac{\chi}{3w}|A_j|^2\,,\tag{12}$$

which—upon substitution into Equation (9)—yields the *discrete nonlinear Schrödinger equation*

$$i\hbar\frac{dA_j}{dt} - E_0 A_j + J(A_{j+1} + A_{j-1}) + \frac{\chi^2}{3w}|A_j|^2 A_j = 0\,.\tag{13}$$

It should be emphasized that every parameter in this system can be (and has been) experimentally determined [45], and the most significant parameters are listed in the following Table 3.

3.3 Quantum theory

Let us take a brief look at Equations (9) and (11) from the perspective of quantum theory. As was noted above, we now consider $|A_j|^2$ to be the probability of finding a quantum of CO oscillator energy on the jth turn of the helix. In general this quantum polaron problem is unsolved [29], so every published result involves one or more simplifying assumptions. It is important for the reader to be aware of this unhappy fact because authors sometimes fail to indicate the limitations that are implied by their approximations. Many of the disagreements that have arisen in this area over the past decade and a half might have been avoided had the scientists involved recognized that they were making incompatible assumptions.

The approximation used by Davydov [15] was to write the total quantum mechanical wave function, ψ, as a product

$$\psi = \Psi \times \Phi\,,\tag{14}$$

where Ψ describes the CO oscillators and Φ describes the longitudinal vibrations of the helix. Introducing appropriate time dependent variables into the structures of Ψ and Φ leads to equations that are identical in form to Equations (9) and (11) but with a quantum mechanical—rather than classical—interpretation of the dynamical variable A_j. The parameters remain the same.

Equation (14) is in the spirit of the Born-Oppenheimer approximation of molecular physics, in which the electronic motion is separated from that of the atomic nuclei through their large difference in natural frequency. Unfortunately for Davydov's approximation, the ratio of CO oscillation frequency to that of the longitudinal sound waves on the helix is only about thirty to one, so the errors involved are larger than those of the Born-Oppenheimer approximation. Michael Collins has studied in detail the errors arising from the assumption of Equation (14) and shown how these errors depend on the anharmonic parameter [9]

$$\alpha \equiv \frac{\chi^2}{3wJ}. \tag{15}$$

Equation (14) implies a soliton binding energy that is proportional to α^2, as in the classical case, but this neglects the negative *self-energy* of the soliton, which arises from a shift in the phonon density of states. The self energy is proportional to α and therefore dominates the α^2 contribution when $\alpha < 1$. From Table 3, α is somewhere between .14 and .64 so the product assumption of Equation (14) leads to a significant *underestimate* of a soliton's binding energy on alpha-helix.

3.4   Thermal effects

As they stand, Equations (9) and (11) do not include the influence of ambient temperature, and this is a serious omission because biological organisms function at a temperature of ~ 300K rather than absolute zero. To see how the effects of ambient temperature should be introduced, recall that the average number of quanta (bosons) in an harmonic oscillator of frequency ω at temperature T is

$$\bar{n} = \frac{1}{\exp(\hbar\omega/kT) - 1}, \tag{16}$$

where k is Boltzmann's constant.

For the CO oscillators described in Equation (9)

$$\frac{\hbar\omega}{kT} \doteq 8$$

so

$$\bar{n} \doteq 0,$$

and the direct influence of physiological temperature can be neglected. For longitudinal oscillations on the helix, on the other hand,

$$\frac{\hbar\omega}{kT} < \frac{1}{4}$$

so

$$\bar{n}\hbar\omega \doteq kT,$$

which implies that the mode (or degree of freedom) has an average energy of kT. Thus a standard thermal noise term can be added to Equation (11) to account for the effects of physiological temperature.

¿From the theory of ordinary differential equations, the general solution of Equation (11) can be divided into two parts:

- The *soliton part*, which is driven by the source term—$\chi(|A_j|^2 - |A_{j-1}|^2)$—on the right hand side, and

- The *thermal part*, which is a solution of Equation (11) with the source term equal to zero. In this "homogeneous" part of the general solution, each mode of oscillation has an average energy equal to kT as is shown in the above discussion.

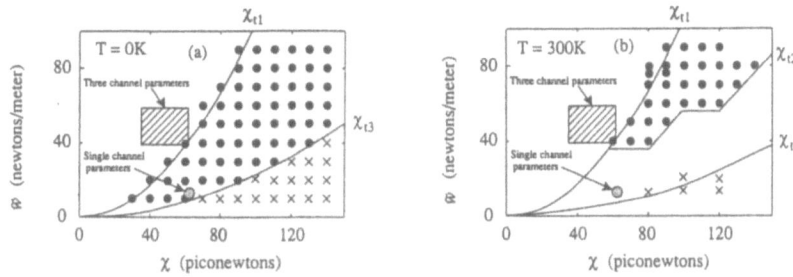

Figure 4: Numerical results obtained by Förner for the propagation of solitons on alpha- helix. A soliton that is launched and propagates across the entire 200 turns of the helix is indicated by a (•) and a pinned soliton by a (×).

It is important to notice that the thermal part of the solution to Equation (11) has an *indirect* effect on Equation (9) because it enters through the $\chi(\beta_{j+1} - \beta_j)A_j$ term. The effect is to scatter the (localized) soliton solution of Equations (9) and (11) into (extended) solutions of Equation (9) that do not have the localized properties of a solitary wave. Thus the effect of temperature is to attenuate the soliton through this indirect mechanism.

Is this effect significant at physiological temperature? As we see from Table 3, the experimental values of $\tilde{w} \equiv 3w$ and χ are somewhat uncertain so it is necessary to investigate how well an alpha-helix soliton propagates over a range of parameter values. The most ambitious study of soliton lifetime in the presence of thermal noise has been carried out by Förner and his colleagues [19, 20, 21, 22, 23, 24, 25, 26], and in these calculations it is assumed that the alpha-helix has 200 turns for a total length of about 0.1 micron, which is the length of the alpha-helix in myosin. Since the helix has 200 degrees of freedom, a random (thermal) energy of $200kT$ is introduced as the thermal part of the solution to Equation (11) before the start of the numerical experiment. At $t = 0$, a soliton is launched at on end of the helix through the initial conditions

$$A_1(0) = 1, \quad \text{and} \quad A_j(0) = 0 \text{ for } j = 2, 3, \ldots, 200; \tag{17}$$

thus the number of Amide-I quanta is equal to unity in these calculations.

Typical results are shown in Figure 4a for $T = 0K$ and in Figure 4b for $T = 300K$ and a rather wide range of the parameters χ and $\tilde{w}$. If a soliton was observed to form and propagate across the entire length of the helix (200 turns), the corresponding point in parameter space is marked by a solid black circle (•). If the Amide-I vibrational energy is observed to be self-trapped but pinned to the lattice, that point is indicated by a cross (×). At other (unmarked) points in the parameter space, solitons did not develop or soliton lifetime was not long enough to permit transit of the entire molecule.

Consider first the calculations for $T = 0K$ that are shown in Figure 4a. There are two critical levels of the exciton-phonon coupling parameter: χ_{t1} and χ_{t3}. The larger of these, χ_{t3}, is the condition for the soliton size to be approximately equal to the lattice spacing. Larger values of χ lead to pinning. The smaller threshold, χ_{t1}, indicates the level of anharmonicity for which the soliton will form with the initial conditions of Equations (17), and this threshold can be calculated from the "inverse scattering transform method" of soliton theory [45]. Thus χ_{t1} depends upon the *shape* of the initial conditions.

Next regard the calculations for $T = 300K$ that are shown in Figure 4b. The threshold level for soliton formation (χ_{t1}) and for pinning (χ_{t3}) remain unchanged (as is expected) but a new threshold level (χ_{t2}) appears that is clearly dependent on temperature. This threshold indicates the value of χ above which the random (thermal) part of the term $\chi(\beta_{n+1} - \beta_n)\phi_n$ becomes large enough to induce significant scattering from soliton to exciton states.

Finally let us consider where "real alpha-helix" lies in this parameter space. From the parameter values given in Table 3, the best available representation of alpha-helix in this model and are indicated as the "three channel parameters" on Figure 4. These parameters lie to the left of χ_{t2} on Figure 4b, and one must conclude that the soliton does not experience significant thermal scattering into exciton states at 300K.

Since the region of the "three channel parameters" lies to the left of χ_{t1} in Figures 4a and 4b, one might be tempted to conclude that the exciton-phonon coupling is too weak for the soliton to form with the initial conditions of Equations (17), but as noted above this depends on the spatial shape of the initial conditions. Extensive numerical simulations indicate that soliton formation is not unexpected on real alpha-helix [45]. In any event, the success or failure of the soliton to become launched from a particular set of initial conditions is unrelated to the effects of ambient temperature.

Also indicated on Figure 4 are the "Single channel parameters" that some have used to characterize alpha-helix. It has been assumed that alpha-helix can be represented by using the parameters of a single channel, which is shown cross-hatched on Figure 3. Overlooking the "factors of three" in Equation (11) leads to erroneous conclusions concerning thermal stability of the alpha-helix soliton.

3.5   Crystalline acetanilide

At this point in the discussion, an appropriately critical biologist might remark: "All this theory is fine, but can you people *show* me a soliton?" Experiments on natural protein are difficult because the structures are rather irregular and measurements are often correspondingly fuzzy. One way around this difficulty is to investigate crystalline acetanilide (ACN)—an optically clear molecular crystal with a peptide structure close to that of natural protein—as was proposed by Careri in 1973 [4]. Infrared measurements in the vicinity of the Amide-I (CO stretching) frequency reveal an anomalous band at 1650 cm^{-1}, which has been interpreted as a self trapped state [5, 18]. Since the current status of this assignment is discussed in detail in the chapter by Barthes, only the following points will be mentioned here:

- The self-trapping parameters for ACN indicate that the CO oscillation is localized near a single molecule. This "small polaron" picture has been evident since the earliest discussions of the self-trapping explanation of the 1650 cm^{-1} band [5, 18]. This point needs to be emphasized because some authors have confused the parameters of ACN with those of alpha-helix in Table 3.

- The "Born-Oppenheimer-like" product wavefunction indicated in Equation (14) works well for ACN. This agreement is discussed in detail in reference [45].

- Since the self-trapped state is localized close to a single molecule of ACN, a perturbation theory in small J is appropriate. This analysis gives a good prediction of the overtone spectrum of the 1650 cm^{-1} band [46].

Parameter	Value	Units
χ^2/W	4.91×10^{-22}	joules
J	7.87×10^{-23}	joules

Table 4: Parameter values for the self-trapped state in crystalline acetanilide.

- The same perturbation theory in small J predicts the temperature dependence of the 1650 cm^{-1} band in ACN [42].

- The precise nature of the phonon modes that lead to self-trapping in ACN has not yet been determined [3], but spectral evidence suggests the participation of several optical phonon modes [42] and perhaps also an acoustic phonon mode [2].

- Experimental studies yield the parameter values of self-trapping of the Amide-I energy in ACN that are listed in Table 4. Note that anharmonicity parameter χ and optical mode spring constant W have not been individually measured, and the composite parameter- $-\chi^2/W$—should be viewed as a sum over several lattice modes.

3.6 Transport of electronic charge

To this point our attention has been directed to Davydov's original idea: the self-trapping and transport of Amide-I vibrational energy [12, 16], corresponding to the conformon of Green and Ji. In 1979 Davydov indicated how the polaron analysis might be generalized to describe a means for self-trapping and transport of electronic charge in protein [13], which corresponds to the conformon of Vol'kenstein and of Kemeny and Goklany. Perhaps the best introduction to this generalization is chapter VI of reference [15].

Davydov's picture of the *electrosoliton* in alpha-helix is based upon two facts: (i) The peptide group (CONH) in the helix has a static electric dipole moment of about 1.55×10^{-29} coulomb-meters, which arises from the static charge distributions on the atomic constituents, and (ii) An electron can be bound to a static electric dipole if it exceeds a minimum value of 5.42×10^{-30} coulomb-meters [47].

Thus Davydov proposed that that an electron could propagate in an exciton band formed by the static dipole potential wells along a channel (see the cross-hatched path of Figure 3) of the alpha-helix.Furthermore the propagation of this electron should be influenced by interactions with acoustic mode phonons on the helix, just as for the self-trapping of vibrational energy. Taking the three channels into account leads to equations that are identical to Equations (9) and (11), but with the following differences in interpretation:

- A_j is the probability amplitude for an electron to be trapped in the jth turn of the helix,

- J is the nearest neighbor longitudinal electronic overlap integral,

- E_0 is the site energy of the electron, and

- χ is an anharmonic coupling constant between an electron and the longitudinal phonons on the helix.

Parameter	Value	Units
χ^2/w	$\sim 8 \times 10^{-20}$	joules
J	$\sim 1.6 \times 10^{-19}$	joules

Table 5: Parameter values for the electrosoliton in alpha-helix.

It is expected that the electrosoliton will transport charge with less loss of kinetic energy (phonon damping) because of its weak residual interaction with acoustic phonons. In this model, w and M are the same as in Table 3, and other values have been estimated by Davydov as in Table 5 [15].

3.7   Protonic solitons

In an often quoted paper by Eigen and De Meyer [17], ice is described as a "protonic semiconductor" in which the mobility of the proton is unexpectedly large. From the work of von Hippel and his colleagues, however, it appears that Eigen and De Meyer were measuring surface currents arising from multicrystallinity and not the bulk properties of pure ice [32, 33, 40]. In a biological context this suggests that proton currents should be expected at interfaces between hydrogen bonded phases, for example between protein and water.

Encouraged by the successes of the soliton paradigm during the 1970's, Davydov, et al. [14], Kashimori, et al. [38], and Yomosa [51] were led independently to a soliton theory of protonic conduction. The basic concept in these works is to model the potential seen by a proton in a hydrogen bond as the double well potential

$$U(\beta) = \varepsilon_0(1 - \beta^2/\beta_0^2)^2 \,, \tag{18}$$

where $\pm\beta_0$ are the two equilibrium positions for the proton, and ε_0 is the height of the potential barrier between them. In a linear chain of such potentials (with periodic spacing R_0), the position of the proton at the jth link is governed by the *nonlinear Klein-Gordon equation*

$$m_p\frac{d^2\beta_j}{dt^2} - K(\beta_{j+1} - 2\beta_j + \beta_{j-1}) = \frac{dU(\beta_j)}{d\beta_j} \,, \tag{19}$$

where m_p is the mass of a proton and the restoring force constant, K, is related to the speed of protonic sound waves, u_0, by

$$u_0 = R_0\sqrt{\frac{K}{m_p}} \,.$$

Equation (19) is well known to have soliton solutions in the form of "kinks" that can travel at any speed between zero and $\pm u_0$ [43]. It is interesting to note that these solitons do not require coordinated motion of the background lattice, although such motion may have a strong effect on the dynamics of the soliton [1, 52].

3.8   Biological applications

We now turn to some biological applications of the foregoing ideas, but this task is approached with hesitation. It is the right (and indeed the *responsibility*) of physical science to propose and examine dynamical theories that might be important in the

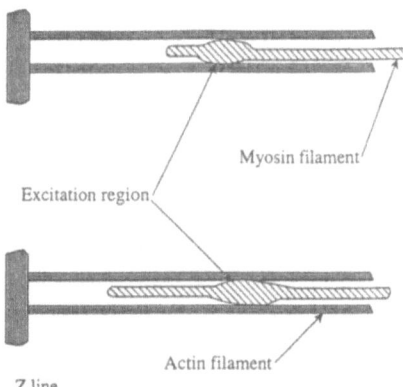

Figure 5: Force generation by solitons in the sliding filament model of skeletal muscle.

realm of biology. Having done this, it becomes the responsibility of biological science to decide which (if any) of these theories might actually play a functional role in a living organism. When a physical scientist encroaches upon the biological realm, he or she is walking on thin ice. It is with this humble thought in mind that the following applications of "solitonics" to biology are sketched.

- **Ionic pumps.** In the resting state of a nerve fiber, which was described in Section 2, the concentration of sodium ions is larger outside the cell than inside, yet the electrostatic potential on the inside is 65 millivolts *below* that on the outside. This strongly nonequilibrium situation (expressed by the polarity of the "sodium batteries" in Tables 1 and 2) drives the leading edge of the nerve impulse, and it is maintained through the action of a "sodium pump" that somehow pushes sodium ions out of the cell against gradients of both concentration and electrostatic potential. Similar phenomena are the pumping of hydrogen ions across purple membrane by the action of light and the pumping of hydrogen ions out of a mitochondrion at the end of the citric acid cycle.

  It is tempting for the physical scientist to suggest that solitonic transport mechanisms might play a functional role in the operation of these mysterious ionic pumps, but one should be aware that the true biological mechanism may be much more complicated.

- **Muscular contraction.** In Davydov's first paper on solitons in protein [12], the main application was to explain the generation of force in the "sliding filament theory" of muscular contraction, and Ji has also discussed the role that the conformon might play in muscular contraction [34]. Davydov asked: What makes the myosin heads attach to actin, pull, and then detach themselves? His explanation can be described in relation to Figure 5, which is redrawn from reference [12]. He suggested that a "herd" of solitons traveling along the bundle of myosin fibers would cause a localized swelling that would, in turn, provide the opportunity for attachment, forcing, and detachment of myosin heads to the actin filaments.

• **The turning wheel.** In 1975 Wyman proposed the "turning wheel" as a general model for the steady state operation of an enzyme [50]. The simplest example is illustrated in Figure 6, which describes the conversion of the ligand (substrate) L into its product L'. In this figure it is assumed that the concentration of L is kept constant and the k's are rate constants in the associated system of ordinary differential equations

$$
\begin{aligned}
[\dot{M_1}] &= -(k_1[L] + k_{-4})[M_1] + k_4[M_2] + (k_L + k_{-1})[M_1L] \\
[\dot{M_2}] &= k_{-4}[M_1] - (k_{-3}[L] + k_4)[M_2] + k_3[M_2L] \\
[\dot{M_1L}] &= k_1[L][M_1] - (k_L + k_{-1} + k_2)[M_1L] + k_{-2}[M_2L] \\
[\dot{M_2L}] &= k_{-3}[L][M-2] + k_2[M_1L] - (k_{-2} + k_3)[M_2L]
\end{aligned}
\tag{20}
$$

where the square brackets denote concentrations of the various species and the "dots" indicate time derivatives. In other words,

$$
[\dot{X}] \equiv \frac{d}{dt}(\text{concentration of species } X).
$$

In this simple example the enzyme is assumed to have only two conformational states, M_1 and M_2, each of which can bind to the ligand L. Wyman assumed "one-step transitions", so the diagonals, $M_1 \longleftrightarrow M_2L$ and $M_2 \longleftrightarrow M_1L$, are not included in his scheme.

For the steady state solution of Equations (20), there is a counterclockwise circulation in the diagram of Figure 6 and the product ligand is produced at the rate

$$
[\dot{L'}] = k_L[M_1L].
\tag{21}
$$

In a real enzyme, of course, the dynamics can be much more complicated. Many of these details have been discussed by Wyman [50] and also by Hill and Chen [30] and some related experimental results have been discussed by Giacometti et al. [27]. Nonetheless two important questions remain:

– What keeps the wheel turning?

– Why should all enzymes be proteins?

Motivated by experimental observations on crystalline acetanilide [5], Careri and Wyman have suggested that a protein soliton might provide an answer to these questions [10]. Their basic idea is illustrated in Figure 6b where the corresponding ordinary differential equations are again as in Equations (20) and (21). The soliton, S, is assumed to be "an energy packet trapped ... inside a portion of a protein matrix, with the possibility of storing energy without dissipation until it decays into heat as a result of a strong perturbation from some other portion of the same protein." The soliton is assumed to be formed as the substrate binds to the ligand L and to have a long lifetime in M_2 but a short one in M_1. This short lifetime provides an explanation for the unidirectional output reaction producing L' and the energy of the soliton pays for the energy difference between L' and L, the remainder being dissipated as heat.

In a later paper [11] Careri and Wyman extended these ideas to a suggestion for a prebiotic mechanism to harvest solar energy that is based upon the overtone bands of crystalline acetanilide [46]. Before the development of chlorophyll a

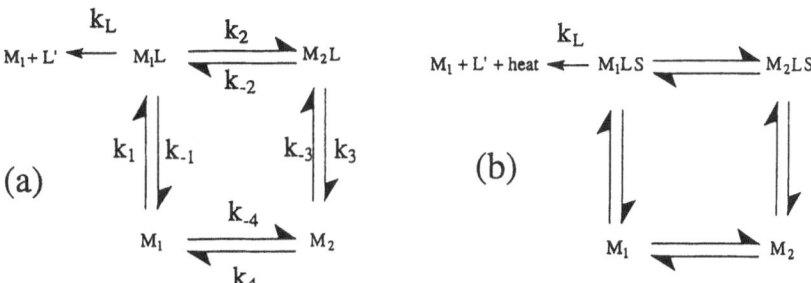

Figure 6: Diagrams related to Wyman's "turning wheel".

"protoenzyme" may have developed with infrared properties similar to ACN. The fundamental band is ruled out of consideration because of the high absorption by water and the first overtone is similarly ruled out because of high protein absorption. But the second overtone falls in a frequency region where water displays a window for infrared transmission.

- **Protonic currents in living systems.** Recent work by Careri and his colleagues has shown that proton conductivity on the surface of protein is an important feature of the living state. Since these results are discussed in detail in Careri's chapter of this book, I merely note that the mechanism of solitonic proton transport presented above may play a role in these observations.

4   CONCLUSIONS

Solitary waves in nerve are well established; they are the familiar "action potentials" of electrophysiology, which balance the rate energy release from the membrane to the power that is consumed by circulating ionic currents. The evidence for energy conserving solitary waves—or solitons—in protein is less secure, but there is confirmation from optical experiments on polypeptide crystals. In any event, vibrational energy, electronic charge and protons do move about in living systems. If associated distortions of the background structures helps to hold these packets of energy or charge together, then solitons can be said to play some role in the functioning of biological systems. For realistic assessments of such effects it is important to pay particular attention to the independent determination of the parameters that appear in the models.

Finally it is interesting to note that an implication of the solitary wave concept for biological theory is related to the concept of *emergence*. The structure of biological science is hierarchical in nature; thus biochemistry emerges from chemistry, biopolymers (DNA, RNA, protein, and lipid membrane) emerge from biochemistry, subcellular structures (the cytoskeleton, mitochondria, and nucleus of a cell) emerge from biopolymers, and so forth up to the structure and dynamics of a fully developed organism. At each level of the hierarchical structure, new entities emerge from the appropriate nonlinear dynamics to provide an atomistic basis for the next higher level. The emergence of solitary waves in nerve and protein serves as a paradigm for this important feature of biological structure. To the extent that we understand it, the complexity of life seems to emerge from several layers of the biological hierarchy.

Acknowledgement

Support from the NSF under Grant No. DMS-9114503 is gratefully acknowledged.

References

[1] Antonchenko V.Ya., Davydov A.S. and Zolotariuk A.V., *Phys. Stat. Sol. (b)* **115** (1983) 631–640.

[2] Alexander D.M. and Krumhansl J.A., *Phys. Rev. B* **33** (1986) 7172–7185.

[3] Barthes M., Almairac R., Sauvajol J.L, Moret J., Currat R. and Dianoux J., *Phys. Rev. B*, **43** (1991) 5223–5227.

[4] Careri G., Cooperative phenomena, Haken H. and Wagner M., editors, Springer-Verlag, Berlin, 1973.

[5] Careri G., Buontempo U., Galluzzi F., Scott A.C., Gratton E. and Shyamsunder E., *Phys. Rev. B* **30** (1984) 4689–4702.

[6] Cole K.S. and Curtis H.J., *Nature* **142** (1938) 209.

[7] Chirgadze Yu.N. and Nevskaya N.A., *Dokl. Akad. Nauk SSSR* **208** (1973) 447–450.

[8] Cole K.S., Membranes, ions and impulses, University of California Press, Berkeley, 1968.

[9] Collins M.A., *J. Chem. Phys.* **88** (1988) 399–404.

[10] Careri G. and Wyman J., *Proc. Nat. Acad. Sci. USA* **81** (1984) 4386–4388.

[11] Careri G. and Wyman J., *Proc. Nat. Acad. Sci. USA* **82**(1985) 4115–4116.

[12] Davydov A.S., *J. theoret, Biol.* **38** (1973) 559–569.

[13] Davydov A.S., *Theor. Mat. Fiz.* **40** (1979) 408–421.

[14] Davydov A.S., Antonchenko V.Ya. and Zolotariuk A.V., Report No. ITP-81-60R, Institute of Theoretical Physics, Kiev 1981.

[15] Davydov A.S., Solitons in molecular systems (2nd edition), Reidel, Dordrecht, 1991.

[16] Davydov A.S. and Kislukha N.I., *Phys. Stat. Sol. (b)* **59** (1973) 465–470.

[17] Eigen M. and De Meyer L., *Proc. Roy. Soc. London A* **247** (1958) 505–533.

[18] Eilbeck J.C., Lomdahl P.S. and Scott A.C., *Phys. Rev. B* **30** (1984) 4703–4712.

[19] Förner W. and Ladik J., Davydov's soliton revisite, ChristiansenP.L. and Scott A.C., editors, Plenum, New York, 1990.

[20] Förner W., *Phys. Rev. A* **44** (1991) 2694–2708.

[21] Förner W., *J. Phys. Cond. Matter* **3** (1991) 4333–4348.

[22] Förner W., *J. Phys. Cond. Matter* **4** (1992) 1915–1923.

[23] Förner W., *J. Phys. Cond. Matter* **4** (1993) 803–822.

[24] Förner W., *J. Phys. Cond. Matter* **5** (1993) 823–840.

[25] Förner W., *J. Phys. Cond. Matter* **5** (1993) 3883–3896.

[26] Förner W., *J. Phys. Cond. Matter* **5** (1993) 3897–3916.

[27] Giacometti G.M., Focesi A., Giardina B. and Wyman J., *Proc. Nat. Acad. Sci. USA* **72** (1975) 4313–4316.

[28] Green D.E. and Ji S., The molecular basis of electron transport, Schultz J. and Cameron B.F., editors, Academic Press, New York, 1972.

[29] Haken H., Quantum field theory of solids, North-Holland, Amsterdam, 1976.

[30] Hill T.L. and Chen Y.D., *Proc. Nat. Acad. Sci. USA* **72** (1975) 4313–4316.

[31] Hodgkin A.L. and Huxley A.F., *J. Physiol.* **117** (1952) 500–544.

[32] von Hippel A., *J. Chem. Phys.* **54** (1971) 145–149.

[33] von Hippel A., Knoll D.B. and Westphal W.B., *J. Chem. Phys.* **54** (1971) 104–123.

[34] Ji S., *Ann. NY Acad. Sci.* **227** (1974) 211–226.

[35] Kemeny G., *J. theoret. Biol.* **48** (1974) 231–241.

[36] Kemeny G. and Goklany I.M., *J. theoret. Biol.* **40** (1973) 107–123.

[37] Kemeny G. and Goklany I.M., *J. theoret. Biol.* **48** (1974) 23–28.

[38] Kashimori Y., Kikuchi T. and Nishimoto K., *J. Chem. Phys.* **77** (1982) 1904–1907.

[39] Landau L.D., *Phys. Z. Sowjetunion* **3** (1933) 664–665.

[40] Maidique M.A., von Hippel A., Knoll D.B. and Westphal W.B., *J. Chem. Phys.* **54** (1971) 150–160.

[41] Pekar I., Untersuchungen über die Elektronentheorie der Kristal, Akademie Verlag. Berlin, 1954.

[42] Scott A.C., Bigio I.J. and Johnston C.T., *Phys. Rev. B* **39** (1989) 517–521.

[43] Scott A.C., *Am. J. Phys.* **37** (1969) 52–61.

[44] Scott A.C., Neurophysics, Wiley, New York, 1977.

[45] Scott A.C., *Physics Reports* **217** (1992) 1–67.

[46] Scott A.C., Gratton E., Shyamsunder E. and Careri G., *Phys. Rev. B* **32** (1985) 5551–5553.

[47] Turner J.E., Anderson V.E. and Fox K., *Phys. Rev.* **174** (1968) 81–89.

[48] Turing A., *Phil. Trans. Roy. Soc. (London)* **B237** (1952) 37–72.

[49] Vol'kenstein M.V., *J. theoret. Biol.* **34** (1972) 193–195.

[50] Wyman J., *Proc. Nat. Acad. Sci. USA* **72** (1975) 3983–3987.

[51] Yomosa S., *J. Phys. Soc. Japan* **51** (1982) 3318–3324.

[52] Yomosa S., *J. Phys. Soc. Japan* **52** (1983) 1866–1873.

Exact two-quantum states of the semiclassical Davydov model and their thermal stability

L. Cruzeiro-Hansson and V.M. Kenkre*

Department of Crystallography, Birkbeck College,
Malet St., London WC1E 7HX, U.K.
* Department of Physics and Astronomy,
University of New Mexico,
Albuquerque, NM 87131, U.S.A.

1   INTRODUCTION

The Davydov model describes a possible mechanism for energy transfer in protein α-helices. These are secondary structures that occur in many proteins when a linear sequence of amino acids is folded into a helix, which is stabilized by hydrogen-bonded spines, i.e. one-dimensional chains with the following structure:

$$\cdots \quad \underset{n-1}{\text{H-N-C=O}} \quad \cdots \quad \underset{n}{\text{H-N-C=O}} \quad \cdots \quad \underset{n+1}{\text{H-N-C=O}} \quad \cdots$$

The Davydov model describes energy transport along these one-dimensional chains. The idea is that the energy liberated in, for instance, the hydrolysis of adenosine triphosphate (ATP), creates an excited vibrational state in the peptide group, called amide I, which is essentially a stretching vibration in the C=O bond. This vibration excitation propagates from one group to the next because of the dipole-dipole interaction between the groups. But it also interacts with the neighboring hydrogen bond, leading to a deformation of the lattice and a lower energy state. This new state, which is constituted by an amide I excitation and its associated hydrogen bond distortion, is called the Davydov soliton [1].

An ansatz-free quantum mechanical simulation of the Davydov system [2] has lent strong support to the existence of such correlations between an excitation and local associated displacement, at least for some values of the parameters of the system. Despite the analysis in that and related papers, the crucial question of whether at biological temperatures the localized solutions last long enough to be biologically useful has remained unsolved. We make a contribution towards the solution of that question by determining the exact two-quantum states of the system and by investigating their thermal stability. We describe the results of that investigation in sections 2 and 3 and make general remarks in section 4.

2   THE DAVYDOV MODEL

The Hamiltonian $\hat{H}$ for the Davydov model is:

$$\hat{H} = \hat{H}_{qp} + H_{ph} + \hat{H}_{int} \tag{1}$$

where $\hat{H}_{qp}$ is the quasiparticle Hamiltonian, which describes the transfer of an amide I excitation between adjacent sites, H_{ph} is the phonon Hamiltonian, which describes the vibrations of the peptide groups n in the one dimensional chain, and $\hat{H}_{int}$ is the interaction Hamiltonian, which describes the interaction of the amide I excitation with the motions of the lattice sites.
The quasiparticle Hamiltonian $\hat{H}_{qp}$ is

$$\hat{H}_{qp} = -V \sum_{n=1}^{N} [(\hat{a}_n^\dagger \hat{a}_{n-1} + \hat{a}_n^\dagger \hat{a}_{n+1})] \tag{2}$$

where $-V$ is the dipole-dipole interaction energy between neighbouring sites and $\hat{a}_n^\dagger(\hat{a}_n)$ is the creation(annihilation) operator for a quantum quasiparticle (the amide I excitation) at site n. N is the number of sites (the peptide groups) in the lattice.
The phonon Hamiltonian H_{ph} is

$$H_{ph} = \frac{1}{2} \sum_{n=1}^{N} [\kappa (u_n - u_{n-1})^2 + \frac{P_n^2}{M}] \tag{3}$$

where u_n is the displacement from equilibrium position of site n, P_n is the momentum of site n, M is the mass of each site and κ is the elasticity constant of the lattice.
Finally, the interaction Hamiltonian $\hat{H}_{int}$ is

$$\hat{H}_{int} = \chi \sum_{n=1}^{N} [(u_{n+1} - u_n) \hat{a}_n^\dagger \hat{a}_n] \tag{4}$$

where χ is an anharmonic parameter arising from the coupling between the quasiparticle and the lattice displacements.

While the lattice variables u_n and P_n are quantum mechanical in the original Davydov model, a number of authors, including Davydov himself, have taken them to be classical and thus represented the displacements and momenta of the sites as real numbers. We make that semiclassical approximation in the rest of the paper. On the other hand, we treat the amide I excitation as a quantum system: its variables are operators, a difference that is marked by the hats above the operators in the above equations. The advantage of this assumption is that exact wavefunctions for the model can be determined without the need for any ansatz (along with its corresponding uncertainties). The disadvantage is, of course, that the range of validity of the classical treatment of the lattice is not known. The investigation of this range of validity is not an easy matter [3]. A usual argument that is often given in support of the semiclassical assumption is that the relatively massive nature of the lattice sites, which are peptide groups of average mass 114 amu, should make their quantum mechanical features negligible.

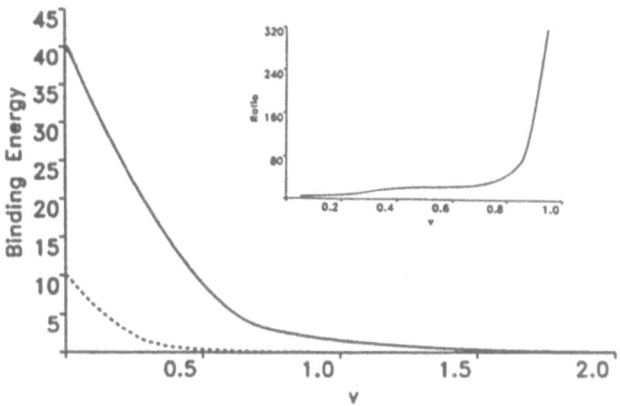

Figure 1: Binding energy of the minimum energy two-quantum states (solid line) and the minimum energy one-quantum states (dotted line) as a function of $v = V\kappa/\chi^2$, for $N=20$. The inset shows the ratio of the two binding energies.

3 TWO-QUANTUM STATES AND THEIR THERMAL STABILITY

In a previous publication, we have studied exact one-quantum states [4]. These are:

$$|\psi(t)> = \sum_{i=1}^{N} \varphi_i(\{u_m\}, \{p_m\}, t)\, a_i^\dagger\, |0> \tag{5}$$

where the dependence of the probability amplitude for an excitation at site i on the displacements and momenta of all the sites is not specified *a priori*. Averaging the Hamiltonian over this state and minimizing the corresponding functional $\mathcal{H}$

$$\mathcal{H}(\{\varphi_n\}, \{u_n\}, t) = <\psi|\hat{h}|\psi> \tag{6}$$

with respect to the probability amplitudes and the displacements, we can find exact minimum energy states of the semiclassical Davydov system for all parameter values [4].

Our interest in the present paper is in the exact *two-quantum* states:

$$|\psi(t)> = \frac{1}{\sqrt{2}} \sum_{i,j=1}^{N} \varphi_{ij}(\{u_m\}, \{p_m\}, t)\, a_i^\dagger a_j^\dagger\, |0> \tag{7}$$

where φ_{ij} is the the probability amplitude matrix for an excitation in sites i and j. Again, averaging the Hamiltonian (1-4) over the two-quantum state (7) and minimizing the corresponding functional with respect to the probability amplitude matrix elements and the displacements leads to the exact two-quantum states. Previously, we have chosen the quantities $v = V\kappa/\chi^2$ and $g = \kappa/\chi$ to parametrize the states of the system [4]. As in the case of the one-quantum states, the minimum energy two-quantum states also turn out to be dependent only on v. In figure 1, the binding energy of localized states, i.e. the difference between the minimum energy of a delocalized state ($-2V$ for one quantum state and $-4V$ for two-quantum states) and the total energy of the state is shown. When the binding energy is zero, the minimum energy states are delocalized and the larger the binding energy, the more stable a localized state is. Figure 1 shows that the two-quantum states are more stable than one quantum states, over a much larger range of parameter values. The dynamical behaviour of a localized solution is given by inserting it in the appropriate equations of motion. For the two-quantum states they are:

$$
\begin{aligned}
i\hbar\frac{d\varphi_{nm}}{dt} &= -V\left(\varphi_{n-1m} + \varphi_{n+1m}\right) + \chi\left(u_{n+1} - u_n\right)\varphi_{nm} + \\
&\quad -V\left(\varphi_{nm-1} + \varphi_{nm+1}\right) + \chi\left(u_{m+1} - u_m\right)\varphi_{nm} + \\
&\quad + \sum_{l=1}^{N}\left[\frac{P_n^2}{2M} + \frac{1}{2}\kappa(u_l - u_{l-1})^2\right]\varphi_{nm}
\end{aligned}
\tag{8}
$$

$$
M\frac{d^2u_n}{dt} = 2\chi\sum_{l}(|\varphi_{ln}|^2 - |\varphi_{ln-1}|^2) + \kappa\left(u_{n+1} + u_{n-1} - 2u_n\right)
\tag{9}
$$

Inserting a minimum energy state in these equations leads to a stationary state and constitutes one way of checking the minimization procedure.

One method of studying the thermal stability of nonlinear structures is by augmenting the above equations of motion into Langevin equations by adding stochastic forces $F(n)$ and friction terms $-\Gamma\frac{du_n}{dt}$ to the right hand side of the equations of motion for the displacements to represent bath interactions with the classical lattice [5-8]. The stochastic forces are usually chosen such that their correlation functions obey the following relation

$$
< F_n(t)\,F_m(t') > = 2M\Gamma k_B T\delta_{nm}\delta(t - t')
\tag{10}
$$

where k_B is Boltzmann constant and T is the temperature. Initial simulations reported the disappearance of one-quantum states in a few picoseconds [7]. A more systematic study has indicated that the picture may be more complex [8]. In our present analysis, we focus on the average state in order to monitor the amount of localization. The average state is calculated by adding the probability of excitation per site of each snapshot so that the site of maximum probability always coincides. At low temperature, we get the minimum energy states, as shown in figure 2. At biological temperatures, comparison of the average states for two-quantum and for one-quantum cases, using the same values of parameters, indicates that the two-quantum state is more stable than the one-quantum state. Although the time scale is beyond the few picoseconds predicted by the original Langevin simulations for the soliton lifetime [7], the difference does not seem to be very significant. However, this is underscored by the problem that Langevin dynamics, in the manner employed, seems to lead to an underestimation of localization in this system. This is a consequence of the fact that Langevin dynamics, when applied to semiclassical systems such as (1-4), leads to a partly classical treatment of the quantum part of the system [5,10]. This is clearly seen in the thermal equilibrium regime, which can be obtained by running very long Langevin dynamics [9]. While the exact thermal average

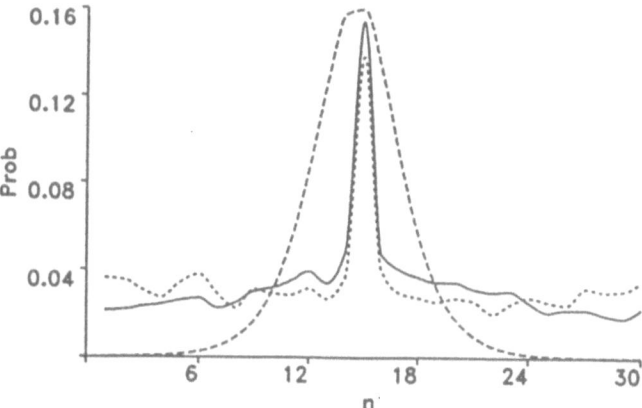

Figure 2: Average probability of excitation per site n for the two-quantum state at $T \approx 0$ (dashed line), at $T=310K$ (solid line) and for the one quantum state (dotted line) at $T=310K$. $V=1.55 \ 10^{-22}$ J, $\kappa=39$ N/m, $\chi=62$pN and $M=114$ m_p. The average is over the first 36 ps.

of an operator $\hat{A}(\{u_n\}, \{\hat{a}_n\})$ is:

$$<< \hat{A} >>= \frac{\int \{du_n\} Tr[e^{-\beta \hat{H}} \hat{A}]}{\int \{du_n\} Tr[e^{-\beta \hat{H}}]} \qquad (11)$$

where $<< \cdots >>$ stands for thermal averaging and $\beta= 1/kT$, it can be shown that Langevin dynamics leads to the following thermal average [5,9]:

$$<< \hat{A} >>= \frac{\int_{2N-unitsphere} \{d\varphi_n^r\} \{d\varphi_n^i\} \int \{du_n\} e^{-\beta <\psi|\hat{H}|\psi>} <\psi|\hat{A}|\psi >}{\int_{2N-unitsphere} \{d\varphi_n^r\} \{d\varphi_n^i\} \int \{du_n\} e^{-\beta <\psi|\hat{H}|\psi>}} \qquad (12)$$

where φ_n^r and φ_n^i are, respectively, the real and imaginary parts of the probability amplitude. Expression (12), where an integration over all possible superpositions of eigenstates is made, violates the postulate of quantum statistical mechanics that the wavefunctions must be taken with a priori random phases [11]. In other words, while in the exact average (11) the nondiagonal terms of the density matrix are zero, in the approximate average (12) they are finite. Thus, Langevin dynamics, which leads to this same result, tends to bring the quantum part of the system into the classical regime as well. A measure of the error involved can be obtained by comparing the exact thermal average (11) with the semiclassical Langevin average (12). In figure 3 such a comparison is made for the interaction Hamiltonian, which is nonzero when the correlation between the excitation and the lattice displacement, which underlie the Davydov soliton, is finite. Figure 3 shows that the approximation is good at low temperatures, when the minimum energy states are predominant. On the other hand, at higher temperatures, the correlation between the excitation and the lattice distortion is underestimated by the semiclassical Langevin average, over all parameter ranges. The underestimation of localization at equilibrium illustrated in figure 3 suggests that such an underestimation might also be occurring in non-equilibrium conditions, as treated by the present form of semiclassical Langevin dynamics. This underestimation should therefore also be

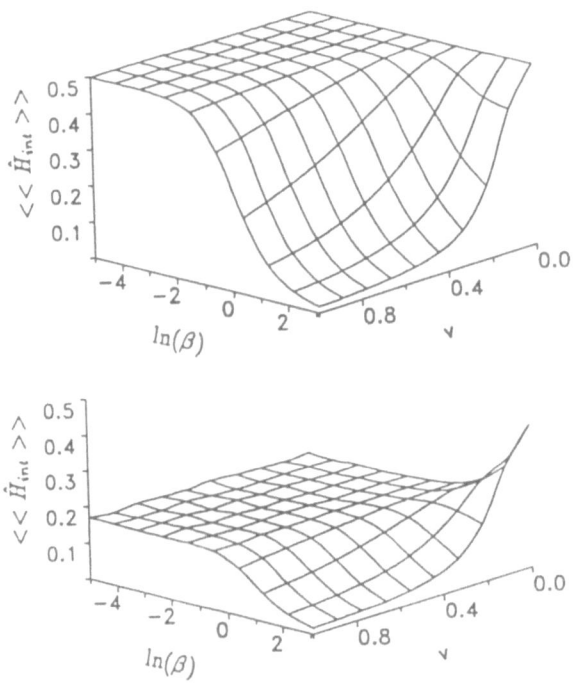

Figure 3: Exact thermal average (a) and semiclassical Langevin average (b) of the interaction Hamiltonian as a function of $v = V\kappa/\chi$ and $\ln\beta$, where $\beta = 1/k_B T$ (in units of χ^2/κ), and $N=2$.

expected to appear in the thermal lifetimes of localized states determined in this fashion. More accurate estimates require the development of other ways of introducing thermal fluctuations in semiclassical systems [10].

4   GENERAL REMARKS

In the above we have discussed the problem of the thermal stability of the two-quantum state. In this section we make general comments, first on the biological relevance of the soliton model and then on experiments to test its validity.

Davydov's first application of the soliton mechanism for energy transfer in protein α-helices was to muscle contraction [1]. Because the Davydov soliton leads to a local compression or to a kink in the α-helix, a Davydov soliton produces a bulge in the myosin tails that travels with it. Davydov expected that this bulge would push the myosin head against actin and provide the dragging force that makes the actin filaments move with respect to myosin. The model was in general agreement with the sliding filament model of muscle contraction and provided a molecular description of the cause of the movement. One important distinct feature of Davydov's model is the emphasis it puts on the *myosin tails*.

Since the 70's, when Davydov's model was first proposed [1], many advances have been made in the field of muscle contraction and energy transduction. These include studies of the structure of the myosin head [12,13], fluorescence assays in which the motion of single actin filaments is monitored [14,15] and kinetic studies [16]. At present, there is some controversy about the size of each cross-bridge movement and about the number of strokes per ATP hydrolysed [15-18]. The Davydov model of an excitation travelling in the myosin tails can explain multiple strokes per ATP hydrolysed and an apparent large step size as a result of multiple smaller steps of the myosin heads affected by each excitation.

Moreover, the soliton idea can be naturally used in contexts more general than muscle contraction or other processes where ATP hydrolysis is used. Indeed, following Green's comments on the unification of the bioenergetics at the 1973 conference, Skulachev [20] has recently proposed on the basis of accumulated data that 1) there are 'three convertible energy currencies', namely, ATP, proton gradients and sodium gradients; 2) all cells store energy in at least two of these 'currencies', one being ATP and the other either a sodium and/or a proton gradient; and finally 3) all energy requirements of the living cell can be satisfied if at least one of the three convertible energy currencies is produced at the expense of external energy sources. The soliton concept, we would like to point out, can enter naturally in the context of all three energy currencies. That chemical reactions such as ATP hydrolysis produce vibrational excitations is well-known. On the other hand, Careri and Wyman have proposed the formation of vibrational excitations when ions bind to proteins [21]. They show that, energetically, because of the large dipole moment associated with an α-helix, the binding of a phosphate on the NH_2 side can be associated with an attractive energy of 13 kcal mol^{-1} (4500 cm^{-1}, 0.554eV), enough for at least two Davydov solitons. Thus, if the action of sodium and proton gradients involves the binding of ions to proteins, the mechanism proposed by Careri and Wyman [21] can lead to vibrational excitations. This shows that all three Skulachev currencies can be related to the creation of vibrational excitations which serve as starting points for the soliton dynamics.

In our opinion, the present state of the studies on the Davydov model suffers from a serious lack of direct experimental evidence that the type of energy transfer that is invoked in the model really occurs in biologically relevant molecules. The theoretical situation is not clear. But even if the model is shown to be theoretically possible, the ultimate proof that it plays a role in nature can come only from experiment. At present, there is little data available [22] although the design of several clear experiments has been discussed to test the existence of Davydov solitons in an α- helix [22,23] and related systems [24]. We would like to make suggestions for two types of experiment. The first is spectroscopic and focuses on the importance of amide I excitations. The implication of the Davydov model is that there are a lot of amide I excitations about when muscle contraction takes place. If in a spectroscopic experiment it is shown that when ATP and Ca are present (the two initiators necessary for muscle contraction) the intensity of the amide I band increases substantially, this will be evidence that the energy transduction mechanism does involve the amide I excitation as postulated by Davydov. The second kind of experiment we suggest involves manipulation of the biologically relevant elements of the system. The main point of the Davydov model for muscle contraction is that the movement of heads is due to a soliton travelling in the tails of myosin. This means that affecting the tails should affect the power stroke. Namely, reducing the length of the tails or making them more rigid should reduce the strength of the contraction. And eliminating the α-helix part of the myosin molecule should completely impair muscle contraction. Thus, an experiment showing that the α-helix part of the myosin has been substantially impaired while the myosin head is still hydrolysing ATP and no muscle contraction takes place would constitute evidence that the tails are important. Such an experiment would lend support to Davydov's idea. If, on the other hand, the myosin heads are the sole actor in muscle contraction, the Davydov model for muscle contraction would be shown not to work.

From a more general perspective than Davydov's soliton theory, one wonders if ATP hydrolysis is able to energise amide I modes of vibration in proteins. One way to answer this question would be to look for anti-Stokes lines in Raman scattering from a variety of water soluble proteins in the presence of ATP. A positive result from measurements of this sort would have implications well beyond Davydov's soliton picture.

Acknowledgement. Very useful discussions with Prof. A.C. Scott are gratefully acknowledged. LCH thanks the BBSRC for the Advanced Fellowship B/93/AF/1646 and for supercomputing facilities at ULCC. This work is also partially support by the NATO Scientific Affairs Division through a grant for collaborative research between the University of New Mexico, USA and Birkbeck College, University of London, UK.

References

[1] Davydov, A.S. and Kislukha, N.I., *Phys. Stat. Sol. B* **59**, (1973) 465-470; Davydov, A.S., *J. Theor. Biol.* **38**, (1973) 559-569; Davydov, A.S. and Kislukha, N.I., *Phys. Stat. Sol. B* **75**, (1976) 735-742; Davydov, A.S., *J. Theor. Biol.* **66** (1977) 379-387; Davydov, A.S., *Physica D* **3** (1981) 1-22; Davydov, A.S., *Int. J. Quant. Chem.* **16** (1979) 5-17; Davydov, A.S., Biology and Quantum Mechanics (Pergamon, New York, 1982); Davydov, A.S., *Sov. Phys. Usp.* **25** (1982) 898-918; Brizhik, L.S. and Davydov, A.S., *Phys. Stat. Sol. B* **115** (1983) 615-630; Davydov, A.S. *Phys. Stat. Sol. B* **138** (1986) 559-576; Davydov, A.S. *J. Biol. Phys.* **18** (1991) 111-125.

[2] Wang, X., Brown, D.W. and Lindenberg, K. *Phys. Rev. Lett.* **62** (1989) 1796-1799.

[3] See e.g. Vitali, D., Allegrini, P. and Grigolini, P. *Chem. Phys.* **180** (1994) 297-318.

[4] Cruzeiro-Hansson, L. and Kenkre, V.M. *Phys Lett A* **190** (1994) 59-64.

[5] Kenkre, V.M. and Grigolini, P. *Z. Phys.* **b 90** 247-253.

[6] Grigolini, P., Wu, H.-L. and Kenkre, V.M. *Phys. Rev. B* **40** (1989) 7045-7053; Kenkre, V.M. *Physica D* **68** (1993) 153-161.

[7] Lomdahl, P.S. and Kerr, W.C. *Phys. Rev. Lett.* **55** (1985) 1235-1238; Lawrence, A.F., McDaniel, J.C., Chang, D.B., Pierce, B.M. and Birge, R.R. *Phys. Rev A* **33** (1986) 1188-1201.

[8] Motschmann, H., Förner, W. and Ladik, J., *J. Phys. Condens. Matter* **1** (1989) 5083-5093; Förner, W., *J. Phys. Condens. Matter* **3**, (1991) 4333-4348; Förner, W., *Phys. Rev. A* **44** (1991) 2694-2708; Förner, W., *J. Comput. Chem.* **13** (1992) 275-313; Förner, W., *J. Phys. Condens. Matter* **4**, (1992) 1915-1923; Förner, W., *J.*

Phys. Condens. Matter **5** (1993) 803-822; Förner, W., *J. Phys. Condens. Matter* **5** (1993) 823-840; Förner, W., *J. Phys. Condens. Matter* **5** (1993) 3883-3896; Förner, W., *J. Phys. Condens. Matter* **5** (1993) 3897-3916.

[9] L. Cruzeiro-Hansson, *Physica D* **68** (1993) 65-67.

[10] Mauri, F., Car, R. and Tosatti, E. *Europhys. Lett.* **24** (1993) 431-436.

[11] Tolman, R.C., The Principles of Statistical Mechanics, (Oxford Univ. Press, London, 1946) pp. 342-356.

[12] Wakabayashi, K., Tokunaga, M., Kohno, I., Sugimoto, Y., Hamanaka, T., Takezawa, Y., Wakabayashi, T. and Amemiya, Y. *Science* **258** (1992) 443-447.

[13] Rayment, I., Rypniewski, W.R., Schmidt-Bäse, K., Smith, R., Tomchick, D.R., Benning, M.M., Winkelmann, D.A., Wesenberg, G. and Holden, H.M. *Science* **261** (1993) 50-58; Rayment, I., Holden, H.M., Whittaker, M., Yohn, C.B., Lorentz, M., Holmes, K.C. and Milligan, R.A. *Science* **261** (1993) 58-65.

[14] Yanagida, T., Arata, T. and Oosawa, F. *Nature* **316** (1985) 366-369; Saito, K., Aoki, T., Aoki, T. and Yanagida, T. *Biophys. J.* **66** (1994) 769-777.

[15] Finer, J.T., Simmons, R.M. and Spudich, J.A. *Nature* **368** (1994) 113-119.

[16] Irving, M., Lombardi, V., Piazzesi, G. and Ferenczi, M.A. *Nature* **357** (1992) 156-158.

[17] Yanagida, T., Harada, Y. and Ishijima, A. *TIBS* **18** (1993) 319-324.

[18] Huxley, H.E. *J. Biol. Chem.* **265** (1990) 8347-8350.

[19] D.E. Green, *Ann. N. Y. Acad. Sci.* **227** (1974) 6-45.

[20] V.P. Skulachev, *Eur. J. Biochem* **208** (1992) 203-209.

[21] G. Careri and J. Wyman, *Proc. Nat. Acad. Sci. USA* **81** (1984) 4386-4388.

[22] Scott, A., *Phys. Rep.* **217** (1992) 1-67 and references therein.

[23] Knox, R.S., Maiti, S. and Wu, P. "Search for Remote Transfer of Vibrational Energy in Proteins", Davydov's Soliton Revisited. Self-Trapping of Vibrational Energy in Protein, Thisted, Denmark, July 30 - August 5, 1989, P.L. Christiansen and A.C. Scott (Plenum, N.Y.,1990) pp. 401-412.

[24] Kenkre, V.M., Rudolph, W. and Scott, A.C., unpublished.

Post-soliton quantum mechanics

D.W. Brown

*Institute for Nonlinear Science, University of California,
San Diego, La Jolla, CA 92093-0402, U.S.A.*

———————————

—————

Schrödinger: *"If one has to stick to this damned quantum jumping, then I regret ever having been involved with this thing!"*

Bohr: *"But we others are very grateful to you that you were, since your work did so much to promote this theory."* [2]

The theory of solitons in quantum systems is plagued by it.s semi-classical character. Indeed, to walk through the theory solitons is to venture into the semi-classical world of Schrödinger's Cat. [3] As nonstationary states in quantum mechanics, solitons should, with the passage of time, evolve into a wave packets that are in some sense superpositions of possible soliton futures, yet the bulk of prevailing theory does not reproduce this quantum property. So how are we to understand the quality of semi-classical results, and how are we to make use of them in circumstances that require a greater consistency with quantum orthodoxy?

We learn from Schrödinger's Cat that while we must accept an other-worldly blurring of possible futures under quantum mechanics, there is a meaningfulness to semi-classical results that survives quantum propagation. Indeed, there can be little doubt that despite their semi-classical character, soliton excitations reflect at least some real properties of anharmonic quantum systems, and that the challenge to progress is to understand how to strip away the semi-classical aspects while retaining the meaningful properties in an improved theory. This paper illustrates methods of energy band theory that can be used to this end where soliton theory meets polaron theory on the field of solid-state physics.

Energy band theory is concerned with determining the energy eigenvalues and energy-momentum eigenfunctions of excitations in translationally invariant systems. The delocalized character of energy-momentum eigenfunctions and the identification of these states with particular momentum labels are essentially symmetry properties of the space in which the problem is presented and beyond this carry no information about the particular systems in which they are found. Indeed, all system-specific information is embedded in the *internal* structure energy-momentum eigenfunctions, and is revealed in correlation functions probing local properties that do not require absolute spatial reference. Indeed, one may take the view that the uniform states are translationally-invariant superpositions of localized functions carrying these properties; we shall refer to such localized functions as "form factors" and denote them $|\zeta>$.

Such a uniform state (unnormalized) may be expressed in the form

$$| \psi(\kappa) \rangle = \sum_n \frac{e^{i\kappa n}}{\sqrt{N}} \, | \zeta(n) \rangle \, ,$$

where $| \zeta(n) \rangle$ is a copy of the basic form factor $| \zeta \rangle$ displaced n lattice sites from the origin, and κ is the total momentum label. In order to construct a proper candidate for an energy-momentum eigenstate, this must be normalized using the quantity

$$\langle \psi(\kappa) | \psi(\kappa) \rangle = \sum_n e^{-i\kappa n} \langle \zeta(0) | \zeta(n) \rangle \, ,$$

which may be understood as a momentum space spectral density of the form factor in that it is the Fourier transform of the quantum autocorrelation function of the form factor in real space. Only if the form factor is orthogonal with all translated copies of itself can the spectral density be uniform in momentum space. Solitons as result from prevailing theories generally are *not* orthogonal to *any* translated copies of themselves, leading to the conclusion that only in the singular case that the soliton is collapsed onto a single lattice site can the spectral density be uniform in momentum space; in all other cases a nontrivial spectral density remains to be considered. Indeed, should the more traditional notion of a broad soliton spanning many lattice sites be the relevant circumstance, a spectral density sharply localized in momentum space should be expected. It is thus that this spectral density carries a "memory" of the soliton profile.

Just as the delocalization of the soliton or other form factor erases memory of the location of the original form factor in real space, the normalization of the uniform state adjusts the weight of each uniform state in such a way as to erase memory of the absolute spatial profile of the form factor, yielding what solid-state physicists call a *Bloch state*:

$$| \Psi(\kappa) \rangle = \frac{| \psi(\kappa) \rangle}{\langle \psi(\kappa) | \psi(\kappa) \rangle^{\frac{1}{2}}} \, .$$

So what remains of the soliton? The properties of the soliton or other localized state that are lifted into the structure of the momentum eigenfunction are those properties that are not directly determined by the shape or locus of the form factor; that is, correlation properties that do not require absolute spatial reference. The momentum eigenfunction construction detailed above can be viewed as peeling off the semiclassical skin from the soliton and retaining only the internal correlation properties that are consistent in principle with a momentum space description.

The Inverse Problem

It is reasonable to ask whether there is any localized state, perhaps analogous to the soliton (or perhaps not), in which the internal correlation properties of the energy-momentum eigenstates may be manifest. This question has the nature of an *inverse problem*. Considering that the energy-momentum eigenfunctions completely specify the possible states of motion of the polaron, the inverse problem seeks to determine the state of the polaron *at rest*. That the resting state of the polaron must have a decomposition over the complete set of energy-momentum eigenstates implies, of course, that the localized state we seek is a non-stationary state under quantum mechanics. In this respect the resting state of the polaron is no different than the orbitals of the hydrogen atom or even the feline shape of Schrödinger's Cat; all describe irreducible local structure that in varying degrees becomes blurred under quantum propagation.

A key consideration in seeking the resting state of the polaron is that *all* the information required for its construction should be available in the Bloch States themselves. Thus when considering a general superposition of the Bloch states of a particular energy band

$$|\Phi(n)> = \sum_\kappa C_n^\kappa |\Psi(\kappa)> \ ,$$

we seek to determine the coefficients C_n^κ in a manner unbiased by external considerations. The localized state should be normalized in order to be interpretable as a proper quantum state; thus, [†]

$$\sum_\kappa |C_n^\kappa|^2 = 1 \ , \quad or \quad C_n^\kappa = \frac{e^{-i\kappa n}}{\sqrt{N}} f_\kappa \ .$$

If we further require the orthogonality of distinct localized states

$$<\Phi(n)|\Phi(n')> = \delta_{nn'} \ ,$$

then it follows that $f_\kappa = 1$. Thus we arrive at the definition of the polaron *Wannier state* $|\Phi(n)>$, by applying the same reasoning as in rigid-lattice band theory. [4] Wannier states form a complete orthonormal set of localized states that are unitarily equivalent to the energy-momentum eigenstates

$$|\Phi(n)> = \sum_\kappa \frac{e^{-i\kappa n}}{\sqrt{N}} |\Psi(\kappa)> \ , \qquad |\Psi(\kappa)> = \sum_n \frac{e^{+i\kappa n}}{\sqrt{N}} |\Phi(n)> \ .$$

Since distinct Wannier states differ only by their location in the lattice, one may, without loss of generality, recognize the existence of only *one* distinct Wannier state. Moreover, though we have referred to these states as "local" functions, we have not forced any local properties upon them; in particular, we have made no assumption that the energy-momentum eigenstates may be based upon soliton form factors. The general properties of Wannier states are independent of the degree to which they are localized, or of other details of their internal structure.

General Properties of Wannier States

It is natural to compare the Wannier state with the soliton-like form factor underlying the whole construction. It is important to note one difference, however, that makes such comparisons somewhat ill-posed. The Wannier state is a *generating function* for the complete set of Bloch states, since any Bloch state can be generated from a single Wannier state merely be delocalizing the Wannier state with the appropriate phase. Moreover, since the Bloch states determine the energy eigenvalues, we may say that *all* the ascertainable properties of the polaron in the state space of the given energy band can be generated from a single Wannier

---

[†] One may consider more general phases θ_n^κ rather than the special choice $\theta_n^\kappa = \kappa n$ we use here, but such considerations are beyond the scope of this paper; see, e.g., Ref. 4.

state. No corresponding claim can be made for the soliton. At best, the κ-by-κ variation usually implemented to optimize the energy band identifies a different soliton-like form factor with every one of the N distinct values of the total momentum; thus, there is no single soliton-like localized state that is representative of the whole set.

Dynamic vs. static structure

Any state in the space associated with the ground state energy band can be expanded in the Bloch basis, the Wannier basis, or over any other complete set; e.g.,

$$|u(t)> = \sum_{\kappa} u_{\kappa}(t) | \Psi(\kappa)> = \sum_{n} u_n(t) | \Phi(n)> .$$

Whatever the expansion, the complete set of properties carried by the state $|u(t)>$ is carried partly by the time-dependent expansion coefficients and partly by the time-independent basis functions. Those properties carried by the expansion coefficients are *dynamic*, while those carried by the basis functions are *static*. Static properties define the structure of the state space in which dynamics takes place.

While it has been common in polaron theory for some exciton-phonon correlation properties to be bound in the static portion of polaron states, it is been essentially the rule in soliton theory for *all* exciton-phonon correlations to be expressed dynamically. This dynamical bias of soliton theory tends to over-weight the importance of nonlinear coherences and to underweight the importance of correlations built into the structure of the quantum state space.

Thermal properties

Thermal properties are essentially dynamic in nature, since temperature is (in the canonical sense) defined through fluctuations in the kinetic energy. The above distinctions between static and dynamic structure translate into distinctions between the temperature-dependent and temperature-independent parts of the polaron density matrix:

$$\hat{\sigma}(T) = Z^{-1} \sum_{\kappa} e^{-E(\kappa)/k_B T} | \Psi(\kappa)><\Psi(\kappa)| \quad , \qquad Z = \sum_{\kappa} e^{-E(\kappa)/k_B T} \quad ,$$

$$\hat{\sigma}(T) = \sum_{mn} \sigma_{mn}(T) | \Phi(m)><\Phi(n)| \quad , \qquad \sigma_{mn}(T) = (ZN)^{-1} \sum_{\kappa} e^{-E(\kappa)/k_B T} e^{i\kappa(m-n)} .$$

This distinction becomes particularly relevant at high temperatures, where, for observables $\hat{O}$

$$< \hat{O} >_{T \to \infty} = <\Phi(0)| \hat{O} | \Phi(0)> .$$

That is, the *only* structure surviving at high temperatures is that carried by the Wannier state.

(This argument applies to the ground state band in isolation from other bands, and breaks down when higher-lying bands can become significantly populated before an essentially uniform density profile can be established over the ground state band. When the argument breaks down, it remains true that the distinction between dynamic and static structure affords static correlations a measure of insulation against thermal fluctuations, since the involvement of many higher energy bands would be required to significantly degrade static correlations.)

Energy band structure

The energy band can be expressed using either Bloch states or Wannier states:

$$E(\kappa) = \langle \Psi(\kappa)| \hat{H} | \Psi(\kappa)\rangle = \sum_n e^{i\kappa n} \langle\Phi(0)| \hat{H} | \Phi(n)\rangle \ .$$

The utility in this lies not in computational facility, but in relating features of energy bands with corresponding characteristics of Wannier states, particularly the spatial distribution of electron density within the Wannier states. It is evident, for example, that the energy band of a broad Wannier state will encompass more harmonics than that of a narrow Wannier state, implying a direct relation between the width of a Wannier state and distortion of the energy band away from its rigid-lattice shape.[†]

Weak Coupling Limit

The Bloch/Wannier apparatus can be applied as well to a free electron in a rigid lattice or a self-trapped electron in a deformable lattice. A scenario at the interface between these two is the weak coupling limit. It is common to think of the weak coupling limit of an electron-phonon system as being equivalent to a free electron in a rigid lattice, since any initial condition in the electronic subspace evolves unperturbed by phonons. This is a statement about dynamics, however, and not a statement about the joint eigenstates and eigenvalues of the subsystems considered as a whole, which is precisely the aim of energy band theory.

For definiteness, we focus our attention on the weak coupling limit of the Holstein molecular crystal model, defined by the quantum Hamiltonian [5]

$$\hat{H} = (E + 2J)\sum_n a_n^+ a_n - J\sum_n a_n^+ (a_{n+1} + a_{n-1}) + \hbar\omega\sum_n b_n^+ b_n - \chi\sum_n (b_n^+ + b_n) a_n^+ a_n \ ,$$

in which a_n^+ creates an electron in the rigid-lattice Wannier state at site n, and b_n^+ creates a quantum of vibrational energy in the Einstein oscillator at site n. This model reduces to noninteracting electron and phonon subsystems when $\chi \to 0$; nonetheless, the determination of the polaron energy band and the associated states requires that we consider the electron in the presence of various numbers of phonons, for all combinations of electron and phonon momenta consistent with each of the allowed values of the total momentum. In this way we build up bands of joint electron-phonon eigenvalues that may be identified with 0, 1, 2, or more phonon quanta.

The zero-phonon band is trivially equal to the free-electron energy band since in the absence of phonons the electron must carry all the momentum (See Figure 1):

$$E_0(\kappa) = 0 \cdot \hbar\omega + E + 2J[1 - \cos(\kappa)] \ .$$

The one-phonon band, however, must account for the fact that a given total momentum may be shared by the electron and the phonon in many different ways, resulting in a band of states $4J$ in width commencing $\hbar\omega$ above the global ground state (See Figure 1):

[†] Note that a transformation $E(\kappa) \to \alpha \cdot E(\kappa) - \beta$ does not affect this relationship; thus, it is a κ–by–κ distortion that is implied and not a simple shift or uniform narrowing.

D.W. Brown

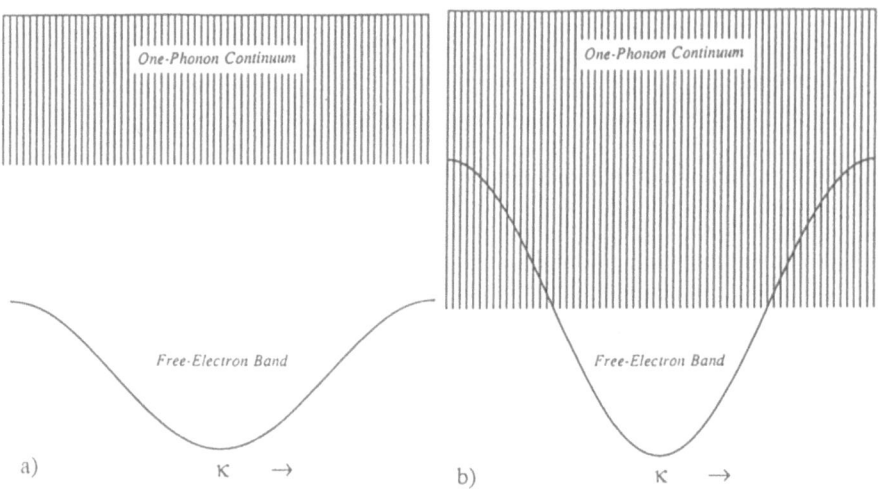

Figure 1. The zero-phonon band and the one-phonon continuum for an electron in a lattice of Einstein oscillators, but in the absence of any electron-phonon coupling.
Panel a) The non-adiabatic case, $4J < \hbar\omega$. Panel b) The adiabatic case, $4J > \hbar\omega$.

$$E_1(\kappa) = 1 \cdot \hbar\omega + E + 2J[1 - \cos(\kappa - q)] \qquad \text{for all } q \in [-\pi, +\pi] \;.$$

Similarly, the two-phonon band commences $2\hbar\omega$ above the global ground state:

$$E_2(\kappa) = 2 \cdot \hbar\omega + E + 2J[1 - \cos(\kappa - q_1 - q_2)] \qquad \text{for all } q_1, q_2 \in [-\pi, +\pi] \;.$$

When $4J < \hbar\omega$, which we call the non-adiabatic regime, the lowest energy at every value of the total momentum is given by the unique energy of the zero-phonon band. The Bloch states in this case are exactly the free-electron Bloch states, and owing to the absence of any distortion of the energy band, the associated Wannier states are identical to the rigid-lattice Wannier states having their electron density localized on a single lattice site.

On the other hand, when $4J > \hbar\omega$, which we call the adiabatic regime, there exists a total momentum κ^* (given by the condition $2J[1 - \cos(\kappa^*)] = \hbar\omega$) above which the zero-phonon states are no longer the minimum energy states. Above κ^*, the joint electron-phonon state having the minimum energy is the unique state in which the electron is at rest ($k = 0$) and all momentum is carried by a single phonon ($q = \kappa$). This implies that the ground-state energy band changes abruptly at κ^* from being free-electron-like below to being free-phonon-like above, and is reflected in the Bloch state

$$|\Psi(\kappa)\rangle = \quad\;\; a^+_{k=\kappa}\,|0\rangle \qquad \text{for } \kappa < \kappa^* \qquad E(\kappa) = E + 2J[1 - \cos(\kappa)] \;,$$
$$a^+_{k=0}\,b^+_{q=\kappa}\,|0\rangle \qquad \text{for } \kappa > \kappa^* \qquad E(\kappa) = E + \hbar\omega \;.$$

These strong distortions are accompanied by changes in the Wannier state

$$|\Phi(n)\rangle = \sum_{\kappa<\kappa^*} \frac{e^{-i\kappa n}}{\sqrt{N}}\, a^+_{k=\kappa}\, |0\rangle + \sum_{\kappa>\kappa^*} \frac{e^{-i\kappa n}}{\sqrt{N}}\, a^+_{k=0}\, b^+_{q=\kappa}\, |0\rangle .$$

The meaning of which is evident in the electron density (See Figure 2):

$$\rho^{el}_n = \langle\Phi(0)|\, a^+_n a_n\, |\Phi(0)\rangle = \frac{\sin^2(\kappa^* n)}{\pi^2 n^2} + \frac{\pi-\kappa^*}{\pi N} .$$

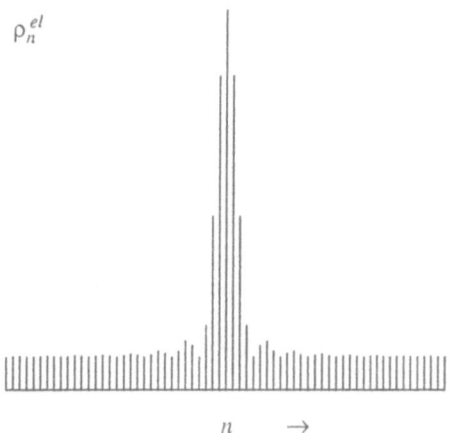

ρ^{el}_n

$n \quad \rightarrow$

Figure 2. Electron density in the Wannier state for the adiabatic situation in Figure 1-b.

This structure contains a localized component having a width of order $1/\kappa^*$ and a weight κ^*/π, and a uniform background having a weight $(\pi-\kappa^*)/\pi$. The phonon number density has a similar spatial distribution displaying a striking electron-phonon correlation *despite* the absence of

$$\rho^{ph}_n = \langle\Phi(0)|\, b^+_n b_n\, |\Phi(0)\rangle = \frac{\sin^2(\kappa^* n)}{\pi^2 n^2} + \delta_{n,0}\frac{\pi-2\kappa^*}{\pi}$$

any dynamic interaction.

Interestingly, this Wannier state is a state of indefinite phonon number; thus, though the Wannier state is *not* an eigenstate of the phonon annihilation operator (i.e., it is not a coherent state), some phonon annihilation operators have nonvanishing expectation values.

Exercise: Determine the time-dependent densities $\rho^{el}_n(t)$, $\rho^{ph}_n(t)$, and the propagator $P(n,t|0,0) = |\langle\Phi(n)|\exp(-i\hat{H}t/\hbar)|\Phi(0)\rangle|^2$ in the weak-coupling limit of the Holstein model; prove $\sum_n P(n,t|0,0) = 1$.

Conclusion

While ruminating on the conundrums of quantum solitons, we have shown how methods of energy band theory are capable of producing localized states exhibiting interesting properties, some of which are seductively reminiscent of the properties of solitons. Are polaron Wannier states quantum solitons? If they are, they are not like any solitons seen before:

- Polaron Wannier states constitute a complete, orthonormal basis for each energy band.

- Distinct polaron Wannier states differ only by a lattice translation.

- The polaron Wannier state constitutes a generating function containing all the information available from the complete set of Bloch states.

- The electron density within each polaron Wannier state generally is localized over a region the size of which is related to the electronic tunneling parameter and the electron-phonon coupling strength.

- The localized electron density is accompanied by a localized phonon energy density.

- Polaron Wannier states are generally states of indefinite phonon number.

- The spatial extent of polaron Wannier states is reflected in distortions of the polaron energy band that may be accessible to spectral measurement.

- All information bearing on the dynamics within a particular polaron band are encoded in the expansion coefficients of the dynamical state in the Wannier basis, and not in the Wannier states themselves.

- Correlations internal to polaron Wannier states are somewhat insulated from thermal perturbations.

- In sufficiently adiabatic systems, polaron Wannier states retain non-trivial properties in the absence of any electron-phonon interaction, and hence in the absence of the non-linearity heretofore associated with solitons.

- The method for constructing polaron Wannier states is rigorous and applies equally well to both integrable and nonintegrable systems.

In work to be published elsewhere, polaron Wannier states for the Holstein model ($\chi \neq 0$) have been computed for variational energy bands at several different levels of theory, confirming the qualitative descriptions presented here and allowing a more complete assessment their structure and properties.

References

1. With apologies to Steven S. Tomsovic and Eric J. Heller, Physics Today **46**, July, 38-46 (1993).

2. Max Jammer, *The Conceptual Development of Quantum Mechanics* (McGraw-Hill, New York, 1966).

3. E. Schrödinger, Naturwissenschaften **23**, 807-12, 823-8, 844-9 (1935).

4. G. H. Wannier, Phys. Rev. **52**, 191 (1937); W. Kohn, Phys. Rev. **115**, 809-821 (1959); Gabriel Weinreich, *Solids: Elementary Theory for Advanced Students* (Wiley, New York, 1965); O. Madelung, *Introduction to Solid-State Theory*, (Springer-Verlag, New York, 1978).

5. T. Holstein, Ann. Phys. (N.Y.) **8**, 325-342 (1959); 343-389 (1959).

Dynamic form factor for the Yomosa model for the energy transport in proteins

A. Neuper and F.G. Mertens

Physikalisches Institut, Universität Bayreuth,
95440 Bayreuth, Germany

----

1. SOLITONS IN MUSCLE FIBERS

During the conference it showed up that there is a need to see the fingerprints of various nonlinear interactions in neutron scattering data, i.e. in dynamic form factors $S(q,\omega)$. This interest was present e.g. in the field of DNA and protein research.

Within the Yomosa model [1, 2] the energy transport, which is necessary for muscle contraction, is described by lattice solitons, in particular by Toda solitons. More recently [3, 4] the initiation of the process by the ATP reaction in the myosin head (see also Fig. 1) has been investigated. During the hydrolysis the contact to the thin filament is cut by the ATP binding to the myosin head. The relaxation of the stretched myosin head leads to a new contact about 5.5 nm further on the thin filament on the next actine unit. This binding becomes firm when the ADP is released. The measured shifts are usually about 11 nm, sometimes even 17 nm. This can be explained by a rapid succession of jumps by 2 or 3 heads [5]. But in our opinion this is in contradiction to the distance histogram shown in [4]. We think the second movement is by the same head, but this time for a different reason. After the head has relaxed there is still a lot of energy stored in the tension of the helix connecting the head with the thick filament. Assuming it behaves like a spring and having not too much friction it might shift the head a second or sometimes even a third time when it overstretches during the relaxation process. This might also be supported by the ratchet mechanism of "molecular motors" [6] and can even happen when a force is applied since the second movement uses momenta conservation.

We now return to the Yomosa model which considers the motion within the hydrogen-bonded chains of the helix *after* the initiation.

2. FORM FACTORS IN THE TODA LATTICE

Among the one-dimensional lattices with non-linear nearest-neighbor interactions the Toda lattice [7, 8, 9] is particularly interesting because of its complete integrability [10, 11, 12]. Many exact results are known here, both for the dynamics at zero temperature

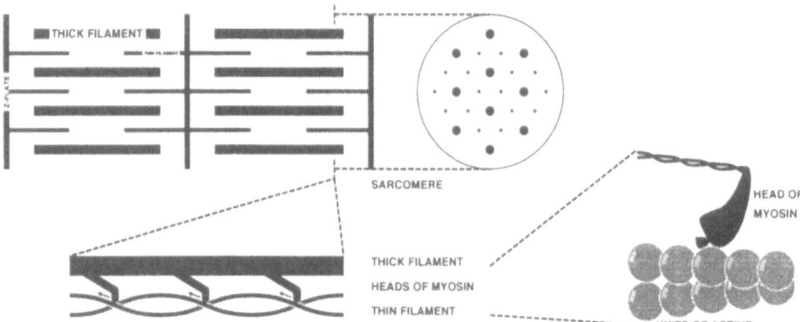

Figure 1: *A schematic view into a muscle fiber*

(e.g. the N-soliton solution) and the static properties at finite temperatures (e.g. the free energy), see [9].

But for the dynamics at finite temperature very little is known exactly. In particular, the dynamic correlation functions are not known. These functions generally are very important for the identification of soliton effects. For systems with non-topological solitons, the knowledge of these functions is absolutely necessary, because here the static quantities do not contain unique informations about the solitons. A good example is the free energy of the Toda lattice: The low temperature expansion yields the free energy of a harmonic lattice plus a power series in the temperature [13, 14]. Many authors have tried to identify the first term of this series with the free energy of an soliton gas [13, 14, 15, 16, 17, 18]. Eventually this idea may be successful by taking into account all phonon and soliton phase shifts due to scattering [19, 20, 21]. On the other hand the power series can be obtained, term by term, from a conventional perturbation theory using the anharmonicities of the Toda potential[13]. This fact explains why an ideal gas of "cnoidal waves" also yields the first term of the above power series[16, 22, 23]. The cnoidal waves are periodic solutions for the Toda lattice, i.e. a generalization of the harmonic phonons[8].

Thus the interpretation by solitons is not unique. (By contrast, the energy gap δE of the topological solitons leads to a free energy $\propto \exp\left(-\delta E/k_B T\right)$, which cannot be obtained by perturbation theory.) For these reasons the dynamic form factors $S(q,\omega)$ of the Toda lattice have been studied extensively by phenomenological theories, approximate theories and computer simulations: The soliton-gas approach [14, 24, 25] predicted a non-Lorentzian soliton resonance slightly above $\omega = q$, in addition to a Lorentzian phonon peak near $\omega_p = \sin q$. Since the two peaks are closely together, they can be expected to appear separately only for a certain range of both q and T. The relative weight of the peaks was not known at that time due to discrepancies in the literature about the soliton density [14, 15, 17, 19, 22].

Diederich[26, 27] used a truncated cumulant expansion; the resulting dynamical equations for the response functions in q-space were solved numerically. The dynamic form factors show a double-peak structure for a small range of q and for a temperature range of about $0.15 \leq T \leq 0.3$. The position, width and tail of the high-frequency peak agree rather well with the soliton peak of [25].

However, in a more recent work Cuccoli et al. [28] used a cumulant expansion up to eighth order and found only a wide single-peak structure. Also the computer simulations so far have not given clear-cut evidence for solitons. Schneider and Stoll[14, 24] used a Langevin-type molecular dynamics (MD). On the one hand snapshots of the particle displacements of the full chain show a large number of narrow and high pulses which seem to correspond to one soliton solutions. (Unfortunately nothing is said about the velocity of the pulses which should be supersonic in the case of solitons.) On the other hand the dynamic form factors exhibit only a single broad peak near the phonon frequency $\omega_p(q)$. However, later a small second peak on the high-frequency side of the phonon peak was seen for some q-values, and this was interpreted as a soliton feature [29].

As the exact free energy can be represented by anharmonic perturbation theory (see above), it should be possible to calculate $S(q, \omega)$ in this way, too. Unfortunately the existing results differ qualitatively: Schneider[29] obtained a highly non-Lorentzian soliton peak plus a δ-function resonance at the high-frequency side of this peak. Diederich[30] obtained a phonon peak which has a pronounced shoulder at the high-frequency side, especially for intermediate q-values.

In view of all these discrepancies between the different results we feel that a new, high-precision simulation is necessary in order to determine unambiguously whether there are soliton signatures in $S(q, \omega)$ and which theories are good approximations. Last but not least, it is interesting to compare different simulation techniques. We have chosen a combined Monte Carlo - Molecular Dynamics method, in contrast to the Langevin method used by Schneider and Stoll[14, 24]. However, they did not really work with a canonical distribution: They took a number of snapshots from the solution of the Langevin equation and used them as initial configurations for molecular dynamics without damping and noise. In this sense their technique is similar to ours, which is microcanonical, too.

Complementary to our simulations we recalculate the dynamic form factor for two reasons: First, in the early papers[14, 24, 25] the soliton-gas approach was used and the correct weights for the soliton and phonon contributions were not yet known; secondly, the mechanical soliton momentum was used instead of the canonical one [14, 24].

2.1 Simulation Results

As mentioned we investigated the Toda lattice with Hamiltonian (1) numerically at different temperatures. We have selected starting configurations by a Monte-Carlo method and performed MD to get dynamic structure factors. Figures 2 and 3 contain more than 500 starting configurations and the resolution in q and ω direction is about 0.01. No smoothing was applied. Looking at the numerical results in Figure 3 one might not like to definitely exclude the possibility of a multi-peak structure. Therefore we show the prediction of the soliton-gas approach [25](see also Sect. 2.3) for comparison, where the phonon and the soliton peak were added with equal weights. We find no similarity, except that the positions of maximum weight are closely together and that there is an asymmetry in both cases. The low frequency side drops like a Lorentzian while the high frequency side shows an exponential decay. We will explain this at the end of the following section, where we find that the shape of the simulation results agrees with the cnoidal-wave calculations.

Including Figure 2 we can also state from the numerical point of view that there is evidence for a broad asymmetric single-peak in the correlations since there is no shoul-

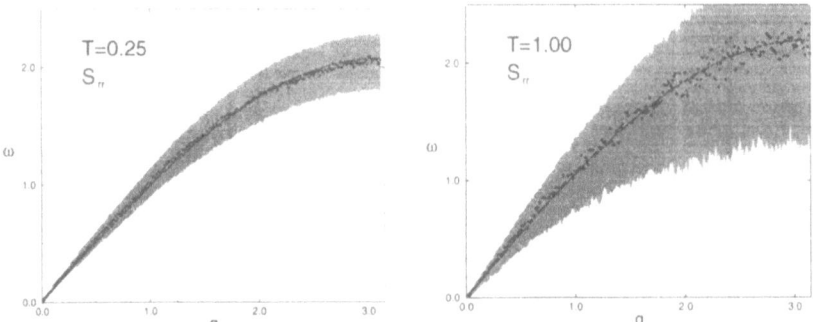

Figure 2: *The peak positions for constant q (circles) of the dynamic structure factor $S_{rr}(q, \omega)$ for two temperatures T. We tried to fit a sine (dashed line). In order to give an impression of the width the area where the amplitude is higher than half of the peak amplitude is shaded.*

derlike irregularity which we can follow for several adjacent q-values. Furthermore, due to the asymmetric potential (see Eqn. (1)) we find the length of the chain varying with temperature. The signature of this variation is found in the dispersion. For $k = \pi$ the change is about 3% at $T = 0.25$ and it is about 8% at $T = 1.00$. This change is not uniform so that fitting a sine in Figure 2 failed. We obtained best fitting results with a $\text{sn}(k|m \approx 0.2)$ nearly without temperature dependence. We hope to explain this also within the cnoidal-wave approach in a later publication.

2.2   Cnoidal-Wave Approach

We now calculate the dynamic force-force form factor $S_{ee}(q, \omega)$ in the Toda lattice, which is described by the Hamiltonian

$$\mathcal{H} = \frac{1}{2} \sum_{n=1}^{N} \dot{y}_n^2 + \frac{1}{2} \sum_{n=1}^{N} \left[e^{-(x_n - x_{n-1})} + (x_n - x_{n-1}) - 1 \right] \tag{1}$$

where x_n are the particle displacements and we set all constants to unity. We choose periodic boundary conditions for a large system $N \longrightarrow \infty$. The periodic solutions, involving a period $\frac{1}{\nu}$ and a wavelength λ, are exactly known [8]

$$e^{-b r_n} - 1 = (2\,K\,\nu)^2 \left\{ \text{dn}^2 \left[2 \left(\frac{n}{\lambda} \pm \nu t \right) K \right] - \frac{E}{K} \right\} \tag{2}$$

and the dispersion relation for these waves as well

$$(2\,K\,\nu)^{-1} = \sqrt{\frac{1}{\text{sn}^2 (2\,K/\lambda)} - 1 + \frac{E}{K}} \quad . \tag{3}$$

For convenience we abbreviate the force between two neighboring particles by $e_n = e^{-b r_n} - 1$, where $r_n = x_n - x_{n-1}$ is the strain. $\text{sn}(u) = \text{sn}(u|m)$ and $\text{dn}(u) = \text{dn}(u|m)$, are Jacobian elliptic functions; $K(m)$ and $E(m)$ are the complete elliptic integrals of first, and second kind, respectively.

We Fourier transform $e_n(t)$ to $e(q,t)$ using a Fourier expansion for $\mathrm{dn}^2(u|m)$ [8, 31]:

$$e(q,t) = \frac{1}{\sqrt{N}} \sum_{n=0}^{N} e^{iqn} e_n = \nu^2 \lambda^2 q \operatorname{csch}\left[\frac{q\lambda K'}{2K}\right] e^{-iq\lambda\nu t} \tag{4}$$

The sum over n leads to a lattice delta function which cancels the one from the Fourier expansion.

Every trajectory of a single particle in phase space corresponds to a certain modulus of the Jacobian elliptic function. Instead of the modulus m one may also use the "soliton" parameter α or the energy E (the complete elliptic integral is denoted by E), as we do. Since we use energy we have to be aware of the fact that $E = E(m, \lambda)$, i.e. energy does not only depend on the form of the humps but also on their number which is less or equal to half of the number of particles. The Complete Elliptic integrals K and K' depend on m and therefore on the energy E. Due to Eqn. (3) the wavelength λ depends on E, too. Finally we find for a certain energy E the dynamic form factor

$$S_{ee}(q,\omega,E) = \frac{1}{2\pi} \int dt\, e^{i\omega t - i\lambda q\nu t}\, e(-q,t) \cdot e(q,0)$$
$$= \delta\left(\frac{2\pi}{q} - \lambda\right) \nu^3 q \left[\lambda^2 \operatorname{csch}\left(\frac{q\lambda K'}{2K}\right)\right]^2. \tag{5}$$

For evaluating the complete dynamic structure factor we need to take into account all possible states. One way to do this is to simplify one integration [22] using the Hamiltonian Equation $\dot{Q} = \frac{\partial H}{\partial P}$ for a fixed energy E, i.e. the integral can be rewritten

$$\int dQ\, \frac{\partial P}{\partial E} = \int dQ\, \frac{dt}{dQ} = \int dt = \frac{1}{\nu} \tag{6}$$

where $\nu = \nu(E)$ the inverse period at a certain energy E. We are now well prepared to integrate Eqn. (5) over all states.

$$S_{ee}(q,\omega) = \int dP \int dQ\, e^{-\beta E} S_{ee}(q,\omega,E)$$
$$= (2\pi)^2 \frac{\omega^2}{q^3} \left[\frac{\partial E}{\partial \lambda} e^{-\beta E} \operatorname{csch}^2\left(\frac{q\lambda K'}{2K}\right)\right]_{\lambda=\frac{2\pi}{q}} \tag{7}$$

where m and therefore the Elliptic Integrals K and K' are determined via the dispersion relation Eqn. (3), an implicit transcendental relation. The energy for a single cnoidal-wave is

$$E = \sum_{l=1}^{\infty} \left[\frac{4}{l}\sinh\left(\frac{\pi l K'}{K}\right) + \omega^2\right] \left[\cosh\left(\frac{\pi l K'}{K}\right)\sin\left(\frac{\pi l}{\lambda}\right)\right]^2 - \ln\left[1 + \omega^2 \sum_{l=1}^{\infty} l\operatorname{csch}\left(\frac{\pi l K'}{K}\right)\right] \tag{8}$$

Again we find the non-trivial dependence on λ and ν. Please note that we use the "harmonic" definition for $\omega = 2\pi\nu$ and not $4K\nu$, as one might expect. Although we have not yet evaluated numerically Eqn. (7) together with Eqn. (8), we can already see (cf. Sect. 2.3) that this will lead to a smooth asymmetric single-peak structure.

Other form factors can be obtained from exact relations, e.g. the displacement-displacement form factor S_{xx} is connected to S_{ee} by [14]

$$\omega^4 S_{xx}(q,\omega) = 4\sin^2\left(\frac{q}{2}\right) S_{ee}(q,\omega) \tag{9}$$

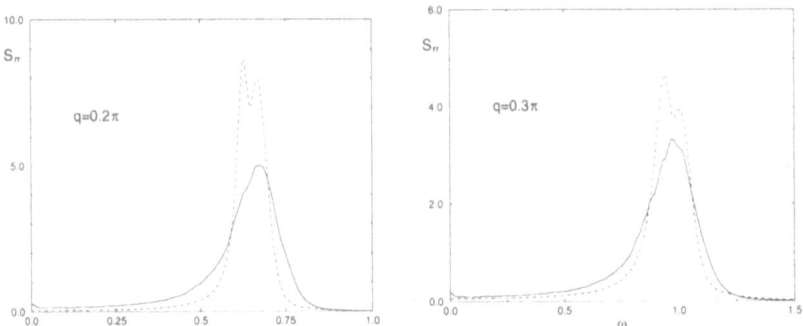

Figure 3: *Normalized correlation functions from simulations (solid line) and soliton-gas approximation (dashed line) at* $T = 0.25$.

And the density-density form factor is related to these structure factors by a relation valid for small q in the low temperature limit [24]

$$S_{\rho\rho}(q,\omega) \approx q^2 S_{xx}(q,\omega) \tag{10}$$

The density-density form factor $S_{\rho\rho}$ is proportional to the strain-strain form factor S_{rr}, which we plotted in Figures 2 and 3.

Other functions, e.g. the energy-energy form factor [24], which are not related to the above group, must be calculated separately by the same method.

2.3   The Soliton Limit

There are two interesting limits of the result (7), namely the soliton and the phonon limit. We will show that for the former case the general result coincides with the soliton-gas approximation [15].

Taking $m \longrightarrow 1$, where the elliptic functions become hyperbolic, implies several other limits:

$$
\begin{array}{lll}
K' \longrightarrow \frac{\pi}{2} & K \longrightarrow \infty & E \longrightarrow 1 \\
\mathrm{sn} \longrightarrow \tanh & \mathrm{cn} \longrightarrow \mathrm{sech} & \mathrm{dn} \longrightarrow \mathrm{sech} \\
\lambda \longrightarrow \frac{\pi K}{\alpha K'} = \frac{\sinh \alpha}{\nu \alpha} & & \nu \longrightarrow \frac{1}{2K} \sinh \alpha
\end{array}
\tag{11}
$$

Since we are now able to eliminate the Complete Elliptic Integrals we are also able to simplify the energy (8) [25]

$$E = \sinh(2\alpha) - 2\alpha \tag{12}$$

Hence the partial derivative with respect to the wavelength is

$$\frac{\partial E}{\partial \lambda} = \frac{\partial E}{\partial \alpha} \bigg/ \frac{\partial \lambda}{\partial \alpha} = [2\cosh(2\alpha) - 2]\frac{\nu}{v'} = 4\frac{\nu}{v'}\sinh^2\alpha \ . \tag{13}$$

We also simplify the argument of the hyperbolic function, i.e. $\mathrm{csch}\left(\frac{q\pi}{2\alpha}\right)$. In Eqn. (7) all dependencies on the modulus or the wavelength are substituted by the correct relation on the soliton parameter α and we obtain finally

$$S_{ee}(q,\omega) = \frac{32}{q^2\pi} e^{-\beta E(\alpha)} \frac{\sinh^5(\alpha)}{\alpha v'} \left(\frac{q\pi}{2\alpha}\right)^2 \mathrm{csch}^2\left(\frac{q\pi}{2\alpha}\right)\bigg|_{\frac{\omega}{q}=\frac{\sinh\alpha}{\alpha}} \tag{14}$$

This is exactly the form from reference [25] except that we have a Boltzmann distribution function whereas they used Fermi statistics. We remark that (14) decays exponentially, whereas the form factor of the harmonic lattice is approximately a Lorentzian [32].

3. SUMMARY

The dynamic form factor of the cnoidal-wave approach can be obtained by adding up a continuum of homotopically altered functions between the phonon and the soliton limit with the correct weights. This will lead to a smooth, asymmetric *single*-peak in the form factor. The weighted addition of both limiting curves, as implied in some references, leads to a two-peak structure within a certain range of temperature which is not realistic, however. Therefore we conclude that the introduction of the soliton concept does not pay for calculating dynamic form-factors of the considered group in thermal equilibrium.

However, if we have no thermal equilibrium, e.g. some high-energy solitons in a low-temperature environment, signatures in the form factor will be seen, as described in the soliton limit. On the other hand, we can consider different dynamic form factors, e.g. the energy-energy form factor, where we also find shapes we would not see without the nonlinearity. But these functions usually cannot be measured in neutron scattering.

Acknowledgments

One of the authors (A.N.) wishes to express his gratitude for the invitation and the support by the Euroconference Programme.

References

[1] Yomosa S., *J. Phys. Soc. Japan* **53** (1984) 3692 – 3698.

[2] Yomosa S., *Phys. Rev. A* **32** (1985) 1752 – 1758.

[3] Svoboda K., Schmidt C. F., Schnapp B. J., and Block S. M., *Nature* **365** (1993) 721 – 727.

[4] Finer J. T., Simmons R. M., and Spudich J. A., *Nature* **368** (1994) 113 – 119.

[5] Howard J., *Nature* **368** (1994) 98 – 99.

[6] Astumian R. D. and Bier M., *Phys. Rev. Lett.* **72** (1994) 1766 – 1769.

[7] Toda M., *J. Phys. Soc. Japan* **22** (1967) 431 – 436.

[8] Toda M., *Prog. Theor. Phys. Suppl.* **45** (1970) 174 – 200.

[9] Toda M., *Theory of Nonlinear Lattices* (Springer, Berlin, 1989).

[10] Hénon M., *Phys. Rev. B* **9** (1974) 1921 – 1923.

[11] Flaschka H., *Phys. Rev. B* **9** (1974) 1926 – 1925.

[12] Manakov S. V., *Zh. Eksp. Teor. Fiz.* **67** (1974) 543 – 555.

[13] Büttner H. and Mertens F. G., *Solid State Comm.* **29** (1979) 663 – 665.

[14] Schneider T. and Stoll E. Classical statistical mechanics of the Toda lattice: Static and dynamic properties. (1980, unpublished).

[15] Mertens F. G. and Büttner H., *Phys. Lett. A* **84** (1981) 335 – 337.

[16] Bolterauer H. and Opper M., *Phys. Lett. A* **83** (1981) 69 – 70.

[17] Yoshida F. and Sakuma T., *Phys. Rev. A* **25** (1982) 2750 – 2762.

[18] Theodorakopoulos N., *Phys. Rev. Lett.* **53** (1984) 871 – 874.

[19] Theodorakopoulus N., "Counting solitons and phonons in the Toda lattice", *Statics and Dynamics of Nonlinear Systems*, Ettore Majorana Centre, Erice, Italy, July 1st – 11th 1983, G. Benedek, H. Bilz, and R. Zeyher Eds., (Springer Verlag, Berlin, 1983), pp. 271 – 277.

[20] Theodorakopoulus N. and Bacalis N. C., "Thermally excited lattice solitons", *Proton transfer in hydrogen-bonded systems*, Heraklion, Crete, May 21st – 25th 1991, T. Bountis Eds. (Plenum Press, New York, 1992) pp. 131 – 138.

[21] Takayama H. and Ishikawa M., *Prog. Theor. Phys.* **76** (1986) 820 – 836.

[22] Bolterauer H. and Opper M., *Z. Phys. B* **42** (1981) 155 – 161.

[23] Bolterauer H. and Opper M., Private communication, (1981, unpublished).

[24] Schneider T. and Stoll E., *Phys. Rev. Lett.* **45** (1980) 997 – 1000.

[25] Mertens F. G. and Büttner H., *J. Phys. A: Math. Gen* **15** (1982) 1831 – 1839.

[26] Diederich S., *Phys. Rev. B* **24** (1981) 3186 – 3192.

[27] Diederich S., *Phys. Rev. B* **24** (1981) 3192 – 3203.

[28] Cuccoli A., Spicci M., Tognetti V., and Vaia R., *Phys. Rev. B* **47** (1993) 7859 – 7868.

[29] Schneider T., "Classical statistical mechanics of lattice dynamic model systems: Transfer integral and molecular dynamics studies", *Statics and Dynamics of Nonlinear Systems*, Ettore Majorana Centre, Erice, Italy, July 1st – 11th 1983, G. Benedek, H. Bilz, and R. Zeyher Eds., (Springer Verlag, Berlin, 1983), pp. 212 – 241.

[30] Diederich S., *Phys. Lett. A* **85** (1981) 233 – 235.

[31] Abramowitz M. and Stegun I. A. Eds., *Pocketbook of Mathematical Functions* (Verlag Harry Deutsch, Thun, 1984).

[32] Mikeska H. J., *Solid State Comm.* **13** (1973) 73 – 76.

Energy and charge transfer in photosynthesis

W. Mäntele[1]

Institut für Biophysik und Strahlenbiologie,
Universität Freiburg,
79104 Freiburg, Germany

1. INTRODUCTION. PHOTOSYNTHESIS: A PHYSICO-CHEMO-BIOLOGIST'S VIEW

Photosynthesis is the most important biological energy conversion process on earth, and no higher forms of life would be possible without the oxygen-containing atmosphere produced by the oxygenic photosynthesis of plants or algae, and more and more endangered by technical consumption of oxygen and combustion to CO_2. Only primitive forms of life, some anaerobic and sulfur bacterial families, would be able to exist, and even some of these successfully perform simple forms of nonoxygenic photosynthesis, i.e. they utilize solar energy without producing oxygen.

Nature has developed a large diversity of photosynthetic organisms, all of which convert a substantial fraction of the incoming photon energy to energy forms which can be used by the organism to grow, to sustain form and function in periods of darkness, or to drive other processes like the movement of bacteria by means of flagellae. This draws a sharp line between processes which draw net energy from light, and sensoric processes in plants or bacteria where light acts as a stimulus for germination, growth, and development, but where metabolic energy input from the organisms is required.

This short review cannot intend to cover all diverse forms of photosynthesis. Instead, the purpose of this tutorial is to show the common physical principles which all organisms utilize to perform the primary steps of solar energy conversion. We shall consider physical mechanisms and patterns which lead from extremely rapid processes such as light absorption (in the order of femtoseconds) to the much slower formation of stabilized redox states of some relevant molecules in the photosynthetic apparatus or the formation of a proton gradient across biological membranes (in the order of milliseconds), either of which can store a large fraction of the absorbed light energy. These processes are usually referred to as *light reactions* in the community of scientists working in photosynthesis, although only the very primary steps

[1] *present address*: Institut für Physikalische und Theoretische Chemie, Universität Erlangen-Nürnberg, Egerlandstrasse 3, 91058 Erlangen, Germany

involve the interaction of light by matter. The so-called *dark reactions*, which involve the fixation of carbon dioxide by plants or the chemical reactions of reduced compounds rich in energy, will not be dealt with here.

A useful concept to classifiy the mechanisms of these *light reactions* is that of following a time scale for the primary processes. Table 1 summarizes the major processes and the approximate time domain where they take place. We shall use here some specific terms for functional units of the photosynthetic apparatus which will be defined in detail below. We shall furtheron see that most of these processes are highly nonlinear and thus are amenable to a description by the physical principles which are the major topic of this summer school.

Table 1: Time scale of primary events in photosynthesis:

Photophysics:

light absorption by chlorophylls, carotenoids, and other pigments	10^{-15} sec
photophysical processes, internal conversions, vibrational relaxations	10^{-13} sec
energy conduction in excited *antenna* complexes	$10^{-12} - 10^{-10}$ sec

Electron transfer and protein electrostatics:

charge separation in *reaction centers*	$10^{-12} - 10^{-9}$ sec
stabilization of separated charges in *reaction centers*	10^{-4} sec

Redox chemistry and protein electrostatics:

coupling of proton transfer to electron transfer	10^{-3} sec
formation of proton gradients across a membrane	10^{-3} sec

Mainly chemistry

formation of *Adenosine-triphosphate* (ATP)	$10^{-3} - 10^{-2}$ sec
formation of stable reduced compounds (NADPH)	10^{-2} sec

2. THE LIPID BILAYER MEMBRANE AS A STRUCTURAL BASIS

Most of the processes listed above are located at lipid bilayer membranes in the photosynthetically active organisms or within distinct compartments formed by such membranes. These membranes are called *photosynthetic membranes*, and are classified by the biologist as *thylakoid membranes* in the case of plants or algae and as *chromatophore membranes* in the case of photosynthetic bacteria. In the latter case, the structural diversity is extreme, but nevertheless the functional principles are the same. Figure 1 shows a cartoon of the plant thylakoid membrane with schematic electron flow: This *linear electron transfer* scheme is inherently complicated and involves spatial and functional coupling of many different components. For the discussion of the major mechanisms, we shall later switch to the much more simple *cyclic electron transfer* in photosynthetic bacteria, and occasionally draw parallels to plant photosynthesis.

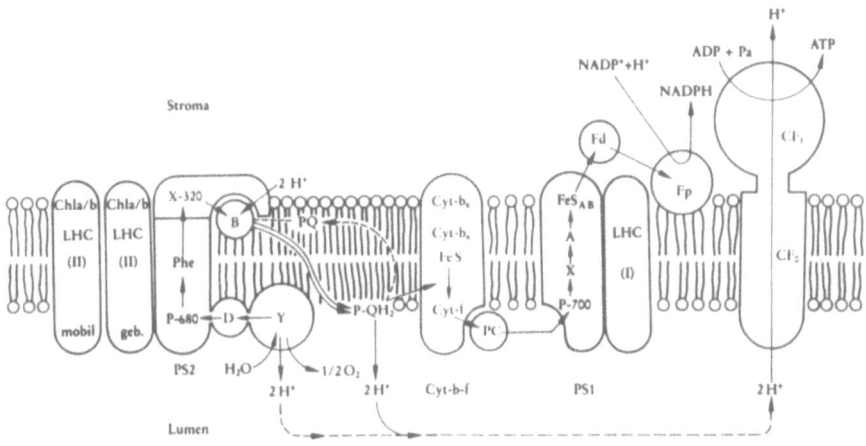

Figure 1: Schematic view of the plant thylakoid membrane (modified from Mäntele, 1990)

Two major membrane-spanning complexes, Photosystem I and Photosystem II, utilize light reactions to drive electrons from protein-bound redox components across the membrane. Photosystem II and the water-splitting enzyme associated with it utilize light to oxidize water. In four subsequent light reactions, four electrons are removed from two water molecules, and four protons as well as one oxygen molecule are released:

$$2H_2O \Rightarrow 4\ H^+ + 4\ e^- + O_2.$$

Photosystem I uses a light reaction to oxidize a water-soluble protein, plastocyanin, and transfers an electron to its acceptor side where stable reduced compounds are formed. A third membrane-spanning complex, Cytochrome b/f, and membrane-soluble quinones interconnect Photosystems II and I and form a Z-like electron transfer pathway. The cytochrome b/f complex acts as a redox-driven proton pump. Protons released from this complex and protons released from the water-splitting enzyme build up a proton gradient across the membrane which discharges through an ATPase and is used to synthesize ATP. In this scheme, the pigment arrays serving to harvest light ("antennae") are not shown.

Why has nature chosen to localize these processes in and at the lipid bilayer membrane ? The answer is relatively simple: First, diffusional processes are constrained to lateral diffusion in the membrane plane and along the membrane surface, guided by interactions with the polar headgroups of the lipids, and can thus be much faster than diffusion in three dimensions, if only water-soluble proteins were used. Second, the membrane provides a sidedness and can form inside or outside compartments to prevent the loss of reaction partners. Third, the membrane defines clear partition coefficients for molecules, which keep hydrophobic electron carriers like quinones inside the membrane bilayer, and water-soluble molecules outside. Fourth, the membrane provides electrical properties as an insulator and as a capacitor which are ideal for intermediate storage of energy in electrical form, i.e. as separated charges, or in an electrochemical gradient formed by different concentration of ions outside. The specific resistance of a pure lipid bilayer membrane is of the order of $10^8\ \Omega\ cm^2$, and may decrease to $1 - 10^4\ \Omega\ cm^2$ in the presence of proteins. This corresponds to a perfect insulator as compared to the specific resistance of a layer of 0.1 Mol/l NaCl with the thickness of the membrane, which is around 10^{-4} Ω cm^2. The specific capacity of a lipid membrane is of the order of 1 μF/cm^2.

> The lipid bilayer membrane is the structural basis of the photosynthetic apparatus and provides inside/outside phases. It acts as a capacitor for short term charge storage. With respect to the inside/outside phases, it orients transmembrane proteins which bind pigments and redox-active cofactors and connect both sides of the membrane. Peripheral or soluble proteins and small membrane-located redox carriers interconnect these transmembrane proteins to form a linear or a cyclic electron transfer system.

3. PIGMENTS USED FOR LIGHT ABSORPTION IN PHOTOSYNTHESIS:

To date, about 60 natural variants of chlorophyll molecules are known and partly characterized in their structure. This is a result of the great variety of photosynthetic organisms. All these chlorophylls have the tetrapyrrole ring structure and a central magnesium atom coordinated by the four nitrogens. The variants are characterized by small differences at the peripheral substituents to the tetrapyrrole system, which nevertheless have a strong impact on the electronic levels and redox properties of the conjugated system. Figure 2a shows the structure of a bacteriochlorophyll a molecule, as well as the parts of the molecule where chlorophyll a, chlorophyll b, and bacteriochlorophyll b differ from it. These four molecules make up the majority of the pigments in plants and photosynthetic bacteria.

Figure 2b shows the structure of the carotenoid spheroidene. Carotenoids are found with different chain length and substituents in almost all native photosynthetic structures. Figure 2c shows the structure of the chromophore of phycocyanin, an accessory pigment found in some blue-green algae. Altogether, these pigments with their electronic levels which are intrinsically different and which are further modified by the binding to proteins make up very efficient light absorbers which cover the spectral range from below 400 nm to above 1000 nm. Within this range, the plant pigments chlorophyll a, chlorophyll b, and carotenoids efficiently harvest light at wavelengths up to approx. 700 nm. At longer wavelengths, plant photosynthetic membranes are transparent, and, provided that our eye would be sensitive for far-red and near-infrared light, the bottom of a dense green forest would appear bright. Below 650 nm, but predominantly between 750 and 900 nm and, in one case, above 1000 nm, photosynthetic bacteria can harvest light. As a consequence of evolution, there is almost no absorption from bacterial photosynthetic systems between 900 and 1000 nm, since an overtone of an infrared water vibrational mode prevents light around 950 nm from penetrating deeper than a few cm into water, the habitat of photosynthetic bacteria.

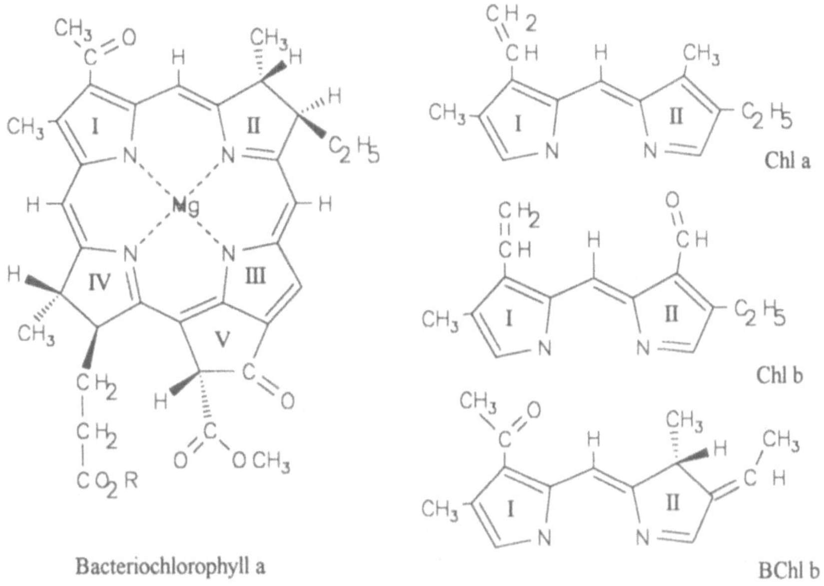

Bacteriochlorophyll a

Chl a

Chl b

BChl b

Figure 2a: **Structures of major chlorophylls**

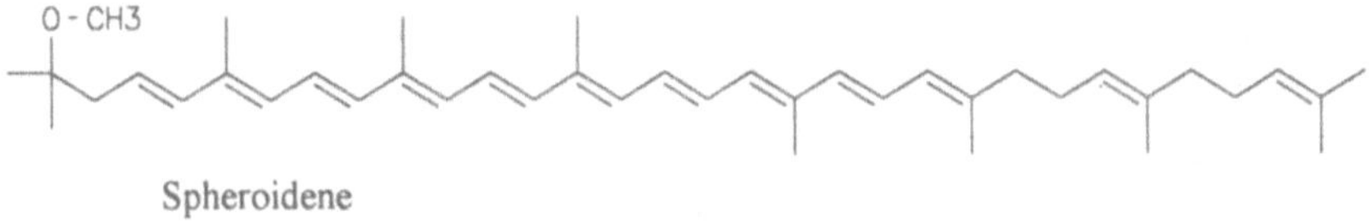

Spheroidene

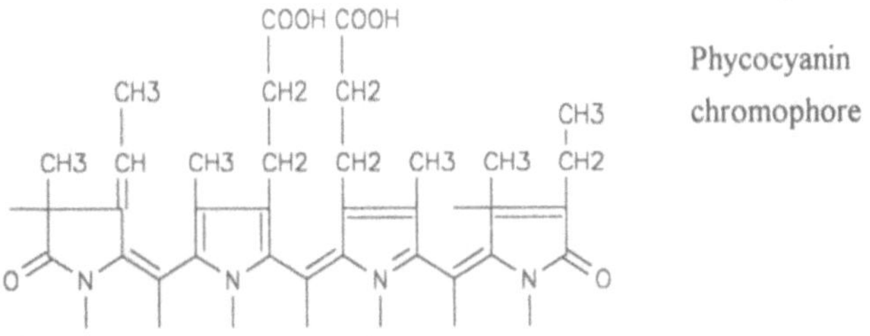

Phycocyanin

chromophore

Figure 2b: **Structure of the carotenoid spheroidene**
Figure 2c: **Structure of the chromophore from the accessory pigment phycocyanin**

Nature uses some of these pigments not only for absorbing light and transfering excited states within **light-harvesting complexes**, but also as protein-bound redox carriers in **reaction centers**. In addition, derivatized pigments, **pheophytins**, are used as electron carriers in reaction centers. Generally, a pheophytin is a magnesium-depleted chlorophyll with two protons bound to opposite pyrrole nitrogens. In photosynthetic reaction centers, only one type of pigment, i.e. chlorophyll a and pheophytin a, or bacteriochlorophyll a and bacteriopheophytin a, or bacteriochlorophyll b and bacteriopheophytin b appear together. Chlorophyll b only appears in a light-harvesting complex.

4. PIGMENTS AND PROTEINS FORM PIGMENT-PROTEIN-COMPLEXES

In all known photosynthetic membranes, the chlorophyll pigments are bound to a protein and form a **pigment-protein complex**. There is no evidence for free, unbound pigment involved in absorbing light and transfering excitation energy. For the sake of completeness, one should mention that in one case under debate there is some evidence for bacteriochlorophyll c forming aggregates without proteins, held together in small vesicles called **chlorosomes** which are bound outside the membrane of green photosynthetic bacteria. The binding of chlorophylls to proteins is **non-covalent**, as can be demonstrated by the extraction of pigments from membranes and pigment-protein-complexes by organic solvents like acetone or tetrahydrofurane.

This binding of chlorophylls to a protein by mechanisms which we shall discuss below seems to fulfil a multiple purpose: First, the protein rigidly **binds and orients** the pigments and thus gives their electronic transition moments a fixed orientation with respect to each other and with respect to the membrane plane. It can also hold two or more pigments closely together to form dimeric or oligomeric structures with modified electronic properties. Second, the protein **determines the spectral properties** of the pigment by acting on the conjugated system and thus lowering the energy gap for electronic transitions. Third, the protein **tunes the redox properties** and can determine whether a specific pigment may be easier to oxidize or to reduce than another.

In order to explain the impact of the protein, or more detailed, the pigment binding site of the protein, on the pigment properties, several possibilities for binding have to be considered. Figure 3 shows a hypothetical fully-bound bacteriochlorophyll a molecule in a binding pocket. This binding pocket can simultaneously provide hydrophobic interactions in the region of the tetrapyrrole ring, and polar interactions or hydrogen-bonds for the carbonyl groups at the periphery of the molecule. In addition, but not shown here, the protein provides a fifth or sixth ligand to the central magnesium atom, whose six coordination sites are not saturated by the four tetrapyrrole nitrogens. This fifth or sixth ligand is typically provided by the nitrogen of a histidine side chain in the protein. Unlike a homogeneously polar or nonpolar organic solvent, the protein binding pocket can thus provide a **heterogeneous environment** for the pigment and influence its conjugated system. This has often led to a description of the protein binding properties for pigments as an **optimized solvent**.

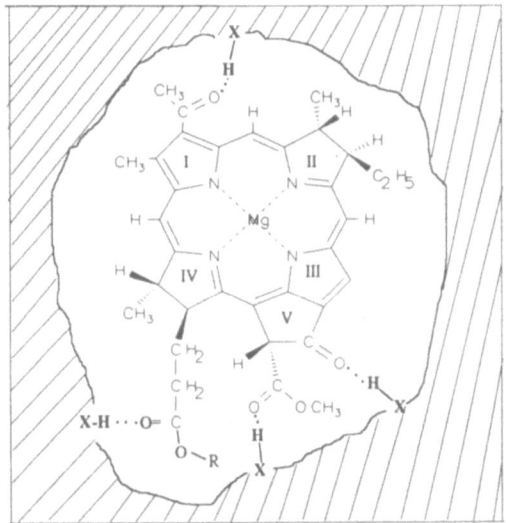

Fig. 3: A hypothetical fully "solvated" bacteriochlorophyll a molecule in its protein pocket. The ligands for the central magnesium are above and below the plane. X denotes H -bonding partners for the C=O groups.

The binding of chlorophyll pigments to a protein has a tremendeous impact on the conjugated system. The lowest energy singlet ($S_0 \Rightarrow S_1$) transition of bacteriochlorophyll a, which is polarized approximately along the line connecting the upper left and the lower right nitrogen of the molecule, gives rise to a strong ($\varepsilon \approx 100.000$ $M^{-1}cm^{-1}$) absorbance band at around 770 nm for bacteriochlorophyll *in vitro* (for example in ethanol) This transition is shifted to 800 nm, 820 nm, 850 nm, or 875 nm for distinct populations of bacteriochlorophyll a in pigment-protein light-harvesting complexes. This shift is even stronger for bacteriochlorophyll b, which absorbs around 790 nm *in vitro*, but at 1020 nm in the light-harvesting complex of the photosynthetic purple bacterium *Rhodopseudomonas viridis*. This corresponds to a lowering of the lowest singlet transition by more than 2800 cm^{-1} (!).

The purpose of this tuning of pigment energy levels is obvious: A set of distinct "antenna" populations with, from population to population, decreasing singlet energy is created which allows an excited state to "hop" without loss and finally end in a "trap", which is the pigment with the lowest S_1 energy level. This trap is either part of the antenna with an energy level close to the reaction center, or the electron donor pigment of the reaction center itself. Hopping may involve hundreds of pigments and may take tens to hundreds of picoseconds.

The binding of the pigments to the protein can also drastically modify the redox midpoint potential of pigments. For example, isolated chlorophyll a in the organic solvent tetrahydrofurane has a redox midpoint potential of approx. 0.4 V (vs NHE, normal Hydrogen electrode). In photosystem I, the two chlorophylls which form the electron donor P_{700} (named according its absorption maximum at 700 nm) exhibit a midpoint potential close to + 0.5 V. In photosystem II, the two chlorophylls thought to form the primary electron donor P_{680} have a redox midpoint potential of approx. + 1 V. On the other hand, the potential for chlorophyll a as an intermediate electron acceptor in photosystem I can be strongly negative. It is obvious that the "tuning" of the midpoint potential by the protein environment provides a means of adapting the light-induced charge transfer to the needs of a redox chain: Extremely positive (and close to the conditions for self-deterioration of a protein) for the purpose of water oxidation, and negative for operation under more reducing conditions.

5. ANTENNA SYSTEMS OF PLANTS AND BACTERIA:

The size of light-harvesting complexes in photosynthetic structures may vary considerably. It is characterized by giving the size of a *photosynthetic unit* (PSU), which is simply the total number of pigments divided by the number of pigments in a reaction center. For native photosynthetic bacteria, the PSU may vary from approx 50 to >200, but antenna-deficient mutants have been created with a PSU of 1. Nevertheless, these mutants can live happily, although they need some extra light, and grow slower. Plants have huge interconnected antenna complexes and may reach PSU sizes of up to 2000. Figure 4 shows antenna sets for plants and photosynthetic bacteria. For the plant antennae, it should be mentioned that the absorbance maxima given may be inaccurate: the difference between two adjacent populations is much smaller than the half-width of the absorbance bands of each population.

Antennae increase the absorption cross section for a photosynthetic reaction center, regulate energy flow within a "lake" of pigments, and dissipate excess excitation by fluorescence or thermally before it can lead to harmful photochemistry.

Chlorophylls or bacteriochlorophylls in light-harvesting complexes ideally perform only $S_0 \Rightarrow S_1$ or $S_0 \Rightarrow S_2$ excitation, deactivation $S_2 \Rightarrow S_1$, and $S_1 \Rightarrow S_1$ energy transfer. Ideally, there is no triplet formation or ionization of pigments, and dissipation via fluorescence ($S_1 \Rightarrow S_0$) occurs in case of high excitation densities, when excitation cannot be taken over by the reaction center.

Reaction centers perform charge separation and stabilization and couple electron transfer and proton transfer.

The "trap", upon $S_0 \Rightarrow S_1$ excitation ("trap*"), becomes a strong reductant and transfers an electron to a series of spatially separated acceptors.

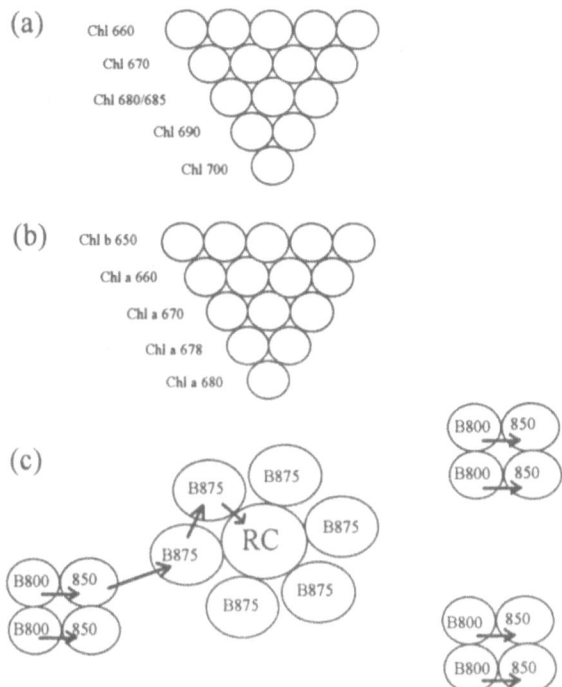

Figure 4: **Arrangement of chlorophylls in light-harvesting complexes of plants and photosynthetic bacteria.**

(4a) Light-harvesting pigments associated with photosystem I (adapted from Richter, 1982). To date, photosystem I has been crystallized, and X-ray structure analysis is in progress (Krauss et al., 1992). Chl 700 represents the primary electron donor of photosystem I.

(4b) Chlorophyll-a/b light-harvesting complex associated with photosystem II (adapted from Richter, 1992). Structural information from this complex is available from electron diffraction of 2-D crystals (Kühlbrandt and Wang, 1991). Chl a 680 represents the primary electron donor of photosystem II

(4c) Schematic arrangement of typical antenna populations for purple photosynthetic bacteria. B800-850 denotes a peripheral light-harvesting complex which is synthesized in a variable stoichiometry as a function of the light available. Some bacteria synthesize a B800-820 complex instead. B875 denotes a light-harvesting complex intimately associated with the reaction center (RC), and synthesized at fixed stoichiometry (unless molecular biologists have messed around with its genes). The corresponding complex for the bacteriochlorophyll b -containing *Rhodopseudomonas viridis* is B1020.

6.   FROM ANTENNAE TO REACTION CENTERS:
        CONVERTING AND STABILIZING LIGHT ENERGY

The photosynthetic electron transfer system of plants, which was shown in fig. 1, is unnecessarily complicated for the description of the principles of light energy conversion and short time storage. Instead, we turn to the much simpler bacterial systems, in particularly those of **purple photosynthetic bacteria**, which have become sort of a model for all photosynthetic systems. For reaction centers from purple photosynthetic bacteria, structures at atomic resolution are known, protein units have been engineered by molecular biologists, and a substantial amount of spectroscopic data is available. Let us thus describe the bacterial photosynthetic reaction center in detail and draw comparisons to plant photosystems I and II whenever it is appropriate.

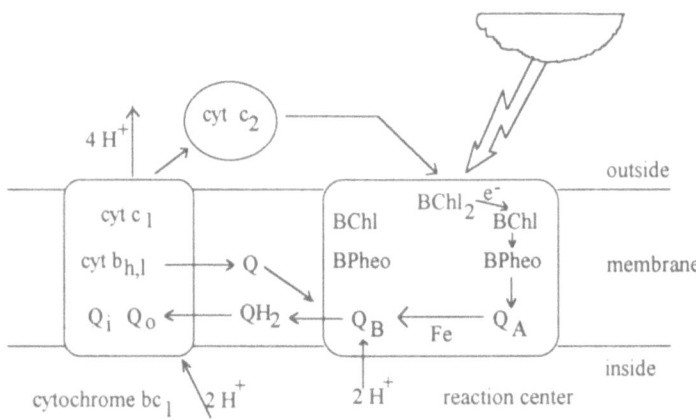

Fig. 5: Schematic view of the cyclic electron transfer system for purple photosynthetic bacteria

Purple photosynthetic bacteria utilize light energy to drive a cyclic electron transfer and to build up a proton gradient across the membrane as shown in figure 5. The photosynthetic apparatus utilizes two transmembrane protein complexes, the photosynthetic reaction center and the cytochrome b/c_1 complex, as well as cytochrome c_2 as a small water-soluble redox carrier protein, and membrane-confined quinones. Purple bacterial reaction centers contain quinone molecules as first and second electron acceptor, as does photosystem II of green plants and algae. It is thus common to refer to this type as *quinone-type reaction center*, in contrast to photosystem I and reaction centers of green bacteria, which contain iron-sulfur complexes as electron acceptors operating at lower redox potentials.

The purpose of the reaction center is to perform a primary charge transfer from an excited pigment molecule to acceptors with increasing distance within picoseconds to nanoseconds, and then to stabilize these separated charges by dispersing reorganization energy. By alignment of these acceptors along a path across the membrane, electron transfer is vectorial, and charge separation creates a transmembrane dipole. Subsequently, protonation reactions are coupled at the acceptor side of the reaction center. Upon two light-driven turnovers, the secondary quinone (Q_B) is converted to a dihydroquinone, or synonymous, a "quinol" according to:

$$Q + 2\,e^- + 2\,H^+ \Rightarrow QH_2$$

This quinol can leave from the reaction center into the membrane phase; the site then takes up a quinone from the *quinone pool*, a total of approx. 20 quinones in a PSU. The second membrane-spanning complex, the cytochrome bc_1, serves as a return path for electrons. The quinol is reoxidized in this complex and the electrons are transferred via the electron carriers cytochrom b (high potential), cytochrom b (low potential), an iron-sulfur complex, and a cytochrome c_1, all within the protein cluster called cytochrome bc_1, to the other side of the membrane. The analogous unit in the electron transfer system of plants is the cytochrome b/f complex shown in fig. 1. A docking site for the water-soluble cytochrome c_2 is suposed to be near the cytochrome c_1 part. Quite analogous to the bc_1 complex, the reaction center exhibits a similar docking site near the primary donor, which has been structurally mapped by getting different cytochrome c and c_2 proteins to react with the reaction center. It is clear that, although both docking sites must be similar and can accept the same molecule, they must be specific enough to render different affinity for the two redox states: a high affinity for oxidized cytochrome c_2 at the bc_1 complex, and a high affinity for reduced cytochrome c_2 at the reaction center. We may call this a simple yet efficient mechanism for protein-protein recognition, which is performed by complementary charge patterns on the surfaces of the protein.

This water-soluble cytochrome c_2 acts as a shuttle between both membrane-spanning complexes and closes the cyclic electron transfer chain. Two protons bound on the lower side to form the quinol and released on the upper side of the cytochrome bc_1 complex, as are two further protons which are vectorially transferred in a so-called *Q-cycle* which involves rereduction, protonation, reoxidation, and deprotonation of the quinone in the cytochrome bc_1 complex and which is driven by excess redox energy. Thus, altogether four protons per two light reactions are actively pumped across the membrane. They build up a proton gradient across the membrane (*cf* the electrical properties of the membrane mentioned above) which discharges through an ATPase (not shown in this scheme) and which is used to synthesize ATP as the stable intermediate in bacterial photosynthesis.

7. INFORMATION ON STRUCTURE AND COMPOSITION OF REACTION CENTERS

Reaction centers can be classified into two main groups, according to the redox components they contain, but also according to the redox potential range in which they operate. The first type is the purple bacterial reaction center, photosystem II, or more general the "quinone-type" reaction center, which contains quinones as electron acceptors and operates at approx. $+ 0.5$ V (bacterial RC) or approx. $+ 1$ V (plant photosystem II) for the oxidizing side, and close to 0 V at the reducing side. The second type is photosystem I or the reaction center of green bacteria, or more general the "iron-sulfur" type of reaction center, which contains iron-sulfur centers as electron acceptors, and operates at moderately positive potentials for the oxidizing side and at negative potentials for the reducing side.

Detailed structural information is only available for purple bacterial RC. The first high-resolution crystal structure of the reaction center from *Rhodopseudomonas viridis* by Deisenhofer, Michel, and Huber, who reveived the 1988 Nobel prize in Chemistry for this work, has unified many biochemical and spectroscopic results available before (BC: before crystallization), and has subsequently (AC: after crystallization) initiated numerous experiments which addressed precise questions with regard to the structure-function relations. Apart from the structure of the *Rhodopseudomonas viridis* reaction center (meanwhile at approx. 2.3 Å), high resolution structures have been worked out for the reaction center of *Rhodobacter sphaeroides*, and structural details are close to 2.5 Å. Small crystals from photosystem II have been reported, but yet await improvement, diffraction to high resolution, and structure analysis. A crystal structure of photosystem I is available at 4 Å resolution (Krauss et al., 1992), but still awaits better resolution and refinement in order to attribute electron density to the approx. 100 chlorophyll molecules which surround the core electron transfer path and act as intrinsic antenna. Unfortunately, structural information is not yet available for the reaction center of green photosynthetic bacteria.

Figure 6 shows the structure of the *Rhodobacter sphaeroides* reaction center. The top part shows the protein with large α-helical sections which run predominantly transmembrane. A the lower part, the protein part is omitted and the arrangement of the pigments, the quinones and the iron is shown schematically. A c2 symmetry axis is indicated which runs through the bacteriochlorophyll a dimer molecule on the top and through the non-heme iron at the bottom, with pigments and quinones arranged on a left and right branch. Only the right branch, however, is active, for reasons which we shall discuss below.

RC Rhodobacter sphaeroides

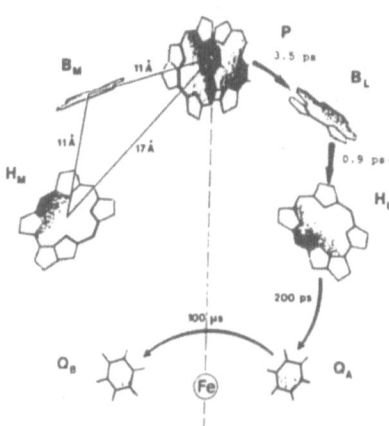

Figure 6: Structure of the *Rhodobacter sphaeroides* reaction center.
The atomic coordinates provided by Allen et al. were used for plotting.
The lower figure shows a schematic arrangement of the pigments.

8. CHARACTERISTICS OF A "QUINONE TYPE" REACTION CENTER

The present knowledge on the purple bacterial photosynthtic reaction center allows to point out the major characteristics - architectural and functional features - of these "biological photocells":

(1) A reaction center is a transmembrane protein, rigid, predominantly α-helical, with a minimum of 2-3 subunits, and a molecular weight of > 100 kDa.

(2) It contains rigidly packed pigments, quinones, and a non-heme iron between two strongly interacting protein subunits. In the bacterial RC, these are the M (medium) and L (low) apparent molecular weight subunits (amino acids in these subunits are coded H or L with a number giving their position in the amino acid sequence). The pigments and other cofactors have little translational or rotational freedom in the protein. Further protein subunits are packed around and are responsible for stability, orientation in the membrane, docking, secondary electron transfer, and coupling of proton transfer. In the bacterial RC, this is the H (high) molecular weight subunit which protrudes from the membrane into the aqueous phase, or the C (cytochrome) subunit of *Rhodopseudomonas viridis* containing hemes involved in electron transport to the primary electron donor.

(3) The set of pigments and redox cofactors bound by the protein moiety defines, by decreasing redox midpoint potential, a pathway for electrons from the donor (placed close to one side of the membrane) to the acceptor(s) placed close to the other side of the membrane.

(4) The electron donor (P) is a molecule which can be excited by light or can act as a "trap" for an excited state (from an antenna), and which is a strong reductant in the excited state to transfer an electron to a remote acceptor..

(5) The intermediary electron acceptors (B, H, Q_A) are molecules which can, by interaction with their protein site, provide enough internal and external reorganization energy to slow down back reactions.

(6) The terminal electron acceptor (Q_B) is a molecule which can couple proton transfer to electron transfer.

(7) Reaction centers of purple bacteria show an almost perfect symmetry for the pigment arrangement, with a c_2 symmetry axis running through the primary electron donor. The asymmetry is on a subtle level, in particular for the pigment-protein and quinone-protein interactions.

9. PRIMARY PROCESSES, FORWARD, AND REVERSE ELECTRON TRANSFER AND THE CORRESPONDING RATES IN THE REACTION CENTER

When the dimeric bacteriochlorophyll (denoted "P" for primary electron donor) in the reaction center gets excited (P^*), it becomes a strong reductant and transfers an electron to a bacteriopheophytin molecule (H) approximately halfways through the membrane. Excitation of P can occur either directly by absorbing a quantum of light at approx 860 nm ($S_0 \Rightarrow S_1$ transition) or at approx 600 nm ($S_0 \Rightarrow S_2$) transition, the latter followed by fast internal relaxation to the lowest (S_1) singlet level. Alternatively, light absorbed by any of the reaction center pigments (bacteriochlorophyll monomers: $S_0 \Rightarrow S_1$ at approx. 800 nm, $S_0 \Rightarrow S_2$ at approx. 600 nm; bacteriopheophytins: $S_0 \Rightarrow S_1$ at approx. 750 nm, $S_0 \Rightarrow S_2$ at approx. 530 nm) or by the surrounding antennae leads to rapid radiationless transfer to the trap P (all wavelengths given are for the *Rhodobacter sphaeroides* reaction center; somewhat different numbers hold for that of *Rhodopseudomonas viridis*).

Electron transfer from P^* to H occurs within approx. 3.5 psec, fast enough to compete with deexcitation via fluorescence. Ultrafast spectroscopic investigations have indicated that this electron transfer occurs sequentially via the monomeric bacteriochlorophyll, forming an intermediate B^- which transfers the electron further to H. At this step, the separated charges (P^+H^-) are approximately 15-20 Å apart, and enough reorganisation energy is dispersed to slow down charge recombination to the order of nsec. From H^-, the electron is further transferred to the primary quinone Q_A within approx. 200 psec. At this step, separation of the charges has proceeded almost across the entire thickness of the membrane (approx. 30 Å). If one considers the differences of redox midpoint potentials (P^+/P and Q_A^-/Q_A) and the shift of their equilibria, one finds that around 45 % of the incoming light energy (1.4 eV for a 860 nm quantum) are stored in this state; 55 % are dispersed as internal and external reorganization energy for the respective cofactors to accomodate the charged radical state. As a consequence, this state is almost infinitely stable compared to the time of formation: in case electron transfer from Q_A to Q_B is blocked or Q_B is removed, charge recombination $P^+Q_A^-$ $\Rightarrow$ P Q_A occurs at a half-time of 60 msec, i.e. 9-10 orders of magnitude slower than charge separation. Electron transfer from Q_A to Q_B occurs along the plane of the membrane and does not lead to further separation of the charges. This transfer is driven by a small Free Energy; the redox midpoint potentials of the primary and the secondary quinone differ by only 30-40 mV. Consequently, the transfer is slow and is controlled by charged residues in the vicinity of Q_B; thus, electron transfer becomes proton-dependent and vice versa - we have reached the stage of coupling between electron and proton transfers.

Once charge separation and stabilization has reached the state $P^+Q_B^-$, charge recombination becomes very slow - in "idling" reaction centers the process $P^+Q_B^-$ $\Rightarrow$ P Q_B proceeds with about 1 sec half-time, and mutants were made where it takes up to one minute. This is remarkably long, in particular because the species involved - a bacteriochlorophyll dimer cation radical and a semiquinone anion - would not be stable in a non-structured solvent environment. In "working" reaction centres, rereduction of P^+ by a secondary electron donor (a soluble cytochrome c_2 for *Rb. sphaeroides* reaction centers, *cf.* Fig 5, and one of the four hemes in the cytochrome c subunit of *Rps. viridis* reaction centres) occurs after $P^+Q_A^-$ charge separation and takes micro-to-milliseconds, further preventing charge recombination from Q_A^- or Q_B^-. In this state, the primary electron donor is ready for a second turnover, which, up to $P^+Q_A^-$ charge separation, occurs in rates quite comparable to the first electron transfer. Electron transfer from Q_A to Q_B in this case ($Q_A^-Q_B^-$ $\Rightarrow$ $Q_AQ_B^{2-}$) shows a different rate and strictly depends on protonation of adjacent amino acid residues. It is probably not a simple two-electron reduction step, but rather a complex interplay of protons and electrons at the Q_B

site. Finally, after the second electron turnover, the quinol leaves the site exchanging with a quinone, a process which take up to tens of msec.

10. WHY DO REACTION CENTERS USE QUINONES AS ELECTRON ACCEPTORS ?

Photosystem II and bacterial reaction centers use quinone molecules as redox-active cofactors on the acceptor side. Figure 7 shows the structure of menaquinone, ubiquinone and of plastoquinone. The reaction center of *Rhodobacter sphaeroides* uses ubiquinone-10 at both sites (Q_A and Q_B), whereas the reaction center of *Rhodopseudomonas viridis* uses menaquinone for Q_A and ubiquinone for Q_B, and photosystem II uses plastoquinone. There are a number of reasons why these molecules are used as terminal electron acceptors:

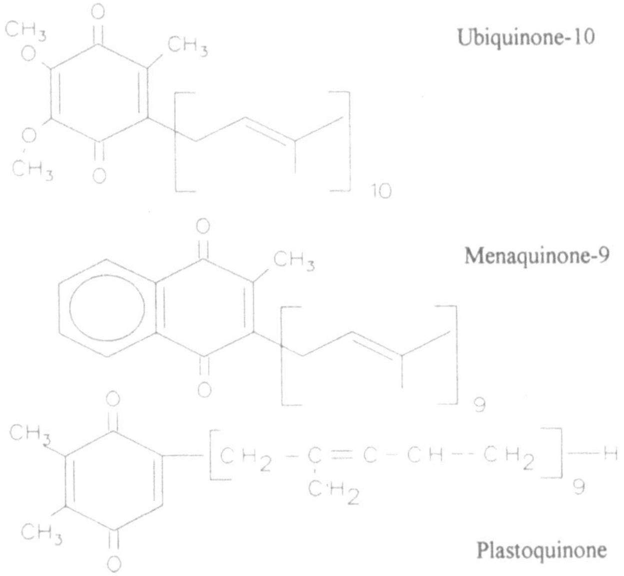

Figure 7: Structure of quinones used by "*quinone -type*" reaction centers
      (a) Ubiquinone-10 from *Rhodobacter sphaeroides* reaction centers
      (b) Menaquinone-9 from *Rhodopseudomonas viridis* reaction centers
      (c) Plastoquinone from Photosystem II of green plants and algae

(1) Quinones have appropriate redox midpoint potentials, which fit the needs of the redox chains for photosystem II or purple bacterial reaction centers. The midpoint potential for single-electron reduction ($Q \Rightarrow Q^{-\cdot}$) is between - 0.1 V and + 0.1 V for most quinones.

(2) Quinones have unique redox-dependent binding properties. If a binding constant k_b is defined which describes the equilibrium between the free quinone and the quinone bound to the protein site, k_b (Q) $\neq$ k_b (Q$^{-\cdot}$) $\neq$ k_b (Q^{2-}) $\neq$ k_b (QH$_2$). While neutral quinone is moderately strong bound, the semiquinone anion is tightly bound (awaiting a second electron), and the quinol can easily leave the site. Whether the species Q^{2-} exists in a stable form at all is presently unclear. Provided that the appropriate site is given, quinones can thus act as one-electron gates or two-electron gates, or leave the site as soon as they are protonated.

(3) Quinones have a unique coupling of redox chemistry to protonation chemistry:

$$Q + 2\,e^- + 2\,H^+ \Rightarrow QH_2$$

This coupling can proceed on two ways within an electron transfer/protonation scheme:

$$
\begin{array}{ccccc}
Q + e^- \Rightarrow & & Q^{-\cdot} + e^- \Rightarrow & & Q^{2-} \\
 & & + H^+ & & + H^+ \\
 & & \Downarrow & & \Downarrow \\
 & & QH + e^- \Rightarrow & & QH^- \\
 & & & & + H^+ \\
 & & & & \Downarrow \\
 & & & & QH_2
\end{array}
$$

In this scheme, electron transfer proceeds from left to right, and proton transfer from top to bottom. The pathway via Q^{2-} is energetically unfavorable, since a negative charge has to be put on an already negatively charged molecule. The diagonal pathway is energetically much more favorable. As we shall see below, intermediate reaction schemes are conceivable in a protein, where the negative charge on the quinone can be shielded by neighboring positive charges.

(4) Provided they have a longer (n > 4-5) isoprene side chain, quinones have a partition coefficient which confines them to the inner part of the membrane.

(5) Quinone molecules are small enough to be exchanged from the binding site with the quinone pool in the membrane without the need of larger protein conformational changes.

The two binding sites of the quinones in the bacterial reaction center differ substantially and point out to the different function of the sites, though stocked with the same quinone molecule in RC of some bacteria. The site of the primary quinone, Q_A, is characterized by the absence of ionizable amino acid residues within at least 8 - 10 Å. On the other hand, there are several aromatic residues nearby. One of them, tryptophan M252, has its aromatic ring approximately parallel to the ring of Q_A. Together with hydrogen-bonding of the quinone carbonyls (an alanine peptide nitrogen and a threonin side chain oxygen are within H-bonding distance, although actual bonding appears to be strictly asymmetrical), this structure reminds of a quinhydrone, i.e. a quinone and a quinol interacting both with their ring structures and hydrogen-bonded with their oxygens. In this structure provided by the Q_A site, the quinone cannot be protonated and thus acts as a one-electron gate.

The site of the secondary quinone, Q_B, contains several ionizable residues in immediate contact with the quinone. Two residues, glutamic acid L212 and aspartic acid L213, are in van-der-Waals contact with the quinone, while further residues (aspartic acid L210, arginine L217, and glutamic acid H173) are all within 10 Å. The ionization state of these residues is not clear at present. Electrostatic calculations on the basis of the known X-ray structure, measurements of electron transfer rates for native RC and for RC where one or more of these residues have been exchanged with non-ionizable residues, and infrared spectroscopic data, however, indicate that some these residues interact strongly and act as a cluster which titrates as an entity (instead of titrating like individual residues). Furthermore, the protein environment of Q_B, which seems to contain single water molecules bound at specific locations, appears to shift the pK_s values of some of these amino acid side chain groups by more than 3-4 pH units.

The role of the ionizable residues near Q_B is to provide protons for the protonation of Q_B upon two electron transfer turnovers. However, it also appears that protonation changes of these residues and a net proton uptake of the RC acceptor side occurs after a single electron turnover, and that the cluster as a whole is more protonated in the state Q_B^- than for neutral Q_B. We shall develop a more detailed scenario below.

11. WHAT IS THE MOLECULAR BASIS FOR EXTERNAL REORGANIZATION ENERGY IN THE REACTION CENTER ?

One can imagine different ways for a protein to solvate a buried charge. First, let us assume that a cofactor with a given electronic distribution in its neutral form is in an energetically favorable equilibrium with its site, inasfar charge-charge and dipolar interactions are concerned. When the cofactor then becomes reduced (or oxidized), its changed charge distribution will lead to a rearrangement of its site. An efficient way of balancing an excess negative or positive charge on the cofactor would be the redistribution of (partial) charges at ionizable amino acid residues in the vicinity. This mechanism can account for the solvation of the secondary quinone Q_B. In the absence of ionizable residues (such as for Q_A), polarization of peptide C=O dipoles can dissipate a substantial fraction of reorganization energy. Reorientation of peptide dipoles is certainly a process too slow for the solvation of the redox cofactors involved in the very fast processes, i.e. for the dimeric and monomeric bacteriochlorophyll.

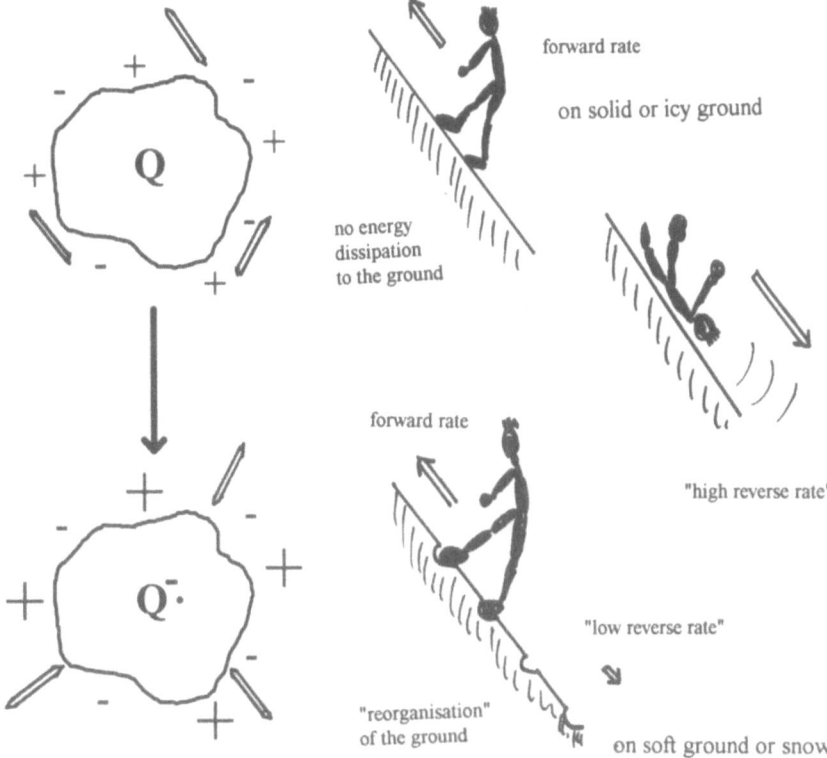

forward rate

on solid or icy ground

no energy
dissipation
to the ground

forward rate

"high reverse rate"

"low reverse rate"

"reorganisation"
of the ground

on soft ground or snow

Figure 8:

Left: Schematic representation of the external reorganization energy upon transformation of a quinone (Q) to a semiquinone anion radical (Q⁻·). Polarization and reorientation of dipoles as well as changes of the charge pattern in the site are shown.

Right: The analogon to electron transfer in the reaction center (following an idea of H. Frauenfelder): A Hillclimber on hard icy ground may proceed rapidly and with some leftover energy to the top of the hill, since he does not "dissipate" energy to the ground. However, he may have a high "reverse rate", sliding down the hill. A hillclimber on soft ground or snow "dissipates" a significant fraction of his available energy to the ground and may arrive on the top quite exhausted. However, his "reverse rate" is very low, thus ensuring that he has a high probability of reaching the hilltop.

One possibility to characterize this molecular reorganization is by use of infrared spectroscopic methods, which monitor changes of bond lengths, bond geometries, bond polarizations, protonation/deprotonation reactions, hydrogen bonding, and of dipole orientation. All methods use difference techniques (see Mäntele, 1993a) to get rid of the large protein background. Vibrational difference (radical-minus-neutral) spectra were obtained the P $\Rightarrow$ P⁺, H $\Rightarrow$ H⁻, and the $Q_A \Rightarrow Q_A^-$ and $Q_B \Rightarrow Q_B^-$ transitions as well as for the donor-acceptor couples in the charge separated state (for an overview, see Mäntele, 1993b). Although by far not completely assigned, these vibrational difference spectra represent the total changes of molecular coordinates upon electron transfer and allow to visualize reorganization energy.

Only small contributions from the protein were detected for the $P \Rightarrow P^+$ and for the $H \Rightarrow H^-$ transitions, and the majority of the signals could be modeled by pigment vibrational modes. This was taken as evidence for mostly internal reorganization energy. It was found that pigment carbonyl modes were shifted in frequency, indicating transitions from the nonsolvated to the sovated form and vice versa. This was taken as evidence that the hydrogen bond pattern around the pigments changes with the charge transfer, a process which could be fast enough to even solvate modified charge distributions within psec.

In contrast to these, a substantial fraction of the IR signals in the $Q_A \Rightarrow Q_A^-$ and $Q_B \Rightarrow Q_B^-$ transition was attributed to the protein site of Q_A and Q_B, respectively. A number of signals was detected in the amide I range (1700 cm^{-1} to 1620 cm^{-1}), which is characteristic for the local and global protein conformation. These signals were taken as evidence for rearrangements in the site, and are probably caused by strengthening, weakening or reorientations of peptide C=o groups. Besides these, changes of protonation state for amino acid side chain groups were detected and taken as evidence for the modification of charge patterns. Overall, the conformational changes in the reaction center were found to be very small, affecting few residues but not entire domains.

12. CONCLUDING REMARKS

The primary reactions of photosynthesis have received increasing attention, since they include the fastest and most efficient energy and electron transfer reactions. Today, for most of the light-induced steps, the sequence of reactions and the reaction partners have been identified. In contrast to electron transfer in solution, the donors and acceptors in the membrane-bound complexes exhibit fixed spatial arrangement and orientation, a prerequisite for testing the theory of electron transfer and the dependence of rates on the free energy or on the medium between donor and acceptor. Moreover, for some complexes this arrangement is exactly known from X-ray crystallography, and the medium properties can be modified, at least to some extent, by site-directed mutagenesis.

The role of the protein in providing reorganization energy has been recognized, too. However, most spectroscopic tools for the observation of this reorganization energy are still limited or rather indirect. Future theoretical and experimental work on the dynamic properties of the protein binding sites of redox-active cofactors will be needed to fully understand stabilization of the redox state of a protein or of charge-separation in the molecular photocells used by nature in photosynthesis.

ACKNOWLEDGEMENTS:

The author would like to thank his coworkers for critical discussions and comments on the manuscript. The assistance of B. Ehlert for the preparation of the text and the figures is gratefully acknowledged.

BIBLIOGRAPY:
Suggested reviews and textbooks:

Breton, J. and Vermeglio, A. eds.
The photosynthetic bacterial reaction center I (1988)
The photosynthetic bacterial reaction center I (1992)
NATO ASI series Vol 149 (I) and Vol. 237 (II), Plenum Press, New York

Breton, J. (1988)
The reaction center of purple photosynthetic bacteria.
ISI Atlas of Science: Biochemistry, pp. 323-328

Clayton, R.K. (1980)
Photosynthesis: Physical mechanism and chemical patterns
Cambridge University Press

Deisenhofer, J., Michel, H. and Huber, R. (1985)
The structural basis of photosynthetic light reactions in bacteria
Trends Biochem. Sci. 243-248

Deisenhofer, J. and Michel, H. (1989) (Nobel lecture)
The photosynthetic reaction centre from the purple bacterium *Rhodopseudomonas viridis*
The EMBO Journal 8, 2149-2170

Norris, J.R. and Schiffer, M. (1990
Photosynthetic reacton centers in bacteria
C & EN, 22-37

Parson, W.W. (1982)
Photosynthetic bacterial reaction centers
Ann. Rev. Biophys. Bioeng. 11, 57-80

REFERENCES GIVEN IN THE TEXT:

Allen, J.P., Feher, G., Yeates, T.O., Komiya, T.O., and Rees, D.C.
Proc. Natl. Acad. Sci. USA **84**, 5730-5734

Krauss, N., Hinrichs, W., Witt, I., Fromme, P., Pritzkow, W., Dauter, Z., Betzel, C., Wilson, K.S., Witt, H.T., and Saenger, W. (1993)
Three-dimensional structure of system I of photosynthesis at 6 Å resolution
Nature **361**, 326-331

Kühlbrandt, W. and Wang, D.N. (1991)
Nature **350**, 130-134

Mäntele (1990)
Biologie in unserer Zeit **20**, 85-93

Richter, G. (1982)
Stoffwechselphysiologie der Pflanzen
Georg Thieme Verlag Stuttgart

The role of nonlinearity in modelling energy transfer in Scheibe aggregates

O. Bang, P.L. Christiansen*, K.Ø. Rasmussen* and Y.B. Gaididei**

Laboratoire de Physique, ENS Lyon, 46 allée d'Italie,
69364 Lyon cedex 07, France
* IMM, T. Univ. of Denmark, Anker Engelundsvej 1,
2800 Lyngby, Denmark
** ITP, Metrologicheskaya Street 14 B,
252 143 Kiev 143, Ukraine

1. INTRODUCTION

Exciton motion in molecular systems is an important field of physics, and has been undergoing active theoretical, and experimental investigations. For a general review on excitons see e.g. Davydov [1]. The field derives its importance from being a part of the general area of energy transfer, and its consequent connection with a variety of disciplines, even outside physics, such as photosynthesis in biology [2]. Here we consider the highly efficient energy transfer observed in a special kind of ordered molecular system, known as Scheibe aggregates [3], and its possible explanation through nonlinear dynamical effects inherent to the system.

Scheibe aggregates are abundantly found in nature, where they function as energy funnels for sunlight to be used in photochemical processes [4], but they may also be produced artificially in the laboratory by the so-called Langmuir-Blodgett (LB) technique [5,6]. Thus one may use LB Scheibe aggregates to study the important process of photosynthesis [7], and especially the initial harvesting, and transfer of energy. The connection to photosynthesis may also be used technologically to identify the combination of molecules resulting in the most efficient energy transfer.

In the last few years, many attempts have been made to use ideas coming from nonlinear physics (solitons, nonlinear energy localization for instance) to model some biological processes like energy storage, and transport in proteins [8], DNA conformational changes, and thermal denaturation [9], etc. Nonlinear excitations, which are exceptionally stable, immune to perturbations, and build-up spontaneously in many systems, appear as interesting candidates to explain some of these molecular processes.

However, the models of protein, and DNA are both one-dimensional (1D), which is what ensures the exceptional stability of their nonlinear excitations. In multi-dimensional systems, most nonlinear excitations becomes unstable, and new phenomena, such as wave-function collapse, comes into the picture. Famous examples of this effect is the Langmuir wave-collapse from plasma physics [10], and self-focussing of laser beams from nonlinear optics [11]. Molecular systems arranged in Scheibe aggregates are inherently multi-dimensional, and thus the modelling becomes quite different from that of protein, and DNA [12].

In the first of the two main sections in this paper we give a review of Scheibe aggregates, explaining what a Scheibe aggregate actually is, how it is produced by the LB technique, and which experimental measurements have been performed on such LB Scheibe aggregates. The LB Scheibe aggregate has many important technological applications, and is worth a study in its own right. However, in the spirit of this conference, we describe, and discuss the initial harvesting, and transfer of energy in the photosynthetic process. Since the antenna assemblies performing this task are arranged in aggregated structures, there is a close connection to the same process in LB Scheibe aggregates. Finally we give an overview of the physical models of Scheibe aggregates, which have been proposed, and studied so far.

In the second main section we consider a 2D nonlinear dynamical model of monolayer Scheibe aggregates. We do not focus on the derivation of the equation describing the dynamics, which may be found else where, but on how aspects from nonlinear physics, such as solitary waves, and wave-function collapse, may be applied in the modelling. Thus, deriving parameter values for a specific LB Scheibe aggregate used experimentally, the applicability of the model is studied. The key subject is the temperature dependent nonlinear coherence time of the excitation, and its comparison with the experimentally measured life time. This allows an estimate of the time-interval in which the nonlinear model may describe the dynamics. Finally we consider the role of wave-function collapse in determining the exciton life time.

2. SCHEIBE AGGREGATES

2.1 Langmuir-Blodgett films and Scheibe aggregates

When a drop of oil hits the surface of water it will spread out in a thin layer, in principle only one molecule thick. This effect is not limited to oil, but may be observed for a number of amphiphilic compounds (hydrophobic in one end, and hydrophilic in the other). This was realised already in 1891 by Pockels [13]. However, it was Langmuir who in 1917 first carried out a systematic study of monolayers of amphiphilic compounds at the air-water interface [5]. Consequently the name Langmuir films was coined to refer to these monolayers. Langmuir continued his studies, especially on the possibility of transfering the films onto a solid substrate, but it was not until 1935 that his colleague, Katherine Blodgett, developed the necessary equipment for the consecutive transfer of Langmuir films onto solid support [6]. Consequently the name Langmuir-Blodgett (LB) films has become generally accepted for films produced in this way. During the next 30 years a number of papers appeared on the topic, but it was Kuhn, and his group in Göttingen that really revived an active interest in the immense possibilities of molecular architecture offered by the LB technique, i.e. the creation of complex molecular assemblies with a predetermined structure.

The technique, in its simplest form, is sketched in Figure 1. Coarsely it consists of a trough of water, a floating movable barrier, and some equipment to measure the surface tension and temperature. When initially dropped into water the amphiphilic compound spreads out in a gaseous phase, where the individual molecules are far apart, and hardly interact (Fig. 1b). The floating barrier is then moved to compress the layer until the molecules are ordered orientationally, and a continuous film is obtained (Fig. 1c). In this two-dimensional (2D) liquid crystal phase the film is transferred onto a solid substrate (one way is shown in Fig. 1d). For experimental details see e.g. the early work of Blodgett [6], the review by Blinov [14], or the conference proceedings [15].

In a particular kind of LB films, the molecules are arranged in so-called Scheibe aggregates, which is a special highly ordered, and compact molecular structure, reflecting itself in a strong narrow absorption band, and an almost coinciding fluorescence band of the film. Absorption, and fluorescence spectra of a typical LB Scheibe aggregate are shown in Figure 2 (left), taken from Ref. [16]. The characteristic Scheibe band, which was first discovered by Scheibe from spectroscopic studies of concentrated dye solutions in 1936 [3], is clearly seen.

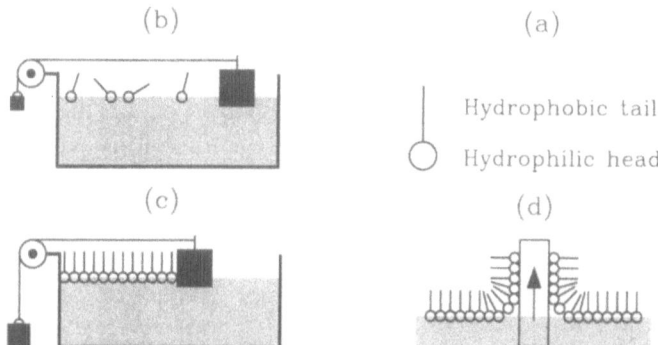

FIG. 1. Simple sketch of the equipment and procedure to produce LB films. (a) Stick diagram representation of the amphiphilic molecule. (b) 2D gaseous phase. (c) 2D liquid crystal phase. (d) One way of depositing the monolayer at the air-water interface onto a solid substrate.

Scheibe aggregate formation of monolayer LB films may be obtained by mixing the dye with inert molecules in the molar ratio 1:1 [16]. The aggregated structure of such a film is also sketched in Figure 2 (right), showing how the chains of the inert molecular filler (in this case octadecane) sticks in the holes left by the hydrophobic chains of the dye (in this case oxacyanine). This closest packing structure results in the largest intermolecular binding energy [16]. In films where the molecular structure is less ordered, and compact the Scheibe band, which gives this kind of LB film its special, and interesting properties, will not be present (e.g. mixing ratios dye:filler other than 1:1).

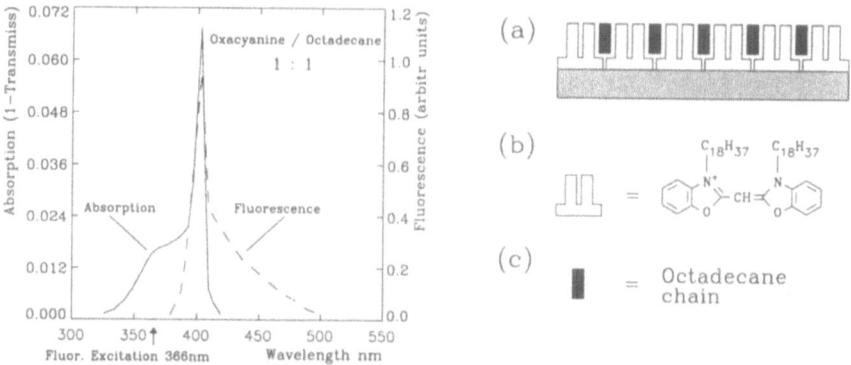

FIG. 2. Left: Absorption and fluorescence spectra of a monolayer LB Scheibe aggregate of an oxacyanine dye mixed with octadecane, acting as inert molecular "filler". Molar mixing ration 1:1, T=300K. Taken from Ref. [16]. Right: (a) Aggregated structure of the film, where octadecane chains sticks in the holes left by the oxacyanine dye molecules, resulting in a closest packing structure. (b) Representation of oxacyanine molecules. (c) Representation of octadecane chains.

2.2 Experiments on Langmuir-Blodgett Scheibe aggregates

Owing to the strong narrow coincident absorption, and fluorescence bands of LB Scheibe aggregates, they have unique energy transfer properties. Furthermore, LB films are particular suited for studies of energy transfer, since the positions, and orientations of the molecules are fixed, and by appropriate planning of the assembly architecture, the distance between donors, and acceptors can be varied systematically.

The confirmation of highly efficient energy transfer over unusually large distances in LB Scheibe aggregates was first made by Möbius and Kuhn in 1979 [17]. They found the efficiency of energy transfer to be proportional to the temperature in the range 20K-300K [17]. In LB films with less ordered, and compact molecular structure than in Scheibe aggregates, the energy transfer is less efficient, and no temperature dependence is observed in the range 20K-300K [17]. In later studies the energy transfer was found to be up to ten times more efficient when donors, and acceptors are situated in the same layer, than if situated in adjacent layers [18].

In this context we are interested in monolayer LB Scheibe aggregates, since they seem to have the best energy transfer properties, and may model the processes of energy harvesting, and transfer in photosynthesis (see next section). Furthermore, this reduces the dimension of the relevant model from 3D to 2D. The main experiments of Möbius, and Kuhn on such films are presented in References [17,18]. For experiments on multilayer LB films consisting of several Scheibe aggregated monolayers see Penner [19], or Sato *et al.* [20].

Theoretically Möbius, and Kuhn found that a coherent exciton model could explain their experimental results [17,18]. In this model it is assumed that on excitation of the aggregate a coherent exciton, extending over a certain number of molecules, is produced. Since the decay rate of a single donor molecule is constant, this results in a radiative decay rate of the exciton, which is proportional to its domain size. Experimentally the radiative decay rate of the exciton was found to be inversely proportional to the temperature, and thus the domain size of the coherent exciton is inversely proportional to the temperature [17,18]. The exciton then moves across the aggregate with its domain size remaining constant. Occasionally it will reach the vicinity of an acceptor molecule, with a certain probability of transfering its energy, which depends on its domain size. This results in the measured temperature dependence of the energy transfer efficiency.

It is important to note that the coherent exciton model put forward by Möbius, and Kuhn is an "average model", and does not state anything about the dynamics. That is, on average the exciton travels with a certain velocity (higher than the speed of sound in the aggregate), since it is supposed to reach an acceptor molecule during its life time, and on average it has a certain domain size. It seems likely that e.g. the domain size will be a decreasing function of time, due to destruction of coherence through thermal fluctuations.

The monolayer LB Scheibe aggregate studied by Möbius, and Kuhn consists of an oxacyanine dye (donors) doped with a thiacyanine dye (acceptors) in the donor:acceptor ratio N, where N may be as high as 10^4. The thiacyanine acceptor molecules are sterically very similar to the oxacyanine donor molecules, and it is therefore assumed to simply substitute a donor molecule in the brickstone work like lattice, in which the donors are arranged. The brickstone work model of the aggregate is depicted in Figure 3 (see also Fig. 2), where also the excitation, and movement of the coherent exciton domain is sketched.

In summary, the main experimental results of Möbius, and Kuhn's work on cyanine Scheibe aggregates is that on illumination of the film

1) a coherent exciton extending over a domain of molecules is excited
2) the domain size is inversely proportional to temperature
3) the exciton stays coherent during its life time
4) the exciton life time is proportional to the temperature
5) the efficiency of energy transfer from donor to acceptor molecules
   is inversely proportional to the temperature.

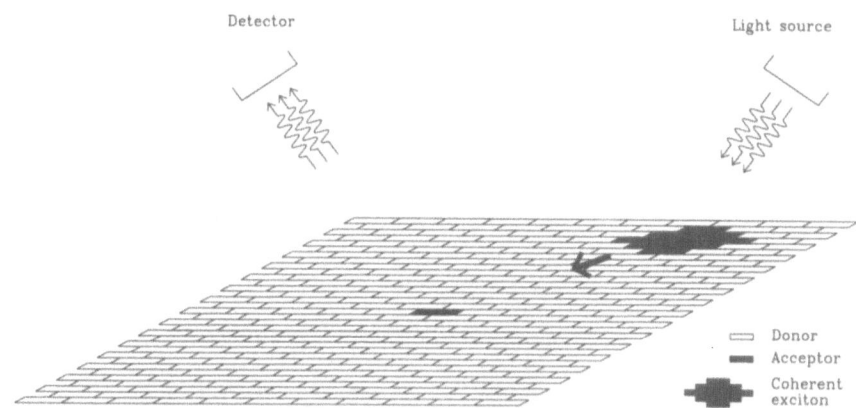

FIG. 3. Brickstone work model of a monolayer LB Scheibe aggregate, where each brick represents a cyanine dye molecule. The excitation of a coherent exciton extending over 9 molecules, and moving over the aggregate towards an acceptor molecule, is sketched.

The temperature-dependent radiative decay rate of excitons in pseudoisocyanine Scheibe aggregates in both solutions, and monolayer LB films, has been studied by Dorn, and Müller [21]. They verified that the fluorescence life time was an increasing function of temperature, indicating a progressive destruction of cooperativity (decreasing domain size of Möbius, and Kuhn's coherent exciton). Similar conclusions were drawn by De Boer, and Wiersma from studies of pseudoisocyanine bromide Scheibe aggregates [22], by Itoh *et al.*, who studied the fluorescence life time of excitons in CuCl microcrystals [23], and by Feldman *et al.* [24], who studied the same effect in GaAlAs quantum wells. Thus the experimental results of Möbius, and Kuhn seem well supported.

2.3 Applications, and the connection with photosynthesis

LB films, and molecular architecture, have many technological applications in electronics (dielectric layers in semiconductors, capacitors, etc.), and especially optoelectronics (photoresistors, photodiodes, photomemory, gratings, 2nd harmonic generation, etc.), which will not be considered in the present context. They may also serve as model systems for fundamental investigations of exciton processes, and structure, and function in 2D, and 3D molecular systems in general. For a general review interested readers are referred to e.g. the review articles of Blinov [14] or Tredgold [25], or Fujihira, and Yamada [26].

The subject of interest here is the highly efficient energy transfer, which may be achieved in multilayer, and especially monolayer LB Scheibe aggregates, [17–20]. Technologically this property is of considerably interest in the photographic industry, where it may be used for (super) sensitization of photographic materials [27]. However, LB Scheibe aggregates are also of great interest for the modelling of various biological processes, since such molecular aggregates abundantly are found in biological systems, where they function as energy funnels for sunlight to be used in photochemical reactions [4]. Thus completely artificial models of photosynthesis may be made by the LB technique [7].

The physical modelling of Scheibe aggregates is at an initial stage, concerned primarily with reproducing the results of Möbius, and Kuhn on excitation, and energy transfer (listed in Sec. 2.2). This corresponds to the initial capture, and transfer of light energy in the photosynthetic process, which occurs in the outer side of the thylakoid membrane [28]. The structure of the thylakoid unit is sketched in Figure 4, representing the membrane by a fluid mosaic model, which is the currently accepted working model [28].

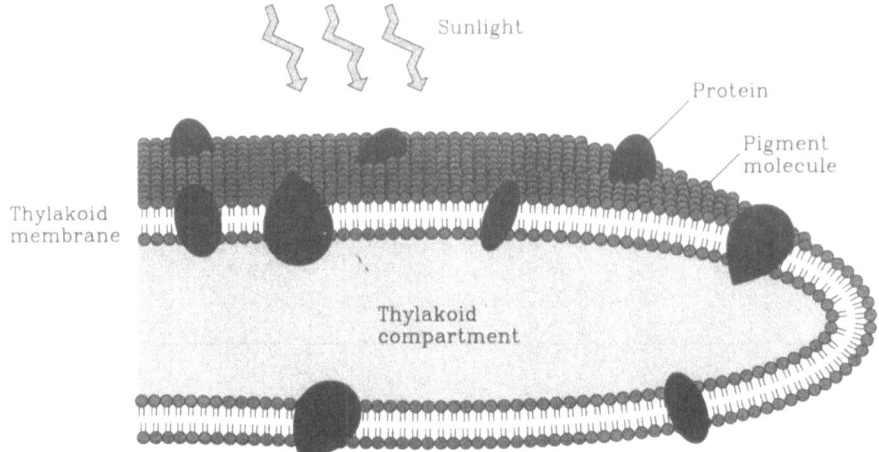

FIG. 4. Cut of the thylakoid unit in chloroplast, where the initial convertion of light energy into chemical energy is performed. The thylakoid membrane is a bilayer of amphiphilic pigment molecules (chlorophyll, carotenoids, etc.), with the inner membrane being hydrophobic, and the outer being hydrophilic. Proteins acting as selective channels or binding sites for filaments of the cytoskeleton, are embedded in the fluid matrix (molecules may move lateratelly) of the bilayer.

The membrane is composed of a bilayer of amphiphilic pigment molecules, with the hydrophobic tails facing each other, exposing only the hydrophilic part to the surrounding fluid (stroma). The hydrophobic core impedes the transport of hydrophilic molecules across the membrane, where as hydrophobic molecules can dissolve in the membrane, and cross it with ease. In each of the layers the pigment molecules may drift, and move around laterally (2D), with a speed averaging about $2\mu m$ per second, where as flip-flopping across the membrane (3rd dimension) is rare [28]. The much larger proteins are embedded in the membrane in a mosaic like structure, and have been shown also to drift around, although much slower than the pigment molecules. They determine the function of the membrane by acting as selective channels for traffic of different kinds of molecules (also hydrophilic), or binding sites for filaments of the cytoskeleton.

Pigments are molecules that by definition strongly absorb visible light. In plants the most important pigment is the chlorophyll (chl) molecule, which occur in the two structural variants, chl A, and chl B. These pigments have two absorption bands, one in the blue/violet region around 450nm, and one in the red/orange region around 670nm. In addition to these structurally distinct types, various spectral forms of chl A has been found, whose maximum absorption of red light is at 660, 670, 678, 685, 690, and 695 to 720 nanometers [28]. The variations are due to different environments of the molecule, e.g. being close to different sorts of proteins. In addition to the chl pigments a number of carotenoid molecules exist with only one absorption band in the blue region. The energy absorbed by these molecules is rapidly transfered to chl, so carotenoids are termed accessory pigments.

Chl A, chl B, and the carotenoids are not dispersed randomly in the thylakoid membrane, but clustered in assemblies of a few hundred pigment molecules. Of the many molecules in each assembly only one pair of chl A molecules can trigger the light reactions by donating their excited electrons to the primary electron acceptor. The location of these specialized chl A molecules in the assembly is called the reaction center. All other pigment molecules in the assembly function collectively as a light-gathering antenna that absorbs photons, and funnels the energy to the reaction center. The antenna complex, together with its reaction center, and primary electron acceptor, is called a photosystem.

Two types of photosystems have been found in the thylakoid membrane, refered to as photosystem I, and II. In this context we will just note, that the two systems are qualitatively similar in structure, but have slightly different absorption spectra, and that their primary electron acceptor feed electrons to different processes. It is however the structure, and function of the antenna which is of interest to us, and thus we concentrate on photosystem I, which is sketched in Figure 5. To the left we see how photons are being absorbed by the antenna complex, and funneled to the reaction center, keeping it constantly in the excited state, where it feeds electrons to the primary acceptor. In photosystem I the reaction center is denoted P700 (P for pigment, and 700 because it absorbs at 700nm). The P700 chl A pair is not different in structure from the many other chl A molecules in the antenna complex, but special due to their location in the thylakoid membrane, where they are bound to specific proteins, and in close proximity to their respective primary electron acceptor.

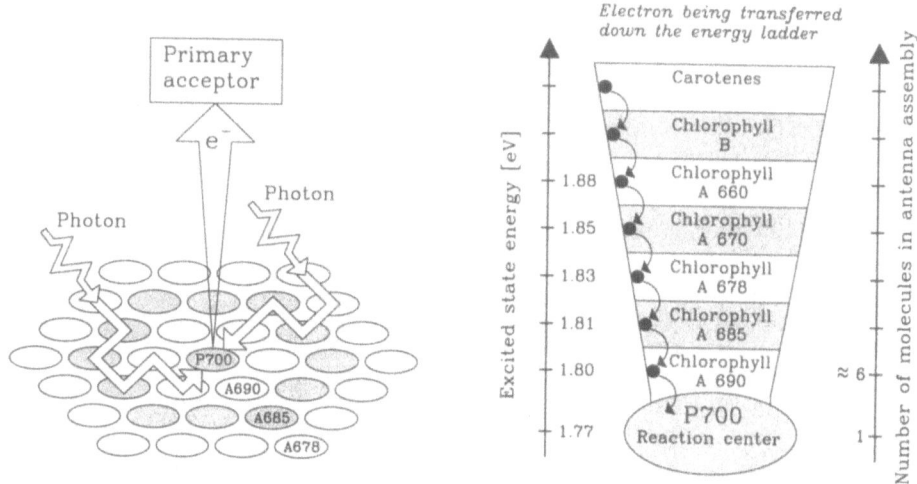

FIG. 5. The energy capture, and transfer process in the plant photosystem I. Left: Photons being absorbed by the large amount of outer pigment molecules (200-300) in an antenna assembly. and transfered to the P700 reaction center, thus keeping it constantly in the excited state, its electrons being transfered to the primary acceptor. Right: Sketch of the electron transport chain, where the directionality is given by the energy gradient.

Note the structure of the antenna complex, where the energy of the excited state of the pigment molecule is progressively lower towards the reaction center, thus giving a degree of directionality or irreversibility to the process, which is essential. In essence the system sacrifice a small amount of energy from each quanta so that nearly all of the quanta can be trapped by the reaction center. This electron transport chain of pigment molecules in photosystem I is sketched to the right in Figure 5.

Let us now elaborate on the connection between the energy harvesting, and transfer in the plant photosystem, and LB Scheibe aggregates. First of all, comparing Figures 1, and 4, we see that the bilayer of amphiphilic pigment molecules in the thylakoid membrane is identical with a two-layer LB film, and may thus be produced artificially. As a matter of fact, the Langmuir films on the surface of water are mentioned as the first artificially produced membranes [28]. However, note that the membrane, as well as the Langmuir film, has a liquid crystal structure, where the molecules are allowed to drift around as mentioned above. This is not the case in the crystalline structure of LB films. However, at physiological temperatures, the time scales of the energy absorption, and transfer to the reaction center in the photosystem, is in the order of 10^{-10}s [29]. Thus the pigments may be regarded as static in this time interval, since their drift velocity only averages about 2μm per second.

Secondly, comparing Figures 2-3, and 5, we see that the molecular structure of the photosystem is identical that of the monolayer LB Scheibe aggregate. In the photosystem the antenna complex consists of several spectral forms of chl A, absorbing at slightly different wave-lengths, and organized to form a unidirectional energy trap. In the LB Scheibe aggregates studied by Möbius, and Kuhn [17] the acceptor was assumed simply to replace a donor in the matrix of donor molecules, thus resulting in a two-component film, and an energy step, in stead of an energy ladder. However, it has not been investigated whether the acceptor molecule alters the absorption spectrum of the nearby donor molecules, thus actually giving a structure as in the antenna complex. Finally, as sketched in Figure 5, textbooks in biology represent the transfer of energy to the reaction center as single electron hopping, conflicting the coherent domain observed in LB Scheibe aggregates.

In any case, since Nature has had considerably more time than mankind to optimize the efficiency of its functions, such as photosynthesis, the structure of the antenna complex in the photosystem could indicate how to produce LB Scheibe aggregates with optimal efficiency of energy absorption, and transfer.

2.4 Physical models

While in the past much work on Scheibe aggregates was aimed at understanding the spectroscopy [30], recently the focus has changed to comprehend their dynamical properties. Thus Mukamel and co-workers [31] have established the existence of a temperature dependent coherence size of the exciton from numerical calculations in 1D, provided that the exciton dephasing time scale is much shorter than the fluorescence life time, and that the exciton-phonon coupling is weak. When the coupling is strong, purely incoherent motion is observed, with an effective coherence size of only one molecule, independent of temperature.

The conclusion that the experimental results may only be reproduced if the exciton-phonon coupling is weak is also reached by Bartnik et al., who studied a purely quantum mechanical model of Scheibe aggregates [32]. Incorporating an acceptor molecule simply as a donor molecule, but with a slightly lower resonant frequency, the acceptor is found to act as an energy trap, provided the parameter values satisfy certain constraints. It is stressed that the 2D structure is special because it leads to rather weak constraints, compared with 1D, and 3D structures. Furthermore, it is proposed that both coherent, and incoherent (diffusive) exciton motion may participate in the energy transfer, initially being coherent, but eventually becoming incoherent.

Huth et al. [33] first proposed a nonlinear model of Scheibe aggregates, where the coherent domain observed experimentally is supposed to arise from nonlinear dynamical effects in the aggregate. Thus the model is based on the cubic nonlinear Schrödinger (NLS) equation [34] with energy transfer occuring through solitary waves. The model was formulated in 1D, where the NLS equation is integrable, and has soliton solutions [34]. However, it was supposed to hold also in 2D with the initial conditions becoming ring waves. In 2D the NLS equation is no longer integrable, and thus the ring wave may collapse at the center [35]. Christiansen and co-workers [36] used this collapse of the ring waves in the nonlinear model to predict the exciton life time at room temperature. For a general review on collapse phenomena in NLS equations see e.g. Rasmussen, and Rypdal [37].

The nonlinear model of Huth *et al.* requires a strong coupling between excitons, and phonons, which contrasts the result obtained by Mukamel, Bartnik, and their collaborators, that the coupling must be weak. It is noted that the definition of weak, and strong coupling is not clear in this connection. Furthermore, the model does not take thermal effects into account. Thus, as for the Davydov soliton in protein [8], two natural questions arises: For how long time can solitary waves exist in Scheibe aggregates at a certain temperature? At which temperatures is it sufficient to explain the dynamics?

In a recent work of O. Bang *et al.* an attempt is made to answer these questions [12]. Here a nonlinear model of monolayer LB Scheibe aggregates is derived from a postulated Hamiltonian, and thermal fluctuations of the molecules are taken into account. In the continuum limit the equation describing the exciton dynamics becomes the 2D NLS equation with multiplicative colored noise, corresponding to destruction of coherence through scattering by the fluctuating phonons. The ground state solitary wave (GS) solution to the 2D NLS equation without noise is assumed to be excited by the impinging photon, modelling Möbius, and Kuhn's coherent exciton. Deriving parameter values for the oxacyanine LB Scheibe aggregate studied by Möbius, and Kuhn [17,18], the temperature nonlinear dependent coherence time of the GS solution is found numerically. From comparison with the experimentally measured life time, the temperature regime is identified, in which this nonlinear model may completely describe the dynamics. Unless the exciton-phonon coupling, and the dispersive dipole-dipole coupling, are significantly lower than the estimated values, this regime seems to be limited to extremely low temperatures ($< 3K$). Thus once again it is found that the exciton-phonon coupling has to be weak.

3. THE NONLINEAR MODEL

3.1 Derivation

The derivation follows closely that of the Davydov model of energy transport in protein [8], except that the present system is two-dimensional, and only the optical phonon branch is considered. Furthermore, thermal fluctuations of the molecules are taken into account. Only the essential steps are given, since details may be found in Reference [12].

Thus the following 3-component Hamiltonian energy operator is considered

$$\hat{H} = \left\{ \sum_n E_0 \hat{B}_n^\dagger \hat{B}_n - \sum_{n \neq p} \sum_p J_{np} \hat{B}_n^\dagger \hat{B}_p \right\} + \left\{ \frac{1}{2} M \sum_n [\dot{u}_n^2 + \omega_0^2 u_n^2] \right\} + \left\{ \chi \sum_n u_n \hat{B}_n^\dagger \hat{B}_n \right\} , \quad (1)$$

where n, and p ranges from 1 to f, f being the total number of molecules. The 1st component of $\hat{H}$ is the exciton energy operator. Here $\hat{B}_n^\dagger$ ($\hat{B}_n$) are creation (annihilation) operators, E_0 is the molecular site energy, and $-J_{np}$ the dipole-dipole interaction energy. The 2nd component is the phonon energy, which is treated classically. Here M is the molecular mass, and $u_n(t)$ represents the elastic degree of freedom at site n. The molecules are approximated by Einstein oscillators, all oscillating at the same frequency, ω_0, and thus only optical phonons are taken into account [38]. The last component is the exciton-phonon interaction energy operator, where $\chi = \mathrm{d}E_0/\mathrm{d}l$ is the exciton phonon-coupling parameter, l being the distance between nearest neighbouring molecules in the Scheibe aggregate. Thus χ measures the amount of energy required to displace a molecule in the aggregate/

Using a nonequilibrium density matrix approach [39] the coupled set of equations describing the dynamics of excitons, and phonons have been found to [12]

$$M\ddot{u}_n + M\lambda\dot{u}_n + M\omega_0^2 u_n = \chi|\phi_n|^2 + \eta_n , \quad (2)$$

$$i\hbar\dot{\phi}_n + \sum_{p \neq n} J_{pn}\phi_p + \chi u_n \phi_n = 0 , \quad (3)$$

where $|\phi_n(t)|^2$ is the probability for finding the excitation at site number n. To describe the interaction of the phonon system with a thermal reservoir at temperature T, damping, λ, and noise, $\eta_n(t)$, have been included in Eq. (2), after its quantum-mechanical derivation. $\eta_n(t)$ is assumed to be gaussian white noise with zero mean, and the autocorrelation function

$$\langle \eta_n(t)\eta_{n'}(t')\rangle = 2M\lambda kT \cdot \delta(t - t')\delta_{nn'} , \qquad (4)$$

where the strength is chosen according to the classical fluctuation-dissipation theorem [40], which assures thermal equilibrium. The damping coefficient, λ, is the linewidth of the infra-red absorption band, k is Boltzmann's constant, and $< \ldots >$ denote ensemble average.

Note that the noise in Eq. (2) is additive, implying that phonons are constantly being created, and destroyed thermally. In contrast the noise in Eq. (3) is effectively multiplicative, implying that only the coherence of the exciton is disturbed by the thermal fluctuations. This is reflected in that the exciton number, N, defined as the total probability for finding the exciton in the system,

$$N\{\phi_n\} \equiv \sum_n |\phi_n(t)|^2 = 1 , \qquad (5)$$

is a conserved quantity in Eq. (3).

Eqs. (2-3) resembles the one-dimensional Davydov equations solved numerically by Lomdahl, and Kerr [41]. Their conclusion was that Davydov solitons do not exist for sufficiently long time at 300K to be of physical interest. In this case we aim at deriving a single equation describing the dynamics of the exciton system.

The solution to Eq. (2) is easily written down

$$u_n(t) = c_1 e^{-\lambda t/2}\cos(\omega_1 t) + c_2 e^{-\lambda t/2}\sin(\omega_1 t) + \qquad (6)$$
$$\frac{\chi}{M\omega_1}\int_0^t |\phi_n(\tau)|^2 e^{-\lambda(t-\tau)/2}\sin(\omega_1[t - \tau])d\tau + s_n(t) ,$$

where c_1, and c_2 are constants, and the perturbed eigen frequency, ω_1, is given by

$$\omega_1 = \omega_0\sqrt{1 - \epsilon^2} , \quad \epsilon = \lambda/(2\omega_0) . \qquad (7)$$

The parameter, ϵ, is normally small for crystals. The noise, $s_n(t)$, which is no longer white, but strongly colored, is most conveniently expressed in Fourier space

$$\tilde{s}_n(\omega) = \tilde{f}(\omega)\,\tilde{\eta}(\omega)/M , \quad \tilde{f}(\omega) = 1/(\omega_0^2 - \omega^2 + i\lambda\omega) , \qquad (8)$$

$\tilde{f}(\omega)$ being the Lorentzian, and tilde denoting the Fourier transformation of a function.

The first two terms on the r.h.s. of Eq. (6) are the solutions of the homogeneous part of Eq. (2). They are fast oscillating, with the perturbed eigen frequency, $\omega_1 \leq \omega_0$. Assuming that $|\phi_n(t)|^2$ varies slowly compared with $\sin(\omega_1 t)$ the integral in Eq. (6) can be evaluated. This assumption also allows us to remove the homogeneous part by averaging over a period of oscillation, and obtain the solution

$$\bar{u}_n(t) \equiv \frac{1}{t_1}\int_t^{t+t_1} u_n(\tau)d\tau = \frac{\chi}{M\omega_0^2}|\phi_n(t)|^2 + \bar{s}_n(t) , \quad t_1 = 2\pi/\omega_1 . \qquad (9)$$

Thus the modulus of the exciton wave function is the slowly varying mean value of the rapidly fluctuating phonons. The averaged noise is still exact. In the averaging it was further used that ϵ is small, by treating $e^{-\lambda t/2}$ as a constant during the time interval t_1.

Averaging Eq. (3) over a period of oscillation as above, still under the assumption of slowly varying $|\phi_n(t)|^2$, leads to the replacement of $u_n(t)$ by $\bar{u}_n(t)$. Inserting Eq. (9) we get

$$i\hbar\dot{\phi}_n + \sum_{p\neq n} J_{pn}\phi_p + V|\phi_n|^2\phi_n = -\sigma_n\phi_n \,, \tag{10}$$

where the nonlinearity parameter, V, is given by

$$V = \chi^2/(M\omega_0^2) \,. \tag{11}$$

The resulting noise, $\sigma_n(t)$, is again given in Fourier space

$$\tilde{\sigma}_n(\omega) = \frac{\chi}{M}\,\frac{\sin(\pi\omega/\omega_1)}{\pi\omega/\omega_1}\,e^{i\pi\omega/\omega_1}\,\hat{f}(\omega)\,\tilde{\eta}_n(\omega) \,. \tag{12}$$

Its specific form, and different approximations, will be considered in Section 3.3.

Without noise Eq. (10) is known as the discrete self-trapping (DST) equation [42], which originally arose from the study of Davydov-like solitons in crystalline acetanilide [43]. It is a simple system of anharmonic oscillators coupled together through linear dispersive interactions. The DST equation has been successfully applied to describe the stretching vibrations of the hydrogen bonds in small polyatomic molecules like water, ammonia, methane, and benzene [44]. Also self-trapping of vibrational energy in hydrogen bonded polypeptide crystals, like acetanilide, and N-methylacetamide, is explained by the DST model [43,45].

Since we are interested in coherent cooperative behavior, and solitary waves, the large number of molecules in the 2D aggregate makes the DST equation impractical to study. Thus, taking only nearest neighbour coupling, J_0, into account, and making the gauge transformation, $\phi_n \to e^{i4J_0t/\hbar}\phi_n$, we take the continuum limit of Eq. (10), and arrive at the 2D NLS equation with noise

$$i\hbar\phi_t + l^2 J_0\nabla^2\phi + l^2 V|\phi|^2\phi = -l^2\sigma\phi \,. \tag{13}$$

Here $|\phi(x,y,t)|^2 = |\phi_n(t)|^2/l^2$ is now a probability density,

$$N\{\phi\} = \int\int |\phi(x,y,t)|^2 dx dy = 1 \,, \tag{14}$$

and $\sigma(x,y,t) = \sigma_n(t)/l^2$ a noise density.

3.2 Parameter values for a monolayer oxacyanine LB Scheibe aggregate

In this section the values of the model parameters are estimated for the monolayer oxacyanine Scheibe aggregate studied by Möbius, and Kuhn [17,18]. The aggregate is represented as a brickstone work [16] (see also Fig. 2), each brick being a molecule, and the molecules are represented as extended dipoles [46], as shown in Figure 6.

From Figure 6 and the experimental data listed in Table I, the distance between nearest neighbouring molecules, l, and the mass, M, of an oxacyanine molecule, are found to

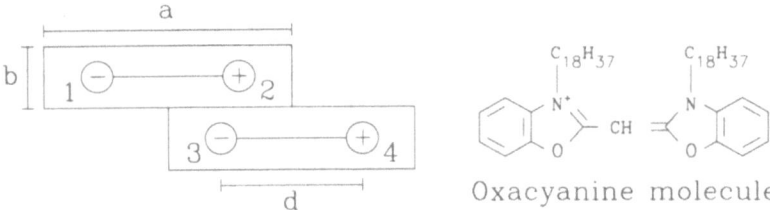

FIG. 6. Left: Brickstone work model of monolayer oxacyanine LB Scheibe aggregate, with each molecule represented as an extended dipole. Right: Oxacyanine molecule.

$$l = \sqrt{a^2/4 + b^2} = 8.72\text{Å} \, , \quad M = 49[C] + 83[H] + 2[N] + 2[O] = 1.2 \cdot 10^{-24}\text{kg} \, , \qquad (15)$$

where $[\cdots]$ denotes the atomic mass.

Since l is comparable with the size of the molecules a point dipole approximation would be inadequate. Instead the molecules are approximated by extended dipoles [16], with the equivalent dipole charge μ_{tr}/d, where μ_{tr} is the transition dipole moment, and d the equivalent dipole length. This gives interaction energies in good agreement with the results from quantum mechanical computations, by which Czikkely $et\ al.$ found that $d = 8.9\text{Å}$, and $\mu_{tr} = 3.13 \cdot 10^{-29}\text{Cm}$ for oxacyanine molecules [46]. The dipole-dipole interaction energy, J_0, can thus be found from static calculations

$$J_0 = -\frac{(\mu_{tr}/d)^2}{4\pi\epsilon_r\epsilon_0} \left[\frac{1}{r_{13}} + \frac{1}{r_{24}} - \frac{1}{r_{23}} - \frac{1}{r_{14}} \right] = 3.6 \cdot 10^{-21}\text{J} \, , \qquad (16)$$

where r_{np} is the distances between the dipole charges in Figure 6.

TABLE I. Physical parameters for the monolayer oxacyanine Scheibe aggregate used in Möbius and Kuhn's experiments [17,18].

Brickstone length	a	15.5Å
Brickstone width	b	4.0Å
Dipole length	d	8.9Å
Transition dipole moment	μ_{tr}	$3.13 \cdot 10^{-29}$Cm
Dielectric constant of the aggregate	ϵ_r	2.25

The Davydov model of energy transport in protein by solitons [8] has been studied for more than two decades, and complete agreement has not yet been reached on the value of the exciton-phonon coupling parameter, χ. The best current estimate is in the range 35-62pN [47]. The application of nonlinear dynamical models in the study of Scheibe aggregates is fairly new, and correspondingly no estimates of χ have been published yet. For sure the compact closest packing structure of Scheibe aggregates implies that a larger energy is required to displace a molecule, than in the case of protein in its natural surroundings. Thus we will consider values in the range of 1pN to 1nN.

The phonon eigenfrequency, ω_0, and the linewidth of the infra-red absorption band, λ, are not known. However, in crystals common values are of the order of $\omega_0 = 10^{12}\text{s}^{-1}$, and $\lambda = 10^{11}\text{s}^{-1}$, which will be used as order of magnitude estimates. The nonlinearity parameter, V, may then finally be found from Eq. (11). All parameters in the model are listed in Table II. Their values are assumed to be temperature independent. Only the strength of the noise, $\sigma(x, y, t)$, will depend on temperature in our model.

TABLE II. Estimated values of the parameters in the nonlinear model of the monolayer oxacyanine Scheibe aggregate used in Möbius and Kuhn's experiments [17,18]. * indicates that the value is an order of magnitude estimate.

Molecular mass	M	$1.21 \cdot 10^{-24} \text{kg}$
Distance between molecules	l	8.72Å
Phonon eigen frequency*	ω_0	10^{12}s^{-1}
Linewidth of infra-red absorption band*	λ	10^{11}s^{-1}
Exciton-phonon coupling parameter*	χ	$1\text{pN} - 1\text{nN}$
Dipole-dipole coupling energy	J_0	$3.6 \cdot 10^{-21} \text{J}$
Nonlinearity parameter*	V	$8.3 \cdot 10^{-25} \text{J} - 8.3 \cdot 10^{-19} \text{J}$

3.3 Approximations of the colored noise

Usually, when noise is simulated numerically, it is defined in Fourier space, and then transformed back to real space [48]. In two dimensions, and for the large number of molecules we want to consider, this would require an excessive amount of storage capacity. However, in the limiting cases of time independent, and pure white noise, this problem may be overcome. The time independent noise is only calculated initially, and the white noise can be generated at each time step by a random noise generator. In Figure 7 $|\tilde{\sigma}_n(\omega)|^2$ is shown, and the two approximations indicated.

We first approximate the power spectrum of the noise in Eq. (10), $\sigma_n(t)$, with a delta function with the strength equal to the integral over the interval $-\omega_1 \leq \omega \leq \omega_1$,

$$|\tilde{\sigma}_n(\omega)|^2 \approx \frac{4\omega_1}{3}|\tilde{\sigma}_n(0)|^2 \delta(\omega) = \frac{8\omega_1 \chi^2 \lambda kT}{3M\omega_0^4}\delta(\omega) , \qquad (17)$$

where $|\tilde{\sigma}_n(0)|^2$ is obtained from Eq. (12). This simply corresponds to disorder in the arrangement of the molecules in the aggregate. Transforming back, and making the continuum approximation the autocorrelation function of the noise density, $\sigma(x, y, t)$, in Eq. (13), is found to

$$\langle \sigma(x, y, t) \sigma(x', y', t') \rangle = D_{disorder} \cdot \delta(x - x')\delta(y - y') . \qquad (18)$$

Here the temperature dependent variance, $D_{disorder}$, is given by

$$D_{disorder} = \frac{4\omega_1 \chi^2 \lambda kT}{3\pi M \omega_0^4 l^2} . \qquad (19)$$

At $T = 300$K, and for $\chi = 0.2$nN the strength of the noise is $\sqrt{l^2 D_{disorder}} = 2.4 \cdot 10^{-21}$J, which is of the same order of magnitude as J_0. With this approximation of the noise an analytical estimate of the nonlinear coherence time cannot be obtained.

In the pure white noise approximation the power spectrum of the noise in Eq. (10), $\sigma_n(t)$, is constant, and given by the mean value in the interval $-\omega_1 \leq \omega \leq \omega_1$

$$|\tilde{\sigma}_n(\omega)|^2 \approx \frac{2}{3}|\tilde{\sigma}_n(0)|^2 = \frac{4\chi^2 \lambda kT}{3M\omega_0^4} . \qquad (20)$$

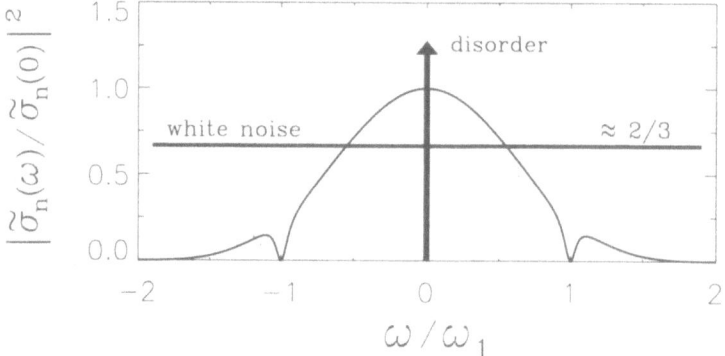

FIG. 7. Normalized power spectrum of the noise, $|\tilde{\sigma}_n(\omega)/\tilde{\sigma}_n(0)|^2$. White noise ($|\tilde{\sigma}_n(\omega)|^2 = 2/3$), and time independent noise ($|\tilde{\sigma}_n(\omega)|^2 = 4\omega_1\delta(\omega)/3$) approximations are indicated. Parameters are taken from Table II. Note that the form the spectrum depends on the specific values of ω_0, and λ.

In this case the autocorrelation function for the noise density in Eq. (13) becomes

$$\langle\, \sigma(x,y,t)\, \sigma(x',y',t')\, \rangle = D_{white} \cdot \delta(x-x')\delta(y-y')\delta(t-t') \,, \tag{21}$$

where the temperature dependent variance, D_{white}, is given by

$$D_{white} = \frac{4\chi^2\lambda kT}{3M\omega_0^4 l^2} \,. \tag{22}$$

Clearly the white noise approximation is poorer than the disorder approximation, since its energy content is infinite, in contrast to the finite energy content in the real noise. Thus it is expected that the white noise will lead to extremely short nonlinear coherence times. However, it represents a limiting case which allows for instructive analytical results to be obtained (see also Secs. 3.5-6).

3.4 Choosing the initial condition

In two (and more) dimensions the NLS equation is not integrable, and possess unstable solutions that may collapse in finite time, the amplitude becoming infinite at some place in space [37]. In describing collapse in the NLS equation its ground state solitary wave (GS) solution plays a vital role. Due to the nonintegrability of the 2D NLS equation, its exact GS solution has not been found. However, its norm, $N \equiv N_s$, has been found numerically to $N_s = 11.7(J_0/V)$ [50], and it has been shown that solutions with $N > N_s$ **may** collapse (can be prevented by a phase factor), while solutions with $N < N_s$ **always** disperses [37]. The GS solution itself is unstable, since a small perturbation of its amplitude will make it either disperse or collapse.

Using a variational approach [51], Anderson *et al.* have derived a good approximation of the GS solution to the 2D NLS equation [52]

$$\phi_s(r,t) = A_0\mathrm{sech}(r/B_0)\, \exp(iJ_0t/\hbar) \,, \tag{23}$$

here given with zero initial velocity, and center at $r = \sqrt{x^2 + y^2} = 0$. The initial amplitude, A_0, and the initial width, B_0, are given by

$$A_0 = \frac{1}{l}\sqrt{\frac{J_0}{V}\frac{12\ln(2)}{4\ln(2)-1}}\ ,\ B_0 = l\sqrt{\frac{2\ln(2)+1}{6\ln(2)}}\ ,$$ (24)

and from Eq. (14) it is indeed found that $N\{\phi_s\} \approx N_s$.

In view of the above it seems obvious to model the initial excitation of the Scheibe aggregate by $\phi_s(r,0)$, since it is the most stable initial condition known for the 2D NLS equation. Furthermore, it represents a localized domain of coherence, as suggested by the experiments of Möbius, and Kuhn [17]. According to Eq. (14) ϕ_s must then be normalized, $N_s = 1$, and thus the following condition must be satisfied by the parameter values

$$N\{\phi_s(r,0)\} = 1 \quad \Rightarrow \quad V/J_0 = 11.7\ .$$ (25)

Using the values of M, ω_0, and J_0 given in Table II Eqs. (25), and (11) gives the specific value of the exciton-phonon coupling, $\chi = \chi_s$, where χ_s is of the order of 0.2nN i.e. a realistic value. The initial condition $\phi_s(r,0)$ will then evolve in time, and propagate over the aggregate, its dynamics being determined by the 2D NLS equation with noise, Eq. (13), as long as it stays coherent. When the coherence is lost the approximations made in the derivation are no longer valid. We note that with the GS solution, $\phi_s(r,0)$, as initial condition the model, given by Eq. (13), can predict the nonlinear coherence time of the exciton in the presence of thermal fluctuations, but cannot predict its radiative life time.

Let us elaborate on what happens if we instead of $\phi_s(r,0)$ have a more general initial condition, and thus allow the parameters V, and J_0 to assume arbitrary values. Then two situations may occur: If V/J_0 is significantly less than 11.7, N will be less than N_s, and the exciton will quickly disperse, with a coherence time much less than can be achieved with $V/J_0 = 11.7$, and $\phi_s(r,0)$ as initial condition. This will e.g. occur if the parameters have the values listed in Table II, and the exciton-phonon coupling is less than χ_s.

If however V/J_0 is significantly larger than 11.7, N will be larger than N_s, and the exciton wave-function will collapse in finite time. This will occur if e.g. the exciton-phonon coupling is larger than the χ_s, which is not unrealistic, as discussed in Section 3.2. The collapse regime is in fact very interesting, since it might lead to an increased nonlinear coherence time, and allow for an estimation of the exciton life time, which is in agreement with experimental results. We will consider this regime in closer detail in Section 3.6.

3.5 Nonlinear coherence time for the ground state solitary wave solution

In References [12], and [49] the 2D NLS equation with noise, Eq. (13), was solved numerically using the disorder, and white noise approximations, respectively. The approximation of the GS solution, ϕ_s, given by Eqs. (23-24), was used as initial condition, and the nonlinear coherence time was defined as the time during which the width of the solution increases by a factor of e. In both cases the temperature dependence of the numerically found nonlinear coherence time followed a power law with a prefactor depending on the parameter values. For the values given in Table II, and $\chi = \chi_s$, the specific expressions are

Disorder   : $t_{coh} = 230\text{fs} \cdot (300\text{K}/T)^{0.24}$, (26)

White noise : $t_{coh} = 25\text{fs} \cdot (300\text{K}/T)^{0.40}$. (27)

In the case of white noise, and the same parameter values, the nonlinear coherence time may be estimated analytically to $t_{coh} = 3.7\text{fs} \cdot (300\text{K}/T)^{1/3}$ [49]. Thus the power law is confirmed analytically with an almost correct power, but a prefactor that deviates by a factor of 7.

The experiments by Möbius, and Kuhn have shown that the exciton stays coherent during its life time [17,18]. Thus, in order for the nonlinear model to be relevant, the predicted nonlinear coherence time must be larger than the life time found experimentally [18]

$$t_{life} = 10^{-10}\text{s} \cdot (T/300\text{K}) . \tag{28}$$

In Fig. 8 (left) the numerically found coherence times given by Eqs. (26-27), and the life time given by Eq. (28), are plotted versus temperature. The temperature where the curves for t_{life}, and t_{coh} for the disorder approximation intersect is denoted T_0. The disorder approximation is used because it has about the same energy content as the real colored noise. We note that the real T_0 then probably will be a little smaller than predicted. At higher temperatures than T_0 the predicted coherence time is smaller than the life time found experimentally. Thus the nonlinear model cannot completely describe the dynamics in this temperature regime. At temperatures lower than T_0 the coherence time is larger than the experimentally found life time. Thus the nonlinear model is relevant in this temperature regime, where it may describe the dynamics of the coherent exciton. With the parameter values given in Table II, and $\chi = \chi_s$, $T_0 \approx 2$K is found from the numerical simulations.

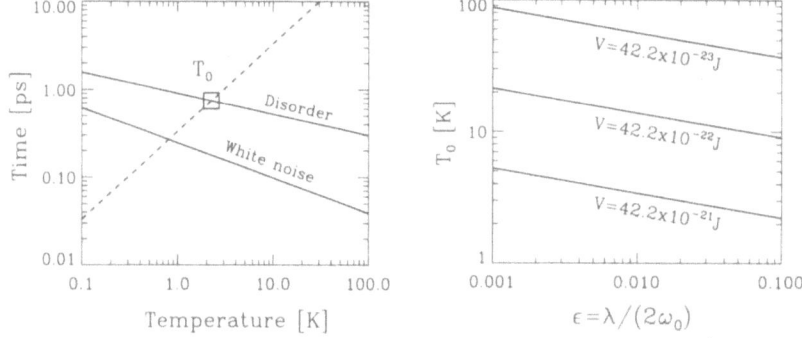

FIG. 8. Left: Temperature dependence of the coherence time (solid curves), given by Eqs. (26-27), and the life time (dashed curve), given by Eq. (28). The approximation of the noise is indicated on the figure. Parameters are taken from Table II, with $\chi = \chi_s$. Right: Temperature, T_0, versus ϵ, as given by Eq. (32), for different values of the nonlinearity parameter, V.

The temperature T_0 depends on the model parameters, since the coherence time can be written more generally as

$$t_{coh} = \alpha \cdot \left(\frac{\hbar}{J_0}\right) \cdot \left(\frac{J_0^2}{l^2 D_{disorder}}\right)^\beta , \tag{29}$$

where $\alpha = 8.07$, and $\beta = 0.24$ are found numerically [12]. Inserting the parameter values given in Table II, with $\chi = \chi_s$, yields Eq. (26). In evaluating the variance $D_{disorder}$ in Eq. (19) the area under the curve in Figure 7 was assumed to be $A \approx 4/3$. However, it depends on the value of the parameter ϵ, and thus Eq. (19) must be replaced by

$$D_{disorder} = \left(\frac{\omega_1 \chi^2 \lambda kT}{\pi M \omega_0^4 l^2}\right) \cdot A(\epsilon) , \tag{30}$$

where the area, $A(\epsilon)$, can be written as a function of $\epsilon = \lambda/(2\omega_0)$ only

$$A(\epsilon) = \int_{-1}^{1} \left[\frac{\sin(\pi x)}{\pi x} \right]^2 \cdot \left[\frac{1}{[1 - (1 - \epsilon^2)x^2]^2 + 4\epsilon^2(1 - \epsilon^2)x^2} \right] dx .$$ (31)

With the parameter values given in Table II, and $\chi = \chi_s$, the area becomes $A \approx 4/3$, and Eq. (30) reduces to Eq. (19).

The dependence of the temperature T_0 on the model parameters can now be found by combination of Eqs. (29-31)

$$\frac{T_0}{300K} = g(\epsilon) \cdot \left(\frac{42.2 \cdot 10^{-21} \text{J}}{V} \right)^{0.61} ,$$ (32)

where the function, $g(\epsilon)$, is given by

$$g(\epsilon) = 8 \cdot 10^{-4} \cdot \left(\epsilon \sqrt{1 - \epsilon^2} A \right)^{-0.19} .$$ (33)

Here the normalization condition, Eq. (25), has been used to eliminate the ratio V/J_0. In Fig. 8 (right) the dependence of T_0 on ϵ is shown for different orders of magnitude of V. It is seen that decreasing ϵ by a factor of 100 only increases T_0 by 3.5K, where as a decrease in the nonlinearity parameter, V, has a much stronger effect.

A value of T_0 around room temperature can e.g. be obtained with an exciton-phonon coupling of $\chi = 4.5$pN. To fulfil the normalization condition, Eq. (25), the dipole-dipole coupling, J_0, must then be reduced by a factor of about 2500. Since J_0 is one of the better determined parameters (see Sec. 3.2), this seems unrealistic. Thus we may conclude that using the GS solution as initial condition nonlinear effects play no important role at biological temperatures. In the following section we consider a more general initial condition.

3.6 The role of collapse

For values of the exciton-phonon coupling parameter higher than χ_s the system may enter the collapse regime (see Sec. 3.4). This would increase the nonlinear coherence time, since the collapse process introduces a contraction effect that counteract the spreading due to thermal fluctuations. The question is then at which temperatures a given initial condition will collapse or disperse, for a given set of parameter values. When the noise is approximated by white noise, a qualitative answer to this question can be obtained analytically [53,54].

Let us assume that we have an initial condition that collapses at zero temperature (no noise), within a certain collapse time. Introducing thermal fluctuations would then tend to counteract the cooperative contraction of the collapse process. Increasing the temperature (or the variance of the noise) would therefore lead to an increase of the collapse time, and above a certain temperature, T_1, collapse would no longer occur.

Using a variational approach [51] the dependence of the collapse time, t_{col}, on the variance of the white noise, D_{white}, has been found to [53]

$$t_{col}(D_{white}) = \frac{t_{col}(D_{white} = 0)}{1 - \frac{l^2}{\hbar J_0} D_{white} \Gamma(\frac{V}{J_0})} ,$$ (34)

where the parameter, Γ, only depends on the ratio V/J_0. Here we clearly see how the collapse is slowed down as the noise becomes stronger. The value of the variance, above which collapse no longer occur, is given by

$$\frac{l^2}{\hbar J_0} D_{white} = \frac{1}{\Gamma(\frac{V}{J_0})} \approx \gamma \left(\frac{V}{J_0} - 11.7\right) , \tag{35}$$

where $\gamma \approx 5.1$ has been found numerically by interpolation [53]. Introducing Eq. (22), and the parameter values in Table II, the ratio V/J_0 correponding to this value of the variance, is found to depend on temperature as

$$\frac{V}{J_0} = \frac{11.7}{1 - T/T_1} , \tag{36}$$

where the limiting temperature, T_1, is much higher than room temperature

$$T_1 = \frac{3\pi\gamma\hbar\omega_0^2}{2k\lambda} \approx 1840\text{K} . \tag{37}$$

We stress that these results are only qualitative due to the approximation of the noise, and other approximations made in the analytical derivation [53].

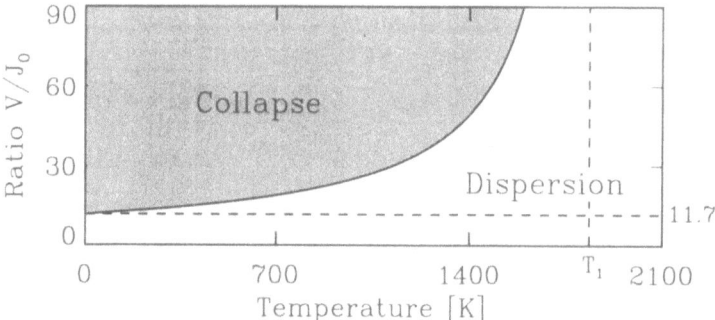

FIG. 9. V/J_0 versus temperature, as predicted by Eq. (36), separating regions of collapse, and dispersion of the exciton wave function.

The ratio V/J_0 as function of temperature, given by Eq. (36), is depicted in Figure 9, and the regimes of collapse, and dispersion indicated. For $V/J_0 \leq 11.7$ the exciton will always disperse, while for $V/J_0 > 11.7$ collapse may occur, depending on the temperature. According to the discussion about the parameter values given in Section 3.2, the strength of the exciton-phonon coupling is not known, and might be higher than χ_s, found when the GS solution is used as initial condition. Assuming a stronger coupling, $\chi = 0.3\text{nN}$, we find that $V/J_0 = 20.7$. Thus the initial pulse will collapse for temperatures below $T \approx 800\text{K}$ according to Eq. (36), and according to Eq. (34) the collapse time will increase with temperature.

In physical systems collapse will be prevented by effects such as dissipation or higher order dispersion or nonlinearities, which were not taken into account in the derivation. For Langmuir waves in plasmas [10], for example, the Landau damping is neglected in the derivation of the NLS equation. However, when the spatial scales of the wave packet becomes too small (comparable to the Debye length), this damping becomes significant, and prevents further growth of the wave amplitude by dissipating its energy [55]. A similar effect in nonlinear optics is the Rayleigh scattering, which becomes important when the wavelength of the electromagnetic field becomes too small [56]. Also, from the results presented here, we see that thermal fluctuations may prevent a collapse from developing.

Thus collapse may be viewed as a strongly nonlinear dissipation mechanism. Since the structure of the singularity near the center of the collapse defines the effectiveness of this dissipation, detailed knowledge about the dynamics of a system before it enters the collapse regime, is essential for the understanding of the properties of the system. The collapse phenomenon has therefore become very important to study.

In our case it is tempting to connect the collapse of the exciton wave function with a strong dissipation of energy, and assume that it dominates the linear radiative dissipation. In this case one may further connect the exciton life time with the collapse time of its wave function. According to Eq. (34) the life time would then increase with temperature, which is exactly what has been found experimentally [17,18,21–24]. If the connection between the collapse time, and life time can be established, it may validate the nonlinear model of Scheibe aggregates.

4. CONCLUSION

To give a basic knowledge of Scheibe aggregates, we have presented a short review of its structure, and special properties, how it may be produced, and why it has a close relation to the biological process of photosynthesis. We have furthermore presented a short résumé of the major attempts to model the exciton dynamics in Scheibe aggregates, and in the spirit of the conference, we have concentrated on the nonlinear model, based on the two-dimensional cubic nonlinear Schrödinger equation with multiplicative noise.

A number of new aspects has been presented. Modelling the initial excitation of the Scheibe aggregate by the ground state solitary wave solution to the noise free equation seems not to be able to describe the full dynamics, especially at room temperature, since the nonlinear coherence time of the solution is too short. Another regime of the model parameters leads to collapse of the exciton wave-function. From priliminary studies this regime appears highly interesting, because it increases the nonlinear coherence time, and may allow for an estimation of the temperature-dependent life time, which is in correspondance with experimental results.

Acknowledgements

We wish to thank H. Kuhn for providing experimental data for the oxacyanine dye LB film, and A.C. Scott for many helpful discussions and valuable suggestions. F. If is thanked for help with the numerical code. O. Bang, acknowledges CEC for financial support with the contract No. SC1-CT91-0705.

References

[1] A.S. Davydov, Theory of molecular excitons, Plenum Press (1971) 1.
[2] R.S. Knox in Primary processes of photosynthesis, ed. J. Barber, North-Holland (1977) 55.
[3] G. Scheibe, *Angew. Chem.* **49** (1936) 563.
[4] G. Feher, M.Y. Okamura in The photosynthetic bacteria, eds. R.K. Clayton and W.F. Sirtronm, Plenum Press (1978) 349.
[5] I. Langmuir, *J. Am. Chem. Soc.* **39** (1917) 1848.
[6] K.B. Blodgett, *J. Am. Chem. Soc.* **57** (1935) 1007.
[7] M. Fujihira in Photochemical processes in organized molecular systems, ed. K. Honda, Elsevier Science Publishers B.V. (1991).
[8] A.S. Davydov, *J. Theor. Biol.* **38** (1973) 559.
[9] M. Peyrard, A.R. Bishop, *Phys. Rev. Lett.* **62** (1989) 2755.
  T. Dauxois, M. Peyrard, C.R. Willis, *Physica D* **57** (1992) 267.
  V. Muto, P.S. Lomdahl, and P.L. Christiansen, *Phys. Rev. A* **42** (1990) 7452.
  V. Muto, *J. Biol. Phys.* **19** (1993) 133.

[10] V.E. Zakharov, *Zh. Eksp. Teor. Fiz.* **62** (1972) 1745 [*Sov. Phys. JETP* **35** (1972) 908].

[11] Y.R. Shen, The principles of nonlinear optics, J. Wiley & Sons Inc. (1984) Chap. 17.

[12] O. Bang, P.L. Christiansen, Y.B. Gaididei, F. If, K.Ø. Rasmussen, *Phys. Rev. E* **49** (1994) 4627.

[13] A. Pockels, Letter to Lord Rayleigh, *Nature* **43** (1891) 437.

[14] L.M. Blinov, *Russ. Chem. Rev.* **52** (1983) 713.

[15] Langmuir-Blodgett Films, ed. G.G. Roberts, Plenum London (1987).

[16] H. Bücher, H. Kuhn, *Chem. Phys. Lett.* **6** (1970) 183.

[17] D. Möbius, H. Kuhn, *Israel J. of Chem.* **18** (1979) 375.

[18] D. Möbius, H. Kuhn, *J. Appl. Phys.* **64** (1988) 5138.

[19] T.L. Penner, *Thin Solid Films* **160** (1988) 241.

[20] T. Sato, Y. Yonezawa, H. Kurokawa, M. Kurahashi, Y. Wada, T. Tanaka, *Thin Solid Films* **210/211** (1992) 172.

[21] H.P. Dorn, A. Müller, *Appl. Phys. B* **43** (1987) 167.

[22] S. De Boer, D.A. Wiersma, *Chem. Phys. Lett.* **165** (1990) 45.

[23] T. Itoh, T. Ikehara, Y. Iwabuchi, *J. Lumin.* **45** (1990) 29.

[24] J. Feldman, G. Peter, E.O. Göbel, P. Dawson, K. Moore, C. Foxon, R.J. Elliott, *Phys. Rev. Lett.* **59** (1987) 2337.

[25] R.H. Tredgold, *Rep. Prog. Phys.* **50** (1987) 1609.

[26] M. Fujihira, H. Yamada, *Thin Solid Films* **160** (1985) 125.

[27] P.B. Gilman, *Phot. Sci. Eng.* **18** (1974) 418.

[28] N.A. Campbell, Biology 3rd ed., Benjamin/Cummings Publ. Company (1993) Chap. 8 and 10. Govindjee, R. Govindjee, *Scientific American* **231** (1974) 68.

[29] W. Maentele, Speech at the Les Houches conference.

[30] P.O.J. Scherer, S.F. Fischer in Time-resolved vibrational spectroscopy, eds. A. Laubereau and M. Stockburger, Springer (1985) 297.

[31] J. Grad, G. Hernandez,, S. Mukamel, *Phys. Rev. A* **37** (1988) 3835. F.C. Spano, J.R. Kuklinski, S. Mukamel, *Phys. Rev. Lett.* **65** (1990) 211.

[32] E.A. Bartnik, K.J. Blinowska, *Phys. Lett. A* **134** 448. E.A. Bartnik, J.A. Tuszynski, *Phys. Rev. E* **48** (1993) 1516.

[33] G.C. Huth, F. Gutmann, G. Vitiello, *Phys. Lett. A* **140** (1989) 339.

[34] V.E. Zakharov, A.B. Shabat, *Zh. Eksp. Fiz.* **61** (1971) 118 [*Sov. Phys. JETP* **34** (1972) 62].

[35] P.S. Lomdahl, O.H. Olsen, P.L. Christiansen, *Phys. Lett. A* **78** (1980) 125.

[36] P.L. Christiansen, S. Pagano, G. Vitiello, *Phys. Lett. A* **154** (1991) 381. P.L. Christiansen, O. Bang, S. Pagano, G. Vitiello, *Nanobiology* **1** (1992) 229.

[37] J.J. Rasmussen, K. Rypdal, *Physica Scripta* **33** (1986) 481.

[38] N.W. Ashcroft, N.D. Mermin, Solid state physics, CBS Publ. Asia LTD (1976) Chap. 22-23.

[39] A.S. Davydov, Quantum mechanics (2nd ed.), Pergamon Press (1965) Chap. 20.

[40] R. Kubo, *Rep. Prog. Phys.* **29** (1966) 255.

[41] P.S. Lomdahl, W.C. Kerr, *Phys. Rev. Lett.* **55** (1985) 1235.

[42] J.C. Eilbeck, P.S. Lomdahl, A.C. Scott, *Physica D* **16** (1985) 318.

[43] J.C. Eilbeck, P.S. Lomdahl, A.C. Scott, *Phys. Rev. B* **30** (1984) 4703.

[44] A.C. Scott, J.C. Eilbeck, *Chem. Phys. Lett.* **132** (1986) 23.

[45] J.H. Jensen, P.L. Christiansen, O. Skovgaard, O.F. Nielsen, I.J. Bigio, *Phys. Lett. A* **117** (1986) 123.

[46] V. Czikkely, H.D. Försterling, H. Kuhn, *Chem. Phys. Lett.* **6** (1970) 297.

[47] A.C. Scott, *Phys. Rep.* **217** (1992) 1.

[48] F. If, M.P. Sørensen, and P.L Christiansen, *Phys. Rev. B* **32** (1985) 1512.

[49] O. Bang, P.L. Christiansen, Y.B. Gaididei, F. If, K.Ø. Rasmussen (unpublished).

[50] K. Rypdal, J.J. Rasmussen, K. Thomsen, *Physica D* **16** (1985) 339.

[51] M. Desaix, D. Anderson, M. Lisak, *J. Opt. Soc. Am.* **8** (1991) 2082.

[52] D. Anderson, M. Bonnedal, M.Lisak, *Phys. Fluids* **22** (1979) 1838.

[53] K.Ø. Rasmussen, Y.B. Gaididei, O. Bang, P.L. Christensen, (unpublished).

[54] P.L. Christiansen, K.Ø. Rasmussen, O. Bang, Y.B. Gaididei, (unpublished).

[55] G.P. Hasegawa, Plasma Instabilities and Nonlinear Effects, Physics and Chemistry in Space **8** Springer Verlag (1975).

[56] G.P. Agrawal, Nonlinear fiber optics, Academic Press (1989) Chap. 1.

Protons in hydrated protein powders

G. Careri, F. Bruni and G. Consolini

*Dipartimento di Fisica, Università di Roma
"La Sapienza", Piazzale Aldo Moro 2,
00185 Rome, Italy*

---

1. ABSTRACT

Previous work from this laboratory has shown that hydrated lysozyme powders exhibit a dielectric behaviour, due to proton conductivity, explainable within the frame of percolation theory. Long range proton displacement appears only above the critical hydration for percolation, when the 2-dimensional motion takes place on fluctuating clusters of hydrogen-bonded water molecules adsorbed on the protein surface. The emergence of biological function, enzyme catalysis, was found to coincide with the critical hydration for percolation.

More recentely, we have evaluated the protonic conductivity of hydrated lysozyme powders, from room down to liquid N_2 temperature. In the high temperature limit a classical isotopic effect can be detected, and the conductivity follows the familiar Arrhenius law for thermally activated hopping. In the low temperature region the conductivity shows a temperature dependance in agreement with prediction by the theory of dissipative quantum tunneling. Below room temperature the static dielectric constant, and the dielectric relaxation time for charge transport showed an increase likely to be identified with the formation of a polaronic-solitonic species as predicted by the theory of proton transport in water chains, a species which displays a larger effective mass and a larger dipole moment than the usual hydrated protonic defects.

The purpose of this paper is twofold. In the first section we present a tutorial report of some previous experimental results on proton displacement in slightly hydrated biological systems at room temperature, to show that in these systems the emergence of biological function coincides with the onset of percolative pathways in the water molecules network adsorbed on the surface of biomolecules. In the second section, we report on preliminary data on the dielectric relaxation of hydrated lysozyme below room temperature, to suggest that protons move along percolative water pathways on the protein surface according to a polaronic-solitonic model recently proposed by theory.

2. PROTON PERCOLATION AND EMERGENCE OF BIOLOGICAL FUNCTION IN NEARLY DRY BIOSYSTEMS

Anhydrous biosystems are very interesting from the biological viewpoint because they display biological function only above a critical water content h_c. Seed germination is a well familiar example of this behaviour. Then one is facing the following problem: can the emergence of biological function at the critical water content h_c be described in the frame of statistical physics as for the case of phase transitions and critical phenomena [1] ?

A general statistical-physical approach, called the percolation model, has been shown to be applicable to a wide range of processes where spatially random events and topological disorder are of intrinsic importance [2]. One of the most appealing aspects of the percolation process is the presence of a sharp transition, where long-range connectivity among the elements of a system suddenly appear at a critical concentration of the carriers. We mention here that for the typical example of a mixture of conducting and non conducting elements, percolation theory expresses the conductivity σ as a function of the critical concentration p_c of the conducting elements as proportional to $(p - p_c)^t$. The theory predicts values for both the critical exponent t and the threshold p_c: these are universal quantities which are only dependent on the dimensionality of the system. Thus one can detect the presence of percolation by measuring the conductivity of a sample as a function of the concentration p [3] of the conducting elements.

Previous work by our group has shown that powders of lysozyme at low hydration display protonic conductivity and that the conduction process follows the percolation model. In this case, single water molecules act as conducting elements, and the conductivity reflects motion of protons along chains of hydrogen-bonded water molecules adsorbed on the surface of the macromolecule, with long-range proton displacement along the extended network created at the percolation threshold [4, 5]. We have been able to detect the critical hydration threshold and exponent of this percolative conduction process in other anhydrous biosystems [6, 7, 8, 9], and the critical hydration for the onset of protonic conductivity coincided with the hydration required to trigger the sample-specific biological function for all samples investigated.

The measurement apparatus consisted of a capacitor assembly mounted on a balance enclosed in a vapor-tight box. The apparatus was designed to monitor simultaneously changing evaporation rates and dielectric properties of the sample inserted between the capacitor plates.

This provided an essentially continuous record of the dielectric properties as a function of sample hydration level h(g/g), expressed as g of H_2O/g of dry weight, and as a function of frequency as well. The dielectric apparatus gives values for the capacitance and the dielectric loss factor of the composite capacitor containing the sample. Using these capacitance data and dielectric theory for a composite capacitor, we evaluated the dc conductivity (σ) of the sample.

The dielectric capacitance of powders of native lysozyme displayed a sharp increase at a water content threshold $h_c = 0.150 \pm 0.016$ g/g, followed by saturation at increasing hydration. Since lysozyme hydration up to one monolayer of water molecules requires $h_m = 0.38 \pm 10\%$ g/g, the experimental critical volume for surface percolation is $h_c/h_m = 0.40 \pm 10\%$, a value very close to 0.45 ± 0.03 predicted by theory. Moreover, the threshold h_c was found to be constant from pH 3 to pH 8, indicating that the local structure of water clusters around ionizable sites of the protein surface is not of primary importance. Thus only the number of water molecules acting as interconnected conductivity sites is relevant; and as a matter of fact, the same threshold is found for both H_2O- and D_2O-hydrated samples, upon consideration of the mass difference between hydrogen and deuterium.

One can easily show that, above the threshold h_c, the conductivity σ must follow the power law $\sigma(h) - \sigma(h_c) = k(h - h_c)^t$, where the exponent t depends on the dimensionality of the system, and k is a constant. Results of this analysis are in very good agreement with the theoretical prediction for a 2D conduction process [10]. The previous analysis of the dielectric data reached independently the same conclusion about the dimensionality of the percolation, and it was based on the close agreement between the measured value of h_c/h_m and the theoretically predicted value for a surface process. The dielectric response at hydration levels near h_c reflects protonic conduction over pathlengths of the order of the diameter of a single macromolecule.

For lysozyme-saccharide complexes a higher value of the percolation threshold has been found, suggesting that the presence of a "foreign body", where the water bridges may not be favorable for proton transfer, must affect the long-range connectivity on the protein surface. This hydration level, $h_c = 0.25$ (g/g), is so close to the critical level for the onset of enzymatic activity in lysozyme powders that it suggests percolation of adsorbed water molecules to be involved in lysozyme catalysis. The value of the exponent t found for the saccharide complex suggests that the protonic conduction remains a surface process.

In Table I we have collected the main results derived from the study of protonic conductivities in anhydrous biosystems [11]. For all samples examined, once the critical exponent t was found close to the theoretical one, the experimental critical hydration for the onset of the protonic percolative process was considered the relevant quantity, to be compared with the hydration content which triggers the sample-specific biological function. A glance at Table 1 shows that indeed these anhydrous biosystems fulfill this comparison quite closely. It is very satisfactory to see that the scaling law for conductivity is followed with such accuracy in biological samples of quite different complexity, ranging from single protein to intact organism.

Table 1: Protonic percolation threshold and emergence of biological function. Adapted from Careri [11].

Anhydrous Biosystem	Dimensionality in scaling law	Hydration induced biological function
Lysozyme-saccharide	2	enzymatic activity
Purple membrane	2	photoresponse
Maize embryo	2	germination
Artemia cyst	3	pre-metabolism

If the previous results will be confirmed for a wider variety of biosystems, one shall be able to generalize these experimental findings by stating that, in order to work as one single functional unit, the components of one anhydrous system must be long range connected by water molecules by a random pathway. Notice that this statement requires concepts and quantities well defined in usual equilibrium conditions, while there is a widespread unjustified belief that the onset of biological function should be described as a non equilibrium phenomenon, a kind of dissipative process like the onset of coherence in a laser above a critical light power.

At this point we may pause and wonder why the percolative model works so well to describe the long range connectivity in water clusters adsorbed on globular proteins. To answer this question two further tutorial consideration should be made, one on the percolative process, and the other on the process of protein hydration.

In our first application of the percolative model we have assumed a 2D network consisting only of nodes connected by one-dimensional links. Inspection of this kind of network, easily generated by computers, show that the conductivity of a cluster is due to an extended back bone, where there are some regions which are singly connected and others which are multiply connected, namely loops to be visualised as street in parallel in a town, and called "blobs". Thus the structure of the infinite cluster just above the critical threshold consists of links (called also red bonds), nodes (crossing point of the links) and blobs (dense region with more than one connection between two points), and also of dead ends which are not relevant for the conduction process. These substructures have fractal dimensions not considered here, and, as we shall see below, are closely similar to the structures formed by water molecules hydrating a protein surface.

Protein hydration is the incremental addition of water to a dry macro-molecule until a dilute solution is obtained, and the hydration level h is usually expressed as the water weight divided by protein dry weight. If we follow the stepwise hydration process of a globular protein, we can detect by different and complementary techniques the events summarized in Table II.

At this point it is easy to realise the one to one correspondence between the fractal sub-structures considered by the percolation model and the structures formed by water molecules detected in the hydration process of a protein surface as follows:

percolation substructures		hydration substructures
blobs	↔	clusters around charged groups
nodes and links	↔	single molecules and chains adsorbed on polar groups

In view of the following section, it is appropriate to point out here the quasi-unidimensionality of the percolative water pathways which support proton transfer along links. This quasi-unidimensionality originates because out of the four hydrogen bonds intrinsic in every ad-sorbed water molecule, one or two are used to bind the molecule to the protein substrate, two are certainly required to bind the molecule to its nearest neighbours waters to allow proton transfer, and hardly one to form a further bond with an occasional water molecule sitting nearby. This quasi-monodimensional water chain must display statistical fluctuations with a renewal time for bonds rearrangements close to the dielectric relaxation time of adsorbed water, which is known to be about 10^{-10} s, thus much shorter than the hopping time for proton diffusion along water clusters, which is close to our observed relaxation time, typically 10^{-6} s.

Therefore, at least near room temperature, dynamic bond percolation can be neglected, and the use of the static site percolation theory is fully justified.

3. LIKELY DETECTION OF PROTONIC POLARONS IN PERCOLATIVE WATER CLUSTERS ADSORBED ON LYSOZYME POWDERS

In a previous work [13] from this laboratory, we have measured the temperature dependence of the protonic conductivity of hydrated lysozyme powder. This was found in good agreement with predictions by the theory of dissipative quantum tunneling. To our surprise, the isotopic effect of samples 5 and 7 (see Table 1 of [13]) was found equal to 1.18 ± 0.05, well below the classical $2^{1/2}$ value already detected at room temperature [5], but this fact was not pointed out in that paper, due to the small uncertainty linked to the linear transformation used to evaluate the conductivities. However, in some recent theoretical work on the proton displacement along H_2O chains (see Peyrard's contribution to this volume), large departure of the isotopic effect from $2^{1/2}$ are predicted. Therefore we decided to revisit this problem with a more sensi-tive experimental apparatus and with a more reliable data processing method, and preliminary findings will be reported here.

The experimental method used in this work is essentially similar to the one used in our previous work [4, 5] , with the following changes:

a) Lysozyme powder was from Sigma, and was used without further treatment.

b) Dielectric measurements were made using a Hewlett-Packard 4284A RCL meter, operating over a set of 50 test frequencies in the range 300 Hz $\leq f \leq 1$ MHz, with a new sample holder consisting of a two layer plane capacitor with the sample inside electrically insulated by means of a Teflon cell.

c) Sample water content was kept constant, and sample temperature spanned over the range 200 to 300 K.

d) Data analysis proved to be the most difficult part of this work to yield a sound description of the several relaxation processes observable over the frequency range used. This analysis was carried out in two steps:

i) Evaluation of the complex dielectric permittivity $\varepsilon''(\omega)$ from the experimental data for the capacity $C(\omega)$ and loss factor $D(\omega)$, collected as a function of temperature T and frequency $f = 2\pi\omega$. This was done according to the following inversion formula,

$$\varepsilon'(\omega) = \beta \left\{ \frac{C_o[C_o - C(\omega)]}{[C_o - C(\omega)]^2 + D^2(\omega)C^2(\omega)} - 1 \right\}$$

(1)

$$\varepsilon''(\omega) = \beta C_o \frac{D(\omega)C(\omega)}{[C_o - C(\omega)]^2 + D^2(\omega)C^2(\omega)}$$

where C_o is the capacitance of the Teflon cell and β is a geometrical factor related to the thickness of sample and Teflon, and to the latter dielectric constant. The real part of the dielectric permittivity $\varepsilon'(\omega)$ is not useful when comparing data of different runs because of its critical dependence on the distance between electrodes, a parameter ill defined in our composite capacitor, and on the filling factor of the sample cell; so we only made use of the imaginary part $\varepsilon''(\omega)$.

ii) $\varepsilon''(\omega)$ was then fitted with the sum of two Cole-Cole relaxations, each one expressed as

$$\varepsilon''(\omega) = (\varepsilon_o - \varepsilon_\infty) \frac{(\omega\tau)^{1-\alpha} \cos(\alpha\pi/2)}{1 + 2(\omega\tau)^{1-\alpha} \sin(\alpha\pi/2) + (\omega\tau)^{2(1-\alpha)}}$$

(2)

to derive the unknowns τ, $(\varepsilon_o - \varepsilon_\infty) = \Delta\varepsilon$, and α for each relaxation process. Here τ is the relaxation time, ε_o and ε_∞ are the limiting low-frequency and high-frequency permittivities, respectively, and α is the Cole-Cole distribution parameter (for a single Debye relaxation, $\alpha = 0$).The dielectric relaxation time τ was not sensitive to ε_o, which is sub-jected to the larger experimental error being an extrapolated value at zero frequency.

In the following, we shall consider the results obtained with 3 lysozyme samples listed in Table III, because these samples displayed sufficiently short relaxation times to be analyzed within the available frequency range.

Table III: examined lysozyme samples and their hydration level.

sample #	hydration	h (g/g)
3	D_2O	0.26 ± 0.01
4	H_2O	0.28 ± 0.01
6	D_2O	0.24 ± 0.01

These samples can be sufficient for a preliminary study of the isotopic effect by comparing #3 with #4, and allow a reproducibility check by comparing #3 with #6.

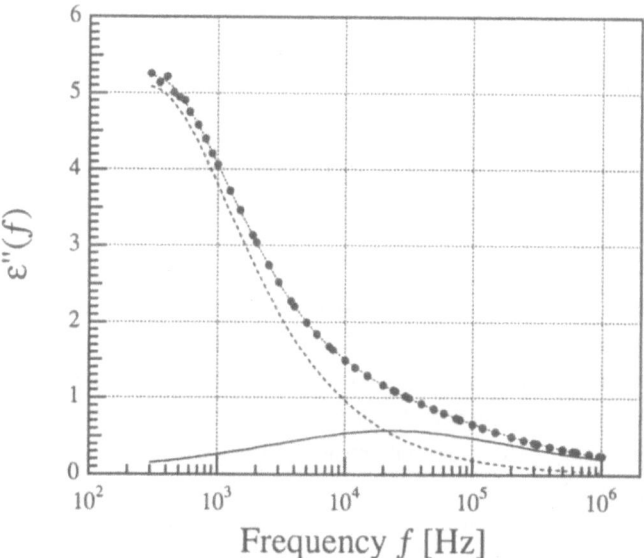

Figure 1: The ε'' relaxation spectra of lysozyme sample #4 at T= 280 K. H_2O content is $h = 0.28$ (g/g). The spectra is resolved into two components.

In Figures 1 and 2 $\varepsilon''(\omega)$ data for samples # 4 and # 3 are plotted at room temperature. Within the frequency region investigated, two relaxation processes can be resolved,

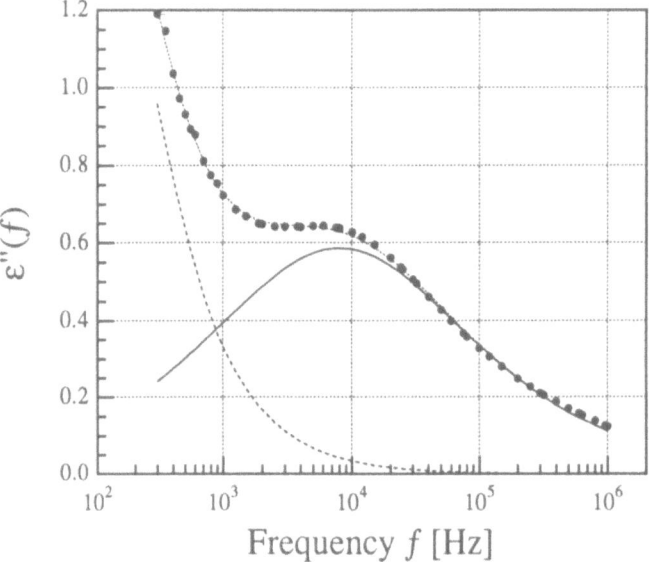

Figure 2: The ε'' relaxation spectra of lysozyme sample #3 at T= 280 K. D_2O content is $h = 0.26$ (g/g). The spectra is resolved into two components.

namely a low frequency large loss, along wiith a smaller one sitting on its shoulder at higher frequencies. The low-frequency relaxation is likely due to a Maxwell-Wagner dielectric dispersion, while the weaker band was assigned to proton relaxation on the hydrated protein surface [14]. Here we shall neglect the larger dispersion because it shifts outside our frequency window with decreasing temperatures, and we shall concentrate on the proton relaxation dispersion only. The analysis of this dispersion, with the method outlined above, yields the temperature dependence of τ, $\Delta\varepsilon$, and α. Since α data display a larger error compared to τ and $\Delta\varepsilon$, only these two latter quantities are shown in Figures 3 and 4. These two sets of data, for samples #3 and #4, can be used to derive the respective isotopic ratio: this is plotted as a function of temperature in Figures 5 and 6.

A glance at the isotopic ratio of both τ and $\Delta\varepsilon$ reveals the occurrence of a "transition" near 278 K from a temperature independent value at high T to a different regime (Figs. 5 and 6). Notice that the steady value of the isotopic ratio of the relaxation times at high temperature is somewhat above the classical value $2^{1/2}$, because the two compared samples are not at the same hydration level. The same holds for the $\Delta\varepsilon$ isotopic ratio, showing a value slightly

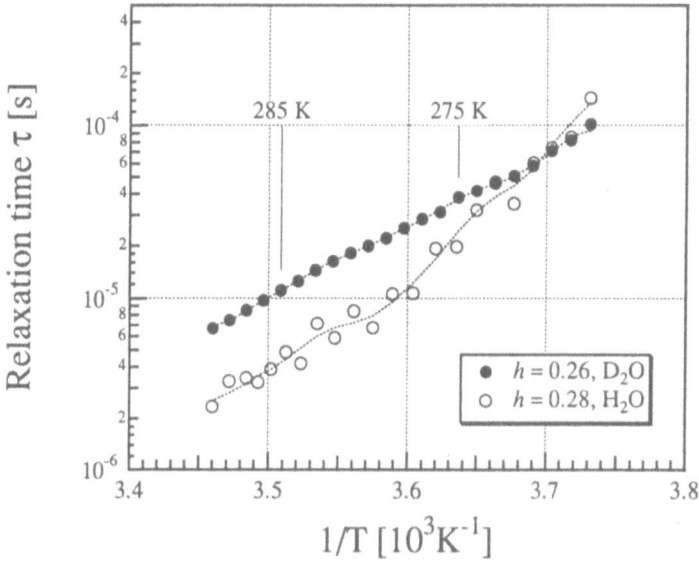

Figure 3: Arrhenius plot of the dielectric relaxation time τ of lysozyme sample #3 ($\bullet$, $h = 0.26$ (g/g)) and lysozyme sample #4 ($\bigcirc$, $h = 0.28$ (g/g)).

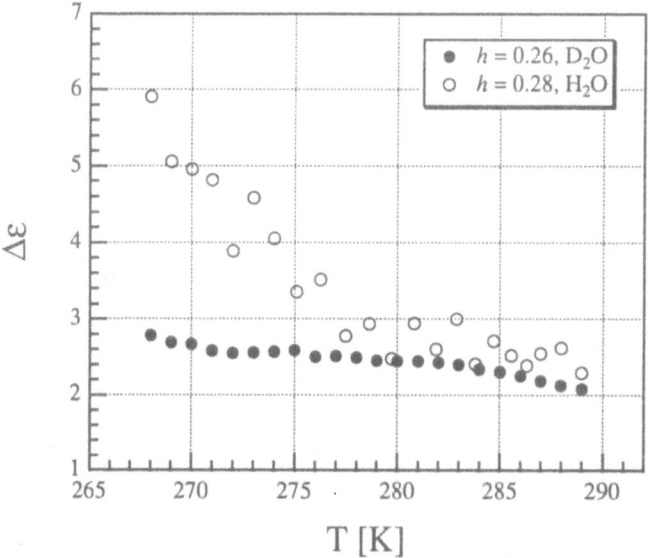

Figure 4: Temperature dependence of $\Delta\varepsilon$ of lysozyme sample #3 ($\bullet$, $h = 0.26$ (g/g)) and lysozyme sample #4 ($\bigcirc$, $h = 0.28$ (g/g)).

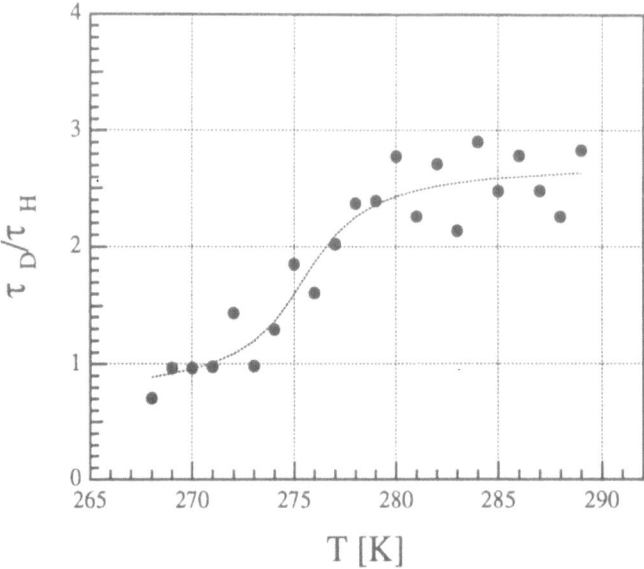

Figure 5: Temperature dependence of the relaxation time isotopic ratio (τ_D/τ_H) of lysozyme samples #3 and #4,

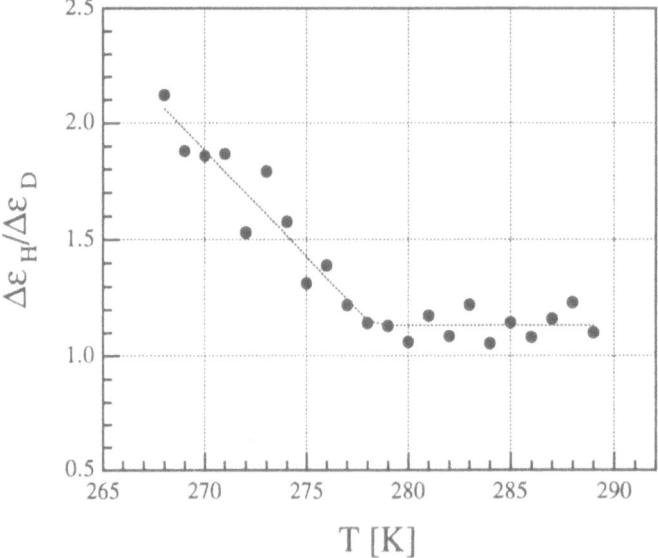

Figure 6: Temperature dependence of the reciprocal of the isotopic ratio of the dielectric dispersion $(\Delta\varepsilon_H/\Delta\varepsilon_D)$ of lysozyme samples #3 and #4,

larger than 1. The ratio $(\Delta\varepsilon_H/\Delta\varepsilon_D)$ is proportional to the dipole moment of the charges displaced by the electric field, if the number density of the charged species remain constant. Then the simplest way to account for these results is to suggest that, below about 278 K, a new species contributes to proton relaxation process for the H_2O-hydrated protein, a new species with a larger effective mass and a larger dipole moment. These two properties are consistent with the current picture of a polaron, namely a stable polarisation of the medium which follows the displaced charge. However, it should be noticed that this is not the only possible explanation, because, in the lower temperature region (not yet investigated) also the D_2O-hydrated protein could display a "transition" similar to the one shown by the H_2O-hydrated protein near 278 K.

We are thus continuing these experiments by extending the frequency range to the lower end and by increasing the set of frequencies, with a new apparatus designed for this purpose. Preliminary data confirm the existence of a transition in the temperature range indicated above, however we do not feel fully confident in our own data as yet. This is because there are always some uncertaincies in the detection of the dielectric properties of non-homogenous samples, and moreover our data analysis may suffer from limitations deriving from the use of the inversion formula needed to transform $C(\omega)$ and $D(\omega)$ into $\varepsilon'(\omega)$ and $\varepsilon''(\omega)$ [15]. Furthermore, the extension of our measurement to lower frequencies may include some contribution from electrode polarization which are difficult to assess. The main reason to report our preliminary findings at this time is to provide some experimental support to the current theoretical work on transport of defects in H_2O monodimensional chains, where the polarization of the medium is found relevant for the formation of a solitonic-polaronic species. This research area is at its very beginning, but if the present early suggestion will be supported by further theoretical and experimental work, the biological significance of these protonic-polaronic species occuring near room temperature should receive appropriate consideration.

References

[1] Careri G., *Nanobiology* **1** (1992) 117-126.

[2] Stauffer D., Introduction to Percolation Theory (Taylor & Francis, London, 1985) pp. 124.

[3] Zallen R., The Physics of Amorphous Solids (Wiley, New York, 1983) pp. 304.

[4] Careri G., Geraci M., Giansanti A. and Rupley J.A., *Proc. Natl. Acad. Sci. USA* **82** (1985) 5342-5346.

[5]   Careri G., Giansanti A. and Rupley J.A., *Proc. Natl. Acad. Sci. USA* **83** (1986) 6810-
6814.

[6]   Rupley J.A., Siemankowski L., Careri G. and Bruni F., *Proc. Natl. Acad. Sci. USA* **85**
(1988) 9022-9025.

[7]   Bruni F., Careri G. and Clegg J.S., *Biophys. J.* **55** (1989) 331-338.

[8]   Bruni F., Careri G. and Leopold A.C., *Phys. Rev. A* **40** (1989) 2803-2805.

[9]   Bruni F. and Leopold A.C., *Physiol. Plant.* **81** (1991) 359-366.

[10] Careri G., Giansanti A. and Rupley J.A., *Phys. Rev. A* **37** (1988) 2703-2705.

[11] Careri G., "Emergence of function in disordered biomaterials", Symmetry in Nature,
(Scuola Normale Superiore, Pisa, 1989) p. 213-215.

[12] Rupley J.A. and Careri G., *Adv. Protein Chem.* **41** (1991) 37-172.

[13] Careri G. and Consolini G., *Ber. Bunsenges. Phys. Chem.* **95** (1991) 376-379.

[14] Hawkes J.J. and Pethig R., *Biochim. Biophys. Acta* **952** (1988) 27-36.

[15] Morgan F.D. and Lesmes D.P., *J. Chem. Phys.* **100** (1994) 671-681.

Nonlinear models of collective proton transport in hydrogen-bonded systems

M. Peyrard

*Laboratoire de Physique, Ecole Normale Supérieure de Lyon,
URA 1325 du CNRS, 46 allée d'Italie,
69007 Lyon, France*

1. INTRODUCTION

Proton transport plays a fundamental role in cellular bioenergetics because pH gradients across membranes are the driving mechanism of many biomolecular reactions. In particular, proton transport is coupled to synthesis and hydrolysis of ATP which is the essential energy storage mechanism in biology. In spite of its importance, the mechanism of proton transport is far from being properly understood and this problem extends beyond biology. Even in physical systems which appear much simpler than biological molecules, such as ice, the mechanism of proton transport is not elucidated completely. The main question is the high proton mobility which is observed experimentally. When the proton-hydroxide permeability of biological membranes has been measured, it has been found to be orders of magnitude greater than expected from sodium-potassium permeabilities [1], showing that protons play a special role. A similar observation had been made previously for ionic mobility in water solutions [2].

This remark prompted Nagle and Morowitz [3,4] to suggest that the hydrogen-bonded chains that were previously invoked by Bernal and Fowler [2] to explain proton transport in water and ice could also be responsible for proton transport in biology. Extending this idea, Antonchenko, Davydov and Zolotaryk [5] and Yomosa [6] proposed independently two models introducing the concept of *soliton* to describe collective effects in proton transport. These pioneering works have been followed by various models based on the same idea, but the experimental detection of the cooperative excitation had not yet been made. The recent experiments reported by G. Careri in this workshop [7] revive the interest of these studies by providing a possible way to detect the cooperative effects. In this paper, we review the soliton models and then propose an extension which attempts to combine the recent knowledge of interatomic potentials provided by ab-initio calculations, ideas from the soliton models, and a treatment of the thermal fluctuations to provide data that can be compared to these experimental results. In his lectures at this meeting, H. Frauenfelder pointed out that only the *experiment* can find interesting things. So what is the role of a *model* especially when it relies heavily on numerical simulations, as it is the case here? Our view is that it can help us to select among several explanations of the experiments, determine the basic mechanisms which dominate the experimental results, and make predictions to be tested in further experiments.

2. THE PHYSICAL PROBLEM AND THE FIRST ANSWERS.

The first models for cooperative proton transport were intended to describe the very high mobility of protons in ice and water. In ice, the number of carriers is very low (about 10^{-6} per molecule at $-10°C$), but their mobility is very high. Measurements of the conductivity and saturation current indicate proton transfers from individual ions as frequent as 10^{13} per sec. or even more, comparable to $k_B T/h$ or even greater [8]. Moreover, as mentioned in the introduction, the theories of ionic mobility in water based on a picture of water as a fluid of definite viscosity and dielectric constant containing ions treated as spherical charged particles subject to resisting forces proportional to their velocity, which account satisfactorily for mobilities of large ions like K^+ or Cl^-, fail to account for the much larger mobilities of the H^+ or OH^- ions [2].

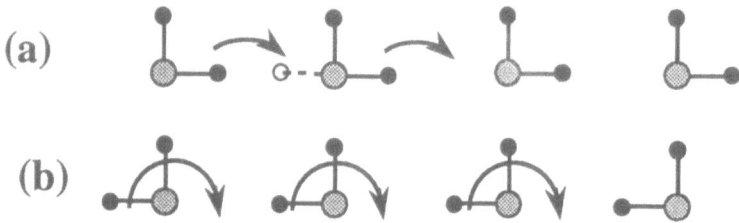

FIG. 1. (a) Schematic picture of proton transfer along a chain of water molecules according to the Bernal Fowler mechanism. (b) Motion of a Bjerrum defect in ice.

These abnormal mobilities of the H^+ and OH^- ions, which are confined to water and related solvents, e.g. methyl alcohol, led Bernal and Fowler to propose the transport of protons along "filaments" as schematized on fig. 1. These filaments are one-dimensional chains of water molecules connected by hydrogen bonds. Such hydrogen-bonded chains exist also in a variety of solid state systems such as ice, carbohydrates, lithium hydrazinium sulfate, imidazole [4] and form within the proteins that are embedded in biological membranes when the protein fold in the hydrophobic lipid environment of the membrane [3]. As shown in fig. 1(a), the transport of a proton along an hydrogen-bonded chain is associated to the existence of two possible positions for the protons involved in an hydrogen bond. Their interaction with the neighboring charges is such that the protons are submitted to a double-well potential. Therefore they can sit close to one or to the other of the two negatively charged entities that they connect. If an excess proton appears at the left end of the chain of water molecules of fig. 1(a), it forms a positive H_3O^+ ion with the first water molecule. Then, by sequential hopping of the successive protons in the chain on their second possible position (shown by the arrows on fig. 1(a)), the H_3O^+ ion moves to the right. In this process a charge is transported along the chain, but each individual proton makes only a short jump. The charge transport can be viewed as the motion of a localized distorsion of the charge density in the system instead of a transport of mass. This explains why the speed of the transport can be much higher for protons than for other ions for which such a mechanism cannot operate. After the transport of a single charge, the chain is left in a state that blocks the passage of another proton by the same process. For a continuous charge flow, the original configuration has to be restored. This can be achieved by a rotation of the water molecules as suggested by Bjerrum for ice [9]. Although the successive rotations involved in the propagation of a Bjerrum defect have also be described as a "soliton" in some models, we shall here concentrate on the Bernal Fowler mechanism, i.e. the first step. For transmembrane proton transport and conduction in water or in a network of water molecules which does not have a fixed structure as in a solid, the contribution of the Bjerrum mechanism is moreover less clear because the hydrogen-bonded channels are not permanent and could well be only temporary channels [10].

Although the Bernal Fowler mechanism provides a concept to understand the fast transport of charge, it does not solve all the questions. In particular, when on computes the barrier for the individual proton jumps that are involved in the process, one find it so high that simple thermally activated hopping would not be sufficient to provide the rate of proton transport observed experimentally. The picture must be completed by an additional feature. The missing ingredient is the distorsion of the heavy-ion lattice (OH^- lattice for a chain of water molecules). If one thinks that the hydrogen bonds, which are perturbed by the proton jumps, are precisely the bonds that connect the heavy ions, it appears rather obvious that the proton jumps will also be accompanied by a motion of the heavy ions. Studies of the proton-heavy-ion interactions, either by ab-initio calculations of the potential energy surface [11], or by spectroscopic measurements of the vibrational frequencies of hydrogen-bonded compounds [12], show that the proton jump tends to induce a decrease in the distance of the two neighboring heavy ions, which in turn causes a lowering of the barrier that the protons have to overcome in their jump. This mechanism, together with the idea that proton jumps can be cooperative, is at the basis of the soliton model proposed by Antonchenko, Davydov and Zolotaryuk.

3. THE A-D-Z SOLITON MODEL FOR PROTON TRANSPORT.

In their model [5], Antonchenko, Davydov and Zolotaryuk (ADZ) include the heavy-ion motion by coupling the proton dynamics to an optical mode of the heavy-ion sublattice. As a consequence, the charge defect of the Bernal Fowler mechanism propagates together with a lattice distorsion, forming a two-component solitary wave.

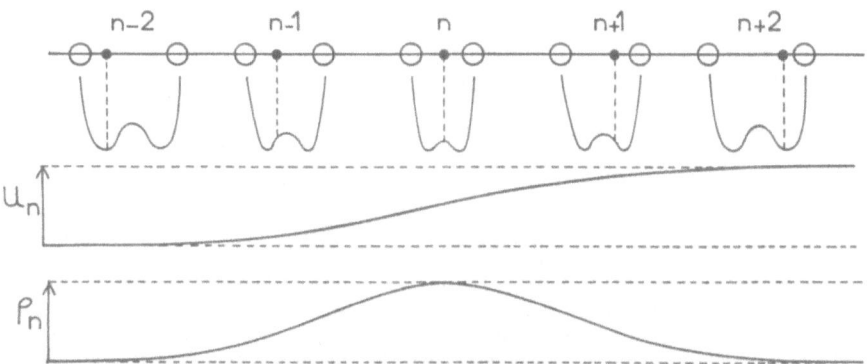

FIG. 2. The ADZ model and a schematic picture of a two component solitary wave showing the displacements u_n of the protons and the relative displacements ρ_n of the heavy ions.

The model consists of two one-dimensional interacting sublattices of harmonically coupled protons (mass m) and heavy ions (mass M) (fig.2). It represents the Bernal Fowler filaments in which only the degrees of freedom that contribute predominantly to proton mobility have been conserved. The two possible positions for a proton involved in an hydrogen bond are described by a double-well potiential expressed by

$$U(u_n) = \epsilon_0(1 - u_n^2/u_0^2)^2 \qquad (1)$$

where u_n denotes the displacement of the nth proton with respect to the center of the oxygen pair, ϵ_0 is the potential barrier, and $2u_0$ is the distance between the two minima of the double-well potential. Thus the proton part of the hamiltonian is

$$H_p = \sum_n \frac{1}{2}m \left(\frac{du_n}{dt}\right)^2 + U(u_n) + \frac{1}{2}m\omega_1^2(u_{n+1} - u_n)^2 , \tag{2}$$

where the last term represents the harmonic coupling with characteristic frequency ω_1 between neighboring protons.

In the heavy-ion part of the hamiltonian, only the relative displacement ρ_n between two neighboring ions linked by an hydrogen bond is considered because it is this quantity that modulates the double well experienced by the protons [11,12]. This restriction means that only an optical mode of the heavy-ions sublattice is taken into account. The heavy-ion part of the hamiltonian is then

$$H_{OH} = \sum_n \frac{1}{2}M \left(\frac{d\rho_n}{dt}\right)^2 + \frac{1}{2}M\Omega_0^2\rho_n^2 + \frac{1}{2}M\Omega_1^2(\rho_{n+1} - \rho_n)^2 , \tag{3}$$

where Ω_0 is the frequency of the optical mode which is involved and the last term describes an harmonic coupling between neighboring heavy-ion pairs. It introduces in the model the dispersion of the optical mode.

The last part in the hamiltonian of the model arises from the dynamical interaction between the two sublattices and describes the modulation of the double-well potential caused by the variation of the distance between the heavy ions that surround a proton:

$$H_{int} = \sum_n \chi\rho_n(u_n^2 - u_0^2) , \tag{4}$$

where χ measures the strength of the coupling.

The expressions of the terms introduced by ADZ illustrate the approach used generally in modeling nonlinear phenomena in biophysics. First the real system is highly simplified by selecting of only a few degrees of freedom. Such a reduction of the full system may appear drastic, but, studies of nonlinear dynamical systems exhibit very often *dominant* degrees of freedom, which govern most of the dynamics, and *slave* degrees of freedom which simply follow the others. The first step in modeling is the identification of the dominant degrees of freedom. The second step in the approach is the selection of the exact expressions for the various interactions in the system. The double-well potential for instance is represented by a polynomial form. For a qualitative study this is sufficient because it provides the appropriate potential shape and moreover, the interest is that it leads to equations of motions which have well known solutions. A similar choice has been made for the interaction term of the hamiltonian. A detailed examination shows that expression (4) gives the expected behavior: the effective barrier for the proton jump is lowered during the jump. Moreover the analytical expression has the great interest that, in the course of the calculations, it leads to terms that can easily be combined with terms coming from $U(u_n)$ to give again a solvable equation. Therefore the model is dictated simultaneously by the physical properties of the real system and by mathematical constraints. Such an approach has its strength because it allows us to get a general insight on the behavior of the system from analytical studies, or through an a-priori knowledge of the properties of nonlinear systems. Of course it has also weaknesses because it may lead to oversimplified models or ad-hoc models built simply because we know how to solve them. These strength and weaknesses show-up clearly in the results that can be derived from the ADZ model.

From the total hamiltonian $H = H_p + H_{OH} + H_{int}$ one can derive an infinite set of coupled nonlinear differential equations for the time evolution of the u_n and ρ_n. In spite of the simplifying assumptions made to establish the model, this set of equation is not solvable analytically. The standard approach to this problem is to assume that the coupling between neighboring sites is sufficiently strong. If this is true, the time evolutions of consecutive sites are very similar and u_n and ρ_n vary smoothly with space. Thus the discrete variables $u_n(t)$, $\rho_n(t)$ can be replaced by two continuous functions of space and time $u(x,t)$ and $\rho(x,t)$. This replaces the infinite set of coupled differential equations by a set of two coupled partial differential equations [5,13]. If one looks for the propagation of a permanent-profile distorsion of the proton density, accompanied by a distorsion of the heavy-ion sublattice, one can look for a solution where u and ρ are functions of a unique dimensionless variable $\xi = (x - vt)/a$ where a is the lattice spacing. This reduces the problem to solving two coupled nonlinear differential equations. Even with all these simplifications the model does not have an analytical solution in the general case. However combining the knowledge deduced from an exact solution which can be determined for a special value of the velocity v, expansions around this value, further approximations in the analytical calculations, and numerical simulations, it is possible to reach a good understanding of the properties of proton transport within the ADZ model [13].

The exact solution, which can be obtained for the particular value $v = v_0 = a\Omega_1$, is given by

$$u(x,t) = u_0 \tanh\left(\frac{x - v_0 t}{L}\right) \tag{5a}$$

$$\rho = \rho_0 \operatorname{sech}^2\left(\frac{x - v_0 t}{L}\right) , \tag{5b}$$

where $\rho_0 = \chi u_0^2 / M\Omega_0^2$ determines the amplitude of the distorsion of the heavy-ion lattice while L, given by

$$\frac{1}{L^2} = \frac{2}{(m\omega_1^2 u_0^2)[1 - v_0^2/(a^2\omega_1^2)]} \left(\epsilon_0 - \frac{\chi^2 u_0^4}{2\Omega_0^2 M}\right) , \tag{6}$$

determines the width of the lattice distorsion. In this expression, the last factor between parenthesis is the barrier that the proton have to overcome which has been corrected by the heavy-ion lattice distorsion. One can notice that, as expected, the distorsion of the heavy-ion lattice reduces the barrier below its value ϵ_0 for a totally rigid heavy-ion lattice. This exact solution has the shape plotted in fig. 2. Although no exact analytical solution is known for $v < v_0$, approximate analytical calculations and numerical solutions show that a permanent profile solution does exist in this velocity domain. This is no longer true at higher velocities. The existence of a maximum velocity for the solution can be understood qualitatively by noticing that, while the light protons can move fast, the heavy ions cannot follow. As a result, instead of a permanent profile localized distorsion, moving at high speed together with the H_3O^+ ion, one observes a permanent leakage of energy from the proton sublattice to non-localized oscillatory modes in the heavy-ion sublattice. This causes a strong "friction" for the proton motions, which appears as a plateau in the mobility curve of the charge carriers [13].

Therefore the calculations performed with the ADZ model show that the dynamics of the heavy ions can have a very strong influence on proton transport in an hydrogen bonded chain. While at low velocity the model shows that the motion of the heavy ions can sustain charge transport by lowering the barrier that the protons have to overcome, at high carrier velocity the dynamics of the heavy ions can on the contrary strongly damp the charge transfer.

In spite of its interest to reach such a conclusion, the ADZ model has many weaknesses. In particular it does not include acoustic motions of the heavy ions. Many improved model have been proposed to correct this problem and provide a more complete description of the hydrogen-bonded chains, but they are generally proceeding along the same lines, in particular, looking for solitonlike solutions without including external factors such as thermal fluctuations, except, to our knowledge, in a a few works where only short time behavior has been examined [14,15] and in the work of Nylund and Tsironis presented in this volume [16,17]. One other major difficulty it that the values of the model parameters are difficult to estimate. It turns out that the parameters are crucial. If the conditions for the validity of the continuum limit approximation are fulfilled, i.e. if the quantity L appearing in the solution (5a) is large with respect to the lattice spacing, the soliton picture is correct, at least in the low velocity range $v < v_0$. In this case the distorsion in the proton density propagates almost freely in the lattice and this would be a very nice explanation of the high proton mobility. Such a situation is however only possible if the motions of the protons is highly cooperative. Data from ab-initio calculations suggest that this is not the case. This would mean that the "solitons" are narrow, with a typical width of the order of the lattice spacing. As discussed in the next section, this would cause a strong pinning of the soliton by the lattice that can only be relieved by thermal fluctuations.

4. IS THERE AN EXPERIMENTAL EVIDENCE OF THE SOLITON?

In order to try to answer this question, it is necessary to have a model for which the potential terms are better controlled and to include explicitly the thermal fluctuations in the calculations. Moreover we must deduce from the model some quantities that can be compared with experimental results. This is not simple because experiments measure generally the conductivity of a sample which includes not only the mobility of the carriers, which can be determined within a model, but also the number of carriers. This is a quantity which is difficult to determine in a calculation especially because it can depend strongly on impurities present in the material (extrinsic carriers) and it can vary significantly with temperature if carriers, weakly trapped by some impurities or defects are released at some temperature. This is why the experiments presented by G. Careri at this meeting are precious because they give the temperature evolution of the isotopic effect versus temperature, i.e. the ratio of the mobilities of hydrogen and deuterium carriers in an hydrogen bonded system [7]. This is a quantity that can be calculated with a theoretical model. This section describes such a calculation and compare the results with the experiments.

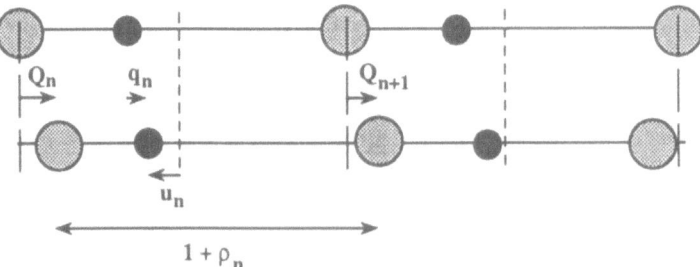

FIG. 3. Our model for the hydrogen bonded. The top figure shows the chain in equilibrium and the bottom one shows the displaced atoms and the corresponding variables.

The model that we consider is closely related to the model used by Savin and Zolotaryuk [15]. It is however slightly modified and some of its parameters are obtained form ab-initio potential energy surface calculations. As in the ADZ model, we consider a one-dimensional diatomic chain (fig. 3). A unit cell contains an heavy ion and a proton. Their displacements are measured in units of the equilibrium lattice spacing, with respect to the equilibrium positions and denoted respectively by Q_n and q_n. It is also convenient to introduce two related variables, the displacement u_n of the nth proton with respect to the middle of the bond linking its two heavy-ion neighbors $u_n = q_n - (Q_{n+1} + Q_n)/2$, and the variation of the distance between these two heavy-ion neighbors $\rho_n = (Q_{n+1} - Q_n) - 1$. Although these two variables make the link with the ADZ model notation, it should be noticed that the present model allows acoustic motions of the heavy ions so that the center of the bond between two heavy-ion neighbors can move with time.

The proton-heavy ion interaction potential can be deduced from the potential energy surface of a short chain of water molecules [11]. The proton transfer which is investigated in such a calculation takes place between two central molecules which are held apart by a fixed distance, i.e. the quantity $1 + \rho$ in our notation is imposed. The geometry of each of the subunits connected by the hydrogen bond is maintained fixed and the position of the proton is varied by changing its displacement u with respect to the center of the bond. The quantum mechanical calculation of the energy of the system is repeated for various values of ρ so that a two-dimensional potential energy surface $E(u, \rho)$ can be constructed. It is then fitted by an analytic function $V(u, \rho)$ which contains some parameters which are optimized to get the best fit. In the study performed by Duan and Scheiner [11], it was found that a combination of two Morse potentials can provide a very good fit of the potential energy surface. The expression of $V(u, \rho)$ is then

$$V(u, \rho) = \frac{\epsilon_0}{(\alpha - 1)^2} \left[\left(\alpha^2 - \frac{1}{2} e^{-b\rho} \right) + \left(\alpha - e^{-b\rho/2} \cosh bu \right)^2 \right] , \qquad (7)$$

where α and b are potential parameters and ϵ_0 sets the energy scale ($\epsilon_0 = 1$ eV in our calculation).

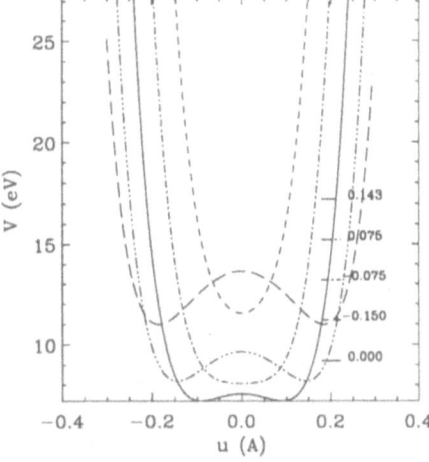

FIG. 4. The total proton-heavy-ion interaction potential, $V(u, \rho) + (1/2)v_0^2 \rho^2$ plotted as a function of u for various values of ρ indicated on the figure ($v_0^2 = 500$).

By construction $V(u, \rho)$ contains the interaction between the proton and heavy ions sublattices and therefore, contrary to the ADZ case, it is not necessary to include explicitly an interaction term in the hamiltonian. The parameters α and b being determined from the fitting with the ab initio potential are no longer arbitrary parameters of the model.

In our dimensionless form their value is $\alpha = 1.466$, $b = 7.315$. However $V(u, \rho)$ does not include the the energy of interaction between the heavy ions since their distance is imposed in the ab initio calculation. As in the ADZ model, this term has to be added afterward. Assuming an harmonic interaction it is written as $v_0^2 \rho^2 / 2$ where v_0 is a constant which measures the interaction between neighboring heavy ions. The total potential which includes this term is shown on figure 4.

This figure shows that the barrier between the two possible proton sites changes when ρ varies and it can even vanish when ρ becomes sufficiently negative. The minimum energy configuration is however obtained when the potential has two minima separated by a barrier of 0.346 eV, which is a value in good agreement with the values generally observed for hydrogen bonds between water molecules.

The hamiltonian of our model is

$$H = \sum_n \frac{1}{2}m \left(\frac{dq_n}{dt}\right)^2 + \frac{1}{2}c_0^2(q_{n+1} - q_n)^2 + V(u_n, \rho_n)$$

$$+ \frac{1}{2}M \left(\frac{dQ_n}{dt}\right)^2 + \frac{1}{2}v_0^2(Q_{n+1} - Q_n)^2 + \frac{1}{2}\Omega^2 Q_n^2 + fQ_n - fq_n . \qquad (8)$$

In this expression f is an external field which is introduced in order to allow the determination of the mobility of the charges through their response to the fiel and c_0^2 measures the strength of the proton-proton coupling. The term $\Omega^2 Q_n^2 / 2$ represents an on-site potential for the heavy ions. It corresponds to the case of water molecules adsorbed on a surface as in the experiments described by G. Careri [7]. This terms prevents the full lattice from sliding freely but we have checked that the properties of the model are only weakly sensitive to the exact value of Ω^2. The case of the parameters c_0^2 and v_0^2 is more critical because they have a strong effect on the dynamical properties of the model. A value for v_0 can be estimated from the shape of the total potential shown in fig. 4 and from structural data because v_0 determines the distance between the two proton equlibrium positions. This distance varies certainly from one system to another, but a value of 0.5 Å is reasonable. Such a distance is obtained for the value $v_0^2 = 500$ that we have selected. The value of c_0 is more difficult to determine. We have obtained our estimation from the structure of a static H_3O^+ ion in the lattice, i.e. from a solution in q_n and Q_n which minimizes the hamiltonian (8).

The minimization is done numerically, in the absence of external field, with boundary conditions such that the protons occupy different equilibrium positions, at the two ends of the chain. This enforces the presence of a charge defect in the chain and the minimization determines its structure. The boundary conditions can be chosen such that the defects is either a rarefaction of the proton density, corresponding to a OH^- ion, or a compression of the proton density corresponding to an H_3O^+ ion. The two types of solutions are shown in fig. 5 (a) and (b).

The general remark that can be made from these results is that the heavy-ion lattice distorsion is small (amplitude 0.054 Å for an OH^- defect), but extends quite far from the center of the defect since it is not negligible even 20 sites away from the center. The distorsion of the proton sublattice also has an extended tail, but its center is very sharp. With the value $c_0^2 = 40$ that we have chosen, the proton jump occurs mainly on one lattice spacing, i.e. the defect is very localized. This is known to be the case in hydrogen-bonded systems like ice and this has motivated our choice for c_0. As a result we are very far from the continuum limit that would lead to a soliton model. Figure 5 shows that even the protons which are the closest to the defect are weakly displaced from their equilibrium position. As they are almost in the bottom of their potential well, moving the defect requires climbing a rather high energy barrier. Figure 5 (c) obtained with a frozen heavy-ion lattice shows that in this case the defect is even narrower giving a higher barrier for translation. Therefore the distorsion of the heavy-ion lattice has the expected effect of promoting charge mobility, but the barrier remains high. Its value can be calculated from the energy difference between a state where the defect has been translated by half a lattice spacing, and the minimum energy state. This difference, known as the Peierls-Nabarro barrier (E_{PN}) in dislocation theory, is listed in table I together with the depinning field of the defect, i.e. the minimum field f_c required to cause its motion along the lattice.

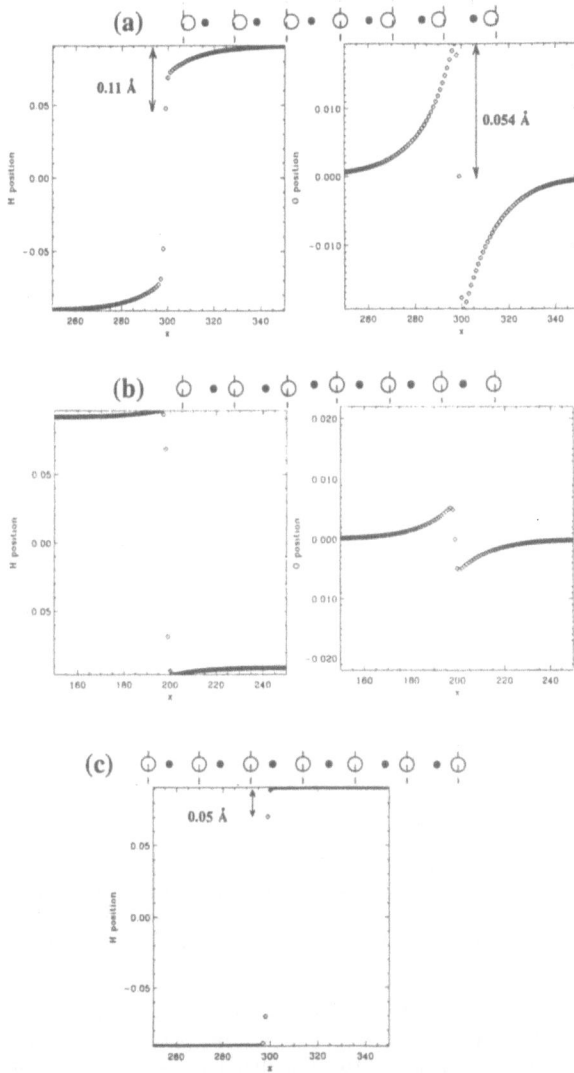

FIG. 5. Static structures of the charged defects obtained by minimization of energy of the hydrogen bonded chains. (a) OH^- ion and (b) H_3O^+ ion. A picture of the chain is shown in both cases below the graph of u_n and ρ_n. On this picture the displacement of the heavy ions has been multiplied by a factor 5 to make it more visible. Figure (c) shows u_n in a case where the heavy ion lattice has been held fixed during the minimization.

TABLE I. Barrier to translation E_{PN} for OH^- and H_3O^+ charge defects calculated by a full minimization (lines "Mobile heavy ions" in the table) or by minimization with heavy ions frozen on the equilibrium positions that they occupy in the absence of defect (lines "Fixed heavy ions" in the table). The table shows also the depinning field f_c required to put the defects into motion at different temperatures. The values which are not filled in the table have not been calculated.

	E_{PN}	$f_c(T = 0)$	$f_c(T = 12K)$	$f_c(T = 120K)$
OH^-				
Mobile heavy ions	0.0337 eV	$2.07\ 10^9$ V/m	$7.0\ 10^8$ V/m	$3.6\ 10^6$ V/m
Fixed heavy ions	0.1156 eV	$7.10\ 10^9$ V/m	$6.9\ 10^9$ V/m	$1.8\ 10^9$ V/m
H_3O^+				
Mobile heavy ions	0.0630 eV	$3.8\ 10^9$ V/m		
fixed heavy ions	0.1156 eV	$7.1\ 10^9$ V/m		

This table shows quantitatively the very strong pinning effect due to lattice discreteness. Although the heavy-ion lattice distorsion tends to reduce this pinning, it is not sufficient to bring the depinning field to the typical values that are found across a biological membrane (10^7 V/m), at least if one does not consider the role of the thermal fluctuations.

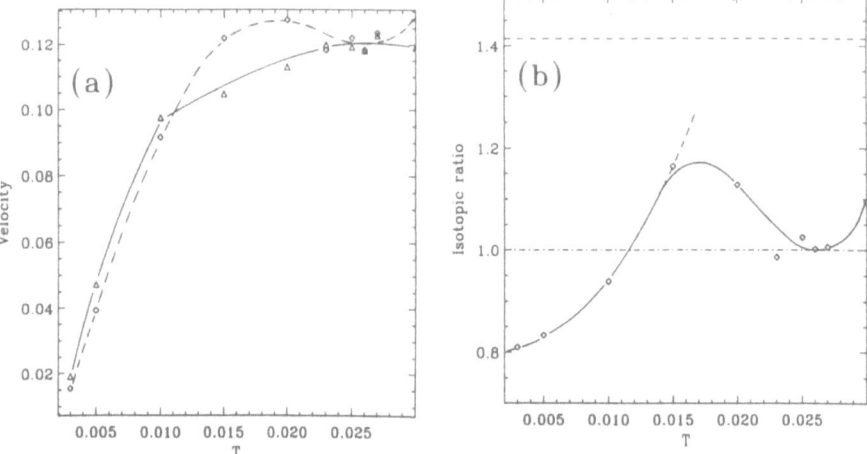

FIG. 6. (a) Averaged equilibrium velocity of a charged defect driven by a field of $1.8\ 10^8$ V/m as a function of temperature. The diamonds are the numerical results for protons and the squares correspond to a lattice where all the protons have been replaced by Deuterium. The lines are only a guide for the eye. (b) Isotopic effect: ratio of the proton and deuterium mobility as a function of temperature.

But the table shows that *thermal fluctuations can drastically reduce the depinning field.* They have a very strong effect, even at very low temperature (12 K) and this points out the importance of including temperature in modeling proton transport. It is rather difficult to use the soliton concept to calculate the mobility of the defects because we are dealing with narrow defects which are only roghly described by the soliton picture. Therefore we have investigated the thermal effects by numerical simulations of the system in contact with a thermal bath. The simulations are performed with the Nose-Hoover method [18,19] which adds to the physical hamiltonian an extra variable which corresponds to the thermal bath. This non-local variable regulates the energy flows between the system and the thermostat and it can be shown that, at equlibrium, the statistical properties of the physical system which is investigated correspond exactly to its properties in the canonical ensemble at the temperature T which is chosen. We are using here a slight variation of the method because we apply an external field in the system which is then out of equilibrium. The mobility of the charge carriers in the thermalized model is obtained by performing many simulations (typically 20) for each value of the field and temperature. Then we average over all these simulations with different initial conditions. We have measured the averaged equilibrium velocity of the charge defect in the presence of an applied field of $1.8\,10^8$ V/m. This value is rather high compared to the field across membranes. In principle the calculations could be performed for smaller fields, but then the time necessary for the defect to reach its equilibrium velocity increases significantly and the computer time exceeds our possibilities.

As expected, fig. 6(a) shows a general tendency to a higher mobility when temperature increases, but the finer shape of the curve is more interesting. First, in the case of protons, we found a dip in the mobility curve around $T = 290$ K which is yet unexplained. This structure does not seem to be a numerical artefact because the corresponding values have been checked by many extra calculations to improve the statistics and rule out the possibility of a simple fluctuation. This feature could perhaps be related to a similar effect observed by Nylund and Tsironis [16], but it requires further studies. The mobility curve shows also clearly two regions. Below 150 K, mobility increases very fast with temperature while, above this temperature, a tendency to saturation is clearly observed. Although further analysis is necessary, we think that this behavior is an indication that a cooperative effect takes place and that the carrier mobility is assisted by the heavy-ion lattice deformation, in agreement with the picture provided by the static solutions shown in fig. 5. This dynamical cooperative effect would lower significantly the barrier that the proton have to overcome as shown by the comparison of the Peirls-Nabarro barrier with mobile or fixed ions listed in table I. To be efficient such a process requires a coherent motion of the heavy ions while the proton moves. This is only possible at low enough temperature. This would explain the saturation effect observed above 150 – 200 K.

This hypothesis is supported by the isotopic effect shown in fig. 6(b). In the low temperature range the mobility of proton and deuterium are found to be similar. In fact we even find that the *heavier carrier is more mobile than the light one*, contrary to standard expectation. This would be understandable if, instead of the motion of an isolated proton, we were observing the motion of a "collective object" involving the proton and the surrounding heavy-ion distorsion. The effective mass of this object would have little to do with the mass of the individual proton or deuterium, hence the anomalous isotopic effect. Then, as temperature is raised above 150 K, the isotopic effect tends to recover the expected value of $\sqrt{2}$ for proton versus deuterium. This is consistent with the hypothesis discussed above that the collective effect is killed by the thermal fluctuations that break the coherence. The tendency toward the value $\sqrt{2}$ is however interrupted by the dip in proton mobility around 290 K that we discussed above.

Although this picture for the charge transport in the hydrogen-bonded chain is consistent with our numerical simulations, it certainly needs to be refined. And in particular the excess mobility of the deuterium has to be explained. It is possible that these heavier ions are more efficient than protons to drive the heavy-ion sublattice.

5. CONCLUSION.

Although we have showed that the *soliton* picture is probably not exactly appropriate to describe proton transport in hydrogen bonded chains because the charge defect is so narrow that the soliton would be pinned by lattice effects, some of the ideas contained in the soliton model could well be valid. In particular, our numerical results with a model for which we have attempted to reduce the arbitraryness of the parameter choice by using results of ab-initio calculations, suggest that a *collective* effect involving a coherent distorsion of the heavy-ion lattice could explain the fast increase of the proton mobility which is found. At higher temperatures, the collective effect seems however to be destroyed by the thermal fluctuations. It is interesting that the results obtained with this model are strongly supported by the results obtained independently by G. Careri and al. who have also observed an anomalous isotopic effect in the low temperature range, which dispapears when temperature is increased [7]. The similarity is not perfect and, in particular, the anomaly in the isotopic effect observed by G. Careri is much larger than the one we found. It is too early to claim that this similarity provides a proof that collective effect play a real role in proton transport, but we have perhaps now an indication that this is the case. However there are other possibilities to explain an anomalous isotopic effect. Quantum effects have for instance been invoked to explain thermally activated diffusion of hydrogen on tungsten surface [20] for which the prefactor to the diffusion rate is higher for deuterium than for hydrogen. A quantum effect is perhaps possible in the experimental system studied by G. Careri and coworkers, but it can be ruled out in our calculations which are purely classical.

In any case, while the width of the charge defect probably rules out a soliton in the strict sense, the possibility that a *"proton polaron"* exists and has been observed should perhaps be taken seriously.

ACKNOWLEDGMENTS

Part of this work has been supported by the CEC grant SC1-CT91-0705. I want to thank G. Careri for having stimulated this work by his experimental results and enthusiasm. I also thank him for precious discussions.

References

[1] D.W. Deamer and J. W. Nichols, *J. Membrane. Biol.* **107** (1989) 91-103.
[2] J.D. Bernal and R.H. Fowler, *J. Chem. Phys.* **1** (1933) 515-548.
[3] J.F. Nagle and H.J. Morowitz, *Proc. Natl. Acad. Sci. USA* **75** (1978) 298-302.
[4] J.F. Nagle, M. Mille and H.J. Morowitz, *J. Chem. Phys.* **72** (1980) 3959-3971.
[5] V. Ya. Antonchenko, A.S. Davydov and A.V. Zolotariuk, *phys. stat. sol. (b)* **115** (1983) 631-640.
[6] S. Yomosa, *J. Phys. Soc. Japan* **51** (1982) 3318-3324.
[7] G. Careri, F. Bruni, and G. Consolini, *Protons in Hydrated Protein Powders.* previous paper in this volume.
[8] Mou-Shan Chen, L. Onsager, J. Bonner and J. Nagle *J. Chem Phys.* **60** (1974) 405-419.
[9] N. Bjerrum, *Science* **115** (1952) 385-390.
[10] J.F. Nagle, *J. of Bioenergetics and Biomembranes* **19** (1987) 413-426.
[11] Xiaofeng Duan and S.Scheiner, *International Journal of Quantum Chemistry: Quantum Biology Symposium* **19** (1992) 109-124.
[12] N.D. Sokolov, M.V. Vener and V.A. Savel'ev, *J. of Molecular Structure* **222** (1990) 365-386.
[13] M. Peyrard, St. Pnevmatikos, and N. Flytzanis, *Phys. Rev. A* **36** (1987) 903-914.
[14] J. Halding and P.S. Lomdahl, *Phys. Rev. A* **37** (1998) 2608-2613.
[15] A.V. Savin and A.V. Zolotaryuk, *Phys. Rev. A* **44** (1991) 8167-8183.
[16] E. Nylund and G. Tsironis, *Phys. Rev. Lett.* **66** (1991) 1886-1889 and E. Nylund, K. Lindenberg and G. Tsironis *J. of Statistical Physics* **70** (1993) 163-181.
[17] G. Tsironis, *Proton-solitons bridge physics with biology,* following paper in this volume.
[18] S. Nose, *J. Chem. Phys.* **81** (1984) 511-519.
[19] W.G. Hoover, *Phys. Rev.* **A31** (1985) 1695-1697.
[20] K.F. Freed *J. Chem Phys.* **82** (1985) 5264-5268.

Proton-solitons bridge physics with biology

G.P. Tsironis

Physics Department, Univ. of Crete and
Research Center of Crete, P.O. Box 1527,
71110 Heraklion-Crete, Greece
and
Computational Physics Laboratory,
Dept. of Physics, Univ. of North Texas,
Denton, TX 76203, U.S.A.

1. INTRODUCTION

Electrons and protons are the "elementary" particles that dominate most physics studies. Electrons are very light particles, are in abundance and are responsible for a very large number of phenomena from the astrophysical scale to condensed matter systems and chemical reactions. In systems such as metals, free electons form bands leading to charge conduction. Protons, on the other hand, are approximately two thousand times heavier, and thus much more sluggish than electrons. In high energy physics, protons are accelerated routinely to very high (cosmic) energies (such as 1 TeV, or an energy approximately one thousand times the energy equivalent to their mass) and are used then as superfast cannonballs for atom smashing experiments. Twelve orders of magnitude below this scale, protons bind with oxygen to form water molecules giving rise to the phenomenon of life, for it is indeed true that most biological phenomena involve water. In this mesoscopic level protons are still mobile because they are part of molecules in the liquid state. Protons can also give way part of their mobility and become intermediaries in the formation of a new building block through "hydrogen bonds" that are very efficient in binding several kinds of atoms together. In this bond, a proton is shared by two oxygen atoms (or other appropriate atoms or molecules) enhanching thus the stability of the complex and assisting in the creation of microscopic order through the formation of periodic structures. The resulting hydrogen-bonded chains (HBC's) can be a few hundred angstroms long in biological settings or even practically infinite as in ice. In order to form HBC's, a proton from one water molecule links weakly with an adjacent molecule forming a weak "hydrogen bond". The resulting structure has a quasi-one-dimensional character defined by the hydrogen bonds themselves.

One of the basic questions that are of interest, especially with regard to the biological implications is that of the means of movement of the protons while forming HBC's. Large proton motion is readily observed, for instance, in proton pumps when protons move through a cellular membrane. Since the protons bind adjacent molecules they cannot be considered "free" (as are electrons in metals, for instance). On the other hand, it is known experimentally that some protons move through the HBC's leading to proton conductivity. The latter must be of drastically different nature compared to electronic conductivity and will certainly depend on the nature and strength of the force that protons experience while participating in a hydrogen bond. It has been established through a variety of direct as well as indirect evidence that the hyrdogen bond is formed by a potential with two minima, usually modeled by a symmetric double-well potential. Each hydrogen bonding proton "lives" in this potential formed as a result of the interaction of two adjacent oxygen or

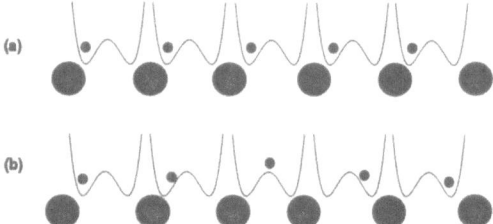

Figure 1: A small segment of a one-dimensional hydrogen-bonded chain with protons (small particles), ions (large particles) and double well potentials approximating the hydrogen bond. Protons as well as ions interact though nearest neighbor interactions that are not shown in the figure. In (a) a typical equilibrium configuration in the left minimum. In (b) a soliton is seen to be a domain wall that separates the protons that are in the left minima from those that are in the right minima. It can also be seen as a defect in an otherwise perfect lattice.

other ions. In addition to this interaction, protons interact with each-other through the dipole-dipole interactions of the atoms that also help energetically in the HBC formation. Finally, the protons interact also with the vibrational modes of the lattice (or chain) that has been formed.

A classical quasi-one dimensional model that includes the essential physical features of an HBC contains the following ingredients[1]-[8]: (i) individual protons are placed in double wells or similar potentials; (ii) adjacent protons interact in the simplest case with harmonic forces; (iii) the lattice of the ions vibrate and (iv) there is an indirect coupling of the protons motion to the ion lattice vibrations. Some of these features are pictorially presented in Figure 1. As with all physics modeling, this picture for the hydrogen-bonded systems attempts to retain all the essential physics without adding "unnecessary" complications. Ingredients (i) and (ii) together have immediate and unavoidable consequences for the proton dynamics itself. In a very large chain having these properties soliton-like waves are formed that can propagate without any dispersion. Technically, the nonlinear waves that are formed are called kinks and antikinks, but hereafter we will be using the term "soliton" to denote loosely a wave entity with particle-like properties. We note here that the initial question related to the dynamical properties of heavy particles such as protons is now addressed as a question involving waves. Waves emerge in large media that involve many coupled constituent parts; when an HBC is large enough, the individuality of each proton is lost and the coupling through (ii) of adjacent protons leads to an essentially continuous, one-dimensional medium. A small amplitude wave in this medium represents small vibrations of the protons from their equilibrium positions; these vibrations can propagate due to the coupling of the neighboring protons leading to low amplitude waves that propagate and disperse in a way not very different from the ripples in the sea. The presence of the local nonlinear potentials however, (responsible for the hydrogen bond itself), change qualitatively the picture when larger amplitude waves are formed. In this case dispersive waves propagate through the chain in the form of solitons. Physically these solitons represent small local *defects* that are formed when a single proton accidentally jumps in the wrong minimum of the double well. As a result of the coupling with adjacent protons, this local error gets somehow amplified leading to an entity that spreads over few lattice sites. If this local defect is spread enough, it can "slide" through the local atom periodicity leading to wave motion but of a local dispersionless nature. When the defect is more localized in only a few sites, the motion still retains some of the soliton propagation aspects but with some additional features. The birth of a proton-soliton in an HBC is thus a consequence of ingredients (i) and (ii).

Once a soliton is formed it interacts with the local lattice vibrations through (iii) and

(iv) leading to a deformation in the local vibrational modes of the ionic lattice. As a result an even more complex entity is formed, not very different from a polaron. When this new entity moves, it carries with it not only the soliton features attributed to the nonlinear proton potential but also the vibrational deformation of the underlying ionic lattice. The dynamics of this composite object is of interest in order to understand the physics of hydrogen-bonded systems.[1],[2],[7].

2. PROTON-SOLITON MOBILITY

Assume that we have an ensemble of particles, such as protons, that are considered classical particles and are moving in a force-free region of space; their mean square displacement from a given origin will grow quadratically in time. This type of motion is termed "ballistic". When, on the other hand, the same collection of particles executes diffusive-type motion, the particle mean square displacement grows linearly in time. Diffusive motion which typically occurs when a collection of particles interacts with a heat bath at high enough temperatures is asymptotically much slower than the ballistic motion. What type of law do the the proton-solitons follow in their dynamics? Do they move ballistically, diffusively or in some other fashion? The answer to this question is quite complicated in its details and depends very much on the specific soliton regimes. An ideal soliton is a "dual" object since it is a wave with particle-like properties. The particle aspect is manifested in the fact that the motion of a soliton that is not interacting with the lattice vibrations is ballistic. The wave nature of the soliton introduces a new feature; viz., the soliton moves in a fashion similar to a relativistic and not a newtonian particle. This means that there is a maximal velocity it can achieve equal to the velocity of the low amplitude-linear waves that can also propagate in the medium.

In order to make these statements more explicit, let us consider the one dimensional hydrogen-bonded chain depicted in Fig. (1a). If we ignore for the moment the motion of the heavy ions we are left with a simple mechanical model of masses (the protons), springs (not drawn in the figure and giving the effective interaction between the nearby protons) and double-well potentials (mimicking the hydrogen bond itself). The Hamiltonian for this system is presented in the Appendix in Eq. (4). For a soliton as long or (better) longer than the one shown in Fig. (1b) the continuum wave approximation is helpfull since it lends itself to an explicit solution for the soliton in that limit shown in Eq. (10). We note that the hyperbolic tangent shape of the kink is reminicent of a "wall" that separates two regions of the hydrogen bonded chain, viz., the one on the left where all the protons are in the left hand side minima of the double well and the one in the right where the protons are all located in the other minimum. The localized nature of this "transition region" is depicted better through the derivative of the kink solution that approximates (in the continuum limit) the relative proton displacements. The latter is given by

$$\frac{du_0}{dz} = \pm\frac{1}{\sqrt{2}}sech^2[\frac{z - z_0}{\sqrt{2}}] \tag{1}$$

We thus see that the soliton strain energy is well localized within a distance equal to about two (in the units of Eq. (1)) from the center point z_0. When this "composit object" moves, it behaves as a particle with an effective mass M_{cl} and follows the laws of relativity. This means that for velocities small compared with c_0, the sound speed of the medium, the soliton moves ballistically like a regular Newtonian particle, whereas for very large velocities (compared to c_0) its inertial properties are modified somehow. What is of interest here is that the effective mass of the kink is actually smaller than the mass of a single proton. The expresion for the soliton mass is derived in the Appendix and is given approximately by

$$M_{cl} \approx 5.963\text{x}10^{-5} \frac{c}{c_0} \frac{b(\text{Å})}{a(\text{Å})} \sqrt{\epsilon_0(eV)} \, m_p \tag{2}$$

where in the parentheses we have inserted the appropriate units. For a typical case where $c_0 \approx 10^5 m/sec$, $a = 2.7\text{Å}$, $b = 0.7\text{Å}$ and $\epsilon_0 = 0.5eV$ we have $M_{cl} \approx 0.01 m_p$. A wide soliton behaves like a light particle and propagates quite efficiently in the quasi-one dimensional hydrogen-bonded chain.

When the hydrogen-bonded chain is placed in a heat bath the individual protons are scattered by the temperature fluctuations and so are the solitons.[9] Since the proton-soliton behaves like a particle we should expect that the stochastic temperature forces (if weak enough so that they do not destroy it) kick it randomly in any direction, thus forcing it to execute diffusive motion. Even though the kink diffuses, it does so in a more complicated manner since its "internal structure" is also affected by the stochastic fluctuations. A manifestation of these phenomena is seen when an external electric field is applied in a hydrogen-bonded chain that has solitons. The kinks respond to the presence of the field and start an accelerating motion that reaches a terminal velocity when the rate of kinetic energy loss from dissipation balances the input rate due to the field.[2] This terminal velocity is seen to be proportional to the applied field (for small enough fields), as expected from the linear response theory. The kink velocity is also seen to have a peculiar nonmonotonic dependence on the temperature of the sample: At low and high temperatures the terminal speed is seen, as expected, to increase with the applied field. There is, however, an intermediate temperature regime where the velocity is seen to decrease with the field.[6],[8]. This novel behavior is directly attributed to the fact that the soliton is not just a point particle but a point-like object with internal structure.

3. QUANTUM TUNNELLING

One of the most interesting questions regarding solitons and their propagation is related to their quantum nature. In the absence of quantum fluctuations the soliton behaves like a classical particle with effective mass M_{cl} that, as we saw previously, is actually smaller than the proton mass in some typical cases of interest. On the other hand, we know that we live in a quantum world and thus, when a proton is considered as a quantum mechanical particle, it has some probability to tunnel through the barrier from one minimum to the other. This tunnelling probability (which is certainly very small since protons are quite heavy particles) is even smaller when the barrier height ϵ_0 or the location $\pm b$ of the minima with respect to the maximum are large numbers. However, when either (or both) ϵ_0 or b decrease the proton tunnelling probability increases and quantum mechanics becomes more important in the problem. In such a case, one ought to address the question of the "fate" of the soliton in the presence of quantum fluctuations. Since tunnelling enhances motion we should expect that a quantum soliton will be even "more mobile". If many of them are present in the system they can form a small band similar to the electronic one with a bandwidth proportional to the (small) tunneling rate.

It is possible to calculate analytically the lowest (proportional to $\hbar$) semiclassical correction to classical kink energy through the ingenious procedure found by Dashen, Hasslacher and Nevue (DHN)[10], [11]. The DHN formula, adopted to a typical hydrogen bonded chain, reads (from the Appendix):

$$E \approx \sqrt{\epsilon_0(eV)} \ \{ \ 1.987 \ b(\mathring{A}) \ - 0.072 \ \frac{1}{b(\mathring{A})} \} \tag{3}$$

where E is the total energy content of the soliton that includes the classical mass term and its lowest order semiclassical correction. In Eq. (3) the units of the variables have been included and the total energy is measured in electron-Volts. We observe that the semiclassical correction enters with a negative sign, reducing thus the energy (or mass) content of the classical proton-soliton. Even though the DHN formula is strictly valid in the weak coupling limit where the barrier height energy is large, we could address the question of the regime in which the term due to quantum fluctuations becomes comparable to the classical mass term. We expect that the breakdown of the DHN expression will give an indication for the parameter regime where quantum fluctuations begin to become important and influence substantially the physics of the problem. As is readily seen from Eq. (3), with the assumption of proton spring frequency of $\omega_0 = 3x10^{14}$, the breakdown regime is close to $b \approx 0.2\text{Å}$. Even though this value is far from the typical hydrogen-bonded values of $b = 5 - 7\text{Å}$, it is certainly in close range and it might be realizable in some systems under pressure. This preliminary discussion of the role of quantum fluctuations shows that quantum effects might be very important in hydrogen-bonded chains and that a more in-depth study is necessary.

4. CONCLUSIONS

We presented a brief qualitative discussion of the possible role of solitons in hydrogen-bonded chains. The existence of kink-like solitons in these chains is unavoidable once we take into account the nonlinear nature of the hydrogen bond and the coupling between the protons. The solitons are seen to be composite or collective objects that behave dynamically like particles. Their internal structure is manifested in some dynamical quantities such as in mobility calculations. Protons could propagate more efficiently in the form of solitons through a hydrogen-bonded chain. This fact can have interesting ramifications in biological HBC's where protons routinely use these chains to move across cellular membranes.

There are many unanswered questions regarding the presence solitons in hydrogen-bonded networks. A very important one involves the influence of quantum fluctuations. It is possible that in some cases quantum fluctuations become important, leading thus to novel phenomena[12]. Another issue that needs to be addressed is the departure from the continuum limit in the case of very narrow domain walls. In this case, even though the soliton picture might be useful, new phenomena arise due to the lattice discreteness. These theoretical issues need to be resolved before there is a clear guide for the experimental observation of proton-solitons.

APPENDIX

Calculation of the classical soliton mass

We start with the following proton Hamiltonian:

$$H = \sum_n \frac{1}{2}m_p(\frac{dy_n}{dt})^2 + \frac{1}{2}m_p\omega_0^2(y_{n+1} - y_n)^2 + \epsilon_0(1 - \frac{y_n^2}{b^2})^2 \qquad (4)$$

where m_p is the mass of the proton, ω_0 is the oscillation frequency of the proton chain, ϵ_0 is the barrier height of the hydrogen bond potential and y_n is the proton displacement from an equilibrium position take to be the center of each potential. We assume that the lattice spacing is a and the the equilibrium positions of the double well are at $\pm b$. The definitions

$$u_n = \frac{y_n}{a}, \quad \alpha = \frac{a}{b}, \quad \beta = \frac{\epsilon_0}{E_0}, \text{ and } \tau = \omega_0 t \qquad (5)$$

with the energy scale $E_0 = ma^2\omega_0^2$ render the dimensionless Hamiltonian

$$H' = \frac{H}{E_0} = \sum_n \frac{1}{2}(\frac{du_n}{d\tau})^2 + \frac{1}{2}(u_{n+1} - u_n)^2 + \beta(1 - \alpha^2 u_n^2)^2 \qquad (6)$$

In the long wavelength limit we can replace the index n by the continuous variable x, or the dimensionless length $\tilde{z} = x/a$, thus having $u_n(t) \approx \tilde{u}(z,t)$ and $u_{n+1} - u_n \approx \frac{\partial \tilde{u}}{\partial \tilde{z}}$. In this fashion we can go from the Hamiltonian of Eq. (6) to the Lagrangian density

$$\tilde{\mathcal{L}} = -\frac{1}{2}(\frac{\partial \tilde{u}}{\partial \tau})^2 + \frac{1}{2}(\frac{\partial u}{\partial \tilde{z}})^2 - \beta(1 - \alpha^2 \tilde{u}^2)^2 \qquad (7)$$

that, with the definitions $u = \alpha\tilde{u}$ and $z = 2\sqrt{\beta}\alpha\tilde{z}$ and by dropping the kinetic energy term (since we are interested primarily in the stationary solutions), can be rewritten as

$$\mathcal{L} = \frac{\tilde{\mathcal{L}}}{4\beta} = \frac{1}{2}(\frac{\partial u}{\partial z})^2 - \frac{1}{2}u^2 + \frac{1}{4}u^4 + \frac{1}{4} \qquad (8)$$

The Euler-Lagrange equation for the static $\mathcal{L}$ is

$$\frac{\partial^2 u}{\partial z^2} + u - u^3 = 0. \tag{9}$$

Equation (9) can be easily solved leading the well known localized soliton (kink and antikink) solution

$$u_0(z) = \pm \tanh[\frac{z - z_0}{\sqrt{2}}] \tag{10}$$

The dimensionless energy of the soliton is given by

$$E_{cl}^d = \int dz \, \mathcal{L}(u_0(z)) = \frac{2}{3}\sqrt{2} \tag{11}$$

whereas, after taking into account all changes of variables, we have in dimensional units:

$$E_{cl} = \frac{4}{3}\sqrt{2}(\frac{b}{a})\sqrt{E_0\epsilon_0} = \frac{4}{3}\sqrt{2\epsilon_0 m_p c^2} \frac{\omega_0 b}{c} = \frac{4}{3}\sqrt{2\epsilon_0 m_p c^2} \frac{b}{a}\frac{c_0}{c} \tag{12}$$

where c is the speed of light. We note that $c_0 = a\omega_0$ is the speed of sound in the chain. Since the soliton solution is relatilvistically covariant, this energy corresponds, through the Einstein formula, to a mass that can be found if we divide the energy of Eq. (12) by the square of the maximum traveling speed in the medium c_0. Using therefore the identities $E_{cl} = M_{cl}c_0^2$ and $E_p = m_p c^2$, where $E_p \approx 1 GeV$ is the energy associated with the rest mass of the proton, we obtain the expression of the kink mass in units of the proton mass:

$$M_{cl}/m_p = \frac{E_{cl}}{E_p}(\frac{c}{c_0})^2 = \frac{4}{3}\frac{b}{a}\sqrt{\frac{2\epsilon_0}{m_p c^2}} (\frac{c}{c_0}) \tag{13}$$

In a typical system (such as ice), $\epsilon_0 \approx 0.5 eV$, $\omega_0 \approx 3.0 \times 10^{14} sec^{-1}$, $a = 2.7 \overset{\circ}{A}$ leading to $c_0 \approx 10^5 m/sec$ and $mc^2 = 1 GeV$. Taking $b \approx 0.7 \overset{\circ}{A}$ we estimate the classical soliton mass to be $M_{cl} \approx 0.01 m_p = 1.0 \times 10^{-2} m_p$. We notice that the soliton mass is considerably smaller than the proton mass and thus protons move more efficiently in the chain in the form of solitons. This result is compatible with the findings of reference [5]

Semiclassical correction

When quantum fluctuations are taken into account the individual protons can tunnel between the two minima of the double well leading to a possibility for soliton tunneling. For large enough barriers, it is possible to use an extension to the WKB approximation and calculate the lowest order semiclassical correction to the classical kink energy. The semiclassical contribution stems essentially from the proton oscillations in the bottom of each well and thus it is proportional to the local oscillation frequency of the minima. In dimensional variables we have for the corrected energy[10][11]

$$E = \frac{4}{3}\sqrt{2\epsilon_0 m_p c^2} \frac{b}{a}\frac{c_0}{c} + \sqrt{6}(-\frac{3}{\pi\sqrt{2}} + \frac{1}{2\sqrt{6}}) \sqrt{\frac{\epsilon_0}{m_p c^2}} \frac{\hbar c}{b}$$

$$= \sqrt{\epsilon_0}\{\frac{4}{3}\sqrt{2}\sqrt{m_p c^2} \frac{\omega_0}{c} b + \sqrt{6}(-\frac{3}{\pi\sqrt{2}} + \frac{1}{2\sqrt{6}}) \frac{\hbar c}{\sqrt{m_p c^2}} \frac{1}{b}\} \tag{14}$$

$$\approx \sqrt{\epsilon_0(eV)} \{ 1.987 \, b(\overset{\circ}{A}) - 0.072 \frac{1}{b(\overset{\circ}{A})}\} \tag{15}$$

where in Eq. (15) we used for $\hbar c = 1973.28 \, eV - \overset{\circ}{A}$ and $\omega_0 = 10^{14} sec^{-1}$. The semiclassical correction enters with a negative sign leading thus to a more mobile soliton due to quantum fluctuations. The expression in Eq. (15) breaks down at $b \approx 0.2\text{Å}$ thus marking a regime where quantum fluctuations are very important. For different parameter values, the onset of this regime might occur for even larger b-values. A more thorough investigation of the quantum nature of the HBC's through Quantum Monte Carlo methods that is in progress will shed more light in this issue[13]. We conclude that quantum fluctuations should not be considered entirely negligible in the modeling of hydrogen bonded systems.

Acknowledgments

I wish to thank David K. Campbell for helpful discussions on the quantum nature of solitons and Bill Deering for critical comments on the manuscript. Partial support from TARP grant 003656-073c is also acknowledged.

References

[1] V. Ya. Antonchenko, A. S. Davydov and A. V. Zolotaryuk, Phys. Status Solidi B **115**, 631 (1983).

[2] M. Peyrard, St. Pnevmatikos and N. Flytzanis, Phys. Rev. A **36**, 903 (1987).

[3] St. Pnevmatikos, Phys. Rev. Lett. **60**, 1534 (1988); G. P. Tsironis and St. Pnevmatikos, Phys. Rev. B **39**, 7161 (1989).

[4] H. Desfontaines and M. Peyrard, Phys. Lett. A **142**, 128 (1989).

[5] A. Zolotaryuk and St. Pnevmatikos, Phys. Lett. A **143**, 233 (1990).

[6] E. S. Nylund and G. P. Tsironis, Phys. Rev. Lett. **66**, 1886 (1991).

[7] St. Pnevmatikos, G. P. Tsironis and A. Zolotaryuk, J. Mol. Liquids **41**, 85 (1989).

[8] E. Nylund, K. Lindenberg and G. P. Tsironis, J. Stat. Phys. **70**, 163 (1993).

[9] M. A. Collins, A. Blumen, J. F. Currie and J. Ross, Phys. Rev. B **19**, 3630 (1979).

[10] R. F. Dashen, B. Hasslacher and A. Neveu, Phys. Rev. D **10**, 4130 (1974).

[11] R. Rajamaran, Phys. Rep. **21**, 227 (1975).

[12] F. Bruni, G. Consolini and G. Careri, J. Chem. Phys. **99**, 538 (1993).

[13] D. K. Campbell, private communication.

Neutron scattering studies of biopolymer-water systems: solvent mobility and collective excitations

H.D. Middendorf

Clarendon Laboratory, University of Oxford,
Oxford OX1 3PU, U.K.

1. INTRODUCTION

Our understanding of biopolymer-water interactions at the molecular level is still rudimentary. The past 10 years have seen much progress in computer simulations of hydrated macro-molecular systems [1-3], and several new or improved experimental techniques have been applied to problems in this area [4-8]. Although we know a good deal now about a number of particular systems (mainly DNA and a few small globular proteins), the structural and dynamical complexity of hydrated biomolecules is such that it is not easy to integrate results from disparate experimental approaches into a coherent picture and to make quantitative contact with theory. The latter is true especially for numerous analytical and simulation studies of transport processes, nonlinear excitations, and critical phenomena. Such studies often depend crucially on a good characterisation of dissipative processes and on accurate model parameters derived from experiments.

One basic reason for the gap between theory and experiment here is that the more commonly used experimental techniques do not adequately cover the important transition region between vibrational and diffusive processes. However, modern neutron scattering techniques are able to provide genuine spatiotemporal information over the relevant parameter domain (see Fig. 1) [9-14]. This domain encompasses the transition from localised, high-frequency vibrations to collective excitations characterised by longer space and time scales, and further to predominantly diffusive modes of motion. Spatially it extends from scale lengths of the order of 1 Å (bond lengths, radius of water molecule) to lengths in the 100 to 1000 Å region (cluster diameters, interstitial dimensions, correlation or persistence lengths). Temporally, this domain is bounded by the highest frequencies of phonon or phonon-like excitations and the slowest random-walk time scales associated with macromolecular diffusion. It is in this space-time domain that living systems have evolved ingenious ways of sustaining complex molecular interaction processes in an aqueous, largely dissipative environment. Of fundamental interest here are the coupling between hard and soft degrees of freedom, the dynamics of hydrogen-bonded molecular components, and the effect of hydration on collective modes.

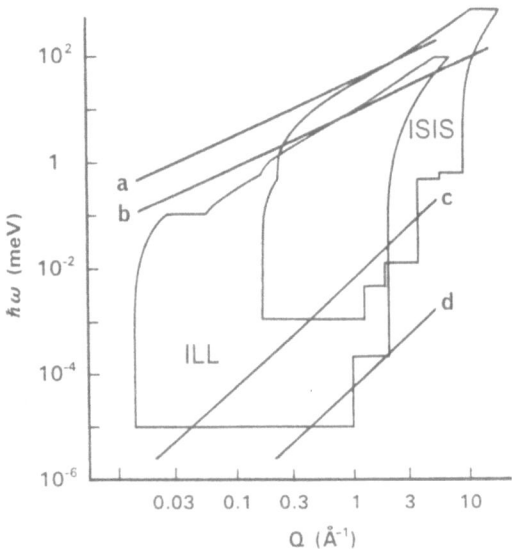

Fig. 1. Envelopes of the Q,ω-domains of commonly used spectrometers at ILL (IN5, IN6, IN10, IN11) and at ISIS (IRIS, TFXA, MARI, HET), for standard operating parameters. Straight lines: momentum-energy relations for acoustic phonons [(a) and (b), $V_{ac} = 3000$ and 1000 m/s] and for diffusion [(c) and (d), $D_s = 10^{-6}$ and 10^{-8} cm^2/s].

2. BASIC RELATIONS AND SCATTERING PROPERTIES

Thermal or cold neutrons with de Broglie wavelengths from 2 to 20 Å have energies between 20 and 0.2 meV, compared to ≈ 10 keV for X-ray quanta. Because of this, it is possible with neutrons to go beyond diffraction and do something that cannot be done with X-rays: to energy-analyse the intensity recorded at each scattering angle, in addition to measuring its variation with angle [9,10]. For unpolarised neutrons, each elementary scattering event (i.e. neutron-nucleus collision) is governed by two conservation equations defining momentum and energy transfer:

$$\hbar Q = \hbar (k_f - k_o) \qquad\qquad \text{(momentum)} \qquad (1)$$
$$\hbar \omega = E_f - E_o = \tfrac{1}{2} m (v_f^2 - v_o^2) = (\hbar^2/2m)(k_f^2 - k_o^2) \qquad \text{(energy)} \qquad (2)$$

Here E, v and $|k| = k = 2\pi/\lambda$ refer to the energy, velocity, wavenumber and wavelength of incident neutrons (subscript o), or neutrons scattered into solid angle element $d\Omega$ around a scattering angle 2θ (subscript f). The neutron rest mass is m; $\hbar$ is Planck's constant divided by 2π. Eqs. (1) and (2) can be solved for Q^2 as a function of either $\varpi = \hbar\omega/E_o$ (cold-neutron spectrometers on reactor sources) or $\hbar\omega/E_f$ (pulsed-source spectrometers). In the former case, the result is

$$Q^2 = k_o^2 [2 + \varpi - 2(1+\varpi)^{1/2} \cos 2\theta]$$
$$= 4k_o^2 (\sin^2\theta + 0.5\,\varpi \sin^2\theta + 0.0625\,\varpi^2 \cos 2\theta + \dots) \qquad (3)$$

For $\varpi = 0$ this reduces to the Q-value for elastic scattering, $Q_o = (4\pi/\lambda_o)\sin\theta$. In *quasi-elastic scattering*, the nondimensional energy transfer ϖ is small and this leads to Doppler-like broadenings of the elastic 'line' by diffusive motions. For *inelastic scattering* proper, the full Eq. (3) is needed to describe the scattering kinematics.

With the exception of spin-echo instruments, all neutron spectrometers provide data in the form of a double differential cross-section, $d^2\sigma/d\Omega\,dE$. For the simplest case of a monatomic assembly of target nuclei this is given by

$$d^2\sigma/d\Omega\,dE \sim N\,(k_f/k_o)\,[\,\sigma_{inc}\,S_{inc}(Q,\omega) + \sigma_{coh}\,S_{coh}(Q,\omega)\,] \tag{4}$$

where σ_{inc} and σ_{coh} are incoherent and coherent scattering cross-sections, respectively. Neutron scattering from any natural (i.e. not covalently deuterated) biopolymer is predominantly incoherent because of the large incoherent cross-section of protons, $\sigma_{inc}^{p} = 79.7 \times 10^{-24}\ cm^2$. This is an order of magnitude larger than the cross-section of all other nuclei of interest. The incoherent component of the scattering relates to single-particle correlations and therefore carries information on diffusive processes or localised excitations. In partially or fully deuterated systems the coherent scattering is important and is governed by two-particle correlation functions describing collective processes such as phonons.

Quantitatively, all analyses of $d^2\sigma/d\Omega\,dE$ data aim to extract or model the dynamic structure factors $S_{inc}(Q,\omega)$ and/or $S_{coh}(Q,\omega)$ [10,14]. These functions represent energy-resolved generalisations of the static structure factors $S_{inc}(Q) \approx 1$ and $S_{coh}(Q) \equiv S(Q)$ measured in neutron diffraction experiments. The connection between $S_{inc}(Q,\omega)$ or $S_{coh}(Q,\omega)$ and the atomic dynamics is established by time Fourier transforms, the so-called 'intermediate' scattering functions $F_{inc}(Q,t)$ and $F_{coh}(Q,t)$. They play a key role in neutron scattering because a subsequent $Q \leftrightarrow r$ Fourier transformation yields two functions depending only on r and t, the van Hove space-time correlation functions. The latter relate the experimentally accessible quantities (mainly S, but also F) to a substantial body of theoretical work on the time-dependent statistical mechanics of systems of interacting particles, thus providing a basis for interpretation that is often more 'direct' than that of other techniques.

Neutron $S(Q,\omega)$ data from experiments on hydrated polypeptides, polynucleotides and polysaccharides are gradually deepening our understanding of biopolymer-water interactions, and quantitative comparisons with simulations are beginning to be made [3]. However, the volume of neutron data on complex systems of this kind is still small owing to the relatively low flux of even the best neutron sources and the fact that world-wide there are only about a dozen high-performance neutron spectrometers. Specifically, the potential of quasi-elastic and inelastic neutron scattering in this area is due to three factors:

(i) The Q,ω-range now accessible by instruments at the major research centres extends over 9-10 decades in energy ($10^{-9} \lesssim \hbar\omega \lesssim 10$ eV or $10^{-5} \lesssim \hbar\omega \lesssim 10^5$ cm^{-1}), and spatially over 3-4 decades ($0.01 \lesssim Q \lesssim 10$ Å^{-1}). There is substantial overlap with complementary information from IR and Raman spectroscopy at intermediate and high frequencies, and with NMR, ESR, fluorescence and microwave techniques towards lower frequencies.

(ii) The theory and practice of hydrogen/deuterium contrast variation, well developed in structural work with neutrons, carry over to spectral analyses of both the incoherent and coherent scattering. By exchange with H_2O/D_2O buffers and/or covalent deuteration, it is possible to create a wide range of contrast and to accentuate or 'fade out' the scattering due to particular constituents.

(iii) The simplicity of point-like nuclear scattering facilitates the interpretation of $S(Q,\omega)$ data in terms of frequency distributions and correlation functions, and makes neutron scattering by far the most informative counterpart to analytical and simulation work.

The structural basis for neutron studies of the dynamics of biological macromolecules is excellent. Diffraction techniques using mainly X-rays, but also neutrons and electrons, have given us superb 3D pictures of the atomic structure of hundreds of biomolecules. Highly resolved Fourier maps also reveal the locations of many closely associated water molecules in the primary hydration shells. The remarkable polymorphism and hydrogen-bonding properties of biopolymers make for a rich diversity of secondary, tertiary and quaternary structures. This structural complexity is reflected in a wide range of dynamical processes, both collective and diffusive. Processes that are individually well known from simpler systems are not only present simultaneously, but may interact to give novel effects. Even in the case of a well-characterised protein crystal we are dealing with a heterogeneous system in which substantial structural and dynamical changes occur over a narrow (5-10 Å), geometrically irregular transition region. This divides, or couples rather, two 'phases' with different properties: the bulk biomolecule, and the bulk-like crystal water or solvent. As with any complex system, it will be useful to examine limiting cases. We can, for example, consider the limits of small ($h \ll 1$) and large ($h \gg 1$) specific hydration, i.e. water content expressed as $h = $ g water / g dry biopolymer. The intermediate region, where h is of order unity, is best represented by protein crystals, but also — over a wider h-range — by biopolymer gels.

To illustrate the scope of neutron experiments on biopolymer-water systems and to show a few representative results, the following sections focus on two systems: a globular, slightly hydrated protein, and a highly hydrated network of polysaccharide fibres. The experimental work discussed in this paper was done mainly at the Institut Laue-Langevin (ILL, Grenoble), the Laboratoire Léon Brillouin (LLB, Saclay), and the ISIS Pulsed Neutron Facility (Chilton near Oxford).

3. GLOBULAR PROTEINS AT LOW HYDRATION

3.1 Hydration dynamics of phycocyanin in the sorption regime

In sorption studies, an amorphous or partially ordered sample is hydrated in steps from a nearly anhydrous state up to a level where there is sufficient water for the molecules to interact with each other and to show specific functional responses (such as enzymatic activity). The initial or reference state is often referred to as 'dry', although a certain number of internal water molecules (in well-defined positions) and some tightly bound surface waters will always be present in a protein. These must be regarded as an integral part of the structure, and amount to residual specific hydrations h_o in the region of 0.03 to 0.08. Depending on the stability of the tertiary structure and the conditions used, a protein may or may not undergo

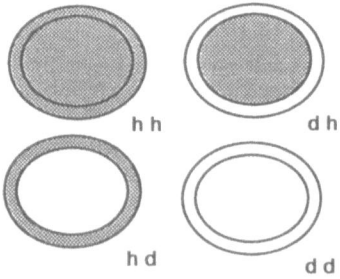

Fig. 2. Globular protein with hydration shell (schematic): the four main hydrogen (h) and deuterium (d) contrast combinations.

observable hydration-induced conformational changes in the range $h_o > h \gtrsim 0.35$.

Gravimetric, calorimetric and spectroscopic techniques have been used extensively to investigate many aspects of biomolecular sorption processes [6,8,9,16]. By measuring sets of contrast-dependent neutron $S(Q)$ and $S(Q,\omega)$ data at a number of hydration levels along a sorption isotherm, it is possible to examine in considerable detail the structural and dynamical changes accompanying the gradual activation of the various degrees of freedom of a protein and its interaction with increasing amounts of closely associated water. Of particular interest here are space and time resolved data that relate to proton transport processes [15] and percolation phenomena [8,16].

So far only one protein has been studied in sorption experiments covering a substantial fraction of the Q,ω-domain shown in Figure 1. This is C-phycocyanin, a light-transducing chromoprotein extracted from blue-green algae (*Synechococcus lividus*) that tolerate mass cultivation in perdeuterated media [17]. These properties made C-phycocyanin a 'guinea pig' protein for neutron studies since the first such experiments at Harwell in the early 1970 s. X-ray crystallography to 2-3 Å resolution has shown its secondary structure to be 62% α-helical [18]. C-phycocyanin can exist in various aggregation states, the basic building block being an αβ-heterodimer of molecular weight 29000-30000. The $(\alpha\beta)_6$ molecule is shaped like an oblate ellipsoid (diameter 110 Å, thickness 40 Å) with a central solvent channel and radial clefts.

A protein that can be biosynthetically deuterated offers much scope for neutron difference spectroscopy using H/D contrast variation [19-25]. If we exclude the more complicated possibilities of partial or selective covalent deuteration, we can measure $S(Q,\omega)$ for four samples as shown schematically in Figure 2. Remembering the fact that hydrogens make up roughly one half of the atoms in a protein, it is possible in such experiments either to focus on the water dynamics at and near the protein surface, or on the effect of hydration on the intramolecular dynamics of the polypeptide globule. For $Q \lesssim 0.1$ Å^{-1}, moreover, the Brownian dynamics of larger entities such as protein clusters and functional assemblies can be probed.

3.2 High-resolution quasi-elastic scattering, $\hbar\omega < 100$ μeV

To begin with the 'dd' case, we can D_2O-hydrate a fully *in vivo* deuterated C-phycocyanin sample (d-PC) in steps from the nearly anhydrous state (residual hydration $h_o \approx 0.05$). At energy transfers $\hbar\omega$ in the 10 to 100 neV range, using the spin-echo spectrometer IN11 at Grenoble, it was found that the normalised intermediate scattering function, $F_{coh}(Q,t)/F_{coh}(Q,0)$, is essentially constant within $0.02 < Q < 0.25$ Å^{-1} for gently dried powder samples of d-PC. When this protein is hydrated slightly, F_{coh} picks up a small time-dependent component which at low Q is mainly due to the activation of restricted Brownian motions of the subunits of d-PC with effective diffusion coefficients of 1 to 2×10^{-6} cm^2/s [19]. At somewhat higher $\hbar\omega$, measuring $S(Q,\omega)$ on backscattering spectrometers covering $0.15 < Q < 5.5$ Å^{-1}, the signals are essentially elastic for small h. With increasing D_2O hydration, up to equivalent monolayer coverage of the protein, it is possible to detect small difference broadenings of 50 to 100 neV, but these are at the limit of observability by backscattering techniques (Fig. 3). Since we are looking mainly at the bulk protein here, the conclusion is that its structure 'loosens up' only slightly with the formation of a network of hydrogen bonds around charged side chains and polar groups at the surface and in the clefts or channels between subunits. For Q values below about 0.6 Å^{-1}, or scale lengths $2\pi/Q > 10$ Å, this is corroborated by $S(Q)$ measurements: total intensity scans using IN11 (which yields S_{coh} separated from S_{inc}) show that upon hydration a broad subsidiary diffraction peak appears in the 0.1 to 0.3 Å^{-1} region (Fig. 4). With increasing h, this peak shifts towards lower Q as a

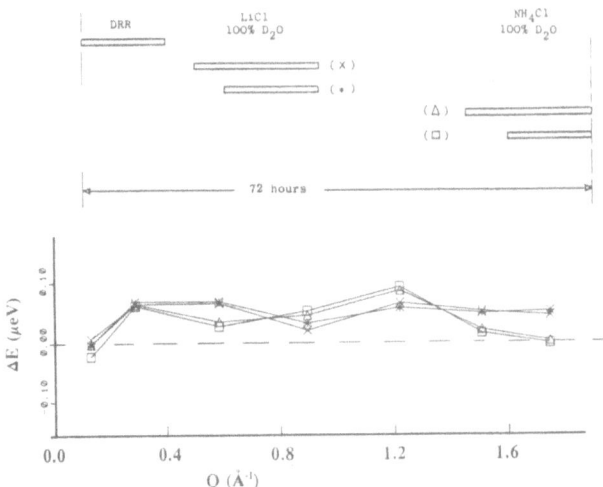

Fig. 3. Lower: Difference line broadenings ΔE as a function of Q for D₂O-hydrated d-PC at two hydration levels, relative to broadening measured for the anhydrous sample. Upper: Hydration history during 72 hours of on-beam exposure while recording quasi-elastic spectra, from dry reference state (DRR) to successive equilibration over saturated LiCl and NH₄Cl solutions (both 100 % D₂O). Different exposure times demonstrate equilibrium reached.

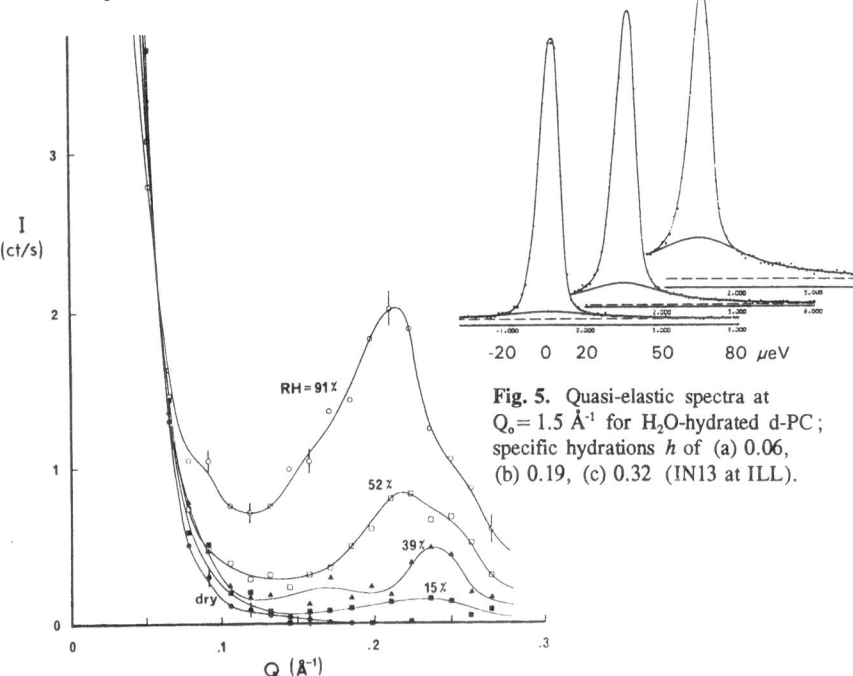

Fig. 5. Quasi-elastic spectra at $Q_o = 1.5$ Å⁻¹ for H₂O-hydrated d-PC; specific hydrations h of (a) 0.06, (b) 0.19, (c) 0.32 (IN13 at ILL).

Fig. 4. Coherently scattered intensity $I \sim S(Q_o)$ for anhydrous d-PC and for H₂O-hydrated d-PC at four relative humidity (RH) levels (T = 22°C).

consequence of the gradual build-up of hydration-induced order in the three-dimensional arrangement of the subunits of d-PC with correlation lengths of 100-200 Å. At higher momentum transfers, as shown by S(Q) measurements over $0.7 < Q < 14$ Å^{-1} at Saclay [25], hydration has little effect on the structure factor of d-PC as long as $h \lesssim 0.25$.

This situation changes considerably upon hydration with H_2O (case 'hd'). It is now feasible, by virtue of the dominant incoherent proton cross-section, to observe quasi-elastic line broadenings which relate mainly to the functionally important outer shell of the protein, i.e. to the single-particle dynamics of the water of hydration together with that of H_2O-hydrated side chains at or very near the surface (primarily those of the amino acid residues Asp, Glu, Lys, and Arg). Concomitant changes in the dynamics of the d-PC interior can be treated as correction terms.

In the sorption regime, i.e. for specific hydrations $h \lesssim 0.5$, it is possible to distinguish at least two components in the scattering, and Figure 5 shows as an example three quasi-elastic spectra measured for H_2O-hydrated d-PC. The quantities of interest are the total intensities S(Q) and the deconvoluted widths $\Delta E(Q)$ of the central and broad components, $S^c(Q,\omega)$ and $S^b(Q,\omega)$, considered separately as well as in relation to each other. The information contained in data of this kind is similar to that derived from quasi-elastic light scattering, except that owing to their much lower wavelengths neutrons are sensitive to a variety of diffusive motions over distances in the 1 to 50 Å range. The broad component seen in Figure 5, i.e. $S^b(Q,\omega)$, relates primarily to scattering from restricted diffusion (slow librational and hindered rotational modes) of water molecules hydrating the polar and charged side chains at or near the surface of d-PC. This component is barely detectable at very low hydrations and becomes measurable when the water uptake has reached a level roughly equivalent to ½ monolayer coverage. The hydration difference broadenings derived from $S^b(Q,\omega)$ measurements are in the 20 to 200 μeV range, corresponding to librational and rotational correlation times of the order of 10^{-10} s.

Lineshape analyses show that the time scales for translational and rotational motions are well separated up to $h \approx 0.5$, or values of water uptake equivalent to about 1.5 monolayers per $\alpha\beta$-subunit of d-PC. In this region the hydration process can be characterised by elastic incoherent structure factors (EISF, [13]) measured on backscattering spectrometers and on cold-neutron time-of-flight spectrometers with resolutions of 10-20 μeV or better. These EISF(Q;h) functions represent an average over the squared Fourier transform of all volume elements swept out by spatially restricted proton trajectories during librational or rotational motions of hydrated sidechains and associated water molecules that are confined to the primary hydration shell for times longer than a few 10^{-9} s.

To observe translational diffusive motions of H_2O molecules sorbed to d-PC, which give rise to very small broadenings of $S^c(Q)$ in Figure 5, one must increase the resolution 20-fold and look at the inner quasi-elastic region ($|\hbar\omega| \lesssim 5 \mu$eV) on a low-Q backscattering spectro-meter such as IN10 at Grenoble. In carefully designed on-beam hydration experiments, this instrument allows line shape changes of about 0.1 μeV to be detected, and hydration difference spectra then reveal an oscillatory Q-dependence of ΔE_{inc} between Q = 0.3 and 1.5 Å^{-1}. With increasing hydration, the first maximum of $\Delta E_{inc}(Q)$ shifts from $Q_{max} \approx 0.4$ to 0.8 Å^{-1} while gaining in intensity by a factor of two, and the following minimum between Q_{min} = 0.7 and 1.1 Å^{-1} becomes progressively more shallow. This oscillatory width function, analysed in conjunction with S(Q) measurements up to $Q \approx 1.8$ Å^{-1}, is consistent with a Chudley-Elliott model [13] in which water molecules perform jump diffusion between a distribution of hydration sites on the surface of a globule. With increasing hydration, the average jump distance decreases from 9 to 6 Å, while the strong Q_{max}-dependence of hydration numbers points to an increased clustering of water molecules at these sites [19].

3.3 Quasi-elastic and inelastic scattering with $\hbar\omega \gtrsim 100$ μeV

Time-of-flight and crystal analyser spectrometers are being used to study the hydration dynamics of globular proteins over wider energy windows that may extend to several 100 meV. On most instruments it is possible to record quasi-elastic spectra simultaneously with the acquisition of inelastic spectra over Q,ω-domains covering the region of dispersive phonon modes together with the librational bands of interest (i.e. those of water between 50 to 100 meV). This dynamic range is achieved at the expense of energy resolution in the quasi-elastic region, which is 1 or 2 orders of magnitude coarser than on backscattering instruments.

Time-of-flight spectra from fully H_2O-hydrated d-PC ($h=0.5$) in the 100 to 333 K region were measured at Grenoble [23,24]. Analysis of the quasi-elastic scattering (resolution 80 to 115 μeV) in terms of the Volino-Dianoux model for restricted diffusion [13] showed that the EISF curve at 333 K resembles that obtained for jump diffusion of water molecules in small spherical volumes of radius $a \approx 3$ Å. Since the structure, distribution and steric freedom of sidechains is known in considerable detail from X-ray crystallography, more quantitative comparisons of EISF models with data for H_2O-hydrated d-PC are feasible but it must be remembered that the information obtained is highly averaged.

Moving further up in energy, we consider the acoustic phonon regime with energy transfers $\hbar\omega$ from ≈ 0.5 to 10 meV. Spatially extended and sufficiently regular structures are required to support phonons with distinct dispersion properties; in globular proteins any such excitations will be highly damped and of short range since α-helices and ß-sheets are normally not longer than 15-20 Å and often couple to a network of hydrogen-bonded water molecules. In a phycocyanin experiment using a triple-axis spectrometer [20], constant-Q scans showed weak side bands roughly symmetrical with respect to the central elastic line. Their energy shifts appeared to increase systematically with Q and hydration; mean-square fits to linear dispersion relations gave propagation velocities V_{ac} between 2200 (nearly dry d-PC) and 3300 m/s (D_2O-hydrated d-PC, $h=0.5$) (Fig. 6). Although the statistics for individual scans were poor, the data as a whole suggest a certain 'stiffening' of structural elements upon hydration. These Brillouin-like side bands have been interpreted as due to short-lived coherent excitations propagating over 10-15 Å in hydrogen-bonded structural elements (such as α-helices) and 'patches' of closely associated water molecules in the primary hydration shell. Excitations of this kind can only be observed with neutrons travelling faster than V_{ac}, i.e. in downscattering experiments (analogous to Stokes scattering in Raman work).

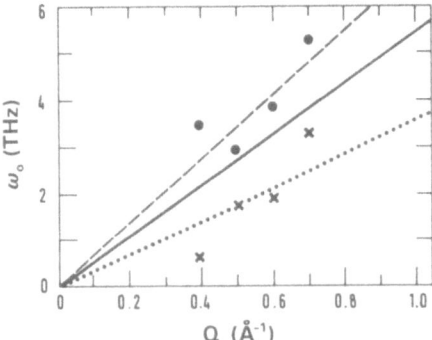

Fig. 6. Dispersion of short-wavelength collective excitations in perdeuterated C-phycocyanin: energy shifts ω_o (1 THz = 4.13 meV) of Brillouin-like side bands for dry ($\times$) and D_2O-hydrated protein ($\bullet$). The solid line gives the dispersion of the 'fast' sound mode in pure D_2O as measured by Teixeira et al. [Phys. Rev. Lett. **59**, 1780 (1985)].

Conventional Fermi-chopper instruments using slow neutrons ($\lambda_o \gtrsim 4$ Å) produce time-of-flight spectra covering the $\hbar\omega \approx 1\text{-}100$ meV region in upscattering (i.e. anti-Stokes scattering), and examples of phycocyanin spectra measured in this way are shown in Figure 7 [20]. It is seen, first of all, that a distinct hydration difference spectrum between the dry and the H_2O-hydrated sample is observed already at a relatively low hydration level (equivalent to about 80 H_2O molecules per $\alpha\beta$-subunit). The prominent peak at 61 meV is due to librational modes of the water of hydration; on partial desorption (not shown here) this peak shifts to higher energy transfers, indicative of tighter 'binding' and some loss of mobility.

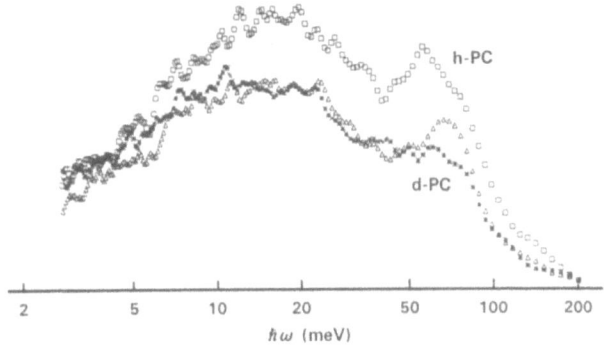

Fig. 7. Time-of-flight spectra (instrument IN5 at ILL, $Q_o = 1.1$ Å^{-1}) of perdeuterated C-phycoyanin : dry ($\triangle$), D_2O-hydrated (*) and H_2O-hydrated ($\square$) (both at $h = 0.045$).

4. HIGHLY HYDRATED SYSTEMS : POLYMER AND BIOPOLYMER GELS

4.1 Structure and dynamics of aqueous gels

Biopolymers, like most hydrophilic polymers, form aqueous gels, i.e. highly hydrated reticulate structures with a water content that in some cases can be as high as 99.8 %. Macroscopically, such gels are jelly-like solids or viscoelastic 'blobs' with unusual transport and rheological properties. They are excellent model systems for investigating cooperative processes in hierarchically structured two-component systems, such as helix formation, collapse transitions, and percolation phenomena [26,27]. Polysaccharide and polyacrylamide gels have important bioanalytical and biomedical applications, especially as electrophoretic media for separating nucleic acids or proteins [28].

From classical X-ray diffraction work, supplemented more recently by computer simulations, we know in atomic detail the primary and secondary structure of slightly hydrated polysaccharide fibres, principally agarose, hyaluronate, and the carrageenans. Our knowledge of the highly hydrated gel state was for a long time limited to results from techniques probing distances $d \gtrsim 1000$ Å (optical techniques) and motions with frequencies $\lesssim 10$ MHz (rheology). In recent years, electron microscopy and neutron diffraction studies of polysaccharide and polyacrylamide gels have provided data on their molecular structure from ≈ 20 Å to values of the order of 10000 Å; the log S(Q) vs. log Q plots for dilute agarose gels are linear over 1-2 decades with fractal exponents between 2 and 2.5 [29]. To understand quantitatively how the electrophoretic and rheological behaviour of aqueous gels is determined by their microscopic properties, it is essential to obtain data linking the space-time domains covered so far, and to examine in particular the transition region from long-range diffusive processes to more localised interactions at the (bio)polymer-water interfaces.

Recent neutron scattering work on biopolymer gels has focussed on agarose and hyaluronate [29-35]. Agarose is a nearly uncharged poly(disaccharide) with 4 OH groups per disaccharide unit. The double-helical secondary structure of agarose is known in atomic detail from X-ray diffraction, and water interactions have been characterised by analytical pair potentials obtained from *ab initio* simulations. In the gel state, hydrogen-bonded bundles of 10-30 double helices form a random 3D network of fairly stiff rods connected through junction zones. Hyaluronic acid (HA), on the other hand, is a highly charged polysaccharide capable of retaining large amounts of water in entangled meshworks with unusual rheological properties. HA possesses an equally high density of potential hydrogen-bonding sites (OH-groups and certain oxygens), but in addition it is a strongly polyanionic molecule due to K^+ or Na^+ ions complexed between 3 or 4 helical poly-saccharide strands. Electrostrictive forces in the primary hydration shell, in particular on water molecules close to these salt ions, may be expected to dominate the microscopic hydration properties of HA.

4.2 Quasi-elastic scattering

These polysaccharide gels behave as viscoelastic materials over scale lengths longer than several microns (i.e. lengths large compared with the maximum polymer network dimension), but microscopically (≤ 1000 Å) they are fluids except for the 'molecular scaffolding' of the polymer network or matrix. From the properties summarised above, it is clear that the rigidity of this 'scaffolding' can vary substantially and that Coulomb interactions may play an important role. The polymer matrix dynamics in agarose and HA gels are very different and couple to the water motions in complex ways that determine the rheological properties of the system as a whole. The dynamics of the matrix as such can be probed by spin-echo and backscattering experiments aimed at determining $F_{coh}(Q,t)$ and $S_{inc}(Q,\omega)$ for energy transfers $\hbar\omega$ in the neV to μeV region, corresponding to times in the μs to ns range. As an example, Figure 8 shows a log-log plot of the quasi-elastic difference widths of the central component for an agarose gel, relative to the same widths measured for a frozen sample which gives nearly elastic lines (as seen at 0.7 μeV resolution). The data show the existence of two distinct regions with a crossover at $Q \approx 0.5$ Å^{-1} where the slope changes from ≈ 3 to 1. This dependence can be interpreted in terms of current theories of the dynamics of polymer chains in solution. A more complete characterisation of the matrix dynamics in the μs to ns range will be possible on the basis of neutron spin-echo measurements using new or upgraded instruments at ILL.

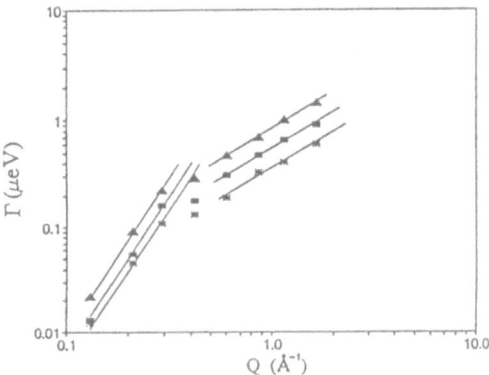

Fig. 8. Log-log plots of the widths $\Gamma(Q)$ (HWHM) of high-resolution quasi-elastic lines for a 0.075 % agarose-H_2O gel, measured on IN10 at ILL for temperatures $T = 283$ K (∗), 303 K (■), 326 K (▲). The continuous lines are linear fits to the low-Q data ($\sim Q^3$) and high-Q data ($\sim Q$).

The translational and rotational mobilities of water trapped by reticulate structures and porous solids are known to be altered substantially relative to those of bulk water, but it is difficult to assess quantitatively how these and related microscopic quantities vary as a function of distance from the (bio)polymer surface. The unique property of agarose to form well-defined gels which are solid-like down to polymer concentrations of around 0.2 % allows a 'perturbative' approach : quasi-elastic spectra can be measured for a concentration series covering the transition from pure H_2O [34] to a 0.2 % gel and further to more concentrated gels of up to 30 % or higher. It is thus possible to systematically probe changes in the molecular dynamics introduced by a random, increasingly dense 3D network of hydrophilic fibres carrying OH groups that mesh with and perturb the transient H-bond networks of water clusters. Because of the dominant proton scattering, the contribution of coherent scattering is negligible in the concentration range of interest here. The spectra obtained are essentially proportional to proton incoherent dynamic structure factors, $S_{inc}(Q, \omega)$, and parameters characterising the microscopic mobility are extracted from multicomponent lineshape analyses on the basis of models assuming different populations of mobile protons (Fig. 9).

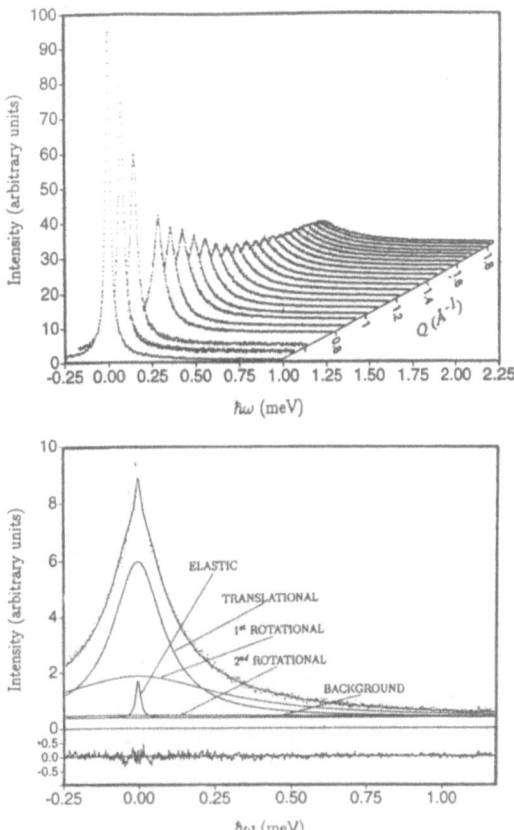

Fig. 9. Upper: Quasi-elastic spectra measured at ISIS (instrument IRIS) for a 1 % agarose-H_2O gel (277 K, momentum transfers Q from 0.3 to 1.5 Å^{-1}). Lower: Example of lineshape analysis for a single spectrum, showing the total fit, the decomposition into elastic, translational and rotational components, and the residual.

The main results of recent quasi-elastic scattering experiments on gels [30-33] may be summarised as follows :

(a) The dynamic properties of water contained by a network of polysaccharide fibres, i.e. of water that can be regarded as a close approximation to 'biological water', vary smoothly during the transition to zero polymer concentration; significant differences between agarose and HA gels are evident in the 0 to 1 % region.

(b) Gel water is less mobile than pure H_2O at the same temperature, and this is manifest in reduced self-diffusion coefficients D_s and residence times τ_o for jump diffusion. Rotational correlation times, on the other hand, are much less affected by the polymer network. The temperature dependence of τ_o suggests a lowered activation energy for translational jump diffusion relative to the bulk water values.

(c) The molecular dynamics of gel water at $T > 270$ K resembles that of pure supercooled water at some corresponding temperature between 250 and 270 K, but quantitatively there are differences in the relative contributions of translational and rotational modes. A 'structural temperature' lower than the actual thermodynamic temperature can be assigned to gel water.

(d) The restricted translational diffusion of biopolymer segments is characterised by effective D_s between 0.5 and 2.5×10^{-8} cm²/s.

On average, the translational broadenings Γ_t vs. Q^2 are well described by a Chudley-Elliott (CE) model with a random distribution of jumps to a shell of nearest-neighbour sites around any given proton. However, the high Q-resolution and good statistics of spectra from a pulsed-source spectrometer (IRIS at ISIS) also revealed small oscillatory components (Fig. 10) modulating the overall $\Gamma_t(Q)$ dependence. More detailed analyses suggest that these oscillations are related to density fluctuations involving on average 20 to 30 water molecules in transient clusters [35]. This interpretation is consistent with molecular dynamics simulations which model water as a continually restructuring gel-like network of H-bonded molecules. The transient clusters in models of this kind are characterised by correlation lengths of a few molecular diameters (6-8 Å). It appears that such effects are enhanced in gel water due a higher degree of transient structuring induced by the network of hydrophilic fibres.

4.3 Inelastic scattering

At energy transfers beyond the quasi-elastic region, i.e. $\hbar\omega \gtrsim 1$ meV, aqueous polymer and biopolymer gels offer opportunities for studying damped phonons in random networks that couple to a strongly hydrogen-bonded liquid. In agarose gels, in particular, the rod-like bundles of double helices support acoustic excitations together with a variety of low-energy segmental and breathing modes, all of them damped in different degrees by the surrounding hydration shell and interstitial water volumes. Results from preliminary experiments at Grenoble have shown significant non-Debye intensity increases in the density of states between ≈ 1 and 20 meV [31,36]. These deviations from classical Debye behaviour resemble those observed in similar measurements for a polymer with variable cross-linking [37], although it is unlikely that they could be due to fracton-like excitations propagating in the random polysaccharide network. More detailed experiments designed to measure Q-dependent frequency distributions as a function of temperature are in progress at ISIS; of greatest interest here are moderately dense agarose gels ($0.2 < h < 0.5$) at temperatures between 255 and 280 K.

Wet-spun fibres of HA are another focus of current neutron work on phonon propagation and damping in hydrated biopolymers. As shown by optical Brillouin scattering and X-ray diffraction, oriented HA fibres exhibit a remarkable hydration-induced phase transition accompanied by a sharp drop in the sound speeds parallel and perpendicular to the helix axes [38]. Neutron $S(Q,\omega)$ data for the principal H/D contrast combinations can be expected to reveal molecular details of this order-disorder transition and its underlying dynamics.

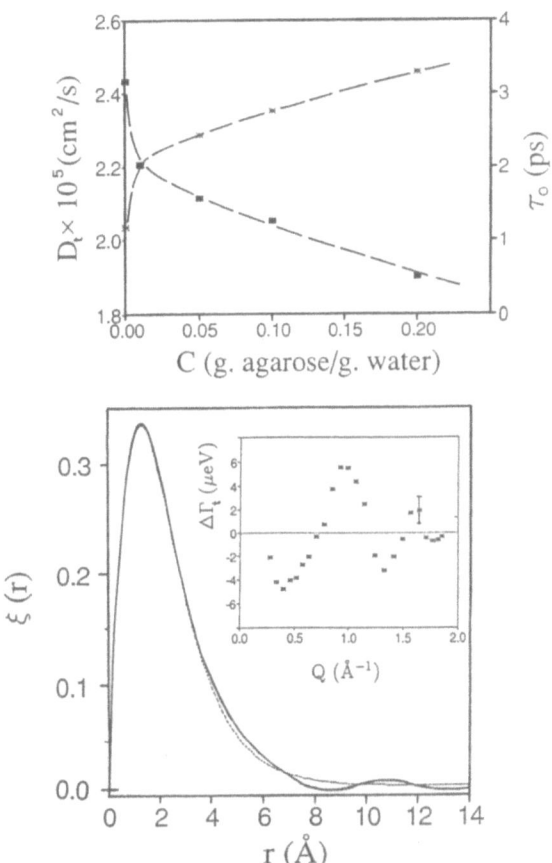

Fig. 10. Upper: Residence times τ_0 for translational jump diffusion (∗) and microscopic diffusivity D_t (■) at 294 K as a function of agarose concentration C. Lower: Radially symmetric jump length distribution obtained by Fourier inversion of experimental $\Gamma_t(Q)$ for a C=0.20 gel at 294 K (continuous line), and of $\Gamma_r(Q) = D_t Q^2/(1+D_t Q^2 \tau_0)$ (random CE model, dotted curve). Inset: Difference linewidths $\Delta\Gamma_t = \Gamma_{exp} - \Gamma_r$ of the translational component Γ_t for a C=0.20 agarose-H_2O gel at 294 K as a function of Q (the bar giving the average error of $\Delta\Gamma_t$).

5. CONCLUSIONS

In summary, the experiments and results discussed here demonstrate how neutron scattering provides new insights into the dynamics of hydrated biomolecules. The ability to do "diffraction-cum-spectroscopy" over a large Q,ω-range corresponding to the space and time scales of biopolymer-water interactions opens new avenues which can be exploited for testing and refining theoretical models and simulations. Outstanding assets of neutron techniques in this context are the capability to probe diffusive as well as cooperative processes over scale lengths from ≈ 1 to a few 100 Å, together with the possibility of varying scattering contrast between the constituents of a heterogeneous system.

The phycocyanin results draw attention to the fact that the 'energy landscape' of a hydrated biomolecule is very complex indeed [39]. Some of the concepts and models used in theoretical work will need to be refined substantially in order to accomodate data from experimental techniques capable of providing genuine spatiotemporal information. We conclude, in particular, that the characterisation of hydration phenomena by two or three discrete relaxation times is too simplistic, and also that two-state models ('bound' vs. 'free' water) are inadequate to describe protein hydration. It is quite clear, moreover, that the study of hydration and exchange processes in biomolecules is inseparable from a quantitative assessment of their intramolecular dynamics. Fully hydrated biomolecules are increasingly the subject of MD simulation studies, and we look forward to more detailed comparisons with $S(Q,\omega)$ data from neutron experiments.

References

[1] Karplus, M. and Petsko, G.A., *Nature* **347** (1990) 631.
[2] Goodfellow, J.M. (ed.), Molecular Dynamics: Applications in Molecular Biology (Macmillan Press, Basingstoke, 1990).
[3] Smith, J.C., *Quart. Rev. Biophys.* **24** (1991) 227.
[4] Kossiakoff, A.A., *Ann. Rev. Biochem.* **54** (1985) 1195.
[5] Saenger, W., *Ann. Rev. Biophys. Biophys. Chem.* **16** (1987) 93.
[6] Goldanskii, V.I. and Krupyanskii, Y.I., *Quart. Rev. Biophys.* **22** (1989) 39.
[7] Wüthrich, K., Otting, G. and Liepinsh, E., *J. Chem. Soc. Faraday Discuss.* **93** (1992) 35.
[8] Rupley, J.A. and Careri, G., *Adv. Protein Chem.* **41** (1991) 37.
[9] Palma, M.U., Parak, F. and Palma-Vittorelli, M.B. (eds.), Water-Biomolecule Interactions, Italian Phys. Soc. Conf. Proceedings, Vol. 43 (Editrice Compositori, Bologna, 1993).
[10] Middendorf, H.D., in Neutrons in Biology (Schoenborn, B., ed.), Vol. 27 of Basic Life Sciences (Plenum, New York, 1983), p. 401.
[11] Middendorf, H.D., in Neutron Scattering in the Nineties (International Atomic Energy Agency, Vienna, 1986), p. 203.
[12] Middendorf, H.D., *Ann. Rev. Biophys. Bioengineering* **13** (1984) 425; *Physica* **B 182** (1992) 415.
[13] Bée, M., Quasielastic Neutron Scattering: Principles and Applications in Solid State Chemistry, Biology and Materials Sciences (Adam Hilger, Bristol, 1988).
[14] Martel, P., *Progr. Biophys. Mol. Biol.* **57** (1992) 129.
[15] Middendorf, H.D., *Solid State Ionics* **65** (1995). In press.
[16] Settles, M., Doster, W., Kremer, F., Post, F. and Schirmacher, W., *Phil. Mag.* **B 65** (1992) 861.
[17] Crespi, H.L., in Stable Isotopes in the Life Sciences (International Atomic Energy Agency, Vienna, 1977), p. 111.
[18] Schirmer, T., Bode, W. and Huber, R., *J. Mol. Biol.* **196** (1987) 677.
[19] Middendorf, H.D. and Randall, J.T., *Phil. Trans. Roy. Soc.* **B 290** (1980) 639.
[20] Middendorf, H.D., Randall, J.T. and Crespi, H.L., in Neutrons in Biology (Schoenborn, B., ed.), p. 381, Vol. 27 of Basic Life Sciences (Plenum, New York, 1983).
[21] Middendorf, H.D. and Randall, J.T., in Structure and Motion: Membranes, Nucleic Acids and Proteins (Clementi, E., Corongiu, G., Sarma, M.H. and Sarma, R.H., eds.), p. 219 (Adenine Press, New York, 1985).
[22] Bellissent-Funel, M-C., Teixeira, J., Chen, S.H., Dorner, B., Middendorf, H.D. and Crespi, H.L., *Biophys. J.* **56** (1989) 713.

[23] Bellissent-Funel, M-C., Teixeira, J., Bradley, K.F., Chen, S.H. and Crespi, H.L., *Physica* **B 180 & 181** (1992) 740.

[24] Bellissent-Funel, M-C., Teixeira, J., Bradley, K.F. and Chen, S.H., *J. Phys. I France* **2** (1992) 995.

[25] Bellissent-Funel, M-C., Lal, J., Bradley, K.F. and Chen, S.H., *Biophys. J.* **64** (1993) 1542.

[26] Bansil, M., *J. Phys. IV France* **3** (1993) C1-225.

[27] Burchard, W. and Ross-Murphy, S.B. (eds.), Physical Networks: Polymers and Gels (Elsevier Science, Amsterdam, 1990).

[28] De Rossi, D., Kajiwara, K., Osada, Y. and Yamauchi, A. (eds.), Polymer Gels, Fundamentals and Biomedical Applications (Plenum, New York, 1992).

[29] Middendorf, H.D., Cavatorta, F. and Deriu, A., *Progr. Colloid Polym. Sci.* **81** (1990) 275.

[30] Deriu, A., Cavatorta, F., Cabrini, D. and Middendorf, H.D., *Progr. Colloid Polym. Sci.* **84** (1992) 461.

[31] Deriu, A., Cavatorta, F., DiCola, D. and Middendorf, H.D., *J. Phys. IV France* **3** (1993) C1-237.

[32] Deriu, A., Cavatorta, F., Cabrini, D., Carlile, C.J. and Middendorf, H.D., *Europhys. Lett.* **24** (1993) 351.

[33] Middendorf, H.D., DiCola, D., Cavatorta, F., Deriu, A. and Carlile, C.J., *Biophys. Chem.* **47** (1994). In press.

[34] Cavatorta, F., Deriu, A., Di Cola, D. and Middendorf, H.D., *J. Phys.: Condensed Matter* **6** (1994) A117.

[35] Teixeira, J., *J. Phys. IV France* **3** (1993) C1-163.

[36] Cavatorta, F., Deriu, A. and Middendorf, H.D., in Structure and Dynamics of Nucleic Acids, Proteins, and Membranes (Clementi, E. and Chin, S., eds.), (Plenum, New York, 1986), p. 75.

[37] Dianoux, A.J., Page, J.N. and Rosenberg, H.M., *Phys. Rev. Lett.* **58** (1987) 886.

[38] Lee, S.A., Oliver, W.F., Rupprecht, A., Song, Z. and Lindsay, S.M., *Biopolymers* **32** (1992) 303.

[39] Frauenfelder, H., Sligar, S.G. and Wolynes, P.G., *Science* **254** (1991) 1598.

Chapter IV

Beyond biological molecules.

Biomolecules like DNA or proteins are only building blocks for more complex systems and there are many steps between atomic physics and a living organism. One of the steps that follows biomolecules is the organised network of protein polymers that form the cytoskeleton of the cells and could be responsible for cell motions as well as information processing. In their paper J. **Tuszynski, B. Trpisova, D. Sept, M. Sataric, and S. Hameroff** show how it is possible to model the assembly or disassembly of microtubules that form the cytoskeleton and how nonlinear excitations could take part in the process. The machinery for the synthesis of proteins is another biological process which has to be considered at the scale of a cell and not of an individual molecule. The transcription of DNA, which was considered in chapter I at the level of a single DNA molecule must be considered at the cell level if one is interested in the evolution of bacteria in the presence of environmental constraints. As shown by **F. Bagnoli, G. Guasti and P. Lio**, this can be done with a statistical models which shows the possibilities of an approach based on ideas of theoretical physics because the authors build a thermodynamics where the "temperature" is proportional to the mutation rate. The last two papers provide other illustrations of the possibilities of fundamental theoretical studies applied to biology related problems. Using techniques of nonlinear dynamical systems, **D. Hennig and H. Gabriel** study the fragmentation of a molecular cluster consisting of an atom weakly bound to a polyatomic molecule. This problem could represent for instance the unbinding of oxygen or CO from myoglobin, which is currently not understood. **C. Kuhn** presents a study of π-electrons in hydrocarbon chains which could be of interest for biological molecules where the π-electrons play an important role.

The cell's microtubules: self-organization and information processing properties

J.A. Tuszyński*,(1), B. Trpisová*, D. Sept*,
M.V. Satarić** and S. Hameroff***

*Department of Physics, University of Alberta,
Edmonton, AB, T6G 2J1, Canada
**Faculty of Technical Sciences,
21000 Novi Sad, Serbia, Yugoslavia
***Department of Anesthesiology,
Univ. of Arizona, Tucson, AZ 85721, U.S.A.

Interiors of living cells are structurally and dynamically organized by networks of protein polymers called the cytoskeleton. Of these filamentous structures microtubules are best characterized and appear to be most fundamental. Bulk cytoplasm of living eukariotic cells contains parallel-networks of individual microtubules which are interconnected by filamentous strands (so called MAPs or microtubule associated proteins). In this paper we report on the main physical features of microtubules focusing our attention on the microtubules dynamics, assembly (disassembly), nonlinear modes suitable for signaling and the spin-glass phase which is ideal for information processing.

1. BACKGROUND INFORMATION

Microtubules (MT's) are the most fundamental filamentous structures that comprise the cytoskeleton [1]. MT's can be viewed as hollow cylinders formed by protofilaments aligned along their axes (see Fig. 1) and whose lengths may span macroscopic dimensions. In vivo, MT's are assemblies of 13 longitudinal protofilaments, each of which is a series of subunit proteins known as tubulin dimers. Each subunit is a polar, 8 nm-long dimer which consists of two slightly different 4 nm-long monomers with a molecular weight of 55 kilodaltons. These two constituent parts are referred to as α and β tubulin. The primary structure of tubulin monomers extracted from porcine brain is well-known [1]. Secondary structure of tubulin has been studied experimentally by both Raman spectroscopy and circular dichroism [2]. In the polymerized state of MT it had been determined that α-tubulin monomers comprise four α-helices, four β-sheets and two random coils, while β-monomers have six α-helices, one β-sheet and seven random coils. Each tubulin dimer possesses an electric dipole moment $\vec{p}$ due to the presence of 18 calcium ions bound within each β monomer. An equal number of negative charges provide an electrostatic balance and are localized near the neighboring α-monomer. Thus, MT's provide an example of an electret substance, i.e. an assembly of oriented dipoles. In this paper we investigate the question of a type of spatial arrangement

---

(1) *on leave at*: Institut für Theoretische Physik I, H. Heine Universität Düsseldorf, 40225 Düsseldorf, Germany

of these dipoles in some detail and the role they may play in information processing at a cellular level.

It was suggested [3–5] that conformational states of tubulin dimers present within MT's may be coupled to charge distribution or dipolar states thereby allowing for cooperative interactions with neighboring tubulin states. Furthermore, this mechanism could lead to the presence of piezoelectric properties which are very common in ferroelectrics. This, in turn, could prove very important in their assembly/disassembly behavior [6,7].

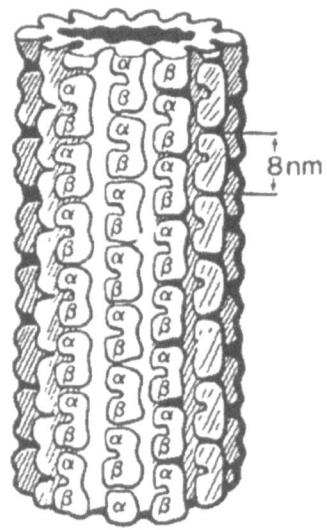

FIG. 1. A schematic illustration of a microtubule.

In addition to energetic factors, information processing appears to be an important aspect of MT behavior. Barnett [8] suggested that filamentous cytoskeletal structures may operate as information strings in analogy to semiconductor word processors. Thus MT's could act as processing channels along which strings of information bits can move transferring messages over substantial distances. Furthermore, MT's form parallel arrays which are interconnected by cross-bridging proteins called microtubule-associated proteins (MAP's). Hence, MT arrays can easily be envisioned as playing the role of parallel-arrayed memory channels. Microtubules participate in a wide variety of dynamic processes in the cell. During cell division they serve as a kind of track for the transport of chromosomes [9]. In the neurons they are the pathway for the transport of neuronal vesicles towards the synapse. In non-dividing cells they form a network in the space between the nucleus and the cell membranes possibly involved in the transport of material from the surface to the center and vice versa.

In this paper we address the question whether the basic functions of MT's are compatible with information processing capabilities. We focus our attention on the dominant physical pattern, i.e. the lattice of dipoles and investigate its ground-state properties. In general, three kinds of geometrical arrangements of dipoles we see in MT's are found to be: (a) random (or paraelectric), (b) ferroelectric (parallel-aligned) and (c) of spin-glass type. Each of the above structures exists under different conditions of temperature, electric field, MAP distribution and MT length giving rise to a possible sensitive state-switching mechanism.

The ferroelectric state appears to be most suitable for assembly/disassembly processes while the spin-glass state is ideally suited for information processing (and thus can be seen as providing the substrate for consciousness) as will be argued in this paper.

2. ASSEMBLY/DISASSEMBLY MODELLING

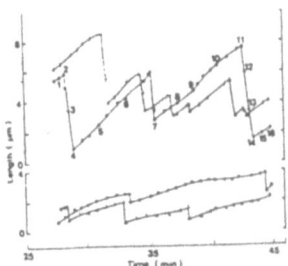

FIG. 2. Direct observation of MTs growing and shrinking [10].

Using dark-field microscopy it was experimentally found [10] that when a single MT is monitored, its two ends grow at different rates. The active "+" end grows faster than the inactive "-" end. However, each end stops growing independently in a stochastic manner and then immediately begins to shorten at a high rate. After the shortening period an MT suddenly stops and restarts the growth phase.

It is believed [1] that the assembly of MTs is an entropy-driven process partly because the unpolarized subunits (monomers) bind and order more water molecules than do the assembled MTs. Both polymerized and unpolymerized monomers bind either GTP (guanosine-threephosphate) or GDP (guanosine-diphosphate) which are analogous to ATP and ADP. The hydrolysis of GTP into GDP releases 0.49 eV of energy. Energy is imparted to the tubulin subunits but it is not required for assembly of MTs. Rather, the hydrolysis of GTP into GDP occurs subsequently within assembled MTs. Utilization of the hydrolysis energy is not well understood, but has been suggested to change bond strengths [11] or contribute to the generation of coherent lattice excitations leading to kink-like orientation of dimer's dipoles.

In their dynamic instability state, MTs are labile structures which can alternate erratically between the growing and the rapidly shrinking phases (see Fig. 2 following [10]). Whether growing or shrinking occurs depends on a variety of factors such as concentration of Ca^+ ions, availability of free GTP tubulin and the rate of hydrolysis of GTP tubulin into GDP tubulin within the assembled MTs. The latter is important since MTs whose ends are comprised of GTP tubulin (GTP "caps") are stable whereas those whose ends consist of GDP tubulin are unstable and disassemble rapidly.

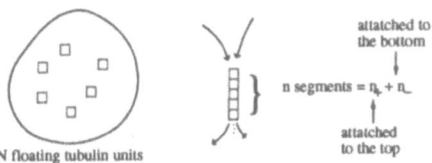

FIG. 3. Illustration of the growth/shrinkage model of MTs developed by the authors.

Consequently, if the addition of free GTP tubulin matches or exceeds the rate of GTP hydrolysis within the MTs, a GTP "cap" protects the growing MT end. As MTs grow and consume free GTP tubulin, the available free GTP tubulin may drop below a critical value and the rate of assembly drops. The two ends of MTs apparently differ in their assembly (disassembly) characteristics because the positively charged beta ends bind GTP tubulin more strongly than do the negatively charged alpha ends. Hotani et al [10] claim that within narrow ranges of the unpolymerized GTP tubulin concentration and the rate of GTP hydrolysis within MTs, and when the dynamic instability is suppressed by the presence of microtubule associated proteins (MAPs), the assembly rate at the beta end may be equal to that of disassembly at the alpha end. This results in a steady state of dynamic MTs of constant length which is referred to as treadmilling. We have developed a simple model of growth and disassembly using the following picture (see Fig. 3).

We use the equation for tubulin conservation

$$N_0 = N + n = N + n_+ + n_- \tag{2.1}$$

and the two equations below describing the evolution at each of the two ends

$$\dot{n}_+ = AF_+(N - n)n - Bn - f^+_{stoch} \tag{2.2}$$
$$\dot{n}_- = AF_-(N - n)n - Bn - f^-_{stoch} \tag{2.3}$$

Note that $f^\pm_{stoch}$ denotes a stochastic force which leads to a random detachment of a whole segment of a microtubule. The force increases in proportion to the length of uninterrupted growth. In the absence of $f^\pm_{stoch}$, the evolution is quite smooth and it can be of 3 generic types depending on the values of model parameter. The three possibilities are: (a) both ends grow monotonically, (b) both ends shrink, and (c) one of the ends grows while the other shrinks.

The results of our numerical simulations are depicted in Fig. 4. Note the striking similarity with the experimental results of Fig. 2. Finally, note that assemblies of MT's tend to synchronize their evolution as is witnessed by the results of coherent oscillations [12].

3. THE DIPOLAR LATTICE AND ITS PHASES

Our basic premise is that the entire MT may be physically viewed as a regular (triangular) array of coupled local dipole moments which interact with their immediate neighbors via dipole-dipole forces. Although Melki et. al [13] showed that tubulin undergoes a conformational change (see Fig. 5a), we will tentatively adopt a simplified view where elastic

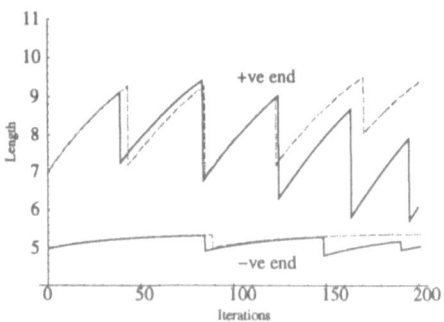

FIG. 4. Results of numerical simulation of MT evolution using eqs. (2.1)–(2.3).

$J_1 = 5.77 \cdot 10^{-21}$ J	$\theta_1 = 0°$
$J_2 = -0.71 \cdot 10^{-21}$ J	$\theta_2 = 58.2°$
$J_3 = 3.40 \cdot 10^{-21}$ J	$\theta_3 = 45.6°$

TABLE I. A summary of the model parameter values used.

degrees of freedom are not explicitly included in the description. However, an appropriate generalization of the physical model poses no technical difficulties.

The starting point in the analysis is to adopt a triangular lattice structure (located on the surface of the MT) with the dimensions and orientations as shown in Fig. 5b and 5c. Each lattice site is assumed to possess a dipole moment $\vec{p} = Q \cdot d$ where $Q = 2e$ and $d \simeq 4$nm and its projection on the vertical axis can only be $+p$ or $-p$. The interaction (dipole-dipole) energy E_{ij} between two neighboring lattice sites (labelled i and j) is, therefore,

$$E_{ij} = \frac{1}{4\pi\epsilon\epsilon_0} \frac{3\cos^2\theta - 1}{r_{ij}^3} p^2 \qquad (3.1)$$

where ϵ_0 is the vacuum permitivity, ϵ the dielectric constant of the medium, r_{ij} is the distance between sites i and j. The angle θ is between the dipole axis and the direction joining the two neighboring dipoles. Fig. 6 illustrates the relevant situation used in our calculations.

In Fig. 6a the signs "+" and "-" refer to dipole-dipole interactions that prefer either a parallel or an antiparallel arrangement of dipole moments, respectively. Above in Table 1, we present the numerical results for the constants J_1, J_2 and J_3 and the corresponding angles which were found based on the known structural data [4]. With the known strong axial anisotropy of interactions we can map this situation onto an anisotropic two-dimensional Ising model on a triangular lattice so that the approximate effective Hamiltonian is now given by

$$H = -\sum_{<nn>} J_{ij} s_i^z s_j^z \qquad (3.2)$$

and the effective spin variable $s_i^z = \pm 1$ denotes the dipole's projection on the vertical MT axis. The exchange constants J_{ij} take the values J_1, J_2, J_3 depending on the choice of dipole pairs.

Due to the fact that $J_2 < 0$ and that there are an odd number (13) of protofilaments, the system exhibits frustration [14] in its ground state. This means that for any closed path, it is impossible to satisfy all bond requirements. Hence, there will always be a conflict (hence the word frustration) between satisfying the energetical requirements of "+"-bonds and "-"-bonds. The ensuing dipolar phase structure is known in the physical literature as a spin-glass phase [15,16]. In a spin-glass (SG), spin orientations are locally "frozen" in random directions due to the fact that the ground state has a multitude of equivalent orientations.

For example, for each triangle reversing the spin on one side with respect to the remaining two leads to an energetically equivalent configuration. Having the number of triangles on the order of the number of lattice sites, i.e.. $N \sim 2 \cdot 10^4$, yields the degeneracy of the ground state on the order of 6^N which is a very large number! This provides a very convenient property from the point of view of encoding information in such a highly degenerate dipolar lattice state. The basic structure of each of the three phases is schematically shown in Fig. 7 where the different phases are contrasted.

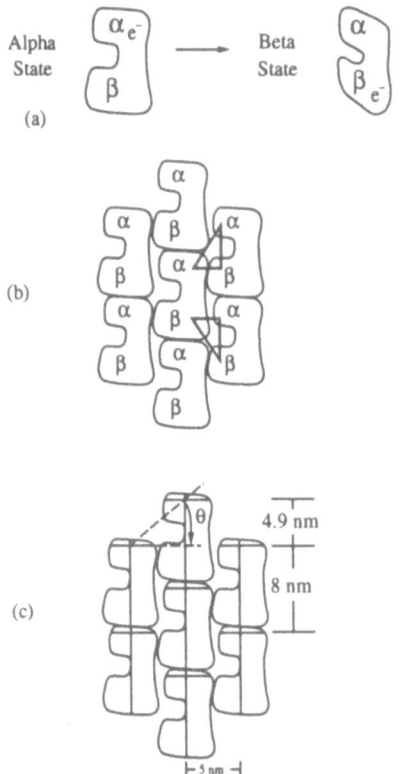

FIG. 5. A graphical description of the structural units in a microtubule: (a) the two electronic states of a dimer, (b) the hexatic unit cell, and (c) the dimensions and angles of a unit cell.

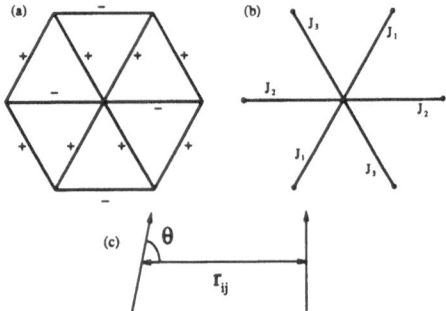

FIG. 6. A schematic illustration of the model parameters used and their meaning.

Relatively small potential barriers separating the various equivalent arrangements of spins in the SG phase may play a very useful role since relaxation times are very long for the various accessible states. Some other properties of the spin-glass phase are: (a) the absence of long-range order between spin sites; (b) the presence of short-range correlations. This means that the system is very "soft" energetically permitting a number of spin arrangements which are relatively stable. However, there is no tendency towards the formation of large correlated spin clusters. This is very attractive from the point of view of computational applications of a MT. A spin-glass state is, in fact, ideal for this purpose since it allows easy formation of local ordered states, each of which carries an information content and is long-lived. On the other hand, it is also not difficult to switch from one state to another through various physical means required to overcome the small potential barriers between individual states. A complete elimination of the SG phase can be achieved by: (a) the application of an electric field along the MT axis. We estimated that fields on the order of 10^4 V/m are sufficient to switch an MT from a spin-glass phase to a ferroelectric state (F). Note that fields up to 10^5 V/m are known to exist at a cell level so this situation is quite feasible. (b) Raising the temperature above the spin-glass transition temperature will drive the system to a disordered (paraelectric) phase. (c) A spin-glass is also sensitive to the boundary conditions. This means, for example, that the presence of domain walls will have a profound effect on the SG phase.

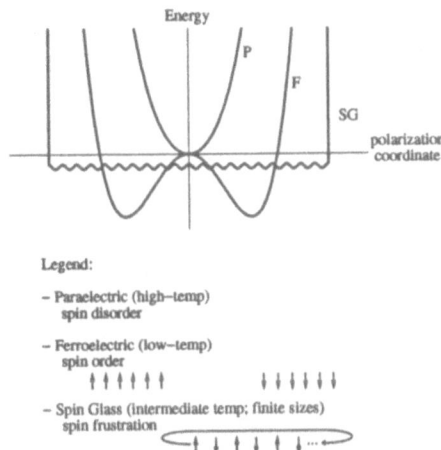

FIG. 7. A comparison of the three distinct dipolar phases of a microtubule.

4. DYNAMICS IN THE FERROELECTRIC PHASE

When a moderate external electric field is applied (or when the temperature is lowered further) a ferroelectric phase is favored which is characterized by long-range order manifested by an alignment of spin directions along the axis. Obviously, from the point of view of information processing potential, this is not a very useful phase. However, it can play a major role in the assembly/disassembly processes as we discuss below.

In the ferroelectric phase, the MT system has a strong uniaxial dielectric anisotropy so that the array of dipole oscillators can be effectively described in terms of only one degree of freedom. In fact Athenstaedt [6] showed experimentally that a tubulin dimer undergoes a conformational change induced by the GTP-GDP hydrolysis in which one monomer shifts its orientation by 29° from the dimer's vertical axis. Thus it was deduced that the single degree

of freedom is the projection on the MT cylinder's axis of the monomer's displacement from
its equilibrium position. The MT cylinder taken as a whole represents a giant dipole. When
the cross section of a MT is viewed using electron microscopy, the outer surfaces of a MT
are surrounded by a "clear-zone" of several nm which apparently represents the orientated
molecules of cytoplasmic water and enzymes [17]. This could be explained by the presence of
an electric field produced by a MT. Therefore, it is assumed that together with the polarized
water surrounding it, a MT generates a nearly uniform intrinsic electric field parallel to its
axis.

The total Hamiltonian regarding the dipole dynamics in a MT is thus given by

$$H = \sum_{n=1}^{N} \left[\frac{1}{2} M \left(\frac{du_n}{dt} \right)^2 + \frac{1}{2} K (u_{n+1} - u_n)^2 - \left(\frac{A}{2} u_n^2 - \frac{B}{4} u_n^4 \right) - c u_n \right]. \tag{4.1}$$

The first term above represents the kinetic energy associated with the longitudinal displacement of constituent dimers, each of which had mass M. The second term arises from
the restoring strain forces between adjacent dimers in the protofilament. The overall effect
of the surrounding dipoles on a chosen site n is supposed to be qualitatively described by the
double well quartic potential (the third term and fourth terms). A and B are model parameters such that $B > 0$ and is temperature independent while A is typically a linear function
of temperature that may change its sign at an instability temperature T_c, ie. $A \simeq \alpha(T - T_c)$.
With $\alpha < 0$, below T_c, $A > 0$ and $u_n = 0$ is a maximum of the potential such that $V(0) = 0$.
Two symmetric local minima exist at $u_n = u_0 = \pm(A/B)^{1/2}$ for which $V_{min} = -A^2/4B$ (Fig.
8a). The anharmonic potential discussed here is meant to approximate the average effect

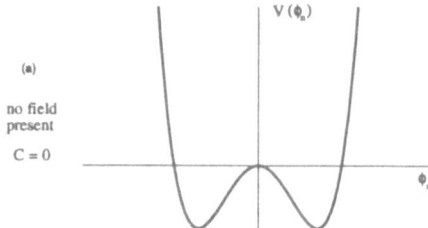

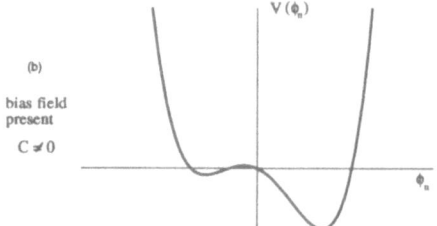

FIG. 8. The effective mean potential: (a) $V = \frac{1}{2}A\phi_n^2 + \frac{1}{4}B\phi_n^4$ when no field is present; (b)
$V = \frac{1}{2}A\phi_n^2 + \frac{1}{4}B\phi_n^4 - C\phi_n$ when a bias field is present.

due to the environment when all the neighboring dimers assume their equilibrium positions. The mobile electrons on each dimer can be localized either more toward the α-monomer or more toward the β-monomer. The latter possibility is associated with a change in the dimer's conformation. Experimental evidence indicates that a conformational distortion of $29°$ from vertical occurs in the β state. Therefore, one can identify the variable u with the amount of β-state distortion when the latter is projected on the vertical axis.

In order to derive a realistic equation of motion for the system described, it is indispensable to include the viscosity of the solvent and introduce the associated damping force. Assuming for simplicity that the solvent is made up of only water molecules the viscosity can be simply taken into account by adding the friction force to the equation of motion with

$$F_v = -\gamma \frac{du_n}{dt},\tag{4.2}$$

where γ represents the damping coefficient that has an appropriate experimental value.

An equation of motion for dipolar oscillations can be thus derived using the Hamiltonian of Eq. (4.1) applying the continuum limit and adding the viscous force of Eq. (4.2). As a result one obtains [11]

$$\psi_{\xi\xi} + \rho\psi_\xi - \psi^3 + \psi + \sigma = 0 \tag{4.3}$$

where $\psi(\xi)$ represents the normalized displacement field $\psi(\xi) = \frac{u(\xi)}{u_o}$, where $u_o = \left(\frac{|A|}{B}\right)^{1/2}$ corresponds to the minimum of the double-well potential (Fig. 8a). The variable ξ is the moving coordinate for the travelling-wave form of the solution

$$\xi = \left[\frac{|A|}{M(v_o^2 - v^2)}\right]^{1/2}(x - vt)\tag{4.4}$$

where v denotes the propagation velocity for dipolar oscillations while $v_o = R_o\sqrt{K/M}$ represents the longitudinal sound velocity through MT. The remaining parameters of Eq. (4.4) are

$$\rho = \gamma v \left[|A|M(v_o^2 - v^2)\right]^{-1/2}\tag{4.5}$$

and

$$\sigma = q\sqrt{B}|A|^{-3/2}E.\tag{4.6}$$

It can be shown the Eq. (4.4) has a unique bounded solution which is given by the formula

$$u(\xi) = u_2 + \frac{u_1 - u_2}{1 + \exp[\beta(u_1 - u_2)\xi]}\tag{4.7}$$

as illustrated graphically in Fig. 9 where

$$u_1 = 2(\frac{A}{3B})^{1/2}\cos\{\frac{1}{3}\arccos[\frac{3qE}{2A}(\frac{3B}{A})^{1/2}]\}\tag{4.8}$$

$$u_2 = -2(\frac{A}{3B})^{1/2}\cos\{\frac{\pi}{3} - \frac{1}{3}\arccos[\frac{3qE}{2A}(\frac{3B}{A})^{1/2}]\}$$

$$\beta = \pm\frac{1}{v_0}(\frac{B}{2M})^{1/2}\tag{4.9}$$

Thus the solution (4.7) is a kink giving the boundary between the two states $u = u_1$ for $\xi \to +\infty$ and $u = u_2$ for $\xi \to -\infty$, if $\beta > 0$. The boundary moves with the unique terminal velocity given by

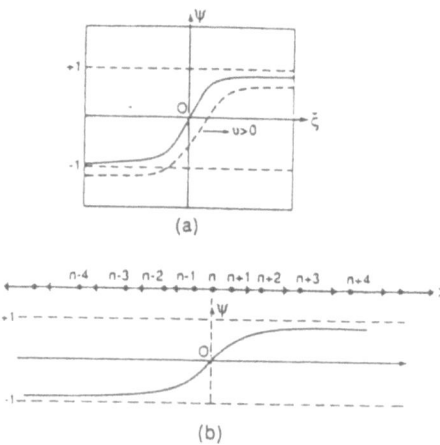

FIG. 9. The form of a kinklike excitation of eq. (4.7). (a) A graphic plot of the function of a bounded kink. (b) The arrangement of dipoles involved in forming a kink.

$$v = \pm\frac{v_0}{\gamma}(\frac{6A}{M})^{1/2}\cos\{\frac{\pi}{3} + \frac{1}{3}\arccos[\frac{3qE}{2A}(\frac{3B}{A})^{1/2}]\} \tag{4.10}$$

and its width Δ is

$$\Delta = \frac{v_0}{(u_1 - u_2)}(\frac{2M}{B})^{1/2}. \tag{4.11}$$

For a long MT we can assume that for points that are sufficiently far from its ends the magnitude of the electric field can be approximated as $E \simeq Q(4\pi\epsilon_0 r^2)^{-1}$, where Q represents the effective charge on the ends of a MT and r is the distance between the selected point and the end of the MT. Taking as an example a moderately long MT consisting of approximately 10^2 dimers with length $L \sim 10^{-6}$m, the effective charge Q is estimated $Q \simeq 1.5 \cdot 10^{-16}$C so that the intrinsic electric field in the vicinity of the middle point of a MT is found to be on the order of $E \sim 10^6 V/m$. Accounting for the dielectric effects of the surrounding water molecules, this value must be reduced to approximately $E \sim 10^5 V/m$.

As far as the potential coefficients A and B are concerned no reliable experimental data exist so a crude estimate is made using come typical values for crystalline ferroelectrics knowing that they do not vary substantially between different compounds. These are typically

$$A \simeq 2 \cdot 10^2 Jm^{-2} \quad ; \quad B \simeq 10^{24} Jm^{-4} \tag{4.12}$$

which then yields

$$\sigma \simeq 2 \cdot 10^{-9} \cdot E. \tag{4.13}$$

It is therefore clear that even for extremely strong electric fields ($E \sim 10^8 V/m$) the inequality $\sigma \ll 1$ holds. The main consequence of the smallness of the electric field is that the propagation velocity (4.9) can be safely approximated as

$$v \simeq \frac{3v_0}{\gamma|A|}(\frac{MB}{2})^{1/2} qE \tag{4.14}$$

and it is generally much smaller than the sound velocity ($v \ll v_0$). Eq. (4.13) links the propagation velocity v and the magnitude of the electric field E. The coefficient of proportionality in this relationship represents the kink mobility

$$\mu = \frac{3v_o}{\gamma|A|}q\left(\frac{MB}{2}\right)^{1/2}. \tag{4.15}$$

Substituting typical numerical data as: $\gamma = 5.6 \cdot 10^{-11}kgs^{-1}(T = 300^\circ K); v_o = 1.7 \cdot 10^3 m/s$(for DNA); $M \simeq 1 \cdot 10^{-22}kg$; and $q \simeq 6 \cdot 10^{-18}C$, yields

$$\mu \simeq 2 \cdot 10^{-5}m^2V^{-1}s^{-1}. \tag{4.16}$$

Thus taking $E = 10^5 Vm^{-1}$ as a representative average of the electric field, the propagation velocity of a kink is on the order of

$$v \simeq 2m/s. \tag{4.17}$$

Assuming a smooth journey from one end of the MT to the other, the average time of propagation for a single kink should be

$$\bar{\tau} = \frac{L}{\bar{v}} \simeq 5 \cdot 10^{-7}s. \tag{4.18}$$

However, increasing the length of the MT will increase $\bar{\tau}$ on two accounts: (a) increasing the numerator in Eq. (4.17) and (b) affecting the mean velocity through the dependence of the electric field on L. Indeed, $E \sim L^{-2}$. Consequently $\bar{\tau} \sim L^3$ leading to a rapid increase of $\bar{\tau}$ with L.

It should be mentioned that the effects of discreteness of the MT's lattice (which are ignored here by using the continuum approximation) may play an important role. For example Kimball [18] performed an analysis of kink dynamics for a discrete lattice inferring that there may exist a threshold value of the external field E required to sustain kink motion. Moreover, it appears that fast-propagating kinks are more stable than slow ones.

It is apparent that the intrinsic electric field governs the rate of propagation of kinks. By adding an external electric field to the Hamiltonian one can introduce a new control mechanism in the MT dynamics. An applied electric field will result in a faster moving population of kinks and thus a greater stability against thermal fluctuations; if, on the other hand, the intrinsic and applied electric fields are oriented in opposite directions, the kink's motion may be slowed or stopped altogether. This then can be seen as a basis for treating MT's as artificial information strings. Each kink within a MT can be viewed as a bit of information whose propagation can be controlled by an external electric field.

Conformational states of tubulin dimers present within MTs have been suggested to be coupled to charge of dipolar states, thereby allowing for cooperative interactions with neighboring tubulin dimer sites [3,4]. This coupling would also lead to the presence of piezoelectric properties which could prove very important in MT signaling, communication and assembly (disassembly) behavior [6,7]. Thus a more accurate model for piezoelectric dynamics could be based on the free energy expansion

$$F = \int dx \, dy \left[\frac{A_2}{2} P^2 + \frac{A_4}{4} P^4 - \vec{E} \cdot \vec{P} + \frac{D_x}{2}(\partial_x P)^2 + \frac{D_y}{2}(\partial_y P)^2 + \frac{1}{2}\rho u_t^2 \right.$$

$$\left. + \frac{1}{2} k_1 u_x^2 + \frac{\alpha}{2} u^2 - u\sigma + \frac{\beta}{4} u^4 + \gamma u P + \frac{1}{2} k_2 u_y^2 \right] \tag{4.19}$$

where $F(P)$ is the polarization energy, $F(u)$ is the elastic energy (axis anisotropy), $F(u, P)$ is the piezoelectric coupling, σ represents the stress field (sensitive to GTP) and E denotes the external electric field (sensitive to pH and membrane dynamics).
Minimization of F with respect to P and u gives:

$$A_2 P + A_4 P^3 - E - D_z \partial_{xx} P - D_y \partial_{yy} P + \gamma u = 0 \tag{4.20}$$

$$\rho u_{tt} - k_1 u_{xx} - k_2 u_{yy} + \alpha u - \sigma + \gamma P + \beta u^3 = 0 \tag{4.21}$$

Taking into account anisotropy (propagation along the x-axis) and considering travelling waves with $\xi = x - vt$ would reduce (4.19) and (4.20) to a system of coupled ODE's. Making an ansatz: $u = aP + b$ with a, b to be determined and adding dissipation terms $\dot{P}$ and $\dot{u}$ would result in an effective damped driven NLKG for the polarization field P. This type of equation has a wealth of behaviors ranging from solitons to period doubling and chaos.

5. STABILITY OF THE SPIN-GLASS PHASE

The next question we wish to address is the range of stability of the various phases identified in the previous section. For an infinite triangular lattice governed by the Hamiltonian of eq.(3.2), only two phases exist: the paraelectric phase (above T_c), the ferroelectric phase (below T_c) and the spin-glass phase is completely absent in this limit. The critical temperature T_c for the $F \to P$ transition is given by the equation

$$\sinh \left(\frac{2J_1}{k_B T_c} \right) \sinh \left(\frac{2J_2}{k_B T_c} \right) + \sinh \left(\frac{2J_1}{k_B T_c} \right) \sinh \left(\frac{2J_3}{k_B T_c} \right) + \sinh \left(\frac{2J_2}{k_B T_c} \right) \sinh \left(\frac{2J_3}{k_B T_c} \right) = -1 \tag{5.1}$$

where k_B stands for the Boltzmann constant. Note that the critical value of T_c depends crucially on the combination of model parameters used, i.e. on $\tilde{Q} \equiv Q^2 d^2/\epsilon$ as shown in Fig. 10. For the realistic (although not known previously) values of Q, d and ϵ, the transition temperature lies between 200 K and 400 K indicating the possibility of the associated phase transition close to room temperature, i.e. at physiological conditions.
Many important factors may affect the value of T_c and thus provide sensitive control mechanisms. Through a coupling to the elastic degrees of freedom (conformational change), the dielectric constant ϵ may be altered by the presence of water molecules surrounding an MT structure thereby decreasing the value of T_c and introducing dipolar disorder. Small structural changes, in particular shifts in the angles between the dimer dipoles, may remove the frustration mechanism effectively preventing the onset of the SG phase. Changes in the opposite direction can enhance frustration favoring the SG over the F-phase effectively switching from the growth mode of operation of MT's to their information processing behavior.
In order to obtain some insight into the above questions, we have performed Monte-Carlo simulations for finite lattices with dimension $13 \times N$ (N is the length of the microtubule in terms of the number of layers). It is clear that as N increases the spin-glass phase is gradually removed (see Fig. 11 for a schematic phase diagram). We also see this effect by directly plotting the mean polarization per site for $N = 26$ (Fig. 12a) and $N = 5000$ (Fig.

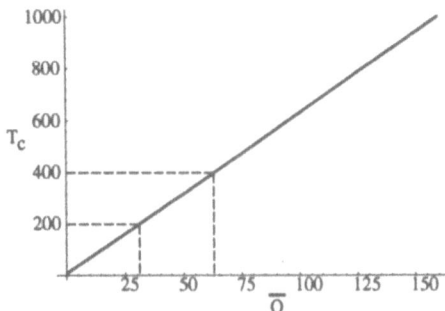

FIG. 10. The dependence of the critical temperature T_c (in K) of an infinite triangular lattice on the combination of model parameters $\bar{Q} \equiv Q^2 d^2 / \epsilon$ (in units $e^2 \cdot 10^{-57}$).

N	T_c (K)	N	T_c (K)
10	240	300	298
20	304	400	318
30	305	500	313
40	296	600	308
50	304	700	304
60	302	800	312
70	311	900	314
80	301	1000	306
90	299	1300	310
100	300	1500	303
200	313		

TABLE II. The dependence of T_c on N for $\bar{Q} = 11.9 \cdot 10^{-56} \, \mathrm{C}^2 \, \mathrm{m}^2$.

12b) as two contrasting examples. In Table 2 we show how the transition temperature T_c changes with N for the choice of model parameters which give $\bar{Q} = 11.9 \cdot 10^{-56} \, \mathrm{C}^2 \, \mathrm{m}^2$.

We conclude that dynamic processes leading to the elongation of MT's could effectively remove the information processing capabilities of MT's by expelling the SG phase. The same can be achieved by raising the temperature above a characteristic value which is length dependent.

We have also examined the effect of external electric fields and MAP's (see Fig. 8 for typical MAP patterns) on the aforementioned transition. The electric field shifts the transition region and makes it broader. A similar effect can be seen by incorporating MAP's as "empty" (i.e. non-polar) lattice sites. The actual magnitude of the shift and broadening depends on the pattern of MAP's chosen and the ratio of MAP's to the total number of lattice sites. Taking the set of parameter values which yields $T_c = (300 \pm 15)$ K for the perfect lattice results in $T_c = (250 \pm 20)$ K for the lattice with MAP's at a ratio of 1:11 while $T_c = (230 \pm 20)$ K is obtained for a ratio of 1:8. This indicates that MAP's substantially lower the transition temperature and make the SG-phase accessible to the MT system at much lower temperatures that those required in the absence of MAP's.

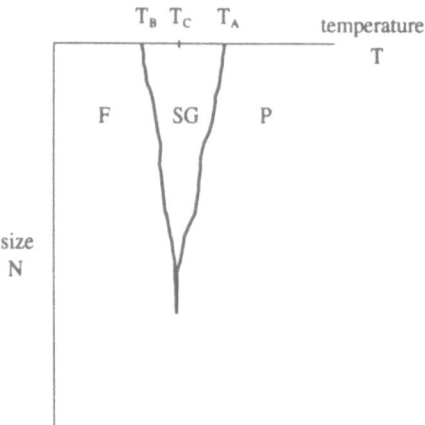

FIG. 11. A schematic phase diagram for a finite-size MT lattice.

6. INFORMATION CAPACITY IN VARIOUS PHASES

Can MTs also transport information by conducting a wave of coherent conformational change? Can MTs store information by showing a mosaic of conformational states? Such functions for MTs have been suggested by Hameroff et al, [3,19].

Some evidence links the cytoskeleton with information processing and cognitive function. For production of MT subunit protein (tubulin) and MT activities is correlated with peak learning, memory and experience in baby chick brains. When baby rats begin their critical learning phase for the visual system (when they first open their eyes), neurons in the visual cortex begin producing vast quantities of tubulin [20]. Tubulin production is drastically reduced when the critical learning phase is over (when the rats are 35 days old). Bensimon and Chernat [21] found that selective destruction of brain MTs by the drug colchicine caused cognitive defects in learning and memory which mimic the clinical symptoms of Alzheimer's disease, in which the cytoskeleton becomes entangled.

Further suggestion for cytoskeleton computation and (or information storage) stems from the spatial distribution of discrete sites (or states) in the cytoskeleton. For example, tubulin subunits in closely arrayed MT have a density of about $10^{17}/cm^3$, which is very close to the theoretical limit for charge separation [22]. Thus the cytoskeleton polymers have maximal density for information representation by charge, and the capacity for dynamically coupling that information to mechanical energy and chemical events via cooperative dipole states.

In order to examine the usefulness of MT's as the cell's information processors we must first evaluate the information capacity within each of the three phases identified. These results can be used to find the optimal conditions for the MT's to function as the substrate for consciousness-related activities. We base the calculations that follow on the standard (Shannon) definition of information I of a statistical system where [23]

$$I = -\sum_{i=1}^{K} p_i \ln p_i. \tag{6.1}$$

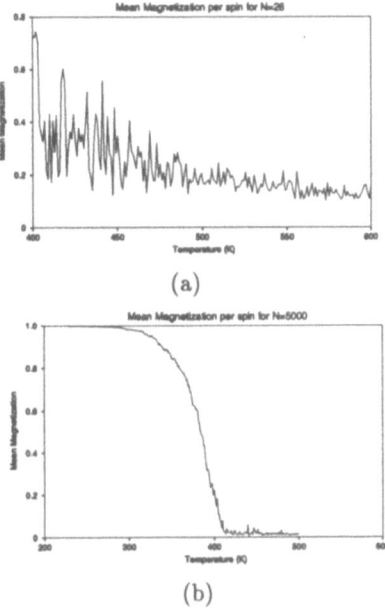

FIG. 12. Plots of the mean polarization per site (y-axis) for (a) N=26 and (b) N=5000

Here, p_i stands for a probability value in state i and, obviously, the probability distribution must satisfy:

$$\sum_i p_i = 1 \text{ with } 0 \leq p_i \leq 1. \tag{6.2}$$

For the ferroelectric and paraelectric phases we adopt the mean-field approximation where each state is characterized by the continuous variable P (mean polarization per site). The energy functional is taken in the Landau form as

$$E = \left(\frac{A}{2}P^2 + \frac{B}{4}P^4\right) N_0 \tag{6.3}$$

where N_0 is the total number of sites in the lattice, $A = a(T - T_c)$ and $B > 0$. As is well-known above the critical temperature, i.e. for $T > T_c$, E is minimized by $P = 0$ while below the critical temperature, i.e. for $T < T_c$ by $P_\pm = \pm\sqrt{-A/B}$. The associated continuous probability distribution $f(P)$ that replaces p_i of eq. (6.1) is the Boltzmann-weighted distribution function in the form:

$$f(P) = Z^{-1}\exp(-\beta E) = f_0 \exp(\alpha P^2 - \gamma P^4) \tag{6.4}$$

where $f_0 = Z^{-1}$ is the normalization, $\beta^{-1} = k_B T$, $\alpha = -\frac{A}{2k_B T}$ and $\gamma = \frac{B}{4k_B T}$. Hence, for $T > T_c$, $f(P)$ is single-peaked at $P = 0$ while for $T < T_c$ it is double-peaked at $P = P_\pm$.

Following Haken [23] we calculate the information capacity in the paraelectric ($P = 0$) and ferroelectric ($P \neq 0$) phases as

$$I = \ln Z - \alpha < P^2 > + \gamma < P^4 > \tag{6.5}$$

where the averages are obtained using:

$$< P^n > \equiv \int_{-\infty}^{\infty} f(P)P^n \, dP. \tag{6.6}$$

We carry out the requisite calculations in a straight-forward manner for both the ferro- and para-electric phases where analytical calculations can be performed. For the SG-phase, however, we assume that the above prescription is valid only within the local domain of coherence (or within the correlation length). Hence, for each domain i we have a local polarization P_i and the associated probability distribution $f_i(P_i)$, essentially analogously to those of eqs. (6.3) and (6.4). Thus for the total system the probability distribution becomes a product of local distributions each of which characterizes a domain of coherence

$$f = \prod_{i=1}^{n} f_i(P_i) \tag{6.7}$$

where n is the number of domains (see Fig. 13).

Note that n depends on temperature and we assume for simplicity that

$$n = 1 + \frac{(N_0 - 1)(T - T_B)}{T_A - T_B} \tag{6.8}$$

in order to interpolate continuously between the ferroelectric and paraelectric phases since $T_B \leq T \leq T_A$. At $T = T_B$, virtually the entire system is uniformly polarized while at $T = T_A$ it is completely depolarized and incoherent. Note, that as a consequence of (4.7) we obtain for the information capacity in the SG-phase

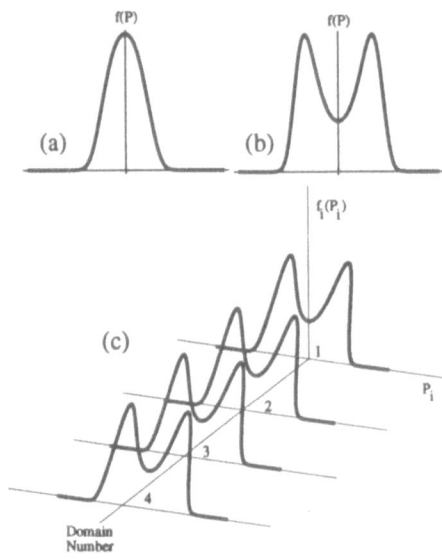

FIG. 13. An illustration of the probability distributions $f(P)$ in the three phases: (a) paraelectric, (b) ferroelectric and (c) spin-glass phase.

$$I = \sum_{i=1}^{n} I_i \tag{6.9}$$

where I_i refers to each individual domain. Our numerical computation clearly indicates that information capacity I is highest at the boundary between the spin-glass and the paraelectric phase (see. Fig. 14) and hence if MT's are to be effective as information processors, they should use this narrow "window of opportunity" at the border area between these two phases. Of course, the actual location of the border area depends on the magnitude of the electric field applied and the concentration of MAP's present.

7. SUMMARY

We have argued in this paper that the spatial arrangement of dipole moments of a microtubule is crucial to its functioning as a dynamic system. Due to the presence of frustration in the dipole-dipole interactions, a spin-glass phase is predicted to arise at low enough temperatures and electric fields. The presence of MAP's will lower the temperature values required for SG-formation. The transition temperature itself decreases in proportion to the MAP ratio. The attractiveness has been recognized earlier [24] and in the present context it lies in maximum computational capabilities offered by a highly degenerate ground state. Moreover,

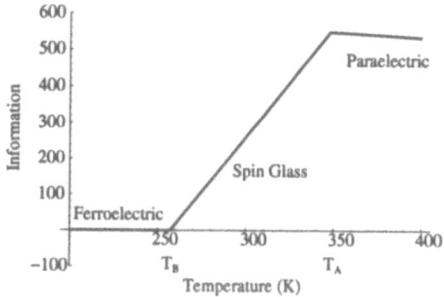

FIG. 14. Plot of the information capacity as a function of temperature.

long relaxation times give relative stability to short-range correlated dipole patterns. Each pattern can be seen as containing binary information encoded in the lattice.

The other possible ordered state is a ferroelectric phase with an almost perfect alignment of dipole moments along the protofilament axis. It is characterized by long-range order and hence its usefulness for information transfer and processing is dubious. However, it eagerly supports the formation of domain walls between the two stable orientations of dipole moments. The application of an external electric field preferentially directs kink-like excitation towards the properly aligned end causing a disassembly of the protofilament due to the energy released by the kink.

We have discussed the various possible control mechanisms (field, distortion, temperature, MAP patterns) that could provide a means by which the MT could select an operating mode between information processing and assembly/disassembly types. This could shed some light on why the MT formation rate is enhanced in particularly important phases of the organism's history and development (learning, division, growth, etc.).

ACKNOWLEDGMENTS

This research was supported by NSERC (Canada), DAAD and the Alexander von Humboldt Foundation.

References

[1] P. Dustin. 1984. *Microtubules.* Berlin:Springer
[2] Dj. Koruga, *Nanobiology* **1**, 5 (1992)
[3] S. R. Hammeroff, *Ultimate Computing.* (Amsterdam:North-Holland, 1987)
[4] S. Rasmussen, H. Karamporsala, R. Vavdyanatu, K. Jensen and S.R. Hammeroff. 1990. *Physica* **D42**: 428
[5] S.R. Hammeroff and R.C. Watt., *J. Theor. Biol.* **98**, 549 (1982)
[6] H. Athenstaedt, *Ann. N.Y. Acad. Sci.* **238**, 68 (1974)
[7] L. Margulis, L. To and D. Chase, *Science* **200**, 1118 (1978)
[8] M. Barnett, *Proceedings of the Third Molecular Electronic Device Conference*, edited by F. Carter. Washington, D.C.:Naval Reseach Laboratory (1987)
[9] Y. Engleborghs, *Nanobiology* **1**, 97 (1992)
[10] T. Horio and H. Hotani, *Nature* **321**, 605 (1986)
[11] M.V. Satarić, J.A. Tuszyński and R.B. Žakula, *Phys. Rev.* **E48**, 589 (1993)
[12] E.M. Madelkow and E. Mandelkow, *Cell Motility and the Cytoskeleton* **22**, 235 (1992)
[13] R. Melki, M.F. Carlier, D. Pantaloni and S.N. Timasheff, *Biochem.* **28**, 9143 (1989)
[14] M. Suzuki, *Prog. Theor. Phys.* **58**, 1151 (1977)
[15] K.H. Fischer, *phys. stat. sol.* (b) **116**, 357 (1983)
[16] K.H. Fischer, *phys. stat. sol.* (b) **130**, 13 (1985)
[17] H. Stebbings and C. Hunt, *Cell Tiss. Res.* **227**, 609 (1982)
[18] J.C. Kimball, *Phys. Rev.* **B21**, 2104 (1980).
[19] S. Hameroff , S.A. Smith and R.C. Watt, *Ann. N.Y. Acd. Sci.* **466**, 949 (1986)
[20] J. Cronly-Dillon, D. Carden and C. Birks, *J. Exp. Biol.* **61**, 443 (1974)
[21] G. Bensimon and R. Chernat, *Pharmacol. Biochem Behavior* **38**, 141 (1991)
[22] F. Gutmann; *Modern Biochemistry*, eds. F. Gutmann and H. Keyzer (Plenum Press, New York, 1986)
[23] H. Haken, *Synergetics: an Introduction* (Berlin:Springer, 1990)
[24] D.L. Stein (ed.), *Spin Glasses and Biology* (Singapore: World Scientific, 1992)

Translation optimization in bacteria: statistical models

F. Bagnoli, G. Guasti* and P. Liò**

Dipto. di Matematica Applicata, Univ. di Firenze,
via S. Marta 3, 50129 Firenze, Italy
* Dipto. di Fisica, Univ. di Firenze,
Largo E. Fermi 2, 50125 Firenze, Italy
** Dipto. di Biologia Animale e Genetica,
Univ. di Firenze, Via Romana 17,
50125 Firenze, Italy

1. INTRODUCTION

Growth rate, in bacteria, is supposed to depend largely on the optimization of the machinery for the synthesis of proteins. A crucial step is the translation from messenger RNA to the protein chain. This phase in bacteria is a highly coordinated and multistep process, being coupled also with the transcription process ([1]). It begins with the interactions of a ribosome with a sequence of the messenger RNA called translation initiation region. This region is characterized by a start triplet, mostly AUG, and normally an upstream Shine-Dalgarno sequence.

Each subsequent displacement of the ribosome along the messenger RNA requires the interaction of a tRNA molecule, carrying a proper aminoacid, with a triplet of bases (codon) on the mRNA. The tRNA molecule, acting as an adapter, translates the information in each codon so that the correct aminoacid is added to the growing protein chain. The attachment of an aminoacid to a tRNA is catalyzed by specific aminoacyl-tRNA synthetases.

Because of the degeneracy of the genetic code (61 codons map into 20 aminoacids) different codons in a mRNA (synonymous codons) specify for the same aminoacid. The codon usage is generally not random and it is very biased in some procaryotic and eucaryotic genes. The codon usage influences the rate of the translation process, being the pairing between codon and the diffusing tRNAs the rate-limiting step compared with the attachment of the aminoacid to the growing protein and the ribosome three-bases movement [2].

Several authors have stressed that in the highly expressed genes the synonymous codons that are translated by the most abundant tRNA species are present in large percentage [3].

Varenne demonstrated experimentally that the presence of codons read by the most rare tRNA species in highly expressed coding regions is associated with pauses in the synthesis of proteins [4].

We can assume that, for a given environment and near to a steady dynamical state, the natural selection favors individuals with higher translation rate. This optimization process is opposed by mutations, that occur randomly along the gene. We focus our interest in synonymous mutations, i.e. mutations that do not change the coded aminoacid. Synonymous mutations are neutral for the functionality of the protein, but affect the translation time.

The dependence of growth rate on tRNA abundances and the dynamics of translation as an evolutionary process in presence of codon mutations, can be faced with statistical mechanics tools. We can interpret the relation between mutations and natural selection as a competition between randomness and order. In this spirit we develop a statistical model from which bacterial evolution emerges as an optimization process under environmental constraints.

2.   A DYNAMICAL MODEL

We assume that the time required by a ribosome to translate a codon is inversely proportional to the abundance of the charged cognate tRNAs in the cell, supposing the absence of spatial gradients (fast diffusion of tRNAs [2]). The dynamics of translation seems to depend also by the differences in the energy of codon anticodon pairing ([5]), but in our models we consider only two types of synonymous codons, namely 0 and 1 with the same interaction energy with their cognate tRNAs; the number per ribosome of conjugate tRNA is n_0 and n_1 respectively. Besides we consider the messengerRNA as a whole, without discriminate an initiation region as in other models [6]. In a first approximation, we assume that the concentration of the charged tRNAs does not vary with the translation process, regardless of the codon composition of mRNAs.

We represent the portion of mRNA that code for a protein (or, alternatively, the whole mRNA) with a vector $\mathbf{c}$; c_i $(i = 1, \ldots, L)$ being the codon at position i and L the length of the coding region. Disregarding fluctuations, the mean translation time per codon τ (in arbitrary time units) is given by

$$\tau_j = \frac{j}{n_1} + \frac{l-j}{n_0} = rj + q;$$
$$r = \frac{n_0 - n_1}{n_0 n_1}; \quad q = \frac{l}{n_0}; \tag{1}$$

where j is the number of ones in the string $\mathbf{c}$.

Assuming that the protein coded by $\mathbf{c}$ is needed in large quantities, the time τ in formula (1) can be considered proportional to the mean duplication time.

Since τ does not depend on the order of the codons in the string, the various strains (different $\mathbf{c}$) can be grouped together according to j. For each group j there are $g_j = \binom{l}{j}$ strains.

The exponential growth state of the mass M_j of bacteria belonging to group j is given by

$$\frac{\mathrm{d}M_j}{\mathrm{d}t} = \nu_j M_j; \tag{2}$$

where $\nu_j = \tau_j^{-1}$.

In the limit of very low mutation rate, we can assume that for each generation there is at most only one synonymous mutation, that changes codon 1 to 0 or vice versa. Let us indicate with μ/l the rate of mutation per codon ($\mu \leq 1$). The average fraction of bacteria in group j that undergo a mutation per unit of time is $\mu\nu_j$. For bacteria in group j, a synonymous mutation changes j of one unit. Including mutations, and working with the mass $m_j = M_j/g_j$ of a single strain in group j, formula (2) becomes

$$\frac{dm_j}{dt} = (1-\mu)\nu_j m_j + \frac{j}{l}\mu\nu_{j-1}m_{j-1} + \frac{l-j}{l}\mu\nu_{j+1}m_{j+1};\tag{3}$$

for $0 \le j \le l$, assuming that $\nu_{-1} = \nu_{l+1} = 0$.

The total mass of the bacterial population is $M = \sum_{k=0}^{l} g_k m_k$. We can derive from eq. (3) the evolution equation for the distribution probability $p_j = m_j/M$ of different strains in the total population. In a natural environment the exponential growth periods are sporadic, generally followed by starvation phases. We mimic this alternation by means of the normalization of distribution, assuming that the strains corresponding to a very low probability are those eliminated by natural selection.

Since mutations do not change the total mass of population, we have, summing up over j in eq. (3)

$$\frac{dM}{dt} = \sum_{k=0}^{l} g_k \nu_k m_k = M \sum_{k=0}^{l} g_k \nu_k p_k = M\bar\nu;\tag{4}$$

obtaining

$$\frac{dp_j}{dt} = \frac{d}{dt}\left(\frac{m_j}{M}\right) = \frac{1}{M}\frac{dm_j}{dt} - \frac{m_j}{M^2}\frac{dM}{dt};$$

and thus a nonlinear evolution equation for the probability:

$$\begin{aligned}
\frac{dp_j}{dt} &= [(1-\mu)\nu_j - \bar\nu]\,p_j + \frac{j}{l}\mu\nu_{j-1}p_{j-1} + \frac{l-j}{l}\mu\nu_{j+1}p_{j+1};\\
\bar\nu &= \sum_{k=0}^{l} g_k \nu_k p_k.
\end{aligned}\tag{5}$$

Before dealing with these equations, let us put some thermodynamical considerations. The scenario is reminiscent of statistical mechanics systems, in which the equilibrium corresponds to the minimum of the free energy, that is a competition, balanced by the temperature, between an energy term (related to the average duplication time) and a disorder one (the entropy).

The maximization of the "entropy" $S = -\sum_{k=0}^{l} g_k p_k \ln p_k$ under the constraints for the normalization of probability distribution $\sum_{k=0}^{l} g_k p_k = 1$ and $\sum_{k=0}^{l} g_k p_k \tau_k = \bar\tau$ (in order to select the most probable probability distribution within those with the same fitness) gives the Boltzmann distribution [7]

$$p_j = C \exp(-\beta\tau_j).$$

The Lagrange multiplier β can be considered as the inverse of an effective "temperature" T, whose interpretation will be discussed in the next section. In the stationary state, $\bar\nu$ is constant in time, and from eq. (4) M (and thus m_j) grows exponentially.

Inserting the *Ansatz*

$$m_j = M_0 \exp(\alpha t - \beta\tau_j)\tag{6}$$

(M_0 is the total mass at time $t = 0$) in the equation (3), and using the relations (1), we get for the asymptotic state

$$(1-\mu)\nu_j - \alpha + \frac{l-j}{l}\mu\nu_{j+1}x + \frac{j}{l}\mu\nu_{j-1}\frac{1}{x} = 0,\tag{7}$$

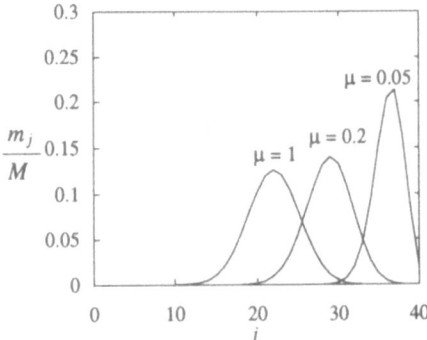

Figure 1: The asymptotic form of m_j/M for $l = 40$ and $a = n_1/(n_1 + n_0) = 0.6$.

where $x = \exp(-\beta r)$. Approximating $\nu_{j+1} \simeq \nu_{j-1} \simeq \nu_j$, with an error of order r/τ_j^2, which is small for l or L large or $n_1 - n_0$ small, eq. (7) becomes a linear equation in j. This equation holds independently of j if

$$
x = \frac{-(1-\mu)lr + \sqrt{(1-\mu)^2 l^2 r^2 + 4\mu^2 q(q + lr)}}{2\mu(q + lr)},
$$

$$
\alpha = \frac{\mu(1 - x^2)}{lrx} = \bar{\nu}.
$$

(8)

Numerical simulations of eq. (5) agree very well with this solution and shows that it is the only stable solution (see Fig. 1 and also [8]).

3. STATISTICAL MECHANICS

The fact that the *Ansatz* (6) is a good approximation for the solution of eq. (3) allows us to identify our model with a kinetic 1D Ising model without interaction among spins, $\tau_j = rj + q$ being the energy $\mathcal{H}(\mathbf{c})$ of strains in group j; $j = \sum_{i=1}^{L} c_i$ the magnetization of the configuration $\mathbf{c}$ and r an external magnetic field. We can stress this analogy by means of the variable change $\sigma_i = 2c_i - 1$, so that

$$
\mathcal{H}(\mathbf{c}) = \frac{r}{2} \sum_{i=1}^{L} \sigma_i + q + \frac{r}{2}
$$

The optimization process induced by the natural selection is interpreted as the minimization of a "free energy", in which the temperature balances the "energy" and the "entropy". The temperature is to be intended here as the influence of the external environment on the asymptotic state of our population, tending to destroy order. In our first model we can calculate explicitly the expression of the temperature, and in the limit of no natural selection ($\varepsilon = n_1 - n_0 \to 0$) we have from eq. (8) $T = 4\mu + O(\varepsilon^2)$. In this case the temperature is proportional to the mutation rate, and this is not surprising (in this case the mutation rate has no influence on the probability distribution, because all groups have the same "energy", but the response to a perturbation depends on it). In the following we shall not develop a dynamical equation, and we start directly from

Table 1: translation times

Table 1: translation times

c_{i-1}	c_i	$(\Delta_i \tau)^{-1}$
0	0	$n_0 - 1$
0	1	n_1
1	0	n_0
1	1	$n_1 - 1$

the statistical interpretation. In this case we do not have an explicit expression for the temperature, that has to be considered an external parameter.

The next step is to take into account that in real bacteria the constancy of tRNAs concentration is not completely fulfilled. From [2] we obtain that in average only 85% of the tRNA pool is acylated, indicating that the reaction rates of the aminoacyl-tRNA synthetases are not substrate limited. Let us suppose that the pool of tRNAs is shared by the ribosomes that are translating a given code c, but it is not affected by the activity of other ribosomes (a local pool). Once that a ribosome uses a tRNA of type k, the number n_k of charged tRNAs is lowered by one unit, until the aminoacyl-tRNA synthetases restore it (our model is valid for $n_k > 1$). If the recovering time is greater than the translation time of a codon, during the translation of a sequence of codons of the same type, the waiting time increases.

Let us assume that the recovering time is a multiple of the translation time (for an average codon), so that we can avoid considering branching in translation. This approximation implies that the translation time for 0 and 1 codons is not very different. The recovery interval δ (1, 2, 3, ... b translation times) induces an antiferromagnetic coupling among spins with range 1, 2, 3, ... b lattice spaces. E.g., for $\delta = 1$, the time $\Delta_i \tau$ needed for the translation of codon c_i depends also on the type of codon c_{i-1}, as shown in table 1.

For $b = 1$, the resulting Hamiltonian is

$$\mathcal{H} = LH_0 - H \sum_{i=1}^{L} \sigma_i - J \sum_{i=1}^{L} \sigma_i \sigma_{i-1},$$

with

$$H = \frac{2d}{(s-2)^2 - d^2},$$

$$J = -\frac{2[d^2 + s(s-2)]}{[(s-2)^2 - d^2](s^2 - d^2)}.$$

Here $s = n_1 + n_0$ and $d = n_1 - n_0$ ($s > d + 2$). The order parameter is the staggered magnetization $\tilde{m} = (1/2) \sum |\sigma_i - \sigma_{i-1}|$. For a completely random configuration $\tilde{m} = 1/2$, for an antiferromagnetic configuration (e.g. the ground state for $H = 0$) $\tilde{m} = 1$, for a ferromagnetic configuration (e.g. the ground state for $J = 0$) $\tilde{m} = 0$. Using standard techniques of equilibrium statistical mechanics one can obtain the explicit expression for the staggered magnetization

$$\tilde{m} = \frac{e^{-4J'}}{\left(\cosh(H') + \sqrt{\cosh^2(H') - 2e^{-2J'}\sinh(2J')}\right)\sqrt{\cosh^2(H') - 2e^{-2J'}\sinh(2J')}};$$

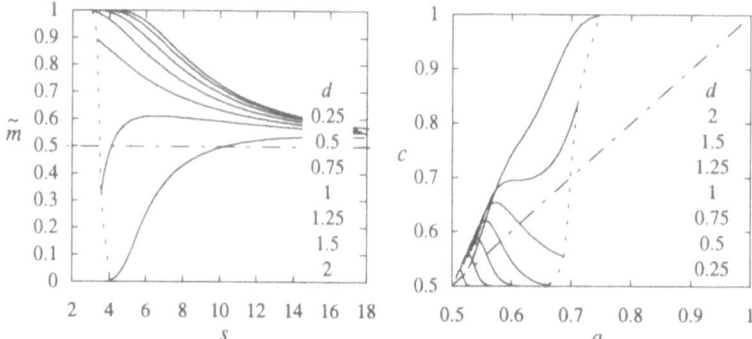

Figure 2: (left) Staggered magnetization $\tilde{m}$ versus total number of tRNA s. (right) Concentration of major codons c versus bias of tRNA pool a. Here $T = 0.05$; the dotted line marks the unphysical region $n_0 < 1$.

and the magnetization m

$$m = \frac{\sinh(H')}{\sqrt{\cosh^2(H') - 2e^{-2J'}\sinh(2J')}}$$

where $H' = H/T$ and $J' = J/T$. In the left part of Figure 2 we show the behavior of $\tilde{m}$ with respect to s for various values of d and $T = 0.05$. For large values of s the effects of a finite size for the pool of tRNAs, and thus the interactions among neighboring codons, vanish. For small s, there is a competition between the tendency to increase the number of major codons, and the disadvantage of having long patches of equal codons. Thus, varying d, we can have either a ferromagnetic ($\tilde{m}$ below 0.5) or an antiferromagnetic ($\tilde{m}$ above 0.5) phase. In the right part of the Figure we report the behavior of the mean concentration of major codons $c = (m + 1)/2$ with respect to $a = n_1/(n_0 + n_1)$, which is a measure of the bias of the pool of tRNAs. The dash-dotted line indicates the stoichiometric ratio between the codon composition and the tRNA pool equal to one. We can observe that, for large values of d, the natural selection favors the elimination of rare codons from the mRNA. For moderate values of d, the optimization process is hinder by large sequences of major codons, and this enhances the presence of rare codons, so that the stoichiometric ratio becomes less that one.

4. CONCLUDING REMARKS

We study the influence of the composition of the tRNA pool in the choice among synonymous codons (major and minor codons) by means of a simplified model. We show that the stationary state of an evolutionary dynamical model for the translation machinery of bacteria can be described by means of statistical mechanics. In the first approximation we consider a constant tRNA pool. In this case our model is equivalent to an Ising model in external field without interactions. The dynamical evolution equations and the statistical mechanics analogies indicates that the asymptotic state of a population in exponential growth is consistent with a Boltzmann distribution, the translation time being the equivalent of the energy. The temperature represents the disordering influence of the external environment, and results to be related to the mutation rate.

The finite efficiency of aminoacyl-tRNA synthetases induces an antiferromagnetic short-range coupling among spins (the codons under translation). This interaction can produce an antiferromagnetic phase, in which synonymous codons tend to be arranged in alternate disposition in order to optimize the translation. In certain conditions the antiferromagnetic interaction increases the presence of rare codons in mRNA above the stoichiometric ratio with the tRNA pool.

The model gives a mathematical description of the the selection pressure for the optimization of the translation process with respect the mutation rate. Bacteria like *E. coli* seem to have a clonal population structure, in the sense that the species undergo periodic selection and the fitter strains are able to replace the other strains in a relatively short time. The fitness regards mainly the exiting of a starvation phase or the high growing rate. In both these two phases tRNAs are lacking. Natural selection forces the cell to select the codons that match the most abundant tRNAs and to avoid stretches of rare codons (due to the finite efficiency of the aminoacyl-tRNA synthetases). At the present we are looking for experimental data supporting the theoretical predictions in the antiferromagnetic phase.

References

[1] Reiss C. (1994), personal communication.

[2] Gouy, M. and Grantham, R. (1980). *Febs Letters* **115**, 151-155.

[3] Ikemura, T. (1981). *J.Mol. Biol.* **146**, 1-21.

[4] Varenne, S., Buc, J., Lloubes, R. and Lazdunsky, C. (1984). *J.Mol. Biol* **180**, 549-576.

[5] Grosjean, H. and Fiers, W. (1982). *Gene* **18**, 199-209.

[6] Liljenstrom, H. and von Heijne, G. (1987). *J. Theor. Biol.* **124**, 43-55.

[7] Landau, L.D. and Lifshitz, E.M. (1958). In: *Statistical Physics*, Pergamon Press, Oxford, UK.

[8] Bagnoli, F. and Liò, P. (1994), submitted to *J. Theor. Biol.*

Dynamics of vibrational dissociation of a pseudo-cluster

D. Hennig and H. Gabriel

Freie Universität Berlin, Fachbereich Physik,
Institut für Theoretische Physik Arnimallee 14,
14195 Berlin, Germany

1 Introduction

In the context of studying phenomena, such as photodissociation of molecules and biomolecular reactions with methods developed in the theory of classical dynamical systems, a variety of the underlying coupled or driven nonlinear oscillator models rely on the use of the Morse oscillator as a main ingredient. These models are mostly Hamiltonian model systems with at least two-degrees-of-freedom based on several assumptions as to the form of the coupling [1],[2], [3],[4],[5], [6]. The model underlying the present study belongs to this category. Its choice was partly motivated by the kicked-oscillator model of ref. [7], in which the strength of the kick was chosen to be proportional to the Morse potential itself. In a sense, our model which merely couples a harmonic oscillator to a Morse potential in a specific way (see Eq. (3)) provides the complimentary autonomous counterpart of the time dependent system studied in [7]. It is a simple model for unimolecular fragmentation, e.g. of a cluster consisting of an atom weakly bound to a polyatomic molecule, the intramolecular dynamics of which being reduced to just one effective harmonic oscillator, responsible for transferring energy to the van der Waals (vdW) bond. We may call this fictitious system a pseudo-cluster.

As to the relation between the onset of chaos and the transport in phase space realizing the molecular fragmentation process we have used tools provided by the theory of dynamical systems [8], [9], [10]. Particularly, we analyze the dissociation dynamics in terms of an area-preserving map.

The paper is organized as follows. Section II presents the model by specifying its Hamiltonian. Section III is devoted to the transformation of the coupled dynamics to an area-preserving map and to a discussion of the onset of extended chaos in the autonomous coupled oscillator system by means of resonance overlap.

2 The model hamiltonian

The simplest model for both, studying migration of vibrational energy in a cluster consisting of a polyatomic molecule and a weakly bound atom as well as investigating the

possibility of breaking one of the weak bonds is a system of nonlinearly coupled oscillators with two degrees of freedom. The model considered in this paper couples a harmonic
oscillator representing a fictitious intramolecular mode to a Morse oscillator which mimics
the intermolecular motion, i.e. the relative motion of the atom and the molecule's center
of mass.

The present model has the Hamiltonian

$$\hat{H} = \tilde{H}_0 + \tilde{H}_1 , \tag{1}$$

where

$$\tilde{H}_0 = \frac{1}{2}(\tilde{P}^2 + \tilde{\Omega}^2 Q^2) + \frac{1}{2\mu}\tilde{p}^2 + D(1 - \exp[-a(\tilde{q} - \tilde{q}_0)])^2 \tag{2}$$

represents the isolated oscillators, a harmonic oscillator with frequency $\tilde{\Omega}$ and displacement coordinate $\tilde{Q}$ and a Morse oscillator for the intermolecular motion. The instan-
taneous (equilibrium) distance of the molecular center of mass from the attached atom
is $\tilde{q}$ ($\tilde{q}_0$), respectively, $\tilde{p}$ is the conjugate momentum and μ a reduced mass. D is the
dissociation energy and a is the range parameter of the Morse potential. The interaction
was chosen as:

$$\tilde{H}_1 = \alpha \tilde{Q}[(1 - \exp[-a(\tilde{q} - \tilde{q}_0)]]^2 \tag{3}$$

with coupling strength given by the real constant α.

The type of coupling introduced by $\tilde{H}_1$ ensure that the interactions vanish at equili-
brium $\tilde{Q} = 0; \tilde{q} = \tilde{q}_0$. For small displacements $(\tilde{q} - \tilde{q}_0)$ $\tilde{H}_1$ reduces to a nonlinear coupling
term of the form $\tilde{Q}(\tilde{q} - \tilde{q}_0)^2$. In the opposite limit of large $\tilde{q}$ the interaction energy
decreases exponentially.

The presence of coupling in the system causes energy migration between the constituents of our pseudo-cluster. Bond breaking occurs, when the amount of energy transferred
to intermolecular vibration exceeds the dissociation threshold. After fragmentation the
constituents move independently of each other. The intramolecular potential energy is
lowered by the dissociation energy needed for breaking the bond. This manifests itself in
a shift of the equilibrium position of the intramolecular harmonic oscillator to $(\tilde{P}_0, \tilde{Q}_0) =$
$(0, -(\alpha)/\tilde{\Omega}^2)$. To simplify the notation let us introduce the following dimensionless quanti-
ties: $\tilde{H}/2D \to H$, $a(\tilde{q} - \tilde{q}_0) \to q$, $\tilde{p}/\sqrt{2\mu D} \to p$, $\tilde{P}/\sqrt{2D} \to P$, $a\tilde{Q}/\sqrt{\mu} \to Q$, $\sqrt{\mu}\tilde{\alpha}/a = \gamma$
and $\tilde{\Omega}/\omega_0 = \Omega$. After time scaling $\omega_0 \tilde{t} \to t$, where $\omega_0 = \sqrt{2a^2 D/\mu}$ is the frequency of the
Morse oscillator in harmonic approximation, we obtain the following system of equations

$$\dot{p} = -(1 + 2\gamma Q)(\exp(-q) - \exp(-2q)) \tag{4}$$

$$\dot{q} = p \tag{5}$$

$$\dot{P} = -\Omega^2 Q + \gamma(1 - \exp(-q))^2 \tag{6}$$

$$\dot{Q} = P. \tag{7}$$

For later use, we introduce action-angle variables (J_1, θ_1) for the harmonic oscillator and
rewrite the Hamiltonian corresponding to Eqs. (4)-(7) as

$$H^\epsilon = H^0 + \epsilon H^1, \tag{8}$$

$$H^0 = \frac{1}{2}p^2 + \frac{1}{2}(1 - \exp(-q))^2 + \Omega J_1$$

$$=: \quad H_M(p,q) + H_H(J_1) \tag{9}$$

$$H^1 \quad = \quad \gamma(1 - \overset{\bullet}{\exp}(-q))^2 \sqrt{2J_1/\Omega} \sin\theta_1$$

$$=: \quad H_1^1(q, J_1, \theta_1; \gamma). \tag{10}$$

The parameter $\epsilon \ll 1$ indicates that we restrict to weak coupling of the two oscillators.

The uncoupled system has two first integrals of motion, the energy (action) of the harmonic oscillator and the "intermolecular" energy of the Morse oscillator and is hence integrable. The phase space of the unperturbed Morse oscillator has a center at $(p,q) = (0,0)$ and a nonhyperbolic fixed point at $(p = 0, \lim q \to \infty)$. The coordinates of the homoclinic orbit corresponding to this fixed point at infinity are given by

$$q^h(t) = \ln[\frac{1 + (t - t_0)^2}{2}], \quad p^h(t) = \frac{2(t - t_0)}{1 + (t - t_0)^2}, \tag{11}$$

where t_0 is the time parametrizing the homoclinic orbit along the unperturbed flow.

The homoclinic orbit forms a separatrix at the dissociation threshold ($E_M = 1/2$), dividing bounded motions for ($E_M < 1/2$) and unbounded motions for ($E_M > 1/2$) motions.

3   Dynamics of the model

3.1   Whisker map

Being interested in the dynamics near the homoclinic orbit, we transform the coupled oscillator dynamics to an area-preserving map, the so called 'whisker map' [11] parametrized in the change of the Morse oscillator energy and the corresponding phase change. In this way, the framework is provided for studying basic features of molecular dissociation in terms of a simple two-dimensional map. In this section we study the dynamics near the separatrix of the Morse oscillator discerning between bounded and unbounded motion, corresponding physically to those solutions which represent either stable pseudo-clusters or fragmented ones.

The main idea of this approach is the investigation of the coupled oscillator dynamics with the help of the solutions of the Morse oscillator, as the degree of freedom of interest, on the separatrix. The derivation of the whisker map is based on a relation between incremental changes in energy and phase of a near-separatrix trajectory during a period of its motion.

The total time rate of energy change of the Morse oscillator is given by

$$\frac{d}{dt}H_M \quad = \quad \frac{\partial}{\partial t}H_M + \{H_M, H\}$$

$$= \quad -2\gamma\sqrt{\frac{2J_1}{\Omega}}\sin(\Omega t + \theta_1^0)\, p\,(e^{-q} - e^{-2q}), \tag{12}$$

from which the energy change during one period

$$\Delta E_M \quad = \quad \int_{-T/2}^{T/2} dt \left(\frac{dH_M}{dt}\right)$$

$$= \quad -2\gamma\sqrt{\frac{2J_1}{\Omega}} \int_{-T/2}^{T/2} dt\, p(e^{-q} - e^{-2q}) \sin(\Omega t + \theta_1^0) \tag{13}$$

is calculated. Since the trajectory is close to the separatrix the integration limits can be extended to $\pm\infty$ and the integral is calculated by inserting the solutions (11) for the homoclinic orbit as well as those for the harmonic oscillator $J_1 = J_1(0)$, $\theta_1 = \Omega t + \theta_1^0$ resulting in

$$\begin{aligned}
\Delta E_M &= 2\pi\sqrt{(2h-1)}\gamma(\Omega-1)\exp(-\Omega)\cos(\theta_1^0 + \Omega t_0) \\
&\equiv -F(\gamma,\Omega,h)\sin(\theta_1^0).
\end{aligned} \tag{14}$$

The value for the action of the harmonic oscillator is determined $J_1 = (h-1/2)/\Omega$, i.e. from the difference between the total unperturbed energy and the separatrix level of the Morse oscillator. We further set $\Omega t_0 = \pi/2$.

$E_M^{n+1} = E_M^n + \Delta E_M$ relates the energy change at the end of the n-th period to that of the $(n+1)$-th period. The corresponding change in the phase $\theta_1^0 \equiv \theta$ during one period of the Morse oscillator $T = 2\pi/\sqrt{|1-2E_M|} = 2\pi/\omega(E_M)$ is determined by

$$\Delta\theta = \Omega T(E_M) = 2\pi\Omega\frac{1}{\sqrt{|1-2E_M|}}. \tag{15}$$

For a description of the dissociation dynamics it is convenient to introduce a normalized energy w_n measuring the decrement from the separatrix level

$$w_n = (1 - 2E_M^n). \tag{16}$$

Trajectories with $w_n < 0$ correspond to solutions bounded in the inner region of the separatrix, whereas trajectories with $w_n > 0$ belong to solutions above the separatrix level.

Finally, the whisker maps becomes

$$\begin{aligned}
w_{n+1} &= w_n - F(\gamma,\Omega,h^0)\sin(\theta_n) \tag{17} \\
\theta_{n+1} &= \theta_n + 2\pi\Omega\frac{1}{\sqrt{|w_{n+1}|}}. \tag{18}
\end{aligned}$$

The main parameter $F(\gamma,\Omega,h)$ determined in equation (14) of the map grows with increasing γ and h^0 and gives an $O(\epsilon)$ estimate of the energy width of the stochastic layer around the broken separatrix of the Morse oscillator. The frequency dependence of the map amplitude $F(\gamma,\Omega,h)$ regulates how the separatrix destruction is affected by the forcing harmonic motion as will be discussed below. For large frequencies Ω, i.e. $\tilde{\Omega} \gg \omega_0$ the exponential decrease of the map amplitude F dominates resulting in small chaotic layers. Further enhancement of the frequency mismatch leads to even vanishing destruction, thus driving the system towards the integrable regime. The most pronounced separatrix destruction appears at $\Omega = 2$, the maximum of the map amplitude F. In Figure 1 (a)-(d) we show the whisker map for various strengths of the coupling parameter γ.

3.2 Resonance overlap

The whisker map illustrates, that the stochastic layer exists whenever $\epsilon \neq 0$, since an infinite numbers of overlapping primary resonance zones are clustered in the neighbourhood

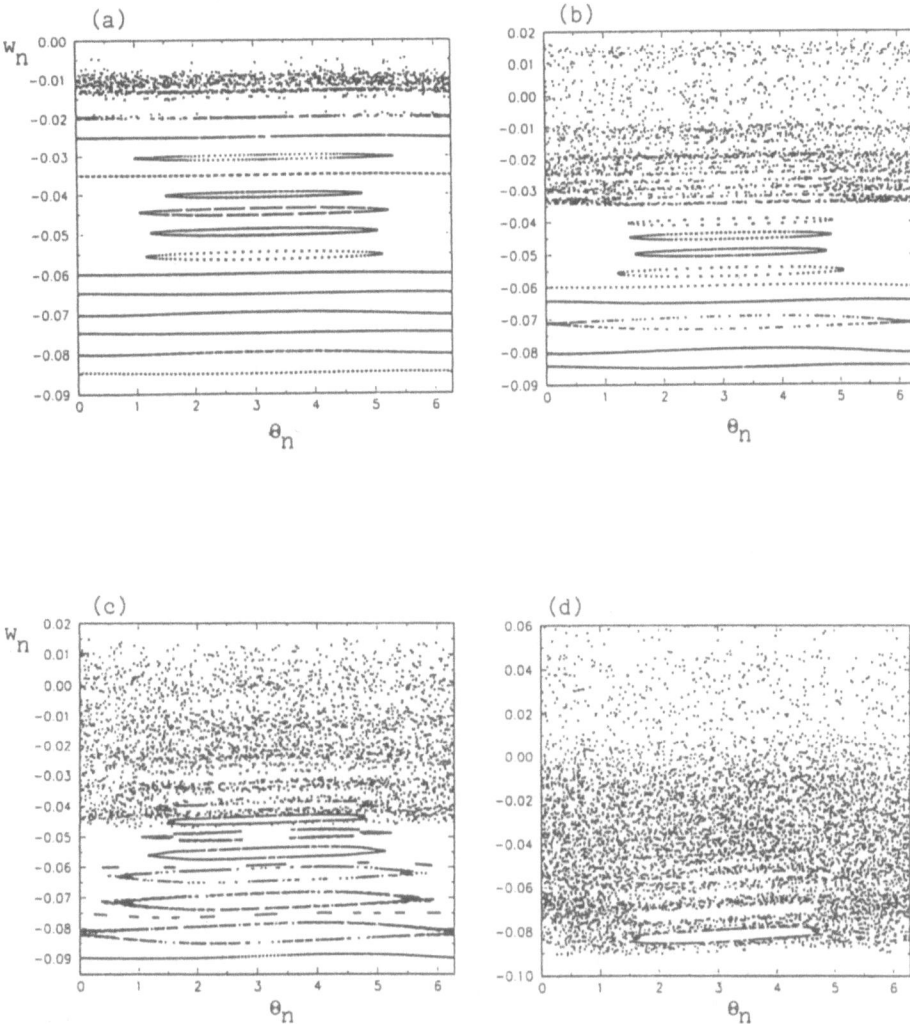

Figure 1: Plots of the whisker map of equations (17) and (18) for parameter values $\Omega = 4$, $h = 1$ and different coupling strengths: (a) $\gamma = 0.0005$, (b) $\gamma = 0.001$, (c) $\gamma = 0.002$ and (d) $\gamma = 0.005$. One clearly sees that with increasing coupling strenghts the chaotic layer penetrates more and more level sets of the Morse oscillator.

of the separatrix of the unperturbed Morse oscillator. The chaotic dynamics allows diffusion of the action of the Morse oscillator from lower to higher values. The corresponding
gain in energy may eventually suffice to break the vdW bond.

The fixed points of the map given in equations (17) and (18) occur at $\theta = m\pi$,
$w^{(m)} = (\Omega/m)^2$, where m is an integer labeling the m-th period-1 resonance state. These
fixed points of the whisker map correspond to resonances $\Omega = m\omega(E_M)$ in the coupled
oscillator system. Note, that for $m \to \infty$ the fixed point approaches the separatrix, i.e.
$w \to 0$. The energy separation between two neighbouring period-1 points is $\Delta w^{(m)} =
w^{(m)} - w^{(m+1)} = (\Omega/m)^2 - (\Omega/(m+1))^2$. At a critical coupling strength γ_{crit} resonances
are destroyed as a result of resonance overlap. This critical coupling strength can be
estimated with the help of Chirikov's criterion [11].

According to this criterion, resonance overlap occurs when the sum of the half-width of
two neighbouring resonances $\delta^{(m)} = \delta w^{(m)} + \delta w^{(m+1)}$ exceeds the distance between them:

$$\delta^{(m)} \geq \Delta w^{(m)}. \tag{19}$$

In order to get an expression for the width of the m-th resonance, we isolate this resonance
and approximate the corresponding single resonance Hamiltonian by those for a pendulum.
To do this it is convenient to pass to action-angle variables (J_2, θ_2) also for the Morse
oscillator yielding

$$H = H^0(J_1, J_2) + \epsilon H^1(J_1, \theta_1, J_2, \theta_2). \tag{20}$$

A particular resonance can be eliminated from the Hamiltonian (20) through a canonical
transformation determined by the following generating function

$$F = (\theta_1 + m\theta_2)I_2 - \theta_1 I_1. \tag{21}$$

Expanding the resulting integrable, single resonance, Hamiltonian $H = H(I_2, \phi_2)$ in the
excursion from the resonance center $\Delta I_2 = I_2 - I_2^r$ we obtain

$$\tilde{H} = -\frac{1}{2}m^2(\Delta I_2)^2 + G_m \cos(\phi_2), \tag{22}$$

where the coefficient is

$$G_m = \gamma\sqrt{(\Omega/2)(I_2^0 - I_1^0)}\sqrt{1 - 2E_M^m}\left[\frac{1 - \sqrt{1 - 2E_M^m}}{1 + \sqrt{1 - 2E_M^m}}\right]^{m/2}, \tag{23}$$

and $(I_2^0 - I_1^0)$ is an integration constant.

For the resonance width we infer from (22):

$$\Delta I_2^m = 2\frac{\sqrt{G_m(\gamma)}}{m}. \tag{24}$$

The critical coupling strength γ_{crit} leading to resonance overlap for the whisker map
is obtained from (19) by using the equality sign and the relation between the action and
energy of the Morse oscillator $E_M(w) = J_2(1 - (1/2)J_2)$ as well as the transformations
(16) and (21). Furthermore, it is possible to estimate a coupling strength which ensures
all period-1 resonance states above an order m to be absorbed in the stochastic layer. As
a result the chaotic layer extends to regions in phase space hosting the separatrix, thus
enabling more and more trajectories to dissociate.

References

[1] M.J. Davis, J. Chem. Phys. **83**, 1016 (1985).

[2] S.K. Gray, S. Rice and D.W. Noid, J. Chem. Phys. **84**, 3745 (1986).

[3] M.J. Davis, J. Chem. Phys. **86**, 3978 (1986).

[4] R.T. Skodjie and M.J. Davis, J. Chem. Phys. **88**, 2429 (1988).

[5] P. Gaspard and S.A. Rice, J. Phys. Chem. **93**, 6947 (1989).

[6] R.E. Gillilan and G.S. Ezra, J. Chem. Phys. **94**, 2648 (1991), and references therein.

[7] S.H. Tersigni, P. Gaspard and S.A. Rice, J. Chem. Phys. **92**, 1775 (1990).

[8] J. Guckenheimer and P. Holmes, *Nonlinear Oscillations, Dynamical Systems, and Bifurcations of Vector Fields* (Springer-Verlag, New York, 1983).

[9] S. Wiggins, *Chaotic Transport in Dynamical Systems* (Springer-Verlag, New York, 1991).

[10] A.J. Lichtenberg and M.A. Lieberman, *Regular and Stochastic Motion* (Springer-Verlag, Berlin, 1982).

[11] B.V. Chirikov, Phys. Rep. **52**, 263 (1979).

The step-potential model for π-electrons in hydrocarbon-systems

C. Kuhn

Laboratoire Léon Brillouin, C.E. de Saclay,
91191 Gif-sur-Yvette cedex, France

I.   Introduction.

In single electron models the many body π-electron Hamiltonian is simplified to a Hamiltonian of independent electrons moving in orbitals to be described as solutions of the 3D-Schrödinger equation of an electron in a potential V(x,y,z) associated with the lattice sites. These orbitals have nodes in the layer plane (orthogonality with the σ-electrons). V(x,y,z) is constructed from atomic contributions [1, 2]. With the particular postulates defining V(x,y,z) the correlation of electrons in the ground state is indirectly considered: it is assumed that V(x,y,z) is the sum of the Slater potentials of the two adjacent charged lattice sites. The contributions of all other sites are neglected: their charges are considered to be shielded by the residual π-electrons while next neigbor sites are unshielded. In this way it is taken care of the interdependence of the electron under consideration with all other electrons. The motion of this electron is considered to be correlated with the motions of the other electrons.

The Schrödinger equations for such potentials V(x,y,z) have been solved in some typical cases [1, 2]. It was shown that a considerable simplification in the shape of this potential still leads to a reasonable description of unbranched π-electron systems. The wave functions can be written as the product of a function $\Psi_\nu(s)$ of the coordinate s along the zig-zag-line connecting the atoms in the chain and a function of the coordinates perpendicular to the chain which is identical for all solutions of physical interest. The functions $\Psi_\nu(s)$ are solutions of the 1D-Schrödinger equation in the potential V(s) where V(s) is the average of V(x,y,z) taken over the coordinates perpendicular to s. V(s) again has been systematically simplified in order to investigate what is crucial to describe the relevant properties of π-electrons [1, 2]. In the case of a chain with bonds fixed to equal bond lengths, the most radical simplification, V(s) = const along the chain, still leads to a reasonable description of the wave functions of an electron in the potential V(x.y,z) if the electron wave is assumed to extend by about one bond length beyond the centers of the sites at the chain ends (length L = a (Z+1), bond length a = 1.40 Å, number of sites Z). In this case $\Psi_\nu(s)$ is simply a sine- function

$$\psi_\nu(s) = \sqrt{\frac{2}{L}} \sin\left(\pi \frac{s}{L} \nu\right) \qquad \nu = 1, 2, 3, \ldots \tag{1}$$

This simplification is appropriate in the case of a cyanine dye with resonating structures

where L = 2na. The N = 2n π-electrons occupy the N/2 = n orbitals of lowest energy.

This simple model [3] (the free electron model) can easily be refined by solving the Schrödinger equation for the given potential V(s) which has troughs at each atom in the chain, but this does not change the wave functions essentially [4]. The troughs can easily be taken into account to check this statement. However, in the case of unequal bond lengths a refinement is important in describing orbital wave functions and energies [1, 2, 4]. V(s) is lower in short bonds because the adjacent nuclei, in the average, are closer to the electron, the Coulomb attraction is larger. We assume, for simplicity, that the potential is constant along a given bond, its value being given by the value of V(s) in the middle of the bond. Numerical evaluation gives the following relation between potential V_i in bond i and bond length d_i [5] :

$$V_i/_{eV} = 39\left(\frac{d_i}{a} - 1\right) - 24\left(\frac{d_i}{a} - 1\right)^2 \tag{2}$$

This model, the step-potential model [4], is the logically simplest extension of the free electron model. It describes this generalized situation sufficiently accurate for the present purposes, and it can easily be refined if desired for its justification. The exact wave functions for a stepwise constant potential are easily obtained by numerical evaluation.

The step potential model allows, in an easy way, to calculate the bond lengths in π-electron systems. Bond lengths and π-electron density in a bond are related: the π-electron cloud attracts the nuclei. Thus, assuming first a σ-bonded molecular skeleton with uniform bond length a, the skeleton is elastically deformed in the field of the π-electron cloud and the cloud is deformed in the changed potential. The process is repeated until self-consistency between bond potential V_i (bond length d_i) and π-electron density ρ_i is reached in each bond:

$$V_i = b(1 - a\rho_i), \quad b = \alpha\frac{h^2}{2ma^2} = 3.8\,eV, \quad \alpha = 1.95 \tag{3}$$

This consideration leads to bond alternation in polyenes

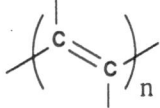

(d_i = 1.34 Å and 1.46 Å for a double bond and a single bond respectively), while the bond lengths in a cyanine dye are essentially equal. Assuming equal bond lengths in polyenes when starting the search for selfconsistency, the total π-electron density is accumulated in bonds 1-2, 3-4, ... causing an instability leading to bond alternation. In cyanine dyes, however, the density accumulations are at atoms 1, 3, 5, ... and not in bonds (Fig. 1). The step potential model leads to equal bond lengths in benzene and Hückel annulenes up to the 14-annulene and bond alternation in all other annulenes and to the expectation that cyanines, with increasing chain length, change from equal bond lengths to alternating bonds and a soliton-like region of equal bonds in their center [4].

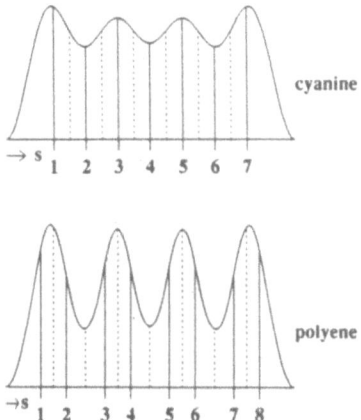

Figure 1: Cyanines and polyenes in the step-potential model. Example $n = 4$.
Charge density maxima at atoms 1,3,5,7 (cyanines), at bonds 1-2, 3-4, 5-6, 7-8 (polyenes).

II. Non-Linear Optics.

To calculate the polarization $P = \mu + \alpha E + \beta E^2 + \delta E^3$ in an applied electric field E parallel to the hydrocarbon chain the Schrödinger equation including the external field and the reaction field produced by the polarized π-electron cloud by the sum of net charges at all atoms along the chain is solved numerically [6]. The linear and non-linear coefficients of the polarization of polyenes, push-pull polyenes, the symmetric cyanines and azacyanines thus obtained [6] were in good agreement with experiments.

III. Kinks in Polyacetylene: Statics and Dynamics.

In an odd numbered ring of a hydrocarbon the two bond length alternation patterns

$$\cdots = C - C = C - C = C - \quad \cdots \quad \text{A-phase}$$
$$\cdots - C = C - C = C - C = \quad \cdots \quad \text{B-phase}$$

unite, forming a topological defect called a kink (Fig. 2) [4, 7] with the bond length alternation pattern $\Delta d_i = (-1)^i (d_i - d_{i-1}) \approx 0.058 \tan((i - i_0)/9.0)$. The kink creates an intergap state, the corresponding wave function is well localized with the envelope $\approx \mathrm{sech}^2((i - i_0)/9.25)$. The weak kink is schematically represented as

$$\cdots - C = C - \overset{\bullet}{C} - C = C - \quad \cdots$$

and the strong kink (the center of defect being shifted by one site), is schematically represented by the mesomeric structures

$$\cdots = C - \overset{\bullet}{C} - C = C - C = \quad \cdots$$
$$\cdots = C - C = C - \overset{\bullet}{C} - C = \quad \cdots$$

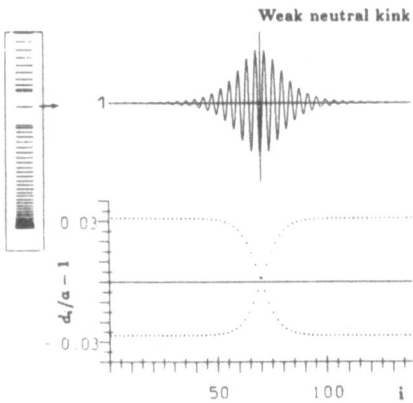

Figure 2: Neutral kink (weak): band gap ΔE=1.34 eV, half filled intergap state (one electron, spin 1/2) and its wave function; dimerisation pattern.

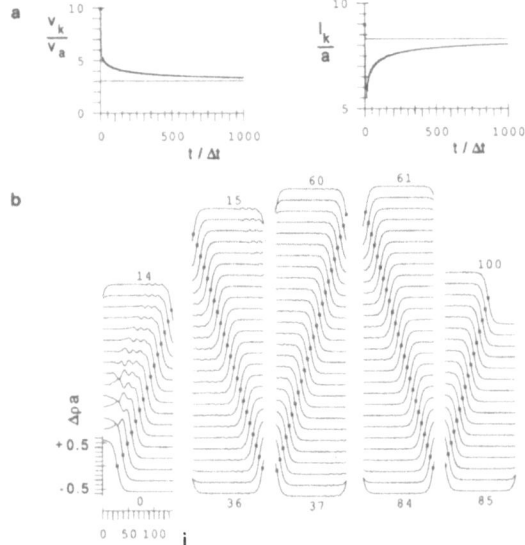

Figure 3: Dynamics of a neutral kink along a ring of 139 sites with an initial velocity $v_k = 10 v_a$ during 1000 time steps $\Delta t = 1.25 \cdot 10^{-15}$ sec. **a)** Time evolution of kink velocity v_k: maximum velocity of $v_\infty = 3.05 v_a$ without energy dissipation into the lattice ($v_\infty = 3.05 v_a$ is given by the electron-phonon coupling α) and of its width l_k: $l_\infty = 8.3\, a$. **b)** Time evolution of charge density alternation $\Delta\rho_i = (-1)^i(\rho_{i+1} - \rho_i)$. Spot-light every $10\Delta t$ indicated by numbers 0-100. $\Delta\rho = 0$ indicated by star. To avoid kink-phonon collision, phonons evolving from fast moving kink ($v_k > v_\infty$) are relaxed.

The dynamics [7, 8] in standard adiabatic approximation follows from Newton's equation. The σ-bonds in the field of the π-electron cloud are compressed according to the π-electron density in each bond. Thus, in analogy to Hooke's law, the strain arising in bond i is proportional to the deviation $\tilde{\rho}_i - \rho_i$ of the actual π-electron density ρ_i from the equilibrium density $\tilde{\rho}_i$. For small deviations of the actual density ρ_i from its equilibrium it is assumed that the equilibrium density $\tilde{\rho}_i(\rho_1, \rho_2, ..., \rho_M)$ is obtained by applying equation (3) for each ρ_i and solving the Schrödinger equation with this configuration of bond potentials. The force caused by this strain acts on the two adjacent sites. The time axis is scaled ($\Delta t = 1.25 \cdot 10^{-15}$ sec) by calculating the frequency v_{max} of the in-phase-streching-mode of the polyene lattice numericaly and compairing it with the experimental value $v_{max} = c \cdot 1400 \, \text{cm}^{-1}$ (vibronic structure of the absorption band of polyenes). The velocity of sound along s is $v_a = 1.85 \cdot 10^4 \, \text{m sec}^{-1}$.

IV. The C$_{60}$ Molecule.

As zeroth-order approximation we can consider these π-electrons in terms of free electrons on the sphere with radius r=3.6Å [9]. The orbitals are spherical harmonics with energies E=$\gamma L(L+1)$, angular momentum quantum number L=0,1,2... and $\gamma = h^2/2mr^2 = 0.3 \text{eV}$. Their occupation for the neutral C$_{60}$ is $(L=0)^2(L=1)^6(L=2)^{10}(L=3)^{14}(L=4)^{18}(L=5)^{10}$ with an open L=5 shell. Figure 5a shows the energy levels of the free π-electron in the network of the C$_{60}$-molecule [10] (Fig.4). In contrast to the free electron on the sphere, where the L=5 shell is 11-times degenerate, we find three seperate levels due to the icosahedral symmetry: the 5-times degenerate 1hu HOMO (highest occupied molecular orbital) of closed shell, the triply degenerate 2t1u LUMO (lowest unoccupied molecular orbital) and the triply degenerate t2u orbital at higher energy. Simlarly, in the case of the L=6 shell, the 13 times degeneracy is removed and we find the triply degenerate 1t1g LUMO+1 and states of higher energy. Relaxing the lattice by electron-phonon coupling, the 60 bonds building up the pentagons become "single bonds" of length 1.45Å, whereas the hexagons become alternating with the "double bonds" of length 1.38Å (Fig.4), the (HOMO-LUMO) gap widens to 2.5eV and the optically allowed (HOMO-LUMO+1) gap widens to 3.1eV (Fig. 5b).

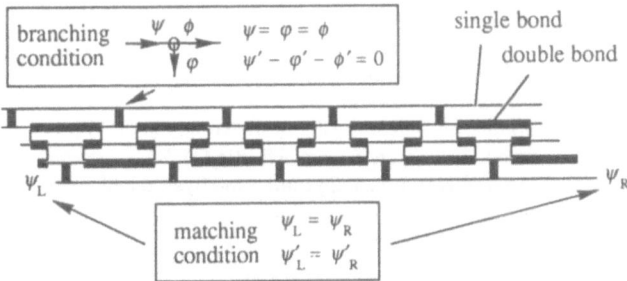

Figure 4: Schematic representation of C$_{60}$-molecule (icosahedral symmetry of ground state). The π-electrons are considered to move along a one-dimensional step-potential forming a network of bond potentials. The one-particle Schrödinger equation is solved numerically by an inverse eigen value problem (matching condition) of a 10*10 matrix (no symmetry restrictions) fulfilling branching conditions at each site.

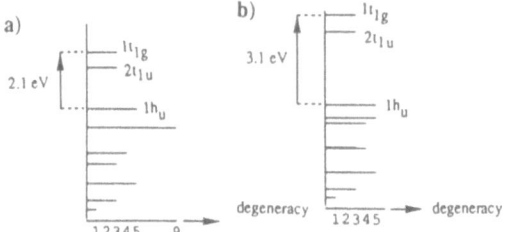

Figure 5: Energy levels vs. degeneracy of C_{60}-molecule calculated by the **a)** free π-electron in icosahedral network, all 90 bonds of length 1.42Å; **b)** nearly free π-electron in the network of the relaxed geometry with bonds of length 1.45Å and 1.38Å (single and double bonds).

References

1.   H.D.Försterling, W.Huber, H.Kuhn, *Int.J.Quant.Chem.* **1**, 225 (1967)
2.   H.Kuhn, W.Huber, G.Handschig, H.Martin, F.Schäfer, F.Bär,
         J.Chem.Phys. **32**, 467 (1960)
       H.Martin, H.Kuhn, F.Bär, W.Huber, G.Handschig
         J.Chem.Phys. **32**, 470 (1960)
       H.Kuhn in "Progress in the Chemistry of Organic Natural Products",
         Edt. L.Zechmeister, Springer Wien **17**, 404 (1959)
       H.D.Försterling, H.Kuhn, *Int.J.Quant.Chem.* **2**, 413 (1968)
3.   H.Kuhn, *Helv.Chim.Acta* **31**, 1441 (1948)
4.   C.Kuhn, *Phys. Rev. B* **40**, 7776 (1989)
5.   H.Kuhn, C.Kuhn, *Chem. Phys. Lett.* **204**, 206 (1993)
6.   C.Kuhn, *Synth. Met.* 41-43, **3681** (1991)
       C.Kuhn in "Electronic Properties of Polymers", Eds. H.Kuzmany,
         M.Mehring, S.Roth, *Springer Series in Solid State Sciences* **107**
         Springer Verlag, Berlin Heidelberg, (1992), pp. 196
7.   W.P.Su, J.R.Schrieffer, A.J.Heeger, *Phys.Rev.B* **22**, 2099 (1980), **1138** (E) (1983)
       H.Takayama, Y.R.Lin-Liu, K.Maki, *Phys. Rev. B* **21**, 2388 (1980)
       A.J.Heeger, S.Kivelson, J.R.Schrieffer, W.-P.Su, *Rev.Mod.Phys.* **60**, 783 (1988)
       D.Baeriswyl, D.K.Campbell, S.Mazumdar, in:
         "Conjugated Conducting Polymers", Edt. H.G.Kiess,
         Springer Series in Solid State Sciences **102**, Springer, Berlin (1992), pp. 7
8.   C.Kuhn, W.F. van Gunsteren, *Solid State Comm.* **87**, 203 (1993)
       C.Kuhn, in "Future Directions of Non-Linear Dynamics in Physical and
         Biological Systems", Edt. P.L.Christiansen et al.,
         Plenum Press, New York (1993), pp. 67
       C.Kuhn, *Synth. Met.* **57**, 4350 (1993)
9.   G.A.Gallup, *Chem.Phys.Lett.* **187**, 187 (1991)
10.   C.Kuhn, in "Electronic Properties of Fullerenes", Eds. H.Kuzmany,
         J.Fink, M.Mehring, S.Roth, *Springer Series in Solid State Sciences* **117**
         Springer Verlag, Berlin Heidelberg, (1993), pp. 131

Conclusion

One recurrent question which showed up during the discussions of the meeting is whether the generalisation and universality of the physicists approach are relevant for biology. The debate over this problem is certainly not over, but, according to their comments after the workshop, both communities have slightly changed their point of view. Clearly the physicists have taken a better measure of the diversity in biology and they will try to introduce a part of it in their models. Deciding how specific a model should be to be relevant is another story! Although biologists are certainly not ready to accept the extreme reductionism of many physical approaches, they have been in contact with the concepts of nonlinear science which have no counterpart in the world of linear excitations, and will perhaps integrate new ideas in their analysis.

In his book cited in the introduction, E. Schrödinger has shown the extraordinary predicting capability of reasoning from basic ideas. Although DNA structure was not known when he wrote the book, he had been able to deduce some of its essential features, such as its property of being an "aperiodic crystal" to code enough information. One question that he could not answer satisfactorily was the origin and the stability of order and coherence in a living organism. It is interesting to notice that self organisation that forms coherent excitations in nonlinear systems answers precisely this question. This supports our expectation that the rich concepts which are provided by nonlinear science will form a basis from which further reasoning along new directions will, one day, lead to great discoveries in biology.